VOLUME ONE HUNDRED AND TWENTY

Advances in PROTEIN CHEMISTRY AND STRUCTURAL BIOLOGY

Inflammatory Disorders - Part B

VOLUME ONE HUNDRED AND TWENTY

ADVANCES IN PROTEIN CHEMISTRY AND STRUCTURAL BIOLOGY

Inflammatory Disorders - Part B

Edited by

Rossen Donev
Head of Research
MicroPharm Ltd
Newcastle Emlyn, Carmarthenshire
United Kingdom

ELSEVIER

ACADEMIC PRESS
An imprint of Elsevier

Academic Press is an imprint of Elsevier
125 London Wall, London EC2Y 5AS, United Kingdom
525 B Street, Suite 1650, San Diego, CA 92101, United States
50 Hampshire Street, 5th Floor, Cambridge, MA 02139, United States
The Boulevard, Langford Lane, Kidlington, Oxford OX5 1GB, United Kingdom

First edition 2020

ISBN: 978-0-12-821322-3
ISSN: 1876-1623

For information on all Academic Press publications visit our website at
https://www.elsevier.com/books-and-journals

Publisher: Zoe Kruze
Acquisitions Editor: Ashlie Jackman
Editorial Project Manager: Leticia Lima
Production Project Manager: Abdulla Sait
Cover Designer: Greg Harris

Typeset by TNQ Technologies

Contents

Contributors

Shivalkar Apurva Abhay
School of BioSciences and Technology, Vellore Institute of Technology, Vellore, Tamil Nadu, India

Indoumady Baskara-Yhuellou
Laboratory for Epigenetics & Environment, Centre National de Recherche en Génomique Humaine, CEA-Institut de Biologie François Jacob, Evry, France

Anupam Bishayee
Lake Erie College of Osteopathic Medicine, Bradenton, FL, United States

R. Bithia
School of BioSciences and Technology, Vellore Institute of Technology, Vellore, Tamil Nadu, India

Biswanath Dinda
Department of Chemistry, Tripura University, Agartala, Tripura, India

Manikarna Dinda
Department of Biochemistry and Molecular Genetics, University of Virginia, Charlottesville, VA, United States

Subhajit Dinda
Department of Chemistry, Dasaratha Deb Memorial College, Khowai, Tripura, India

Marco Gattorno
Autoinflammatory Diseases and Immunodeficiencies Centre, IRCCS Istituto Giannina Gaslini, Genova, GE, Italy

C. George Priya Doss
School of BioSciences and Technology, Vellore Institute of Technology, Vellore, Tamil Nadu, India

Joanna L. Giles
MicroPharm Ltd, Newcastle Emlyn, Carmarthenshire, United Kingdom

Rishin Haldar
School of Computer Science and Engineering, Vellore Institute of Technology, Vellore, Tamil Nadu, India

Prajakta Ingale
National Centre for Cell Science, NCCS Complex, SP Pune University Campus, Pune, Maharashtra, India

Ritika Kabra
National Centre for Cell Science, NCCS Complex, SP Pune University Campus, Pune, Maharashtra, India

Balu Kamaraj
Department of Neuroscience Technology, College of Applied Medical Sciences, Imam Abdulrahman Bin Faisal University, Jubail, Saudi Arabia

Rekha Khandia
Department of Genetics, Barkatullah University, Bhopal, MP, India

Maignana Kumar
Department of Pharmacology, Chettinad Academy of Research and Education, Kelambakkam, Tamil Nadu, India

John Landon
MicroPharm Ltd, Newcastle Emlyn, Carmarthenshire, United Kingdom

Shuang Liang
Department of Oral Immunology and Infectious Diseases, University of Louisville School of Dentistry, Louisville, KY, United States

Elena Lukina
Discipline of Health Sciences, College of Nursing and Health Sciences, Flinders University, Bedford Park, SA, Australia

R. Magesh
Department of Biotechnology, College of Biomedical Sciences Technology & Research, Sri Ramachandra Institute of Higher Education and Research (DU), Chennai, Tamil Nadu, India

Pinky D. Mandal
School of BioSciences and Technology, Vellore Institute of Technology, Vellore, Tamil Nadu, India

Eileen McGowan
School of Life Sciences, Faculty of Science, University of Technology Sydney, Sydney, NSW, Australia

Ashok Munjal
Department of Genetics, Barkatullah University, Bhopal, MP, India

Ahmed Shaikh Nishaat Laeeque
School of BioSciences and Technology, Vellore Institute of Technology, Vellore, Tamil Nadu, India

Riccardo Papa
Autoinflammatory Diseases and Immunodeficiencies Centre, IRCCS Istituto Giannina Gaslini, Genova, GE, Italy

Paolo Picco
Autoinflammatory Diseases and Immunodeficiencies Centre, IRCCS Istituto Giannina Gaslini, Genova, GE, Italy

Oktawia J. Polak
MicroPharm Ltd, Newcastle Emlyn, Carmarthenshire, United Kingdom

Arup Roy
Chemical Science & Technology Division, CSIR - North East Institute of Science and Technology, Jorhat, Assam, India

Srivarshini Sankar
School of BioSciences and Technology, Vellore Institute of Technology, Vellore, Tamil Nadu, India

Shailza Singh
National Centre for Cell Science, NCCS Complex, SP Pune University Campus, Pune, Maharashtra, India

R. Siva
School of BioSciences and Technology, Vellore Institute of Technology, Vellore, Tamil Nadu, India

Olga A. Sukocheva
Discipline of Health Sciences, College of Nursing and Health Sciences, Flinders University, Bedford Park, SA, Australia

D. Thirumal Kumar
School of BioSciences and Technology, Vellore Institute of Technology, Vellore, Tamil Nadu, India

Jörg Tost
Laboratory for Epigenetics & Environment, Centre National de Recherche en Génomique Humaine, CEA-Institut de Biologie François Jacob, Evry, France

S. Udhaya Kumar
School of BioSciences and Technology, Vellore Institute of Technology, Vellore, Tamil Nadu, India

Charles Emmanuel Jebaraj Walter
Department of Biotechnology, Sri Ramachandra Institute of Higher Education and Research (Deemed to be University), Chennai, India

Huizhi Wang
VCU Philips Institute for Oral Health Research, Department of Oral and Craniofacial Molecular Biology, Virginia Commonwealth University School of Dentistry, Richmond, VA, United States

Weizhe Xu
State Key Laboratory of Oral Diseases & National Clinical Research Center for Oral Diseases & Department of Cariology and Endodontics, West China Hospital of Stomatology, Sichuan University, Chengdu, Sichuan, China; Department of Oral Immunology and Infectious Diseases, University of Louisville School of Dentistry, Louisville, KY, United States

Hatem Zayed
Department of Biomedical Sciences, College of Health and Sciences, Qatar University, Doha, Qatar

Wei Zhou
Department of Endodontics, Ninth People's Hospital, School of Medicine, Shanghai JiaoTong University, Shanghai, Pudong, China

CHAPTER ONE

The expanding pathways of autoinflammation: a lesson from the first 100 genes related to autoinflammatory manifestations

Riccardo Papa[a,*], Paolo Picco[a] and Marco Gattorno[a]
[a]Autoinflammatory Diseases and Immunodeficiencies Centre, IRCCS Istituto Giannina Gaslini, Genova, GE, Italy
*Corresponding author: E-mail: papariccardo86@gmail.com

Contents

Advances in Protein Chemistry and Structural Biology, Volume 120
ISSN 1876-1623
https://doi.org/10.1016/bs.apcsb.2019.11.001

Abstract

AutoInflammatory Diseases (AIDs) are a group of innate immune system disorders characterized by sterile inflammation without evidence of pathogenic autoantibodies or auto-reactive T lymphocytes. An expanding spectrum of genes and molecular pathways are associated with AIDs.

Inflammasomopathies are secondary to dysregulation of multi-protein complexes, called inflammasomes, leading to an excessive maturation and secretion of IL1β and IL18. Patients present with persistent or recurrent systemic inflammation, abdominal and chest pain, skin rashes and are sensible to IL1 inhibitors.

Unfolded proteins response causes a small number of AIDs that we propose to call immuno-proteinopathies, characterized by recurrent fevers and deep tissues inflammation.

Other inflammatory conditions can occur in case of abnormalities of actin polymerization and the term of immuno-actinopathies is proposed.

Generalized pustular psoriasis is a marker of autoinflammation mainly affecting the keratinocytes. Specific treatment targeting the p40 subunit of IL12 and IL23 or IL-17 are usually effective.

Granulomatous inflammation characterizes AIDs related to NOD2 signaling defects.

Defects in the ubiquitin-proteasome system cause a group of relopathies and some interferonopathies related to defect of the proteasome function (CANDLE syndrome).

Gain of function of proteins regulating the production of type I interferons lead to severe inflammatory conditions, called interferonopathies. The JAK/STAT inhibitors are usually effective in these latter conditions.

In conclusions, the identification of the main intracellular pathways involved in rare monogenic AIDs allows not only the proper classification of different conditions, but also highlight a pivotal role of possible novel therapeutic targets for the future.

No man ever steps in the same river twice.

Heraclitus

1. Introduction

Autoinflammatory diseases (AIDs) are a heterogeneous group of innate immune system disorders, characterized by sterile inflammation without evidence of pathogenic autoantibodies or auto-reactive T lymphocytes (Gattorno et al., 2008; Masters, Simon, Aksentijevich, & Kastner, 2009). Originally, AIDs were characterized by recurrent fever episodes with urticarial rash and serositis, with possible severe long term complications as mental impairment, infertility and AA amyloidosis (Moghaddas & Masters, 2018; Papa & Lachmann, 2018). After the discovery of the first disease-causing gene, namely *MEFV* for Familial Mediterranean Fever (FMF) (Centola et al., 1998; French FMF Consortium, 1997), an expanding spectrum of monogenic disorders have been associated with errors of the intra-cellular pathways primarily involving the innate immunity (Manthiram, Zhou, Aksentijevich, & Kastner, 2017; Ozen, 2018). All these disorders are mediated by the overproduction of various inflammatory cytokines, such as interleukin (IL)-1, IL-18, and type I interferons (IFNs). According to the mainly intracellular pathways involved in the pathogenesis of these rare conditions, treatments with specific inhibitors of the cytokines involved or their downstream mediators have proven to be very effective in most patients (Havnaer & Han, 2019).

Due to the clinical overlap of symptoms among different monogenic AIDs, the diagnostic approach is usually rather complex. Recently developed technologies for the molecular analysis, as well as the next generation sequencing, allow the simultaneous analysis of various genes associated with a given group of inherited disorders. Next generation sequencing panels are increasingly used in patient with undefined AID with variable success (Rusmini et al., 2016; Papa et al., 2019; Rowczenio & Lachmann, 2019; Kluk et al., 2014; Rowczenio et al., 2019). Furthermore, geneticists try to design appropriate workflow for gene variants interpretation (Van Gijn et al., 2018).

Interesting, inherited errors in the regulatory system of the immune response can result not only in AIDs, but also in primary immunodeficiencies (PIDs). These latter conditions are characterized by autoimmune manifestations caused by defective tolerance to self-antigens inducing organ dysfunctions. However, a growing number of PID can also present with a clear inflammatory phenotype overlapping the clinical manifestations observed in AIDs. These group of diseases are named immune dysregulation (Bousfiha et al., 2018; Toubi & Vadasz, 2019). Thus, the clinical presentation may not favor a correct molecular diagnosis in patients affected by immune disorders.

The knowledge of the underlying dysfunctional pathways may aid clinicians detecting the pivotal cytokine causing inflammatory symptoms, thus providing the best target for treatments even in patients without diagnosis.

Here, we review the genes associated to diseases possibly presenting with a clear inflammatory phenotype, and we propose a pathways-oriented classification of monogenic conditions associated to an inflammatory phenotype.

2. Autoinflammatory pathways

Discovery of AID-causing genes reveals key roles of proteins involved in the common inflammatory response. The number of genes related to AID (Table 1) and inflammatory manifestations associated with disorders alternatively classified (Table 2) are constantly increasing and consent to highlight at least six main autoinflammatory pathways.

2.1 Inflammasomopathies

Enlarging number of AID are caused by constitutive activation of inflammasomes. Inflammasomes are multiprotein intracellular complexes that promotes the maturation and secretion of pro-inflammatory cytokines, as IL-1β and IL-18, in response to pathogenic microorganisms and sterile stressors (Deuteraiou, Kitas, Garyfallos, & Dimitroulas, 2018; Shin et al., 2019). Three of the four best-known recurrent fever syndromes follow this pathway: FMF, mevalonate kinase deficiency (MKD) and cryopirinopathies, also named cryopyrin-associated periodic syndromes (CAPSs).

2.1.1 NLRP3 inflammasome

CAPS are linked to gain of function (GOF) mutations of the nucleotide-binding oligomerization domain-containing protein (NOD)-like receptor family pyrin domain containing 3 (NLRP3) gene causing a constitutive activation of the NLRP3 inflammasome (Table 1). Three clinical phenotypes with increasing severity have been identified: familial cold autoinflammatory syndrome type 1 (FCAS1), Muckle-Wells syndrome (MWS) and Chronic infantile neurological, cutaneous and Articular syndrome (CINCA). Patients with CAPS suffered of systemic inflammation, fever and urticarial rash. Patients with CINCA also present chronic aseptic meningitis with consequent brain atrophy, mental retardation and sensor neural deafness, in addition to osteoarticular manifestations (Kuemmerle-Deschner et al., 2017; Marchica, Zawawi, Basodan, Scuccimarri, & Daniel, 2018; Hui, Johnson, Greemberg, Penney, & Ramsey, 2019; Kilic et al., 2019; Nakamichi et al., 2019).

The discovery of the underlying pathogenic mechanisms led to the use of IL-1 inhibitors for the treatment of these patients (Benedetti, Miettunen, Kallinich, Horne, & Brik; Kaneko, Kurata, Yamamoto, Morikawa, & Masumoto, 2019). Although the treatment response is often optimal, some patients respond only partially, requiring relevant dose escalation. Moreover, some manifestations (for example, the bone abnormalities) are refractory to IL-1 inhibition (Benedetti et al., pp. 1−3).

Table 1 Genotypes considered as consistently associated with AIDs, according pathways description in the text.

Gene	Protein	Mechanism	Effect	Disease	Inheritance	Main symptoms
NLRP3	NLRP3	GOF	Constitutive activation of NLRP3 inflammasome	CAPS (FCAS1, MWD, CINCA)	AD	Urticaria, recurrent fever, CNS inflammation, bone overgrowth
LPIN2	LPIN2	LOF	Increased potassium efflux through P2X7R resulting in activation of TLR4 and NLRP3 inflammasome	MJDS	AR	CRMO, congenital dyserythropoietic anemia, recurrent fever, Sweet syndrome
MEFV	Pyrin	LOF	Activation of pyrin inflammasome	FMF, PAAND	AR or AD	Recurrent fever, serositis, rash, AA amyloidosis
PSTPIP1	PSTPIP1	GOF	Dysregulation of cytoskeleton resulting in activation of pyrin inflammasome	PAPA, PAMI	AD	Sterile abscesses, pyoderma gangrenosum arthritis
MVK	MVK	LOF	Deficit in geranylgeranylation of RhoA resulting in activation of pyrin inflammasome	MKD	AR	Recurrent fever, lymphadenopathy, vomiting, rash
NLRC4	NAIP-NLRC4	GOF	Constitutive activation of NLRC4 inflammasome	AIFEC, FCAS4	AD	Recurrent MAS, enterocolitis, urticaria, recurrent fever, CNS inflammation

(*Continued*)

Table 1 Genotypes considered as consistently associated with AIDs, according pathways description in the text.—cont'd

Gene	Protein	Mechanism	Effect	Disease	Inheritance	Main symptoms
NLRP1	NLRP1	GOF	Constitutive activation of NLRP1 inflammasome	NAIAD	AR or AD	Hyperkeratotic ulcers, vitiligo, arthritis
CEBPE	C/EBPe	GOF	Cytoskeleton abnormalities	CAIN	AD	Recurrent fevers, sterile abscesses, pyoderma gangrenosum
TNFRSF1A	TNFR1A	Misfolding	Activation of unfolded-protein response	TRAPS	AD	Recurrent fever, serositis, rash, AA amyloidosis
TRNT1	TRNT1	LOF	ER stress due to mature tRNA deficiency resulting in activation of unfolded-protein response	SIFD	AR	Immunodeficiency, recurrent fevers, panniculitis
COL1A1	COL1A1	Misfolding	Activation of unfolded-protein response	ICH	AD	Infantile cortical hyperostosis
WDR1	WDR1	LOF	Dysregulation of cytoskeletron result in activation of pyrin inflammasome	PFIT	AR	Thrombocytopenia, immunodeficiency, fever
ARPC1B	ARPC1B	LOF	Dysregulation of cytoskeletron resulting in ARP2/3-dependent F-actin polymerization deficiency	PLTEID	AR	Thrombocytopenia, immunodeficiency, vasculitis, fever

CDC42	CDC42	GOF	Dysregulation of cytoskeletron	NOCARH	AD	Fever, cytopenia, hemophagocytosis, rash, lymphedema, macrothrombo-cytopenia
CARD14	CARD14	GOF	Constitutive activation of NFkB in keratinocyte	CAMPS	AD	Pustolar psoriasis, Pityriasis rubra pylaris
IL36RN	IL36RA	LOF	Deficit of IL-36 receptor agonist	DITRA	AR	Pustolar psoriasis
IL1RN	IL1RA	LOF	Deficit of IL-1 receptor agonist	DIRA	AR	Pustulosis, CRMO
NOD2	CARD15	GOF	Constitutive activation of NOD2 signaling	Blau syndrome	AD	Arthritis, uveitis, rash
PLCG2	PLCg2	GOF	Increased production of inositol trisphosphate, diacylglycerol and calcium	APLAID, PLAID, FCAS3	AD	Colitis, ocular inflammation, cold-induced urticaria, infections
LACC1	LACC1	LOF	Deficit in endogenous fatty acid oxidation, peroxisomal dysfunction and decreased ROS production	LACC1 deficiency	AR	Arthritis, colitis, oral ulcers
RBCK1/RNF31/ *SHARPIN*	HOIL1/HOIP/ SHARPIN	LOF	Deficit in linear ubiquitination	LUBAC deficiency	AR	Fever, lymphangiectasia
OTULIN	OTULIN	LOF	Deficit in hydrolisis of linear Met1-linked ubiquitin chains	ORAS	AR	Lypodistrophy, arthralgia, GI inflammation

(*Continued*)

Table 1 Genotypes considered as consistently associated with AIDs, according pathways description in the text.—cont'd

Gene	Protein	Mechanism	Effect	Disease	Inheritance	Main symptoms
TNFAIP3	A20	LOF	Deficit in hydrolisis of Lys63-linked ubiquitin chains	HA20	AD	Arthritis, GI ulcers, ocular inflammation
PSMG2/A3/B4/8/9/10	PSMβ5i	LOF	Deficit in proteasome assembly	CANDLE	AR or AD	Lypodistrophy, fever, panniculitis, CNS calcification
POMP	POMP	LOF	Deficit in proteasome assembly	PRAID	AD	Neutrophilic dermatitis, autoimmunity, immunodeficiency
USP43	USP43	LOF	Deficit in proteasome assembly	USP43-AID	AR	Recurrent fevers, erythema nodosum, lipodystrophy and polyarthritis
TMEM173	STING	GOF	Constitutional activation of DNA sensor	SAVI	AD	Polyarthritis, interstitial lung disease, vasculitis
COPA	COPA	LOF	Deficit in retrograde Golgi-to-ER transport resulting in ER stress and Th17 upregulation	COPA	AD	Polyarthritis, interstitial lung disease, vasculitis
CECR1	ADA2	LOF	Deficit in growth factor ADA2 activity and stimulation of inflammatory macrophage response	DADA2	AR	Panarteritis nodosa, CNS stroke, livedo reticularis, immunodeficiency

Table 2 Genotypes associated with autoinflammatory manifestations.

Gene	Protein	Mechanism	Effect	Disease	Inheritance	Main symptoms
PSEN1	Presenilin1	LOF	Catalytic component of gamma-secretase	HS	AD	Sterile abscesses
PSENEN	Presenilin2	LOF	Component of gamma-secretase	HS	AD	Sterile abscesses
NCSTN	Nicastrin	LOF	Component of gamma-secretase	HS	AD	Sterile abscesses
NLRP7	NLRP7	GOF	Activation of NLRP7 inflammasome	VEO-IBD	AD	Ulcerative colitis, hydatidiform mole
NLRP12	NLRP12	GOF	Deficit in inhibition of ASC filaments nucleation resulting in accelerated activation of inflammasome	FCAS2	AD	Urticaria, recurrent fever
F12	FXII	GOF	Increased FXII amyloidotic activity	F12-associated FCAS	AD	Cold-induced urticaria
ADGRE2	ADGRE2	GOF	Deficit in inhibition of mast cells degranulation	Vibratory Urticaria	AD	Vibratory-induced orticaria
WAS1	WAS	LOF	Cytoskeleton abnormalities due to CDC42 signaling	WAS	XLR	Macrothrombocytopenia, eczema, recurrent infections
FHL1/PRF1/UNC13D/STX11/STXBP2	FHL1/Perforin/Munc13-4/synthaxin11/synthaxin-binding-protein2	LOF	Defective cytotoxicity	Familial HLH	AR	Recurrent MAS with fever, cytopenias, hepatosplenomegaly

(Continued)

Table 2 Genotypes associated with autoinflammatory manifestations.—cont'd

Gene	Protein	Mechanism	Effect	Disease	Inheritance	Main symptoms
HAVCR2	HAVCR2	LOF	Defective cytotoxicity	SPTCL	AR	HLH, panniculitis
RAG1/2	RAG1/2	LOF	Defective V(D)J recombination	CCHIDG	AD	HLH, granulomas, eosinophilia
STAT1	STAT1	GOF	Increase transcription of acute phase response genes	CVID	AD	Mucocutaneous candidiasis, recurrent infections, autoimmunity
STAT3	STAT3	GOF	Increase transcription of acute phase response genes	CVID, ADMIO1	AD	Autoimmunity, lymphoproliferation, recurrent infections, short stature
CTLA4	CTLA4	LOF	T cells costimulation defects	CVID, CHAI	AD	Hypogammaglobulinemia, cytopenia, lymphoproliferation, autoimmunity
LRBA	LRBA	LOF	Ligand-receptor complexes endocytosis dysregulation	CVID	AR	Hypogammaglobulinemia, interstitial lung disease, enteropathy, recurrent infections
TNFRSF11A	TNFR11A	GOF	Increased RANK signaling	TRAPS-like disease	AD	Recurrent fever, amelogenesis imperfecta
FBLIM1	FBLP1	LOF	Deficit of IL-10/STAT3/NR4A2-dependent inhibition of ERK1/2 phosphorylation resulting in increased RANKL production	Monogenic CRMO	AR	CRMO

SH3BP2	SH3BP2	GOF	Increased response to RANKL stimulation in myeloid cells and osteoclast	Cherubism	AD	Cyst-like overgrowth of the lower head bones
GALNT3	GALNT3	LOF	Deficit of O-glycosylation of FGF23 resulting in accumulation of fragmented FGF23 in osteocytes	HFTC1	AR	Cortical hyperostosis, CRMO, periarticular calcinosis, hyperphosphatemia
TGFB1	TGFb	GOF	Decreased osteoclastic resorption and increased osteoblastic bone formation	Camurati-Engelman disease	AD	Symmetrical long bone hyperostosis, cytopenias, recurrent fevers
ALPL	TNSALP	LOF	Deficit of tissue non specific alkaline phosphatase resulting in hyperprostaglandinism and demineralization	Hypophosphatasia	AR or AD	CRMO, fractures, premature loss of teeth
SLC29A3	ENT3	LOF	Defect of macrophage colony-stimulator factor clearance resulting in proliferation	PHID	AR	Recurrent fever, hyperpigmentation with hypertrichosis, dysmorphic features, hepatosplenomegaly, sensorineural deafness

(*Continued*)

Table 2 Genotypes associated with autoinflammatory manifestations.—cont'd

Gene	Protein	Mechanism	Effect	Disease	Inheritance	Main symptoms
FATP1	FATP1	LOF	Defect of long chain fatty acid transportation in classically activated macrophages resulting in increased pro-inflammatory glucose metabolism	MRS	AD	Recurrent facial nerve paralysis, emifacial swelling, fissured tongue
AP1S3	AP1	LOF	Deficit of TLR3-dependent IFN1β production resulting in hypersecretion of IL-1β	AP1 deficiency	AD	Pustolar psoriasis
IL10/ IL10RA/B	IL-10/IL10R	LOF	Deficit of IL-10 inhibition and STAT3 phosphorylation	VEO-IBD	AR	Colitis, failure to thrive
TRIM22	TRIM22	LOF	Deficit in K63-linked polyubiquitination of NOD2	VEO-IBD	AR	Colitis, severe perianal disease, infections
XIAP	XIAP	LOF	Impaired NOD2 signaling and lymphocytes apoptosis	VEO-IBD	XL	Colitis, failure to thrive, lymphoproliferation
NFIL3	NFIL3	LOF	Excessive IL-1β and TNFα production in macrophages	NFIL3 deficiency	AR	Oligoarthritis, autoimmune thyroiditis, positive HLA-B27, ANA, and pANCA
LYN	LYN	GOF	Costitutive activation of tyrosin kinase	LAID	AD	Vanishing bile duct disease, vasculitis, cytopenia

SKIV2L	SKIV2L	LOF	Loss of cytosolic RNA exosome function in RNA turnover	THES	AR	Severe diarrhea, immunodeficiency, trichorrhexis nodosa, facial dysmorphism
PDGFRB	PDGFRB	GOF	Activation of STAT3	Penttinen syndrome	AD	Lypoatrophy, scar-like skin lesions, hair abnormalities, hypereosinophilia
JAK1	JAK1	GOF	Activation of STAT1/3	Hypereosinophilic syndrome	AD	Severe dermatitis, hepatosplenomegaly, hypereosinophilia
TREX1	DNAse3	LOF	Deficit of exonuclease resulting in accumulation of nucleic acid	AGS1, RVCL	AR or AD	CNS calcification, SLE, chilblains
RNASEH2A/B/C	RNH2A/B/C	LOF	Deficit of RNAse resulting in accumulation of nucleic acid	AGS4/3/2	AR	CNS calcification
SAMHD1	SAMH	LOF	Deficit of nuclease resulting in accumulation of nucleic acid	AGS5	AR	CNS calcification, chilblain, oral ulcers, deforming arthropathy, cerebral vasculopathy with erly onset stroke
ADAR1	DSRAD	LOF	Deficit of RNA deaminase resulting in accumulation of nucleic acid	AGS6, DSH	AR/AD	CNS calcification, dyschromatosiss symmetrica hereditaria

(*Continued*)

Table 2 Genotypes associated with autoinflammatory manifestations.—cont'd

Gene	Protein	Mechanism	Effect	Disease	Inheritance	Main symptoms
IFIH1	MDA5	GOF	Activation of RNA sensor	AGS7, SMS	AD	Aortic calcification, glaucoma, psoriasis, skeletal abnormalities with dental dysplasia
DDX58	RIG1	GOF	Activation of dsRNA sensor	SMS2	AD	Aortic calcification, glaucoma, psoriasis, skeletal abnormalities with acro-osteolysis
C1qA/C1qB/C1qC/C1r/C1s/C2/C3/C4a	C1q/C1r/C1s/C2/C3/C4a	LOF	Deficit of immune complex clearance resulting in activation of autoreactive B cells	Monogenic SLE	AR	SLE-like
DNASE1/DNASE1L3/DNASE2	DNAse1/DNAse1L3/DNAse2	LOF	Deficit of deoxyribonuclease resulting in accumulation of nucleic acid	Monogenic SLE	AD or AR	SLE-like
ACP5	TRAP	LOF	Accumulation of phosphorylated osteopontin in peripheral dendritic cells	SPENCDI	AR	Scleroderma, bone dysplasia, spasticity
ISG15	ISG15	LOF	Reduced intracellular IFNα/β-inducer USP18 accumulation	MSMD	AR	CNS calcification, mycobacterial infections
USP18	USP18	LOF	Reduced intracellular IFNα/β-inducer USP18 accumulation	PTORCH2	AR	CNS calcification, liver dysfunction, cytopenia

POLA1	POLA1	LOF	Deficit of DNA polymerase alpha resulting in deficit of cytosolic RNA:DNA hybrids and costitutive activation of IFN regulatory factor and NFkB	XLPDR	XLR	Recurrent pneumonia, diffuse reticulate hyperpigmentation, hypohydrosis, colitis, uretral stenosis, corneal inflammation, failure to thrive
PNPT1	PNP	LOF	Mitochondrial double-stranded RNA accumulation resulting in interferon upregulation	Mitochondrial disease	AR	Infantile encephalopathy, movement disorder
NGLY1	N-glicase1	LOF	Accumulation of misfolded glycoprotein resulting in interferon upregulation	Mitochondrial disease	AR	Infantile encephalopathy, movement disorder, hypo/alocrimia, thyroid and adrenal insufficiency

In recent years, it has been shown that the secretion of IL-1 and IL-18 mediated by inflammasomes needs the cleavage of gasdermine D induced by caspase-1 and translocation of the N-terminal fragment towards the plasma membrane (Xiao et al., 2018; Zhao, Shi, & Shao, 2018). At this level, the gasdermine D forms pores by binding to membrane phospholipids through which cytokine secretion occurs. A continuous activation of gasdermine D and the excessive formation of pores at the level of the plasma membrane determines the inflammatory form of programmed cell death, called pyroptosis, which is characterized by cellular edema, loss of the plasma membrane integrity with consequent release of inflammatory cytokines in the extracellular environment. Human CAPS murine models demonstrated a close correlation between NLRP3 inflammasome activation, gasdermine D maturation and subsequent pyroptosis, confirming that gasdermine D is an important effector of this pathway. Recently, a novel knock-in murine model of CAPS demonstrated the development of AA amyloidosis and partial response to proton pump inhibitors (Bertoni et al., 2019). Thanks to the demonstration of the NLRP3 involvement in atherosclerosis and other common adult diseases (Moossavi, Parsamanesh, Bahrami, Atkin, & Sahebkar, 2018; Wang et al., 2019), studies about molecules able to regulate the NLRP3-dependent inflammation are greatly increasing during the recent years (Birnbaum, Tran, Bajaj, & Ye, 2019; Hong et al., 2019; Hu, Zhang, Yi, Zhou, & Mi, 2018; Jiang et al., 2018; Pellegrini, Fornai, Antonioli, Blandizzi, & Calderone, 2019; Pu et al., 2019; Wang et al., 2019; Wu et al., 2018) and intramolecular/intracellular regulators have also been discovered (Hafner-Bratkovic et al., 2018; Poudel & Gurung, 2018). Of particular interest, lipin-2 protein acts as a magnesium-dependent phosphatidate phosphatase enzyme which catalyzes the conversion of phosphatidic acid to diacylglycerol during triglyceride, phosphatidylcholine and phosphatidylethanolamine biosynthesis in the reticulum endoplasmic membrane. Lipin-2 restricts overphosphorylation of protein kinases with extracellular regulation (ERK), p38, and c-Jun NH2-terminal Kinase (JNK), thus inhibiting the activation of the NLRP3 inflammasome (Lorden et al., 2017). Mutations of the *LPIN2* gene coding for lipin-2 protein are related to the Majeed syndrome (MJDS), a rare AID characterized by recurrent fever, congenital dyserythropoietic anemia and sterile multifocal osteomyelitis, whose patients respond to IL-1 inhibition (Table 1) (Roy et al., 2019).

2.1.2 *Pyrin inflammasome*

FMF and the pyrin-associated autoinflammation with neutrophilic dermatosis (PAAND) are monogenic AIDs both related to variants in different domains of the *MEFV* gene and associated with an excessive activation of the pyrin inflammasome (Table 1) (Hong et al., 2019; Moghaddas et al., 2017). Pyrin consists of four domains with different functions (Heilig & Broz, 2018). The N-terminal domain is necessary for the formation of the pyrin inflammasome. Other domains seem to have a self-inhibitory role. In particular, mutations in the C-terminal domain have been associated with FMF. Other FMF-related mutations affect a region of the protein close to the N-terminal domain. In the stationary conditions, this protein region is maintained phosphorylated by kinases dependent on molecules that are sensitive to any cytoskeletal alterations. In case of aggression by pathogens not previously encountered by the immune system, the perturbation induced in the organization of the cytoskeleton is able to activate the pyrin inflammasome causing the secretion of IL-1β and guaranteeing an adequate inflammatory response (Heilig & Broz, 2018; Schnappauf, Chae, Kastner, & Aksentijevich, 2019). Colchicine is able to stabilize the cytoskeleton of the cell, limiting the pro-inflammatory effect of a pyrin that is more prone to activation due to specific *MEFV* mutations (Jamilloux et al., 2018; Balci-Peynircioglu et al., 2018).

On the other side, constitutive activation of the pyrin is able to affect cytoskeleton-dependent properties of neutrophils. In particular, neutrophilic dermatosis, peripheral neutropenia, and sterile abscesses have been reported in response to mild stimuli in patients with PAAND, the pyogenic sterile arthritis, pyoderma gangrenosum, acne (PAPA) syndrome, and the PSTPIP1-associated myeloid-related proteinemia inflammatory syndrome (PAMI), these two latter conditions both related to *PSTPIP1* gene mutations (Table 1) (Hashmi et al., 2019; Klotgen et al., 2018).

Hidrosadenitis supporativa (HS), also known as acne inversa, is a long-term skin disease characterized by recurrent inflammatory sterile abscesses. HS manifestations may occur in patients with PAPA syndrome (Saito et al., 2018) and genes related to familial forms of HS are newly recognized AID-related genes (Table 2) (Vossen, van der Zee, & Prens, 2018). In this group of disorders, treatment with IL-1 inhibitor may be effective, despite anti-tumor necrosis factor (TNF) or IL-6 inhibitors are the first line biological therapy for HS (Armyra, Kouris, Markantoni, Katsambas, & Kontochristopoulos, 2017; Negus, Ahn, & Huang, 2018; Magalhaes et al., 2019).

Homozygous or compound heterozygous mutations in the *MVK* gene resulting in a loss of function (LOF) of the mevalonate kinase enzyme greater than 1% cause MKD. Patients with MKD usually present with recurrent febrile episodes associated with cervical lymphadenopathy, abdominal pain with vomiting and diarrhea, arthralgia and cutaneous manifestations; atypical manifestations may occurs (Coppola, Becken, Van Mater, McDonald, & Panayotti, 2019; Pietrasanta et al., 2019). Less than 0.5% of the enzymatic activity determines the more severe disease manifestation, also call mevalonic aciduria, characterized by the early development of systemic inflammation with facial dysmorphisms, growth delay, convulsions and progressive cerebellar ataxia. Patients with an intermediate phenotype can present at birth with persistent inflammation. The pathophysiology of inflammation in MKD is related to the loss of isoprenoid lipids synthesis and, in particular, with the geranylgeranyl diphosphate deficiency. The geranylgeranyl diphosphate is necessary for the prenylation of small GTPases, as well as RhoA, whose inactivation causes constitutive activation the pyrin inflammasome, IL-1 release and inflammation.

On this basis, the main therapeutic attempts for patients with MKD were focused on the IL-1 inhibition, despite the response is partial in 70% of patients. More recently Canakinumab, an anti-IL1β monoclonal antibody, has shown superior efficacy, with a complete response in more than 50% of patients (De Benedetti et al., 2018). These patients may require higher dose and frequency of drug administration compared to patients with other periodic fevers (Zhuang et al., 2018). The involvement of other pro-inflammatory cytokines, including IL-6 and TNF is increasingly emerging in the pathogenesis of MKD. This evidence may explain the limited efficacy of IL-1 blockade and supports alternative pharmacological treatments. Tocilizumab is a humanized monoclonal antibody against the IL-6 receptor that binds to the soluble and membrane receptor of this cytokine, inhibiting IL-6 mediated signaling. Limited experience on the use of tocilizumab in MKD has been reported in the literature (Rafiq, Lachmann, Joensen, Herlin, & Brogan, 2018). However, blockade of IL-6 should represent the third-line treatment before considering hematopoietic stem cell transplantation (Giardino et al., 2015).

2.1.3 Other inflammasomes

Patients with autoinflammation and infantile enterocolitis (AIFEC) carry a heterozygous GOF variant of the *NLRC4* gene and may present with fatal episodes of macrophage activation syndrome (MAS). An exaggerated IL-18 production is the *primu movens* alternative to IL-1, whose inhibition is not effective in these patients. Murine models lacking IL-18 inhibition or with constitutively active IL-18 highlighted the role of toll-like receptor 9 and intestinal mucosa in hyperferritinemic syndromes (Girard-Guyonvarc'h et al., 2018; Weiss et al., 2018). Interestingly, according to these evidences, a patient with AIFEC resistant to treatment with humanized anti-IFN gamma was treated with transplant of maternal stool with benefit (Benedetti et al., 2019; Bracaglia et al., 2019). In healthy individuals, the IFN regulatory factor 8 is a critical regulator of NLRC4 inflammasome during the host defense against Gram-negative bacteria (Karki et al., 2018). The nucleotide-binding domain and leucine-rich repeat receptor family apoptosis inhibitory proteins are the cytosolic receptors that recognize conserved bacterial proteins, as flagellin. IFN regulatory factor 8 is required for the transcription of genes encoding this group of cytosolic receptors that activate the NLRC4 inflammasome, leading to cell death and IL-1/IL-18 secretion.

Patients with C/EBPe-associated autoinflammation and immune impairment of neutrophils (CAIN) present recurrent fevers lasting 4–5 days, every 2–4 weeks, characterized by abdominal pain and oral aphthosis (Goos et al., 2019). Sterile abscesses of the tongue, paronychia, and hemorrhagic diathesis not related to platelets abnormalities are associated. Laboratory tests show high acute phase reactants and defects in segmentation of neutrophils. The genetic cause is a homozygous variant of the *CEBPE* gene, coding for the transcription factor CRP1 or C/EBPe. The reduced ability of binding transcriptional repressors results in over-expression of many NLRP3-related inflammatory genes and hypersecretion of IL-1 and IL-18 by neutrophils. Patients do not suffer of serious long-term complications, presenting an average life span of more than 70 years. However, the reduced quality of life can suggest using IL-1 or IL-18 inhibitors.

On the other side, the evidences of NLRP12 and NLRP7-related AID are scanty and many different inflammatory manifestations have been associated with *NLRP12* and *NLRP7* gene variants (Basaran et al., 2018; Ghosh, Mishra, Shah, Patel, & Shetty, 2019; Jacob et al., 2019; Kostik et al., 2018). The high prevalence of variants of unknown significance in the general population may suggests to be very careful in considering a pathogenic role of the NLRP12 or NLRP7 variants (Table 2).

Rare variants of the *NLRP1* gene cause LOF of the auto-inhibitory N-terminal domain of the NLRP1-inflammasome that acts in keratinocytes preventing the development of the NLRP1-associated autoinflammatory disease with arthritis and dyskeratosis (NAIAD) (Grandemange et al.,

2017; Yu, Moecking, Geyer, & Masters, 2018). The anthrax lethal factor metalloprotease and small-molecule DPP8/9 inhibitors are both able to cleavage the N-terminal domain of the NLRP1, activating the NLRP1-dependent inflammatory response (Chui et al., 2019; Lacey & Miao, 2019; Zhong et al., 2018). NLRP1 inflammasome seems to have an antibacterial role in mucosal innate immunity, preventing the development of inflammatory bowel disease (IBD) (Li et al., 2018; Tye et al., 2018).

Patients with mutations in the *F12* or *ADGRE2* gene also present autoinflammatory urticarial manifestations (Table 2).

2.2 Immuno-proteinopathies

Proteinopathies are a group of diseases where proteins are abnormally associating and aggregating, due to pathological conformational changes. Intracellular deposition of misfolded proteins causes endoplasmic reticulum stress and pro-inflammatory reaction called the unfolded proteins response (UPR). Here, we propose to collect AIDs causing UPR activation under the term of immune-proteinopathies to distinguish them from the neurologic diseases related to misfolded proteins accumulation.

The TNF receptor-associated periodic syndrome (TRAPS) is one of the best-known monogenic hereditary recurrent fever syndromes, resulting from an autosomal dominant variant of the *TNFRSF1A* gene (McDermott et al., 1999). TRAPS patients usually present recurrent episodes of fever, abdominal, chest or limb pain, rash, ocular symptoms, and up to 10% develop systemic AA amyloidosis (Lachmann, Papa, & Minden, 2014). On demand anti-inflammatory drugs in case of flares and regular blood and urine tests to check high acute phase reactants and proteinuria during the symptoms free intervals are required. NSAIDS or steroids are effective in the short term, but colchicine, DMARDs and biologics may be needed as maintenance therapy (ter Haar et al., 2015). Initially, it was hypothesized that TNF receptor mutations are able to increase the binding affinity or induce a state of constitutive activation even in the absence of TNF, but both hypotheses were incorrect (Kimberley, Lobito, Siegel, & Screaton, 2007). Under physiological conditions, the extra-cytosolic portions of TNF receptor undergo a cleaving mediated by the transmembrane glycoprotein called ADAM17 or TNF-converting enzyme 23. Following the shedding, a soluble pool of TNF binding proteins prevents the TNF contact with the receptors on the cell membrane, blocking the transmission of the pro-inflammatory signal. The presence of a receptor-shedding defect in TRAPS patients has been demonstrated in 1999 (McDermott et al.,

1999). On this basis, a therapeutic role for the TNF receptor fusion protein was hypothesized but later not confirmed (Jesus et al., 2008). Subsequent studies have shown that the pathogenic mechanisms related to TRAPS are much more complex. Mouse model showed that TRAPS mutations cause a defect in the transport of TNF receptor towards the cell membrane, leading to its accumulation within the endoplasmic reticulum and consequent formation of aggregates able to induce UPR independently of the extracellular TNF binding. A protein sensor was identified in the endoplasmic reticulum of hepatocytes, called cAMP-responsive-element-binding protein H that is capable of directly increasing the expression of acute phase proteins, including the serum amyloid A. The accumulation of the TNF receptor in the cytoplasm and Golgi has been also associated with nuclear factor kappa-B (NFkB) hyper activity. Furthermore, the exaggerated inflammatory response observable in the TRAPS patient also appears to be supported by the activation of the p38 and JNK proteins. Together with ERK, p38 and JNK constitute the family of the mitogen-activated protein kinases that are crucial in the recruitment and activation of leukocytes at the inflammatory site. The intra-cytosolic accumulation of the unfolded protein is able to inhibit the phosphatases that normally block mitogen-activated protein kinases activity (Greco et al., 2015; Harrison et al., 2018; Tsuji et al., 2019). This data has been demonstrated by the fact that homozygous knock-in mice for TNF receptor mutations are protected from any inflammatory insult. It is therefore likely that the induction of the inflammatory response in TRAPS requires the simultaneous presence of the mutated receptor and the wild type receptor on the membrane, as indeed happens in TRAPS subjects who have 50% of normal membrane receptors.

The UPR is also believe to be the relevant in the sideroblastic anemia with B-cell immunodeficiency, periodic fever and developmental delay (SIFD), as well as in infantile cortical hyperostosis (ICH), also known as Caffey disease (Table 2).

Patients with SIFD present with persistent systemic inflammation since first weeks of life, associated with mucocutaneous and musculoskeletal manifestations, as septal panniculitis and mononuclear myofasciitis (Giannelou et al., 2018). The need for blood transfusions increase during flares (Barton, Kausar, Kerr, Bitetti, & Wynn, 2018; Mariani & Gong, 2018). SIFD is caused by homozygous or compound heterozygous LOF mutation in the *TRNT1* gene, coding for the tRNA nucleotidyltransferase cytidine-cytidine-adenosine-adding 1 that is essential for maturation of tRNAs, which accumulation causes higher spontaneous production of reactive oxygen species and NLRP3-dependent inflammatory cytokines in cultured patients cells secondary to augmented endoplasmic reticulum stress response. Activation of UPR causes the abnormal B cell maturation leading to Ig deficiency (Kumaki et al., 2019). Furthermore, mutant cells showed deficient clearance of the ubiquitin-scaffold protein p62, which is a substrate for the ubiquitine-proteasome system (UPS), which primary defects are associated with other AIDs, as described later in details. Anti-TNF treatments are more effective then IL-1 inhibitors in these patients.

Caffey disease is a familial disorder manifesting in the late fetal period or infancy with excessive periosteal bone formation (Kirby & Wright, 2019). Signs and symptoms regress spontaneously within months. Caffey diseases is caused by the c.3040C>T mutation of the *COL1A1* gene. The consequent p.Arg836Cys substitution leads to increased disulfide crosslinking within or between collagen fibrils, causing the UPR activation. The *COL1A1* mutations in the intron 5 and 6 have been associated with Ehlers-Danlos syndrome, and exon mutations predominately cause osteogenesis imperfecta (Lu, Zhang, Wang, Ren, & Han, 2019).

2.3 Immuno-actinopathies

Abnormal regulation of the cytoskeleton assembly determines complex clinical pictures that associate inflammatory manifestations and immunodeficiency. Actin is a major component of cytoskeletron architecture, and then we propose immuno-actinopathies as name to refer to this group of conditions in which there is a clear overlap between pure autoinflammation and immune-dysregulation.

Wiskott Aldricht syndrome (WAS) is an X-linked immunodeficiency characterized by the clinical triad of microthrombocytopenia, eczema, and recurrent infections (Table 2). The syndrome has a wide clinical spectrum ranging from autoimmunity to atopy and autoinflammation (Candotti, 2018). Autoinflammatory manifestations typically present early in life, are often refractory to therapy, and are associated with a worse clinical prognosis and an increased risk of developing a malignancy (Moratto et al., 2011). Deficiency of WAS protein causes defects in chemotaxis and podosomes formation, leading to abnormal autophagosome and inflammasome function and thus favoring the onset of autoinflammatory manifestations (Lee et al., 2017).

The autoinflammatory periodic fever, immunodeficiency, and thrombocytopenia (PFIT) syndrome is caused by mutation in the actin-regulatory

WDR1 gene, coding for the WD-repeat protein 1 (Table 1) (Standing et al., 2017). The protein plays a critical role in dynamic reorganization of the cytoskeleton (Ono, 2018). Patients present periodic fevers lasting 3–7 days, every 6–12 weeks, with severe acute phase reactants, hyperferritinaemia, and oral mucosal inflammation, causing an acquired microstomia. Severe infections, recurrent perianal ulceration, and pyoderma gangrenosum were also observed. Platelets abnormalities characterize also the platelet abnormalities with eosinophilia and immune-mediated inflammatory disease (PLTEID) caused by bi-allelic mutations of the *ARPC1B* gene (Brigida et al., 2018; Kahr et al., 2017; Kuijpers et al., 2017; Somech et al., 2017; Volpi et al., 2019). The disease is characterized by systemic inflammation with lympho-proliferation and immunodeficiency resembling the Wiskott-Aldrich syndrome, with early onset vasculitis, severe infections, eczema. Laboratory tests show microthrombocytopenia, eosinophilia, T lymphopenia with reduced T naïve cells and high serum levels of immunoglobulin (Ig)E and IgA. Functional studies can detect the defective migration and proliferation of T lymphocytes. A functional fMLP/phalloidin test can differentiate symptomatic patients from asymptomatic mutation carriers (Kopitar et al., 2019). Patients with PLTEID present low production of ARPC1B protein, determining a defect of the protein complex adepts to regulate the elongation of actin filaments in the blood cells.

A new monogenic AID characterized by IL-18 hypersecretion related to cytoskeletal abnormalities has diagnosed in four patients presented with Neonatal onset cytopenia with autoinflammation, rash, and hemophagocytosis (NOCARH) (Gernez et al., 2019). Patients present with fever, hepatosplenomegaly and urticarial rash. Growth retardation and facial dimorphisms similar to those are present in patients with cryopirinopathy or Noonan syndrome were noted (Martinelli et al., 2018). Laboratory tests showed increased acute phase reactants, high serum concentration of IL-18 and transfusion-dependent cytopenia. Whole exome sequencing highlights three stop-codon variants of the *CDC42* gene, coding for the Cell division control protein 42 homolog (CDC42) protein. The variants involve the C-terminal domain that normally allows CDC42 binding to lysosomal membrane containing phosphatidyl-inositol-4,5-bisphosphate, necessary for actin assembly. Patients respond to IL-1 inhibitor with complete control of inflammatory symptoms and growth re-catch. However, severe infections occurred during the follow-up, in association with persistence of increased IL-18 concentration.

An increased IL-18 concentration also characterizes familial forms of hemophagocytic lymphoistiocytosis (HLH) and other diseases with immunodysregulation (Table 2).

2.4 Interleukin receptors signaling defects

2.4.1 IL-1R and IL-36R signaling defects

Generalized pustular psoriasis (GPP) is an extremely rare type of psoriasis, presenting with a whole body-covered pus-filled blisters that can be associated with systemic inflammation (Ly, Beck, Smith, Thibodeaux, & Bhutani, 2019). Three AIDs are characterized by GPP caused by inflammation of the resident cells of the skin, namely keratinocytes: the IL-1 receptor antagonist deficiency (DIRA), the IL-36 receptor antagonist deficiency (DITRA), and the caspase recruitment domain (CARD) 14-mediated psoriasis (CAMPS). A patient with coexisting mutations of the *CARD14* and *IL36RN* genes have been also described (Sawabe et al., 2019). Furthermore, the AP1 deficiency, secondary to specific polymorphisms of the *AP1S3* gene, seems to modify symptoms severity in patients with DITRA (Mahil et al., 2016; Setta-Kaffetzi et al., 2014; Yamaguchi et al., 2009). Anti IL-1 treatment is the first-line treatment in patients with DIRA, that usually presents more significant signs of systemic inflammation then patients with DITRA or CAMPS, as persistent fever and osteitis (Gomez-Garcia et al., 2018; Hospach et al., 2019; Sozeri, Gerceker-Turk, Yildiz-Atikan, Mir, & Berdeli, 2018). Anti IL-1 treatment is less effective in patients with DITRA who dramatically respond to Ustekinumab. The availability of recombinant IL-36 receptor antagonist will represents the treatment of choice in this condition (Bonekamp et al., 2018; Cherqaoui, Rossi-Semerano, Piram, & Kone-Paut, 2018; Bonekamp et al., 2018).

2.4.2 IL10R signaling defects

Patients with rare forms of very early-onset (VEO)-IBD may carry mutations of genes coding for IL10 or IL-10 receptor subunit A and B. Furthermore, two siblings presenting with early-onset ANA-positive multidrug-resistant oligoarthritis carried LOF variants of the *NFIL3* gene that has been related to the IL-10 signaling defects in mice and IBD susceptibility in humans.

2.5 Relopathies

Hyper activation of the NFkB/Rel, leading to inflammatory manifestations, is the final pathway involved in many AIDs, as well as the CARD14 and

CARD15/NOD2 signaling defects, and the newly discovered group of disorders with abnormal ubiquitination (Steiner, Harapas, Masters, & Davidson, 2018).

2.5.1 CARD14 and CARD15/NOD2 signaling defects

CARD14 is a proinflammatory signaling molecule whose expression is predominantly restricted to epidermal keratinocytes (Israel & Mellett, 2018; Satoh, Mellett, Contassot, & French, 2018). The N-terminal domain acts as NFkB activator thought blocking the activity of the inhibitor of kappa B kinase complex. The C-terminal domain seems to have an auto-regulatory activity. Dominant GOF mutations in the *CARD14* gene are associated with psoriasis and LOF mutations are related to severe variant of atopic dermatitis (Peled et al., 2019; Zotti, Polvere, Voccola, Vito, & Stilo, 2018). Patients with CAMPS may present with a wide range of psoriasis-like manifestations, from palmoplantar pustulosis to papulosquamous eruptions, requiring high doses of Ustekinumab (Craiglow et al., 2018; Signa et al., 2019; Fu et al., 2019; Li et al., 2019; Ohnishi et al., 2018).

Patients carrying a variant of the CARD15/*NOD2* gene can be affected by Blau's syndrome, early onset sarcoidosis, or be more prone to develop Crohn's disease. The number of NOD2 variants considered likely benign or of unknown significance according the classification of the International study group for systemic autoinflammatory diseases, freely available at the Infevers web-site (https://infevers.umai-montpellier.fr/web/), are numerous and therefore caution should be used in confirming clinical diagnosis especially in case of the presence of intronic variants (Touitou et al., 2013; Venkataraman & Rivas, 2019; Yao, Li, & Shen, 2019).

The *NOD2* gene coding for a cytosolic pattern recognition receptor and is composed of three domains.

Variants in the C-terminal leucine-rich-repeat domain, which is essential for the muramyl-dipeptide sensing ability of NOD2, are associated with susceptibility to Crohn's disease in relation to their ability to disrupt intestinal mucosal homeostasis (Irazoki, Hernandez, & Cava, 2019; Girardelli et al., 2018a,b).

Intriguingly, stimulation of innate immunity by administration of NOD2 agonists represents a new way to prevent and treat infectious diseases and gastrointestinal cancer (Ferrand et al., 2019; Pashenkov, Murugina, Budikhina, & Pinegin, 2019). On the other hand, NOD2 variants in the central nucleotide binding and oligomerization domain of the protein, which acts as auto-inhibitor domain, cause Blau's syndrome. The classic triad of clinical manifestations are boggy arthritis, panuveitis, and granulomatous dermatitis, despite atypical sites of granulomas have been reported (Cropley, Ashrafy, & Weltman, 2019; Imayoshi, Ogata, & Yamamoto, 2018; Jindal et al., 2019). The severity of articular involvement, even if isolated, or multi-drug resistant uveitis may require anti IL-6 treatment (Toral-Lopez, Gonzalez-Huerta, Martin-Del Campo, Messina-Baas, & Cuevas-Covarrubias, 2018; Wang et al., 2018; Yamasaki et al., 2019; Holzinger, Foell, & Kessel, 2018; Lu et al., 2018; Sarens et al., 2018; Holland & Rosenbaum, 2018; Naik, Annamalai, & Biswas, 2018; Millington et al., 2019; DeSouza & Shah, 2019).

Patients with rare forms of very early-onset (VEO)-IBD may carry mutations of genes whose encoded protein affects the NOD2 signaling and, in particular, those with TRIM22 and XIAP deficiency (Table 2).

2.5.2 Ubiquitination defects

The ubiquitin-proteasome system (USP) allows clearance of abnormal intracellular proteins. Proteasome is a intracellular multi-protein complex able to degrade molecules bended to ubiquitin chains thanks to the action of the Linear ubiquitin chain assembly complex (LUBAC) (Peltzer et al., 2018; Taraborrelli et al., 2018). Negative regulation of the NFκB pathway, which is essential to stop the inflammatory response, depends on the level of ubiquitination of proteins associated with TNF receptor 1 signaling. LUBAC consists of at least three proteins: ring finger protein 31 (RNF31/HOIP), RanBP-type and C3HC4-type zinc finger containing 1 (RBCK1/HOIL1) and SHANK-associated RH domain interacting protein (SHARPIN/SIPL1). The clinical manifestations presented by patients with the *HOIL1* gene variants are amylopectinosis, chronic inflammation and severe bacterial infections (Steiner et al., 2018). HOIP deficiency has been described in a patient with recurrent fevers, chronic steatorrhea and severe bacterial infections (Boisson et al., 2015). A second HOIP patient showed early-onset seronegative polyarthritis that was resistant to steroids, methotrexate and TNF inhibitors (Oda et al., 2019). Generalized epilepsy and eczematous dermatitis were also associated. The skin biopsy showed an extensive $CD4^+$ cells infiltration. Laboratory tests showed high acute phase reactants, low number of memory B cells with hypogammaglobulinemia, and no evidence of immune response to vaccination against Pneumococcus. SHARPIN deficiency was recently described by Oda et al. and characterized by colitis, polyarthritis, mumps and hepato-splenomegaly with partial response to anti IL-6 therapy (ISSAID, 2019).

On the other hand, A20 and OTU deubiquitinase with linear linkage specificity (OTULIN) removes linear polyubiquitin from proteins that have been modified by LUBAC (Heger et al., 2018). Patients with OTULIN-related autoinflammatory syndrome (ORAS) showed relapsing episodes of nodular panniculitis with neutrophil infiltrate and fevers. Reduced growth parameters and painful swollen joints, as well as elevated immunoglobulin levels and serum autoantibodies were reported. Patients with A20 haploinsufficiency (HA20) manifest a Behçet's disease-like syndrome with oral and genital ulcers and, rarely, uveitis. Transmission is dominant, symptoms occurs at early age, and severe inflammatory manifestations, as well as inflammatory bowel disease, arthritis, and thyroiditis, can be associated (Berteau et al., 2018; Hori et al., 2019; Li et al., 2019; Tsuchida et al., 2019). Treatment with TNF and IL-1 inhibitors is partially effective (Damgaard et al., 2019). However, other factors seems to be involved in the inflammatory pathways of relopathies, and combined biologics therapy may be required (Polykratis et al., 2019; Talpin et al., 2019; Vetters et al., 2019).

The interest of neurologists in LUBAC complex is increasing due to the evidence of an impact into the valosin containing protein-related diseases (van Well et al., 2019).

2.6 Interferonopathies

Interferonopathies are a group of monogenic AIDs characterized by severe inflammatory manifestations caused by high secretion of IFNα/β. The type I IFN-dependent inflammatory response is a highly conserved innate immune pathway against viral infections. To the date, several monogenic diseases have been included in this group (Table 2). High production of type I IFN-related mRNAs, called IFN signature, is commonly used biomarker in these patients (Kim et al., 2018a; Lamot, Niemietz, & Brown, 2019; Pescarmona et al., 2019). The discovery of the underlying molecular mechanism leading to the IFNα/β secretion led to the development of target drugs called JAK/STAT inhibitors (Sanchez et al., 2018; Fragoulis, McInnes, & Siebert, 2019).

2.6.1 Proteasome-associated autoinflammatory syndromes

The UPS is also involved in a group of interferonopathies called proteasome-associated autoinflammatory syndromes (PRAASs).

The Chronic atypical neutrophilic dermatosis with lipodystrophy and elevated temperature (CANDLE) is the PRAAS paradigm causing by

LOF mutations in the *PSMB4*, *PSMB8*, *PSMB9*, *PSMA3* genes that encodes the proteasome subunits (Torrelo, 2017). Clinically, patients with CANDLE present with fever, neutrophilic panniculitis that leads to lipodystrophy, joint contractures, myopathy, and aseptic lymphocytic meningitis. During the first decade of life, almost half of the patients develop systemic arterial hypertension, metabolic syndrome, autoimmunity and hepatic steatosis (Al-Mayouf et al., 2018; Yamazaki-Nakashimada et al., 2019). Intellectual disability, generalized lipodystrophy and joint contractures have been reported as long-term outcomes in these patients. Similar clinical picture has been described in patients carrying mutations of proteins that regulate the catalytic activity of proteasome. In particular, an Algerian patient with early onset recurrent fevers, polymorphic rash and hepatosplenomegaly, carrying homozygous LOF mutations of the *PSMB10* gene, has been described as the last proteasome subunit-related PRAAS (ISSAID, 2019).

Proteasome maturation protein (POMP) is a chaperone for proteasome assembly and is critical for the incorporation of catalytic subunits into the proteasome. Heterozygous frame shift variants of the *POMP* gene result in a truncated protein that perturbs proteasome assembly by a dominant-negative mechanism. POMP-related autoinflammation and immune dysregulation disease (PRAID) has been discovered in two unrelated individuals with early-onset combined immunodeficiency, neutrophilic dermatosis, and autoimmunity (Poli et al., 2018).

Heterozygous LOF mutations of the *PSMG2* gene, coding for the PAC2 protein, whose mutations result in a lower proteolytic activity of the UPS, has been found by Jesus et al. in patient with early-onset recurrent fevers, panniculitis, edema of the lower extremities (de Jesus et al., 2019). Hepatosplenomegaly, generalized lymph adenomegaly, transfusion-dependent hemolytic anemia, and myositis were associated.

Homozygous LOF mutations of the *USP43* gene, coding for a protein capable in deubiquitinating proteasome subunits, cause early onset recurrent fevers, lipodystrophy and polyarthritis in a Tartar patient (ISSAID, 2019). The lack of such intracellular regulation also causes a reduced proteolytic activity of the proteasome.

2.6.2 Other autoinflammatory interferonopathies

The STING associated vasculopathy with onset in infancy (SAVI) is caused by GOF mutations in the STING viral sensor (Ergun, Fernandez, Weiss, & Li, 2019; Kumar, 2019). Patients show interstitial lung disease associated with peripheral ulcerative vasculitis of extremities, rarely leading to amputation. Interstitial lung disease also characterize patients affected by COPA syndrome that usually present early-onset symmetric polyarthritis at the disease onset (Volpi et al., 2018; Noorelahi, Perez, & Otero, 2018).

Patients with PRAAS, SAVI and COPA syndrome poorly respond to DMARDS or biological drugs that inhibit primary inflammatory cytokines (IL-1, TNF and IL-6). Baricitinib, leading to the block of STAT phosphorylation, has been proven some efficacy, despite viral respiratory infections and poliomavirus reactivation may developed, rarely leading to acute renal failure (Kim et al., 2018; Volpi et al., 2019). New target therapies are under studying (Hansen et al., 2018).

2.7 Others

The adenosine deaminase (ADA) 2 deficiency (DADA2) is a complex systemic AID in which vasculopathy that histologically resembles panarteritis nodosa, dysregulated immune function, and/or cytopenias predominate (Caorsi et al., 2017; Grossi et al., 2019). ADA2 acts as a growth-factor for monocytes and endothelial cells, altering the balance between pro and anti-inflammatory macrophages and disrupting the integrity of vasa wall (Zhu et al., 2017). Mutations of the *CECR1* gene affect the sites of N-linked glycosylation of the ADA2, altering the enzyme trafficking and secretion (Lee et al., 2018). An abundance of circulating low-density granulocytes that are prone to spontaneous neutrophil extracellular traps formation seems also to play a role in DADA2 manifestations (Carmona-Rivera et al., 2019).

A loss of the auto-inhibitor domain functionality characterizes the disorders related to PLCG2 gene. Patients with the auto-inflammation and phospholipase Cg2 (PLCg2)-associated antibody deficiency and immune dysregulation (APLAID) syndrome, carrying GOF heterozygous mutation that causes a constitutive activation of the PLCG2 protein. The syndrome is characterized by early-onset blistering skin lesions, posterior uveitis, IBD, and cutis laxa, and has been recently described in a second family (Neves et al., 2018; Zhou et al., 2012). On the contrary, patients carrying small intragenic deletions of the auto-inhibitor SH domains, show episodes of cold-induced urticarial rash due to PLCG2 activation in case of a low temperature exposition. Higher calcium efflux in B cells in response to lower temperature exposition is a useful functional tests in patients with PLCg2-associated antibody deficiency and immune dysregulation (PLAID) syndrome, also known as FCAS type 3 (Ombrello et al., 2012). Polymorphisms of the protein regulatory domain, causing mild enzymatic hyperactivity, seem protective against the development of beta-amyloid related neurological diseases (Conway et al., 2018; Magno et al., 2019; Sims et al., 2017; van der Lee et al., 2019). Furthermore, a mild hypogammaglobulinemia with consequent recurrent sinopulmonary infections have been reported in patients with PLCG2 variants and are related to abnormalities in B cells by Bruton's tyrosine kinase receptor activation. In fact, variants of the PLCG2 gene have been reported in tumor cells of patients with chronic lymphocytic leukemia resistant to Bruton's tyrosine kinase inhibitors (Quinquenel et al., 2019; Kanagal-Shamanna et al., 2019).

The LACC1 gene codes for the laccase domain-containing 1 (LACC1) oxidoreductase, that associates with fatty acid synthase on peroxisomes promoting fatty-acid oxidation, and NOD2-signaling complex, inducing mitochondrial production of the reactive oxygen species, cytokine secretion and bacterial clearance by macrophages (Lahiri, Hedl, Yan, & Abraham, 2017; Wang et al., 2018). Variants of this gene has been associated to Crohn's disease, leprosy, and an early-onset multi-drugs-resistant systemic and polyarticular form of juvenile monogenic arthritis (Karacan et al., 2018). LACC1-deficient mouse produces high level of TNF and IL-17, thus suggesting a possible new therapeutic approach for these patients (Skon-Hegg et al., 2019; Rabionet et al., 2019).

2.7.1 RANKL signaling defects

Frameshift GOF mutations of the *TNFRSF11A* gene have been associated to autosomal dominant recurrent fever syndrome by Jeru et al. However, *TNFRSF11A* autosomal recessive LOF mutations are known to cause the osteoclast-poor osteopetrosis with hypogammaglobulinemia (Guerrini et al., 2008), and heterozygous duplications have been associated with the familial expansile osteolysis or early-onset Paget disease of bone (Ralston & Taylor, 2019; Xue et al., 2019). The reason why the *TNFRSF11A* GOF mutations do not affect bones remains unclear. The gene codes for the type I transmembrane Receptor Activator of NFkB (RANK), an osteoclast differentiation factor that mediates essential signals for osteoclastogenesis. The ligand of RANK (RANKL) activates brain regions involved in thermoregulation, inducing fever (Hanada et al., 2009).

On the other hand, variants of the Filamin-binding LIM protein 1, coded by the FBLIM1 gene, have been associated with a hereditary form of chronic recurrent multifocal osteomyelitis (CRMO). The protein localizes at cell junctions and may link cell adhesion structures to the actin cytoskeleton. Authors demonstrated an increased production of RANKL in these patients.

Cherubism is a rare autosomal dominant disorder of the jaws caused by mutation of the *SH3BP2* gene (Kadlub et al., 2018; Sidorowicz, Kubasiewicz-Ross, & Dominiak, 2018). Patients present with a childhood onset bilateral and symmetric proliferative fibro-osseous lesions limited to the mandible and maxilla. SH3BP2 is an adapter protein involved in lymphocyte activation, osteoclast differentiation and bone remodeling. In these patients, hyperactive macrophages secrete a high level of TNFα that drives systemic inflammation, stimulates secretion of RANKL by stromal cells, and ultimately results in bone loss.

3. Conclusions

Throughout history, the doctors classified diseases according main clinical manifestation, the most involved organ, or the name of physician who firstly described each condition. In light of the little advantage of this method, during the last century, thanks to the development of new gene sequencing technologies, reclassification of hereditary human diseases has been proposed, according to the locus of causative mutations. However, a symptom may be secondary to different gene variants and many variants of a gene cause numerous symptoms, thus the gene-targeted classification of monogenic disorders does not seem to provide a useful taxonomy for physicians. On the contrary, molecular pathways cannot only be useful in classifying monogenic disease, but also highlight roles of proteins that can represent future therapeutic targets to better control the deregulated machinery. In fact, thanks to advances in understanding of cell biology, many drugs selectively inhibiting a target protein can treat efficaciously specific symptoms and limit adverse effects. The availability of these molecules makes early diagnosis even more important in order to avoid the development of complications and long-term outcomes.

List of abbreviations

ADA	Adenosine deaminase
ADMIO1	Infantile-onset multisystem autoimmune disease 1
AGS	Aicardi-Goutieres syndrome
AID	Autoinflammatory disease
AIFEC	Autoinflammation with infantile enterocolitis
APLAID	PLCg2-associated antibody deficiency and immune dysregulation

CAIN	C/EBPe-associated autoinflammation and immune impairment of neutrophils
CAMPS	CARD14-mediated psoriasis
CANDLE	Chronic atypical neutrophilic dermatosis with lipodystrophy and elevated temperature
CAPS	Cryopyrin-associated periodic syndrome
CARD	Caspase recruitment domain
CCHIDG	Combined cellular and humoral immune defects with granulomas
CDC42	Cell division control protein 42 homolog
CHAI	CTLA4 haploinsufficiency with autoimmune infiltration
CINCA	Chronic infantile neurological, cutaneous and Articular syndrome
CNS	central nervous system
CRMO	Chronic recurrent multifocal osteomyelitis
CVID	Common variable immunodeficiency
DADA2	ADA2 deficiency
DIRA	IL-1 receptor antagonist deficiency
DITRA	IL-36 receptor antagonist deficiency
DSH	Dyschromatosis symmetrica hereditaria
FCAS	Familial cold autoinflammatory syndrome
FMF	Familial Mediterranean fever
GOF	Gain of function
GPP	Generalized pustular psoriasis
HA20	Haploinsufficiency of A20
HFTC1	Hyperphosphatemic familial tumoral calcinosis
HLH	Hemophagocitic lymphoistiocytosis
HS	Hidrosadenitis supporativa
IBD	Inflammatory bowel disease
ICH	Infantile cortical hyperostosis
IFN	Interferon
Ig	Immunoglobulin
IL	Interleukin
LACC1	Laccase domain containing 1
LAID	LYN associated autoinflammatory disease
LOF	Loss of function
LUBAC	Linear ubiquitin chain assembly complex
MAS	Macrophage activation syndrome
MJDS	Majeed syndrome
MKD	Mevalonate kinase deficiency
MRS	Merkelson-Rosenthal syndrome
MSMD	Mendelian susceptibility to mycobacterial disease
MWS	Muckle-Wells syndrome
NAIAD	NLRP1-associated autoinflammation with arthritis and dyskeratosis
NFkB	Nuclear factor kappa-B
NGS	Next generation sequencing
NLRP	NOD-like receptor family pyrin domain containing
NOCARH	Neonatal onset cytopenia with autoinflammation, rash, and hemophagocytosis
NOD	Nucleotide-binding oligomerization domain-containing protein
ORAS	OTULIN-related autoinflammatory syndrome
OTULIN	OTU deubiquitinase with linear linkage specificity
PAAND	pyrin-associated autoinflammation with neutrophilic dermatosis
PAMI	PSTPIP1-associated myeloid-related proteinemia inflammatory syndrome

PAPA	Pyogenic sterile arthritis, pyoderma gangrenosum, acne
PFIT	Periodic autoinflammatory fever with immunodeficiency and thrombocytopenia
PHID	Pigmentary hypertrichosis and non-autoimmune insulin-dependent diabetes
PID	Primary immunodeficiency
PLAID	PLCg2-associated antibody deficiency and immune dysregulation
PLCg2	Phospholipase Cg2
PLTEID	Platelet abnormalities with eosinophilia and immune-mediated inflammatory disease
POMP	Proteasome maturation protein
PRAAS	Proteasome-associated autoinflammatory syndromes
PRAID	POMP-related autoinflammation and immune dysregulation disease
PTORCH	Pseudo-TORCH syndrome
RANK	Receptor Activator of NFkB
RANKL	RANK ligand
RBCK1	RanBP-type and C3HC4-type zinc finger containing 1
RNF31	Ring finger protein 31
RVCL	Retinal vasculopathy with cerebral leukodystrophy
SAVI	STING-associated vasculopathy with onset in infancy
SHARPIN	SHANK-associated RH domain interacting protein
SIFD	Sideroblastic anemia with B-cell immunodeficiency, periodic fever and developmental delay
SLE	Systemic lupus erythematosus
SMS	Smith-Magenis syndrome
SPENCDI	Spondyloenchondrodysplasia with immune dysregulation
SPTCL	Subcutaneous panniculitis, T cell lymphoma
THES	Trichohepatoenteric syndrome
TNF	Tumor necrosis factor
TRAPS	TNF-receptor associated periodic fever syndrome
UPR	Unfolded protein response
UPS	Ubiquitin-proteasome system
VEO	Very early onset
WAS	Wiskott Aldricht syndrome
XLPDR	X-linked reticulate pigmentary disorder

Acknowledgments

RP would like to thank patients and families, nurses and doctors of the Paediatric Rheumatology Clinic at the Giannina Gaslini Institute for supporting his studies and researches. RP would like to thank especially PP and MG for the continuous support and guidance during the Paediatrics residency program.

References

Al-Mayouf, S. M., et al. (2018). Monogenic interferonopathies: Phenotypic and genotypic findings of CANDLE syndrome and its overlap with C1q deficient SLE. *International Journal of Rheumatic Diseases, 21*, 208–213.

Armyra, K., Kouris, A., Markantoni, V., Katsambas, A., & Kontochristopoulos, G. (2017). Hidradenitis suppurativa treated with tetracycline in combination with colchicine: A prospective series of 20 patients. *International Journal of Dermatology, 56*, 346–350.

Balci-Peynircioglu, B., et al. (2018). Potential role of pyrin, the protein mutated in familial mediterranean fever, during inflammatory cell migration. *Clinical and Experimental Rheumatology, 36*, 116–124.

Barton, C., Kausar, S., Kerr, D., Bitetti, S., & Wynn, R. (2018). SIFD as a novel cause of severe fetal hydrops and neonatal anaemia with iron loading and marked extramedullary haemopoiesis. *Journal of Clinical Pathology, 71*, 275–278.

Basaran, O., et al. (2018). C3 glomerulopathy in NLRP12-related autoinflammatory disorder: Case-based review. *Rheumatology International, 38*, 1571–1576.

Benedetti, F. De, et al. (2019). OP0204 EMAPALUMAB, an interferon gamma (IFN-Y)-BLOCKING monoclonal antibody, in patients with macrophage activation syndrome (MAS) complicating systemic juvenile idiopathic arthritis (SJIA). *Annals of the Rheumatic Diseases, 78*, 178 LP–178.

Berteau, F., et al. (2018). Autosomic dominant familial Behcet disease and haploinsufficiency A20: A review of the literature. *Autoimmunity Reviews, 17*, 809–815.

Bertoni, A., et al. (2019). A novel knock-in mouse model of cryopyrin-associated periodic syndromes with development of amyloidosis: Therapeutic efficacy of proton pump inhibitors. *The Journal of Allergy and Clinical Immunology*. http://dx.doi.org/10.1016/j.jaci.2019.05.034.

Birnbaum, Y., Tran, D., Bajaj, M., & Ye, Y. (2019). DPP-4 inhibition by linagliptin prevents cardiac dysfunction and inflammation by targeting the Nlrp3/ASC inflammasome. *Basic Research in Cardiology, 114*, 35.

Boisson, B., et al. (2015). Human HOIP and LUBAC deficiency underlies autoinflammation, immunodeficiency, amylopectinosis, and lymphangiectasia. *Journal of Experimental Medicine, 212*, 939–951.

Bonekamp, N., et al. (2018). High-dose ustekinumab for severe childhood deficiency of interleukin-36 receptor antagonist (DITRA). *Annals of the Rheumatic Diseases, 77*, 1241–1243.

Bonekamp, N., Caorsi, R., Frenkel, J., & Gattorno, M. (2018). Response to: 'Standard dose of ustekinumab for childhood-onset deficiency of interleukin-36 receptor antagonist' by Cherqaoui et al. *Annals of the Rheumatic Diseases, 77*, e89.

Bousfiha, A., et al. (2018). The 2017 IUIS phenotypic classification for primary immunodeficiencies. *Journal of Clinical Immunology, 38*, 129–143.

Bracaglia, C., et al. (2019). OP0255 Microbiota transplant to control inflammation in a NLRC4-related disease patient with recurrent hemophagocytic lymphohistiocytosis (HLH). *Annals of the Rheumatic Diseases, 78*, 208 LP–208.

Brigida, I., et al. (2018). T-cell defects in patients with ARPC1B germline mutations account for combined immunodeficiency. *Blood, 132*, 2362–2374.

Candotti, F. (2018). Clinical manifestations and pathophysiological mechanisms of the Wiskott-Aldrich syndrome. *Journal of Clinical Immunology, 38*, 13–27.

Caorsi, R., et al. (2017). ADA2 deficiency (DADA2) as an unrecognised cause of early onset polyarteritis nodosa and stroke: A multicentre national study. *Annals of the Rheumatic Diseases, 76*, 1648–1656.

Carmona-Rivera, C., et al. (2019). Deficiency of adenosine deaminase 2 triggers adenosine-mediated NETosis and TNF production in patients with DADA2. *Blood, 134*, 395–406.

Centola, M., et al. (1998). Construction of an approximately 700-kb transcript map around the familial mediterranean fever locus on human chromosome 16p13.3. *Genome Research, 8*, 1172–1191.

Cherqaoui, B. J., Rossi-Semerano, L., Piram, M., & Kone-Paut, I. (2018). Standard dose of ustekinumab for childhood-onset deficiency of interleukin-36 receptor antagonist. *Annals of the Rheumatic Diseases, 77*, e88.

Chui, A. J., et al. (2019). N-terminal degradation activates the NLRP1B inflammasome. *Science, 364*, 82–85.

Conway, O. J., et al. (2018). ABI3 and PLCG2 missense variants as risk factors for neurodegenerative diseases in Caucasians and African Americans. *Molecular Neurodegeneration, 13*, 53.
Coppola, T., Becken, B., Van Mater, H., McDonald, M. T., & Panayotti, G. M. (2019). A case report of mevalonate kinase deficiency in a 14-month-old female with fevers and lower extremity weakness. *BMC Pediatrics, 19*, 245.
Craiglow, B. G., et al. (2018). *CARD14*-associated papulosquamous eruption: A spectrum including features of psoriasis and pityriasis rubra pilaris. *Journal of the American Academy of Dermatology, 79*, 487–494.
Cropley, A., Ashrafy, A. H., & Weltman, M. (2019). An original description of granulomatous liver cirrhosis in Blau syndrome. *Digestive Diseases and Sciences*. http://dx.doi.org/10.1007/s10620-019-05682-8.
Damgaard, R. B., et al. (2019). OTULIN deficiency in ORAS causes cell type-specific LUBAC degradation, dysregulated TNF signalling and cell death. *EMBO Molecular Medicine, 11*.
De Benedetti, F., et al. (2018). Canakinumab for the treatment of autoinflammatory recurrent fever syndromes. *New England Journal of Medicine, 378*, 1908–1919.
De Benedetti, F, Miettunen, P., Kallinich, T., Horne, G. & Brik, R. Genetic phenotypes impacting efficacy and safety of canakinumab in patients with colchicine-resistant FMF, TRAPS and Hids / Mkd : Results from cluster study. 1–3.
de Jesus, A. A., et al. (2019). Novel proteasome assembly chaperone mutations in $PSMG_2$/PAC_2 cause the autoinflammatory interferonopathy CANDLE/$PRAAS_4$. *The Journal of Allergy and Clinical Immunology, 143*, 1939–1943.e8.
DeSouza, P. J., & Shah, R. (2019). Characterization of Blau syndrome panuveitis with wide-field fluorescein angiography. *American Journal of Ophthalmology Case Reports, 14*, 92–94.
Deuteraiou, K., Kitas, G., Garyfallos, A., & Dimitroulas, T. (2018). Novel insights into the role of inflammasomes in autoimmune and metabolic rheumatic diseases. *Rheumatology International, 38*, 1345–1354.
Ergun, S. L., Fernandez, D., Weiss, T. M., & Li, L. (2019). STING polymer structure reveals mechanisms for activation, hyperactivation, and inhibition. *Cell, 178*, 290–301.e10.
Ferrand, A., et al. (2019). NOD2 expression in intestinal epithelial cells protects toward the development of inflammation and associated carcinogenesis. *Cellular and Molecular Gastroenterology and Hepatology, 7*, 357–369.
Fragoulis, G. E., McInnes, I. B., & Siebert, S. (2019). JAK-inhibitors. New players in the field of immune-mediated diseases, beyond rheumatoid arthritis. *Rheumatology, 58*, i43–i54.
A candidate gene for familial mediterranean fever. *Nature Genetics, 17*, (1997), 25–31.
Fu, F., et al. (2019). Rare CARD14 missense variants associated with palmoplantar pustulosis (PPP) in the Chinese Han population. *European Journal of Dermatology, 29*, 99–100.
Gattorno, M., et al. (2008). Diagnosis and management of autoinflammatory diseases in childhood. *Journal of Clinical Immunology, 28*, 73–83.
Gernez, Y., et al. (2019). Severe autoinflammation in 4 patients with C-terminal variants in cell division control protein 42 homolog (CDC42) successfully treated with IL-1beta inhibition. *The Journal of Allergy and Clinical Immunology*. http://dx.doi.org/10.1016/j.jaci.2019.06.017.
Ghosh, K., Mishra, K., Shah, A., Patel, P., & Shetty, S. (2019). Novel deleterious sequence change in the NLRP12 gene in a child with the autoinflammatory syndrome, joint hypermobility and cutis laxa from India. *Mediterranean Journal of Hematology and Infectious Diseases, 11*, e2019018.
Giannelou, A., et al. (2018). Aberrant tRNA processing causes an autoinflammatory syndrome responsive to TNF inhibitors. *Annals of the Rheumatic Diseases*, 612–619. http://dx.doi.org/10.1136/annrheumdis-2017-212401.

Giardino, S., et al. (2015). Long-term outcome of a successful cord blood stem cell transplant in mevalonate kinase deficiency. *Pediatrics, 135*, e211–e215.

Girardelli, M., et al. (2018a). Genetic profile of patients with early onset inflammatory bowel disease. *Gene, 645*, 18–29.

Girardelli, M., et al. (2018b). Novel NOD2 mutation in early-onset inflammatory bowel phenotype. *Inflammatory Bowel Diseases, 24*, 1204–1212.

Girard-Guyonvarc'h, C., et al. (2018). Unopposed IL-18 signaling leads to severe TLR9-induced macrophage activation syndrome in mice. *Blood, 131*, 1430–1441.

Gomez-Garcia, F., et al. (2018). Scoping review on use of drugs targeting interleukin 1 pathway in DIRA and DITRA. *Dermatologic Therapy, 8*, 539–556.

Goos, H., et al. (2019). Gain-of-function CEBPE mutation causes noncanonical autoinflammatory inflammasomopathy. *The Journal of Allergy and Clinical Immunology*. http://dx.doi.org/10.1016/j.jaci.2019.06.003.

Grandemange, S., et al. (2017). A new autoinflammatory and autoimmune syndrome associated with NLRP1 mutations: NAIAD (NLRP1-associated autoinflammation with arthritis and dyskeratosis). *Annals of the Rheumatic Diseases, 76*, 1191–1198.

Greco, E., et al. (2015). The novel S59P mutation in the TNFRSF1A gene identified in an adult onset TNF receptor associated periodic syndrome (TRAPS) constitutively activates NF-kappaB pathway. *Arthritis Research and Therapy, 17*, 93.

Grossi, A., et al. (2019). ADA2 deficiency due to a novel structural variation in 22q11.1. *Clinical Genetics, 95*, 732–733.

Guerrini, M. M., et al. (2008). Human osteoclast-poor osteopetrosis with hypogammaglobulinemia due to TNFRSF11A (RANK) mutations. *The American Journal of Human Genetics, 83*, 64–76.

Hafner-Bratkovic, I., et al. (2018). NLRP3 lacking the leucine-rich repeat domain can be fully activated via the canonical inflammasome pathway. *Nature Communications, 9*, 5182.

Hanada, R., et al. (2009). Central control of fever and female body temperature by RANKL/RANK. *Nature, 462*, 505–509.

Hansen, A. L., et al. (2018). Nitro-fatty acids are formed in response to virus infection and are potent inhibitors of STING palmitoylation and signaling. *Proceedings of the National Academy of Sciences of the United States of America, 115*, E7768–E7775.

Harrison, S. R., et al. (2018). Inositol-requiring enzyme 1-mediated downregulation of microRNA (miR)-146a and miR-155 in primary dermal fibroblasts across three TNFRSF1A mutations results in hyperresponsiveness to lipopolysaccharide. *Frontiers in Immunology, 9*, 173.

Hashmi, S. K., et al. (2019). PSTPIP1-associated myeloid-related proteinemia inflammatory syndrome: A rare cause of childhood neutropenia associated with systemic inflammation and hyperzincemia. *Pediatric Blood and Blood Cancer, 66*, e27439.

Havnaer, A., & Han, G. (2019). Autoinflammatory disorders: A review and update on pathogenesis and treatment. *American Journal of Clinical Dermatology*. http://dx.doi.org/10.1007/s40257-019-00440-y.

Heger, K., et al. (2018). OTULIN limits cell death and inflammation by deubiquitinating LUBAC. *Nature, 559*, 120–124.

Heilig, R., & Broz, P. (2018). Function and mechanism of the pyrin inflammasome. *European Journal of Immunology, 48*, 230–238.

Holland, G. N., & Rosenbaum, J. T. (2018). The challenge of Blau syndrome. *American Journal of Ophthalmology, 187*, xviii–xix.

Holzinger, D., Foell, D., & Kessel, C. (2018). The role of S100 proteins in the pathogenesis and monitoring of autoinflammatory diseases. *Molecular and Cellular Pediatrics, 5*, 7.

Hong, C.-Y., Zhang, H.-D., Liu, X.-Y., & Xu, Y. (2019). Attenuation of hyperoxic acute lung injury by *Lycium barbarum* polysaccharide via inhibiting NLRP3 inflammasome. *Archives of Pharmacal Research*. http://dx.doi.org/10.1007/s12272-019-01175-4.

Hong, Y., et al. (2019). Autoinflammation due to homozygous S208 MEFV mutation. *Annals of the Rheumatic Diseases, 78*, 571–573.

Hori, T., et al. (2019). Autosomal dominant Hashimoto's thyroiditis with a mutation in TNFAIP3. *Clinical Pediatric Endocrinology, 28*, 91–96.

Hospach, T., et al. (2019). Scoping review of biological treatment of deficiency of interleukin-36 receptor antagonist (DITRA) in children and adolescents. *Pediatric Rheumatology Online Journal, 17*, 37.

Hu, Q., Zhang, T., Yi, L., Zhou, X., & Mi, M. (2018). Dihydromyricetin inhibits NLRP3 inflammasome-dependent pyroptosis by activating the Nrf2 signaling pathway in vascular endothelial cells. *BioFactors, 44*, 123–136.

Hui, A., Johnson, L. B., Greemberg, R., Penney, L., & Ramsey, S. E. (2019). Severe cryopyrin-associated periodic syndrome first characterized by early childhood-onset sensorineural hearing loss – case report and literature review. *International Journal of Pediatric Otorhinolaryngology, 120*, 68–72.

Imayoshi, M., Ogata, Y., & Yamamoto, S. (2018). A case of sporadic Blau syndrome with an uncommon clinical course. *Case Reports in Rheumatology, 2018*, 6292308.

Irazoki, O., Hernandez, S. B., & Cava, F. (2019). Peptidoglycan muropeptides: Release, perception, and functions as signaling molecules. *Frontiers in Microbiology, 10*, 500.

Israel, L., & Mellett, M. (2018). Clinical and genetic heterogeneity of CARD14 mutations in psoriatic skin disease. *Frontiers in Immunology, 9*, 2239.

10th Congress of international society of systemic auto-inflammatory diseases (ISSAID). *Pediatric Rheumatology, 17*, (2019), 18.

Jacob, N., et al. (2019). Generalized cytokine increase in the setting of a multisystem clinical disorder and carcinoid syndrome associated with a novel NLRP12 variant. *Digestive Diseases and Sciences, 64*, 2140–2146.

Jamilloux, Y., et al. (2018). Familial mediterranean fever mutations are hypermorphic mutations that specifically decrease the activation threshold of the pyrin inflammasome. *Rheumatology, 57*, 100–111.

Jesus, A. A., et al. (2008). TNF receptor-associated periodic syndrome (TRAPS): Description of a novel TNFRSF1A mutation and response to etanercept. *European Journal of Pediatrics, 167*, 1421–1425.

Jiang, C., et al. (2018). Acrolein induces NLRP3 inflammasome-mediated pyroptosis and suppresses migration via ROS-dependent autophagy in vascular endothelial cells. *Toxicology, 410*, 26–40.

Jindal, A. K., et al. (2019). A young female with early onset arthritis, uveitis, hepatic, and renal granulomas: A clinical tryst with Blau syndrome over 20 years and case-based review. *Rheumatology International*. http://dx.doi.org/10.1007/s00296-019-04316-6.

Kadlub, N., et al. (2018). Molecular and cellular characterizations of human cherubism: Disease aggressiveness depends on osteoclast differentiation. *Orphanet Journal of Rare Diseases, 13*, 166.

Kahr, W. H. A., et al. (2017). Loss of the Arp2/3 complex component ARPC1B causes platelet abnormalities and predisposes to inflammatory disease. *Nature Communications, 8*, 14816.

Kanagal-Shamanna, R., et al. (2019). Targeted multigene deep sequencing of Bruton tyrosine kinase inhibitor-resistant chronic lymphocytic leukemia with disease progression and Richter transformation. *Cancer, 125*, 559–574.

Kaneko, N., Kurata, M., Yamamoto, T., Morikawa, S., & Masumoto, J. (2019). The role of interleukin-1 in general pathology. *Inflammation and Regeneration, 39*, 12.

Karacan, I., et al. (2018). LACC1 gene defects in familial form of juvenile arthritis. *The Journal of Rheumatology, 45*, 726–728.

Karki, R., et al. (2018). IRF8 regulates transcription of naips for NLRC4 inflammasome activation. *Cell, 173*, 920–933.e13.

Kilic, H., et al. (2019). Spectrum of the neurologic manifestations in childhood-onset cryopyrin-associated periodic syndrome. *European Journal of Paediatric Neurology, 23*, 466–472.
Kim, H., et al. (2018a). Development of a validated interferon score using nanostring technology. *Journal of Interferon and Cytokine Research, 38*, 171–185.
Kim, H., et al. (2018b). Pharmacokinetics, pharmacodynamics, and proposed dosing of the oral JAK1 and JAK2 inhibitor baricitinib in pediatric and young adult CANDLE and SAVI patients. *Clinical Pharmacology and Therapeutics, 104*, 364–373.
Kimberley, F. C., Lobito, A. A., Siegel, R. M., & Screaton, G. R. (2007). Falling into TRAPS—receptor misfolding in the TNF receptor 1-associated periodic fever syndrome. *Arthritis Research and Therapy, 9*, 217.
Kirby, K., & Wright, J. E. (2019). *Infantile cortical hyperostosis (Caffey disease).*
Klotgen, H.-W., et al. (2018). The expanding spectrum of clinical phenotypes associated with PSTPIP1 mutations: From PAPA to PAMI syndrome and beyond. *The British Journal of Dermatology, 178*, 982–983.
Kluk, J., et al. (2014). Chronic atypical neutrophilic dermatosis with lipodystrophy and elevated temperature syndrome: A report of a novel mutation and review of the literature. *The British Journal of Dermatology, 170*, 215–217.
Kopitar, A. N., et al. (2019). Flow cytometric determination of actin polymerization in peripheral blood leukocytes effectively discriminate patients with homozygous mutation in ARPC1B from asymptomatic carriers and normal controls. *Frontiers in Immunology, 10*, 1632.
Kostik, M. M., et al. (2018). Multigene sequencing reveals heterogeneity of NLRP12-related autoinflammatory disorders. *Rheumatology International, 38*, 887–893.
Kuemmerle-Deschner, J. B., et al. (2017). Diagnostic criteria for cryopyrin-associated periodic syndrome (CAPS). *Annals of the Rheumatic Diseases, 76*, 942–947.
Kuijpers, T. W., et al. (2017). Combined immunodeficiency with severe inflammation and allergy caused by ARPC1B deficiency. *The Journal of Allergy and Clinical Immunology, 140*, 273–277.e10.
Kumaki, E., et al. (2019). Atypical SIFD with novel TRNT1 mutations: A case study on the pathogenesis of B-cell deficiency. *International Journal of Hematology, 109*, 382–389.
Kumar, V. (2019). A STING to inflammation and autoimmunity. *Journal of Leukocyte Biology, 106*, 171–185.
Lacey, C. A., & Miao, E. A. (2019). NLRP1 – one NLR to guard them all. *The EMBO Journal, 38*, e102494.
Lachmann, H. J., Papa, R., & Minden, K. (2014). The phenotype of TNF-receptor associated autoinflammatory syndrome (TRAPS) at presentation: A series of 158 cases from the Eurofever/EUROTRAPS international registry. *Annals of the Rheumatic Disease, 73*.
Lahiri, A., Hedl, M., Yan, J., & Abraham, C. (2017). Human LACC1 increases innate receptor-induced responses and a LACC1 disease-risk variant modulates these outcomes. *Nature Communications, 8*, 15614.
Lamot, L., Niemietz, I., & Brown, K. L. (2019). Comparable type I interferon score determination from PAXgene and Tempus whole blood RNA collection and isolation systems. *BMC Research Notes, 12*, 511.
Lee, P. P., et al. (2017). Wiskott-Aldrich syndrome protein regulates autophagy and inflammasome activity in innate immune cells. *Nature Communications, 8*, 1576.
Lee, P. Y., et al. (2018). Disrupted N-linked glycosylation as a disease mechanism in deficiency of ADA2. *The Journal of Allergy and Clinical Immunology, 142*, 1363–1365.e8.
Li, J.-Y., et al. (2018). Characterization of an NLRP1 inflammasome from zebrafish reveals a unique sequential activation mechanism underlying inflammatory caspases in ancient vertebrates. *The Journal of Immunology, 201*, 1946–1966.

Li, G.-M., et al. (2019). Expanding the spectrum of A20 haploinsufficiency in two Chinese families: Cases report. *BMC Medical Genetics, 20*, 124.
Li, L., et al. (2019). Variants of CARD14 are predisposing factors for generalized pustular psoriasis (GPP) with psoriasis vulgaris but not for GPP alone in a Chinese population. *The British Journal of Dermatology, 180*, 425–426.
Lorden, G., et al. (2017). Lipin-2 regulates NLRP3 inflammasome by affecting P2X7 receptor activation. *Journal of Experimental Medicine, 214*, 511–528.
Lu, L., et al. (2018). Blau syndrome with good reponses to tocilizumab: A case report and focused literature review. *Seminars in Arthritis and Rheumatism, 47*, 727–731.
Lu, Y., Zhang, S., Wang, Y., Ren, X., & Han, J. (2019). Molecular mechanisms and clinical manifestations of rare genetic disorders associated with type I collagen. *Intractable and Rare Diseases Research, 8*, 98–107.
Ly, K., Beck, K. M., Smith, M. P., Thibodeaux, Q., & Bhutani, T. (2019). Diagnosis and screening of patients with generalized pustular psoriasis. *Psoriasis (Auckland, New Zealand), 9*, 37–42.
Magalhaes, R. F., et al. (2019). Consensus on the treatment of hidradenitis suppurativa – Brazilian Society of dermatology. *Anais Brasileiros de Dermatologia, 94*, 7–19.
Magno, L., et al. (2019). Alzheimer's disease phospholipase C-gamma-2 (PLCG2) protective variant is a functional hypermorph. *Alzheimer's Research and Therapys, 11*, 16.
Mahil, S. K., et al. (2016). AP1S3 mutations cause skin autoinflammation by disrupting keratinocyte autophagy and up-regulating IL-36 production. *Journal of Investigative Dermatology, 136*, 2251–2259.
Manthiram, K., Zhou, Q., Aksentijevich, I., & Kastner, D. L. (2017). The monogenic autoinflammatory diseases define new pathways in human innate immunity and inflammation. *Nature Immunology, 18*, 832–842.
Marchica, C., Zawawi, F., Basodan, D., Scuccimarri, R., & Daniel, S. J. (2018). Resolution of unilateral sensorineural hearing loss in a pediatric patient with a severe phenotype of Muckle-Wells syndrome treated with anakinra: A case report and review of the literature. *Journal of Otolaryngology – Head and Neck Surgery, 47*, 9.
Mariani, R., & Gong, S. (2018). Bone marrow histopathologic findings in SIFD syndrome: Beyond the erythroid lineage. *Blood, 132*, 1459.
Martinelli, S., et al. (2018). Functional dysregulation of CDC42 causes diverse developmental phenotypes. *The American Journal of Human Genetics, 102*, 309–320.
Masters, S. L., Simon, A., Aksentijevich, I., & Kastner, D. L. (2009). Horror autoinflammaticus: The molecular pathophysiology of autoinflammatory disease (*). *Annual Review of Immunology, 27*, 621–668.
McDermott, M. F., et al. (1999). Germline mutations in the extracellular domains of the 55 kDa TNF receptor, TNFR1, define a family of dominantly inherited autoinflammatory syndromes. *Cell, 97*, 133–144.
Millington, G. W. M., et al. (2019). Sporadic Blau syndrome treated with adalimumab. *Clinical and Experimental Dermatology*. http://dx.doi.org/10.1111/ced.14016.
Moghaddas, F., & Masters, S. L. (2018). The classification, genetic diagnosis and modelling of monogenic autoinflammatory disorders. *Clinical Science, 132*(17), 1901–1924.
Moghaddas, F., et al. (2017). A novel pyrin-associated autoinflammation with neutrophilic dermatosis mutation further defines 14-3-3 binding of pyrin and distinction to familial mediterranean fever. *Annals of the Rheumatic Diseases, 76*, 2085–2094.
Moossavi, M., Parsamanesh, N., Bahrami, A., Atkin, S. L., & Sahebkar, A. (2018). Role of the NLRP3 inflammasome in cancer. *Molecular Cancer, 17*, 158.
Moratto, D., et al. (2011). Long-term outcome and lineage-specific chimerism in 194 patients with Wiskott-Aldrich syndrome treated by hematopoietic cell transplantation in the period 1980–2009: An international collaborative study. *Blood, 118*, 1675–1684.

Naik, A. U., Annamalai, R., & Biswas, J. (2018). Uveitis in sporadic Blau syndrome: Long-term follow-up of a refractory case treated successfully with adalimumab. *Indian Journal of Ophthalmology, 66*, 1483—1485.

Nakamichi, S., et al. (2019). A rare case of cryopyrin-associated periodic syndrome in an elderly woman with NLRP3 and MEFV mutations. *Internal Medicine, 58*, 1017—1022.

Negus, D., Ahn, C., & Huang, W. (2018). An update on the pathogenesis of hidradenitis suppurativa: Implications for therapy. *Expert Review of Clinical Immunology, 14*, 275—283.

Neves, J. F., et al. (2018). Novel PLCG2 mutation in a patient with APLAID and cutis laxa. *Frontiers in Immunology, 9*, 2863.

Noorelahi, R., Perez, G., & Otero, H. J. (2018). Imaging findings of Copa syndrome in a 12-year-old boy. *Pediatric Radiology, 48*, 279—282.

Oda, H., et al. (2019). Second case of HOIP deficiency expands clinical features and defines inflammatory transcriptome regulated by LUBAC. *Frontiers in Immunology, 10*, 479.

Ohnishi, H., et al. (2018). Genetic background and therapeutic response in generalized pustular psoriasis patients treated with granulocyte and monocyte adsorption apheresis. *European Journal of Dermatology, 28*, 108—111.

Ombrello, M. J., et al. (2012). Cold urticaria, immunodeficiency, and autoimmunity related to PLCG2 deletions. *New England Journal of Medicine, 366*, 330—338.

Ono, S. (2018). Functions of actin-interacting protein 1 (AIP1)/WD repeat protein 1 (WDR1) in actin filament dynamics and cytoskeletal regulation. *Biochemical and Biophysical Research Communications, 506*, 315—322.

Ozen, S. (2018). What's new in autoinflammation? *Pediatric Nephrology*. http://dx.doi.org/10.1007/s00467-018-4155-4.

Papa, R., & Lachmann, H. J. (2018). Secondary, AA, amyloidosis. *Rheumatic Disease Clinics of North America, 44*, 585—603.

Papa, R., et al. (2019). Next generation sequencing panel in undifferentiated autoinflammatory diseases identify patients with colchicine-responder recurrent fevers. *Rheumatology*. http://dx.doi.org/10.1093/rheumatology/kez270.

Pashenkov, M. V., Murugina, N. E., Budikhina, A. S., & Pinegin, B. V. (2019). Synergistic interactions between NOD receptors and TLRs: Mechanisms and clinical implications. *Journal of Leukocyte Biology, 105*, 669—680.

Peled, A., et al. (2019). Loss-of-function mutations in caspase recruitment domain-containing protein 14 (CARD14) are associated with a severe variant of atopic dermatitis. *The Journal of Allergy and Clinical Immunology, 143*, 173—181.e10.

Pellegrini, C., Fornai, M., Antonioli, L., Blandizzi, C., & Calderone, V. (2019). Phytochemicals as novel therapeutic sStrategies for NLRP3 inflammasome-related neurological, metabolic, and inflammatory diseases. *International Journal of Molecular Sciences, 20*.

Peltzer, N., et al. (2018). LUBAC is essential for embryogenesis by preventing cell death and enabling haematopoiesis. *Nature, 557*, 112—117.

Pescarmona, R., et al. (2019). Comparison of RT-qPCR and nanostring in the measurement of blood interferon response for the diagnosis of type I interferonopathies. *Cytokine, 113*, 446—452.

Pietrasanta, C., et al. (2019). When neonatal inflammation does not mean infection: An early-onset mevalonate kinase deficiency with interstitial lung disease. *Clinical Immunology, 205*, 25—28.

Poli, M. C., et al. (2018). Heterozygous truncating variants in POMP escape nonsense-mediated decay and cause a unique immune dysregulatory syndrome. *The American Journal of Human Genetics, 102*, 1126—1142.

Polykratis, A., et al. (2019). A20 prevents inflammasome-dependent arthritis by inhibiting macrophage necroptosis through its ZnF7 ubiquitin-binding domain. *Nature Cell Biology, 21*, 731—742.

Poudel, B., & Gurung, P. (2018). An update on cell intrinsic negative regulators of the NLRP3 inflammasome. *Journal of Leukocyte Biology*. http://dx.doi.org/10.1002/JLB.3MIR0917-350R.
Pu, Z., et al. (2019). Systematic understanding of the mechanism and effects of arctigenin attenuates inflammation in dextran sulfate sodium-induced acute colitis through suppression of NLRP3 inflammasome by SIRT1. *American Journal of Translational Research, 11*, 3992—4009.
Quinquenel, A., et al. (2019). Prevalence of BTK and PLCG2 mutations in a real-life CLL cohort still on ibrutinib after 3 years: A FILO group study. *Blood, 134*, 641—644.
Rabionet, R., et al. (2019). Biallelic loss-of-function LACC1/FAMIN mutations presenting as rheumatoid factor-negative polyarticular juvenile idiopathic arthritis. *Scientific Reports, 9*, 4579.
Rafiq, N. K., Lachmann, H., Joensen, F., Herlin, T., & Brogan, P. A. (2018). Tocilizumab for the treatment of mevalonate kinase deficiency. *Case Reports in Pediatrics, 2018*, 3514645.
Ralston, S. H., & Taylor, J. P. (2019). Rare inherited forms of Paget's disease and related syndromes. *Calcified Tissue International, 104*, 501—516.
Rowczenio, D. M., & Lachmann, H. J. (2019). How to prescribe a genetic test for the diagnosis of autoinflammatory diseases? *Presse Medicale, 48*, e49—e59.
Rowczenio, D., et al. (2019). Current practices for the genetic diagnosis of autoinflammatory diseases: Results of a European molecular genetics quality network survey. *European Journal of Human Genetics*. http://dx.doi.org/10.1038/s41431-019-0439-9.
Roy, N. B. A., et al. (2019). Majeed syndrome: Description of a novel mutation and therapeutic response to bisphosphonates and IL-1 blockade with anakinra. *Rheumatology*. http://dx.doi.org/10.1093/rheumatology/kez317.
Rusmini, M., et al. (2016). Next-generation sequencing and its initial applications for molecular diagnosis of systemic auto-inflammatory diseases. *Annals of the Rheumatic Diseases, 75*(8), 1550—1557. http://dx.doi.org/10.1136/annrheumdis-2015-207701.
Saito, N., et al. (2018). Novel PSTPIP1 gene mutation in pyoderma gangrenosum, acne and suppurative hidradenitis syndrome. *The Journal of Dermatology, 45*, e213—e214.
Sanchez, G. A. M., et al. (2018). JAK1/2 inhibition with baricitinib in the treatment of autoinflammatory interferonopathies. *Journal of Clinical Investigation, 128*, 3041—3052.
Sarens, I. L., et al. (2018). Blau syndrome-associated uveitis: Preliminary results from an international prospective interventional case series. *American Journal of Ophthalmology, 187*, 158—166.
Satoh, T. K., Mellett, M., Contassot, E., & French, L. E. (2018). Are neutrophilic dermatoses autoinflammatory disorders? *British Journal of Dermatology, 178*, 603—613.
Sawabe, Y., et al. (2019). Case of generalized pustular psoriasis with coexisting mutations in IL36RN and CARD14. *The Journal of Dermatology*. http://dx.doi.org/10.1111/1346-8138.14909.
Schnappauf, O., Chae, J. J., Kastner, D. L., & Aksentijevich, I. (2019). The pyrin inflammasome in health and disease. *Frontiers in Immunology, 10*, 1745.
Setta-Kaffetzi, N., et al. (2014). AP1S3 mutations are associated with pustular psoriasis and impaired toll-like receptor 3 trafficking. *The American Journal of Human Genetics, 94*, 790—797.
Shin, J. I.l, et al. (2019). Inflammasomes and autoimmune and rheumatic diseases: A comprehensive review. *Journal of Autoimmunity, 102299*. http://dx.doi.org/10.1016/j.jaut.2019.06.010.
Sidorowicz, W., Kubasiewicz-Ross, P., & Dominiak, M. (2018). Familial cherubism: Clinical and radiological features. Case report and review of the literature. *European Journal of Paediatric Dentistry, 19*, 213—217.

Signa, S., et al. (2019). Whole exome sequencing approach to childhood onset familial erythrodermic psoriasis unravels a novel mutation of CARD14 requiring unusual high doses of ustekinumab. *Pediatric Rheumatology Online Journal, 17*, 38.

Sims, R., et al. (2017). Rare coding variants in PLCG2, ABI3, and TREM2 implicate microglial-mediated innate immunity in Alzheimer's disease. *Nature Genetics, 49*, 1373–1384.

Skon-Hegg, C., et al. (2019). LACC1 regulates TNF and IL-17 in mouse models of arthritis and inflammation. *The Journal of Immunology, 202*, 183–193.

Somech, R., et al. (2017). Disruption of thrombocyte and T lymphocyte development by a mutation in ARPC1B. *The Journal of Immunology, 199*, 4036–4045.

Sozeri, B., Gerceker-Turk, B., Yildiz-Atikan, B., Mir, S., & Berdeli, A. (2018). A novel mutation of interleukin-1 receptor antagonist (IL1RN) in a DIRA patient from Turkey: Diagnosis and treatment. *Turkish Journal of Pediatrics, 60*, 588–592.

Standing, A. S. I., et al. (2017). Autoinflammatory periodic fever, immunodeficiency, and thrombocytopenia (PFIT) caused by mutation in actin-regulatory gene WDR1. *Journal of Experimental Medicine, 214*, 59–71.

Steiner, A., Harapas, C. R., Masters, S. L., & Davidson, S. (2018). An update on autoinflammatory diseases: Relopathies. *Current Rheumatology Reports, 20*, 39.

Talpin, A., et al. (2019). A20 in dendritic cells restrains intestinal anti-bacterial peptide expression and preserves commensal homeostasis. *PLoS One, 14*, e0218999.

Taraborrelli, L., et al. (2018). LUBAC prevents lethal dermatitis by inhibiting cell death induced by TNF, TRAIL and CD95L. *Nature Communications, 9*, 3910.

ter Haar, N. M., et al. (2015). Recommendations for the management of autoinflammatory diseases. *Annals of the Rheumatic Diseases, 74*, 1636–1644.

Toral-Lopez, J., Gonzalez-Huerta, L. M., Martin-Del Campo, M., Messina-Baas, O., & Cuevas-Covarrubias, S. A. (2018). Familial Blau syndrome without uveitis caused by a novel mutation in the nucleotide-binding oligomerization domain-containing protein 2 gene with good response to infliximab. *Pediatric Dermatology, 35*, e180–e183.

Torrelo, A. (2017). CANDLE syndrome as a paradigm of proteasome-related autoinflammation. *Frontiers in Immunology, 8*, 927.

Toubi, E., & Vadasz, Z. (2019). Innate immune-responses and their role in driving autoimmunity. *Autoimmunity Reviews, 18*, 306–311.

Touitou, I., et al. (2013). The expanding spectrum of rare monogenic autoinflammatory diseases. *Orphanet Journal of Rare Diseases, 8*, 162.

Tsuchida, N., et al. (2019). Haploinsufficiency of A20 caused by a novel nonsense variant or entire deletion of TNFAIP3 is clinically distinct from Behcet's disease. *Arthritis Research and Therapy, 21*, 137.

Tsuji, S., et al. (2019). Functional analysis of a novel G87V TNFRSF1A mutation in patients with TNF receptor-associated periodic syndrome. *Clinical and Experimental Immunology*. http://dx.doi.org/10.1111/cei.13365.

Tye, H., et al. (2018). NLRP1 restricts butyrate producing commensals to exacerbate inflammatory bowel disease. *Nature Communications, 9*, 3728.

van der Lee, S. J., et al. (2019). A nonsynonymous mutation in PLCG2 reduces the risk of Alzheimer's disease, dementia with lewy bodies and frontotemporal dementia, and increases the likelihood of longevity. *Acta Neuropathologica, 138*, 237–250.

Van Gijn, M. E., et al. (2018). New workflow for classification of genetic variants' pathogenicity applied to hereditary recurrent fevers by the International Study Group for Systemic Autoinflammatory Diseases (INSAID). *Journal of Medical Genetics, 55*(8), 530–537.

van Well, E. M., et al. (2019). A protein quality control pathway regulated by linear ubiquitination. *The EMBO Journal, 38*.

Venkataraman, G. R., & Rivas, M. A. (2019). Rare and common variant discovery in complex disease: The IBD case study. *Human Molecular Genetics*. http://dx.doi.org/10.1093/hmg/ddz189.

Vetters, J., et al. (2019). The ubiquitin-editing enzyme A20 controls NK cell homeostasis through regulation of mTOR activity and TNF. *Journal of Experimental Medicine*. http://dx.doi.org/10.1084/jem.20182164.

Volpi, S., et al. (2018). Type I interferon pathway activation in COPA syndrome. *Clinical Immunology, 187*, 33–36.

Volpi, S., et al. (2019). A combined immunodeficiency with severe infections, inflammation, and allergy caused by ARPC1B deficiency. *The Journal of Allergy and Clinical Immunology, 143*, 2296–2299.

Volpi, S., et al. (2019). Efficacy and adverse events during Janus kinase inhibitor treatment of SAVI syndrome. *Journal of Clinical Immunology, 39*, 476–485.

Vossen, A. R. J. V., van der Zee, H. H., & Prens, E. P. (2018). Hidradenitis suppurativa: A systematic review integrating inflammatory pathways into a cohesive pathogenic model. *Frontiers in Immunology, 9*, 2965.

Wang, D., et al. (2018). Missense variants in HIF1A and LACC1 contribute to leprosy risk in Han Chinese. *The American Journal of Human Genetics, 102*, 794–805.

Wang, L., et al. (2018). S100A12 and S100A8/9 proteins are biomarkers of articular disease activity in Blau syndrome. *Rheumatology*. http://dx.doi.org/10.1093/rheumatology/key090.

Wang, S., Yuan, Y.-H., Chen, N.-H., & Wang, H.-B. (2019b). The mechanisms of NLRP3 inflammasome/pyroptosis activation and their role in Parkinson's disease. *International Immunopharmacology, 67*, 458–464.

Wang, Z., et al. (2019). Cardamonin from a medicinal herb protects against LPS-induced septic shock by suppressing NLRP3 inflammasome. *Acta Pharmaceutica Sinica, 9*, 734–744.

Weiss, E. S., et al. (2018). Interleukin-18 diagnostically distinguishes and pathogenically promotes human and murine macrophage activation syndrome. *Blood, 131*, 1442–1455.

Wu, X., et al. (2018). Nicotine promotes atherosclerosis via ROS-NLRP3-mediated endothelial cell pyroptosis. *Cell Death and Disease, 9*, 171.

Xiao, J., et al. (2018). Gasdermin D mediates the pathogenesis of neonatal-onset multisystem inflammatory disease in mice. *PLoS Biology, 16*, e3000047.

Xue, J.-Y., et al. (2019). TNFRSF11A-Associated dysosteosclerosis: A report of the second case and characterization of the phenotypic spectrum. *Journal of Bone and Mineral Research*. http://dx.doi.org/10.1002/jbmr.3805.

Yamaguchi, A., et al. (2009). The microRNA-regulated SBP-Box transcription factor SPL3 is a direct upstream activator of LEAFY, FRUITFULL, and APETALA1. *Developmental Cell, 17*, 268–278.

Yamasaki, Y., et al. (2019). S100A12 and vascular endothelial growth factor can differentiate Blau syndrome and familial mediterranean fever from systemic juvenile idiopathic arthritis. *Clinical Rheumatology, 38*, 835–840.

Yamazaki-Nakashimada, M. A., et al. (2019). Systemic autoimmunity in a patient with CANDLE syndrome. *Journal of Investigational Allergology and Clinical Immunology, 29*, 75–76.

Yao, Q., Li, E., & Shen, B. (2019). Autoinflammatory disease with focus on NOD2-associated disease in the era of genomic medicine. *Autoimmunity, 52*, 48–56.

Yu, C.-H., Moecking, J., Geyer, M., & Masters, S. L. (2018). Mechanisms of NLRP1-mediated autoinflammatory disease in humans and mice. *Journal of Molecular Biology, 430*, 142–152.

Zhao, Y., Shi, J., & Shao, F. (2018). Inflammatory caspases: Activation and cleavage of gasdermin-D in vitro and during pyroptosis. *Methods in Molecular Biology, 1714*, 131–148.

Zhong, F. L., et al. (2018). Human DPP9 represses NLRP1 inflammasome and protects against autoinflammatory diseases via both peptidase activity and FIIND domain binding. *Journal of Biological Chemistry, 293*, 18864–18878.

Zhou, Q., et al. (2012). A hypermorphic missense mutation in PLCG2, encoding phospholipase Cγ2 , causes a dominantly inherited autoinflammatory disease with immunodeficiency. *The American Journal of Human Genetics, 91*, 713–720.

Zhu, C., et al. (2017). CECR1-mediated cross talk between macrophages and vascular mural cells promotes neovascularization in malignant glioma. *Oncogene, 36*, 5356–5368.

Zhuang, L., et al. (2018). Dosage considerations for canakinumab in children with periodic fever syndromes. *Clinical Pharmacology and Therapeutics*. http://dx.doi.org/10.1002/cpt.1302.

Zotti, T., Polvere, I., Voccola, S., Vito, P., & Stilo, R. (2018). CARD14/CARMA2 signaling and its role in inflammatory skin disorders. *Frontiers in Immunology, 9*, 2167.

CHAPTER TWO

Roles of *Porphyromonas gingivalis* and its virulence factors in periodontitis

Weizhe Xu[a,d], Wei Zhou[b], Huizhi Wang[c] and Shuang Liang[d,*]

[a]State Key Laboratory of Oral Diseases & National Clinical Research Center for Oral Diseases & Department of Cariology and Endodontics, West China Hospital of Stomatology, Sichuan University, Chengdu, Sichuan, China

[b]Department of Endodontics, Ninth People's Hospital, School of Medicine, Shanghai JiaoTong University, Shanghai, Pudong, China

[c]VCU Philips Institute for Oral Health Research, Department of Oral and Craniofacial Molecular Biology, Virginia Commonwealth University School of Dentistry, Richmond, VA, United States

[d]Department of Oral Immunology and Infectious Diseases, University of Louisville School of Dentistry, Louisville, KY, United States

*Corresponding author: Email: Shuang.liang@louisville.edu

Contents

Advances in Protein Chemistry and Structural Biology, Volume 120
ISSN 1876-1623
https://doi.org/10.1016/bs.apcsb.2019.12.001

Abstract

Periodontitis is an infection-driven inflammatory disease, which is characterized by gingival inflammation and bone loss. Periodontitis is associated with various systemic diseases, including cardiovascular, respiratory, musculoskeletal, and reproductive system related abnormalities. Recent theory attributes the pathogenesis of periodontitis to oral microbial dysbiosis, in which *Porphyromonas gingivalis* acts as a critical agent by disrupting host immune homeostasis. Lipopolysaccharide, proteases, fimbriae, and some other virulence factors are among the strategies exploited by *P. gingivalis* to promote the bacterial colonization and facilitate the outgrowth of the surrounding microbial community. Virulence factors promote the coaggregation of *P. gingivalis* with other bacteria and the formation of dental biofilm. These virulence factors also modulate a variety of host immune components and subvert the immune response to evade bacterial clearance or induce an inflammatory environment. In this chapter, our focus is to discuss the virulence factors of periodontal pathogens, especially *P. gingivalis*, and their roles in regulating immune responses during periodontitis progression.

1. Periodontitis

Periodontitis is one of the most prevalent inflammatory diseases. In US, 47% of adults, with 70% of the adult population aged 65 years and older, are afflicted with periodontitis (Eke, Dye, Wei, Thornton-Evans, & Genco, 2012; Eke et al., 2015; Papapanou, 2012). Worldwide, 538 million people suffer from severe periodontal disease and 276 million of whom lost their teeth. The figures are expected to rise as population grows and ages (Kassebaum et al., 2017). In 1999, the annual expenditure on the therapy of periodontal disease was estimated to exceed $14 billion in the USA (Brown, Johns, & Wall, 2002). Apart from being the major factor causing loss of teeth, periodontitis is associated with various systemic diseases like atherosclerosis, diabetes, rheumatoid arthritis and adverse pregnancy outcomes like low birth weight (Khader, Dauod, El-Qaderi, Alkafajei, & Batayha, 2006; Kinane, Stathopoulou, & Papapanou, 2017; Madianos, Bobetsis, & Offenbacher, 2013). Periodontal disease may exacerbate these diseases through increased pro-inflammatory mediators or direct infiltration of periodontal pathogens in the circulation (Khader et al., 2006).

2. Periodontal pathogens and their virulence factors

Periodontitis is induced by microbial biofilm and caused by host-mediated inflammation, which leads to collateral tissue damage and clinical attachment loss (Hajishengallis, Darveau, & Curtis, 2012). Oral microbiota contains the microorganisms found in the oral cavity, which are estimated to be more than 700 different species with distinct subspecies (Aas, Paster, Stokes, Olsen, & Dewhirst, 2005; Dewhirst et al., 2010). A balance between the resident microbiota and host immune response maintains a state of homeostasis, while the disruption of this balance contributes to oral diseases like periodontal disease (Lamont, Koo, & Hajishengallis, 2018). Periodontitis is one typical disease of biofilm-associated infection (Bergstrom, 2006; Hajishengallis & Lamont, 2012; Socransky, Haffajee, Cugini, Smith, & Kent, 1998). Oral bacterial biofilms are among the first to be discovered in human and have been studied extensively. From the time that oral bacterial biofilm was first discovered by James Leon Williams in 1897 as a mass of bacteria adhering to the tooth surface, various periodontal pathogens, including *Aggregatibacter actinomycetemcomitans, Fusobacterium nucleatum, Tannerella forsythia* (formerly *Bacteroides forsythus), Porphyromonas gingivalis, Prevotella intermedia* and *Treponema denticola,* have been extensively investigated (Aruni, Dou, Mishra, & Fletcher, 2015). These bacteria and their pathogenic roles in periodontitis have been explained in various models, including red complex, specific and non-specific microorganisms, keystone pathogen, and polymicrobial synergy and dysbiosis models (Hajishengallis & Lamont, 2012).

Red complex model recognizes bacteria including *P. gingivalis, T. Denticola* and *T. Forsythia,* popularly known as the "red complex" in gingival plaques, as the major pathogens that are responsible for periodontitis progression (Socransky et al., 1998). The "red complex model" laid solid foundation for the concept that periodontitis is a multi-microbe disease, and leads to other newly emerged models following the new technology development and new discoveries. Most recently, the etiology of periodontitis has shifted in favor of the theory of polymicrobial synergy and dysbiosis (the 'PSD model') (Hajishengallis, Darveau & Curtis, 2012; Hajishengallis & Lamont, 2012), which proposes that disease is initiated by a synergistic polymicrobial community, in which different members fulfill distinct roles that result in a combined effect on the oral microbial dysbiosis and host immune disruption. Dysbiosis or imbalance in the ecosystem of microorganisms further leads to change in the host-microbe crosstalk that causes destructive inflammation, bone loss, and periodontitis (Abusleme et al., 2013).

During the formation of the pathogenic community, 'keystone pathogens' modulate the immune response to impair host immune surveillance and tip the balance from homeostasis to dysbiosis. Keystone pathogens can also promote the pathogenesis through the enhancement of direct interaction among the oral pathogens. Virulence factors are integral molecules expressed by an organism at various stages of its life cycle with capacity to damage the host. Formation of the pro-inflammatory environment and induction of oral microbiome dysbiosis require diverse virulence factors expressed by the bacteria, such as proteolytic enzymes, capsule, lipopolysaccharide, fimbriae, and certain other surface structures/ligands (How, Song, & Chan, 2016).

Our and others' reports showed that even in relatively low numbers, *P. gingivalis* can interfere with host immune responses and disrupt the symbiosis between the local oral bacteria (Hajishengallis, 2014). It acts as the keystone pathogen in the dysbiosis of oral microbiome and periodontitis development. In this chapter, we will discuss various periodontal pathogens, but focus on *P. gingivalis* and its virulence factors for their important roles in host immune response modulation, which are critical for the pathogen colonization, oral microbial dysbiosis, and periodontitis pathogenesis.

3. *Porphyromonas gingivalis* and its virulence factors

P. gingivalis is a Gram-negative, anaerobic, rod-shaped bacteria forming black colonies on blood agar. It was detected in 85.75% subgingival plaque of chronic periodontitis patients (How et al., 2016). Prevalence of *P. gingivalis* is related to the severity of periodontal disease and has been identified by researchers as one of the major causative agents in the periodontitis pathogenesis. Recently, the keystone pathogen theory postulates that even at low abundance, *P. gingivalis* is capable of inducing chronic periodontitis by remodeling the commensal bacterial community to promote a state of dysbiosis, which leads to disease (Olsen, Lambris, & Hajishengallis, 2017). *P. gingivalis* has been widely investigated, partially benefited from the well-characterized genomics of different *P. gingivalis* strains and the accessibility of a great variety of mutants (Hajishengallis, Darveau, et al., 2012; Naito et al., 2008; Nelson et al., 2003). *P. gingivalis* is known to produce a wide array of virulence factors that could cause tissue destruction on their own or act through other mediators to induce inflammation (Hajishengallis, 2014).

P. gingivalis and *Streptococcus gordonii* are able to interact to form communities and subsequent colonization of the dental plaque. *P. gingivalis* benefits from its interaction and coaggregation with other bacteria in the subgingival plaque for its destructive effects on periodontal tissues (Kuboniwa et al., 2017). *P. gingivalis* can efficiently modify the host immune response and create an environment favorable to its own and other pathogens' continued persistence (Hajishengallis & Lambris, 2011). Virulence factors of *P. gingivalis* play important roles in the coaggregation, biofilm formation, and oral microbial dysbiosis. They can either directly act on other bacteria or induce an optimal environment. One example of direct interaction is that *P. gingivalis* binds to GAPDH and surface proteins on the surface of *S. gordonii*, which ultimately favors the formation of *P. gingivalis* biofilms on the tooth surface (Maeda, Nagata, Yamamoto, et al., 2004).

4. *Porphyromonas gingivalis* LPS

4.1 Structure of *Porphyromonas gingivalis* LPS

Lipopolysaccharide (LPS) is a pathogen associated molecular pattern (PAMP) or more recently identified as microbe-associated molecular pattern (MAMP) (Medzhitov, 2001). LPS is also a major component of the cell wall (i.e. outer cell membrane) of Gram-negative bacteria, including periodontal pathogen *P. gingivalis* (Darveau, Arbabi, Garcia, Bainbridge, & Maier, 2002). LPS was well known for its toxicity and the ability to cause unwanted host inflammation, which gave it the name endotoxin. LPS from *E. coli* was found to be able to facilitate the host inflammatory response by stimulating Toll-like receptor 4 (TLR4) (Lien et al., 2000; Poltorak et al., 1998). *P. gingivalis* LPS, like those of *E. coli* and other Gram-negative bacteria, comprises the outer leaflet of the bacteria and is composed of lipid A, core oligosaccharide that forms the backbone of LPS, and O-specific polysaccharide (Schromm et al., 2000). Acting as the core factor of LPS in the induction of immune responses, lipid A moiety confers toxicity and is structurally composed of a phosphorylated β (1-6) D-glucosamine disaccharide backbone and multiacyl chains acylated by fatty acids at specific positions on the backbone (Dixon & Darveau, 2005). LPS from different bacterial species are structurally different, depending on the fatty acid acyl chain composition of lipid A. The structure difference in LPS may explain the distinct mechanisms through which the host cells recognize bacterial species.

P. gingivalis LPS plays a strong pathogenic role in periodontal tissues. The virulence properties of *P. gingivalis* LPS is determined by its lipid A component. *P. gingivalis* LPS contains multiple forms of lipid A (Darveau et al., 2004), the chemical structure of which was determined to be a glucosamine beta-(1-6) disaccharide 1-monophosphate acylated by 3-hydroxy-15-methylhexadecanoic acid and 3-hexadecanoyloxy-15-methylhexadecanoic acid at the 2- and 2′-positions, respectively (Ogawa, 1993). Host cells respond to *P. gingivalis* LPS lipid A and produce inflammatory responses in gingival tissues, thereby creating a favorable environment for pathogens to sustain and ultimately cause periodontal disease progression (Herath et al., 2013).

While the basic chemical composition of *P. gingivalis* LPS is similar to that of *E. coli* endotoxin, *P. gingivalis* lipid A structure exhibits different variations of acylation, tetra-acylation and penta-acylation. As a consequence, *P. gingivalis* LPS activates distinctive signaling pathways and initiates differential immune responses. Based on its molecular weight, *P. gingivalis* LPS with penta-acylated lipid A is named *P. gingivalis* $LPS_{1690,}$ for its molecular weight of 1690 Da. Similarly, *P. gingivalis* LPS with tetra-acylated lipid A is called $LPS_{1435/1449}$ (Curtis et al., 2011). Basic mechanism for the production of distinctive *P. gingivalis* LPS remains controversial, which can be attributed to different *P. gingivalis* strains and environmental factors, such as the hemin levels, phosphate availability, and incubation temperatures (Darveau et al., 2002) (Rangarajan, Aduse-Opoku, Paramonov, Hashim, & Curtis, 2017).

4.2 Immune responses triggered by *Porphyromonas gingivalis* LPS

P. gingivalis LPS-induced TLR-specific immune regulatory roles have been extensively investigated. While *E. coli*-LPS can activate TLR4 but not TLR2 (Takeuchi et al., 1999), the TLR activations triggered by *P. gingivalis* LPS are much more complicated (Coats et al., 2009; Liu, Desta, Raptis, Darveau, & Graves, 2008). The leading signaling pathway activated by *P. gingivalis* LPS is controversial. Some reports showed that TLR4 exerts a dominant function, while others demonstrated that TLR2 is the major receptor (Wang & Ohura, 2002). Reports proved that TLR2 is required for alveolar bone loss caused by *P. gingivalis* infection in animal models. *P. gingivalis* stimulation led to the up-regulation of TLR2 expression and pro-inflammatory cytokine production *in vitro* (Papadopoulos et al., 2013). TLR2 is also critical for *P. gingivalis* LPS-activated nitric oxide, TNF-α, and IL-6 production by macrophages (Holden et al., 2014). The signaling pathways downstream of TLRs are also controversial. Some reports showed that NF-κB pathway is critical in TLR-mediated pro-inflammatory signaling and the production of cytokines, chemokines, and adhesion molecules (Darveau et al., 2004). NF-κB pathway is also important in alveolar bone loss. Its activation by LPS from non-periodontal pathogen *E. coli* through TLR4 is enough to cause periodontitis (Li et al., 2014). On the other hand, other pathways including the mitogen-activated protein kinase (MAPK) pathway are also reported to be involved in inflammation and bone loss (Bainbridge & Darveau, 2001).

Recent discovery revealed that the binding and activation of specific TLRs by LPS are determined by the structure of different types lipid A, which is the core component and effector of the LPS molecule. Even small changes in lipid A structure can influence the immune responses (Dixon & Darveau, 2005). Lipid A heterogeneity in *P. gingivalis* LPS led to significantly different innate immune responses and pro-inflammatory cytokine production (Herath et al., 2011). When binding to TLR4, penta-acylated *P. gingivalis* LPS_{1690} is similar to classical hexa-acylated *E. coli* LPS in certain ways (Bozkurt, Hakki, Hakki, Durak, & Kantarci, 2017). For example, both *P. gingivalis* LPS_{1690} and *E. coli* LPS elevated the LBP protein expression in human oral keratinocytes. *P. gingivalis* LPS_{1690} acts differently from traditional *E. coli* LPS in several ways. While *P. gingivalis* LPS_{1690} activated NF-κB and p38/MAPK pathways, *E. coli* LPS regulated through NF-κB, p38/MPAK and JNK signaling pathways (Ding, Wang, Darveau, & Jin, 2013). *P. gingivalis* LPS_{1690} may also bind to TLR2. Through TLR2 activation, *P. gingivalis* LPS_{1690} significantly enhanced the transcription of NF-κB in mouse cementoblasts. Comparing to *P. gingivalis* LPS_{1690}, tetra-acylated $LPS_{1435/1449}$ stimulation relied even less on TLR4, and more on TLR2 (Nemoto, Darveau, Foster, Nogueira-Filho, & Somerman, 2006). Indeed, $LPS_{1435/1449}$ acts as a TLR4 antagonist and TLR2 agonist (Herath et al., 2011, 2013). Stimulation with either *P. gingivalis* LPS_{1690} or $LPS_{1435/1449}$ elevated TLR2 expression on the cell surfaces of hGFs, but $LPS_{1435/1449}$ induced higher TLR2 expression than LPS_{1690}. In contrast, *P. gingivalis* LPS_{1690}, but not $LPS_{1435/1449}$, significantly increased TLR4 and MD-2 expression in hGFs. *P. gingivalis* LPS_{1690} activated mainly through NF-κB, while *P. gingivalis* $LPS_{1435/1449}$ primarily induced the p38/MAPK and ERK1/2 signaling pathways.

The structure of lipid A in LPS also determines its binding to a newly found intracellular LPS receptor, caspase-11. *E. coli* LPS can also bind to mouse caspase-11 (homolog of human caspase-4), induce caspase-11 oligomerization, and activate the caspases (Kayagaki et al., 2013). On the other hand, underacylated lipid IVa and tetra-acylated LPS, such as *Rhodobacter sphaeroides* LPS (LPS-RS), cannot induce caspase-11 oligomerization and activation, despite that they can bind to caspase-11 (Shi et al., 2014). Tetra-acylated *P. gingivalis* lipid A was found to evade activation of the noncanonical inflammasome, which enhanced the intracellular survival of *P. gingivalis* in macrophages. In contrast, *P. gingivalis* LPS_{1690}, which contains TLR4 agonist lipid A, induced noncanonical inflammasome activation and led to the production of IL-1β and deprived the survival capability of *P. gingivalis* in macrophages (Slocum et al., 2014). *P. gingivalis* might exploit its TLR4 antagonistic lipid A to evade host clearance by interfering with the activation of TLR4 and caspase-11. Indeed, the lipid A phosphatase, which is responsible for lipid A alterations, was shown to be critical for *P. gingivalis* colonization and even the formation of oral microbiome, indicating the importance of the lipid structure in the processes (Zenobia et al., 2014).

5. Fimbriae

5.1 Types and compositions of fimbriae

P. gingivalis develops fimbriae, or pili, which are thin filamentous surface appendages that protrude from the outer membrane. *P. gingivalis* fimbriae have been among the focuses of research in periodontal pathogens and periodontitis pathogenesis. Fimbriae enhance the biofilm formation, bacterial motility, bacterial adhesion to host cells, and bacterial invasion into the cells. *P. gingivalis* expresses two forms of fimbriae, the long fimbriae (also called major fimbriae) comprised of FimA subunits, and the short fimbriae (or minor fimbriae) with Mfa1 subunit proteins. Both FimA and Mfa1 are critical for host immune responses and pathogenesis of periodontitis, even though their amino acid composition and antigenic properties are different (Amano, 2010). With optimal staining, fimbriae can be observed under microscopes, while different *P. gingivalis* strains showed variance in surface morphology and capability to adhere to host cells (Zheng, Wu, & Xie, 2011).

Long fimbriae are 5 nm wide and extend up to 3 μm long, with sizes varying from 41 to 49 kDa (Dickinson, Kubiniec, Yoshimura, & Genco, 1988; Handley & Tipler, 1986; Lee, Sojar, Bedi, & Genco, 1991). FimA is the shaft subunit, with accessory proteins FimC, FimD, and FimE as minor components of the long fimbriae (Enersen, Nakano, & Amano, 2013; Watanabe et al., 1996). FimE is required for assembly of FimC and FimD onto the distal tip of FimA fibers (Nishiyama et al., 2007), while FimB has been suggested as fimbria anchors and regulators of fimbrial length (Nagano, Hasegawa, Murakami, Nishiyama, & Yoshimura, 2010). Deficiency in FimB expression was reported to lead to the formation of abnormal longer fimbriae, which easily shed from the surface (Nagano et al., 2010).

FimA proteins vary in sizes and amino terminals. Fimbriae can be classified into six types based on nucleotide sequences of the *fimA* gene (type I, Ib, II, III, IV, and V) (Fujiwara, Morishima, Takahashi, & Hamada, 1993). Strains 381, ATCC 33277, and HG565 are type I strains and abundantly fimbriated that showed significant adhesive capability to host tissues and cells. Strains A7A1, SA2, BH18/10, OMZ314, and OMZ409 belong to type II, while strain BH6/26 is a type III *P. gingivalis*. Type IV strains, such as W50 and W83, are poorly fimbriated. HNA99 belongs to type V (Enersen, Olsen, Kvalheim, & Caugant, 2008). When tested for the *fimA* genotypes of *P. gingivalis*, type II and I are the most widely distributed ones in plaque samples from periodontitis patients (Enersen et al., 2008; Nagano, Hasegawa, Iijima, Kikuchi, & Mitani, 2018).

Short fimbriae, composed of mainly Mfa1 structural subunit proteins, differ from 60 to 500 nm in their length (Enersen et al., 2013). Similar to long fimbriae, short fimbriae also contain accessory proteins, which are named Mfa2-5 (Hasegawa et al., 2009, 2016). Mfa2 is an anchor subunit as well as an assembly and elongation terminator (Hasegawa et al., 2009). Mfa3 is localized at the distal tip of the short fimbriae, implying its role as a ligand to host cell receptors (Hasegawa et al., 2013, 2016). Interestingly, Mfa3, Mfa4, and Mfa5 work closely together, by which the absence of any one of these three accessory proteins would lead to lack of all three in the mature fimbriae (Hasegawa et al., 2013, 2016).

5.2 Roles of fimbriae in biofilm formation

As we mentioned earlier, fimbriae play important roles in many of the adhesive properties of *P. gingivalis*. *P. gingivalis* fimbriae bind to host tissues and cells through a wide variety of host molecules and oral substrates, which include proline-rich proteins and glycoproteins, statherins, fibrinogen, fibronectin, and lactoferrin (Amano, 2003). Besides their adhesive function to host tissues and cells, fimbriae are also important for the interaction of *P. gingivalis* with other oral bacteria (Lamont & Jenkinson, 2000), which is indispensable for biofilm formation. Biofilm formation is a complex process, during which the early colonizers first attach to the tooth surface, followed by the attachment of later colonizers (Kuboniwa & Lamont, 2010; Whittaker, Klier, & Kolenbrander, 1996). Fimbriae have been reported to mediate the coaggregation of *P. gingivalis* with other organisms including *Actinomyces viscosus* (Goulbourne & Ellen, 1991), *T. denticola* (Hashimoto,

Ogawa, Asai, Takai, & Ogawa, 2003), *S. gordonii* (Maeda, Nagata, Nonaka, et al., 2004), and *Streptococcus oralis* (Maeda, Nagata, Yamamoto, et al., 2004). The molecules bound by the *P. gingivalis* fimbriae have been under investigation. Some manifested molecules include the dentilisin of *T. denticola* and glyceraldehyde-3-phosphate dehydrogenase (GAPDH) of *S. gordonii* and *S. oralis* (Hashimoto et al., 2003; Maeda, Nagata, Yamamoto, et al., 2004; Park et al., 2005). Interestingly, *P. gingivalis* long fimbriae can also bind to human GAPDH, which imply one pathway in host cell invasion and induction of host responses (Sojar & Genco, 2005). Aside from the major components, the minor components of long fimbriae are also critical for the adherence capability of *P. gingivalis*. It was reported that the mutants lack of fimC-, fimD-, and fimE were less efficient to bind to GAPDH of *S. oralis* as well as fibronectin and type I collagen (Maeda, Nagata, Yamamoto, et al., 2004; Nishiyama et al., 2007).

On the other hand, the roles and working mechanisms of short fimbriae in biofilm formation are likely different from those of long fimbriae. For instance, *P. gingivalis* short fimbriae can bind to SspA and SspB proteins of *S. gordonii* (Lamont, Hsiao, & Gil, 1994). Ssp proteins are members of the antigen I/II family of streptococcal surface proteins. Interestingly, while the antigen I/II proteins are highly conserved among the streptococcal species, *P. gingivalis* can only adhere to SspA/B proteins of *S. gordonii*, but not antigen I/II from *Streptococcus mutans*. Research showed that the *P. gingivalis* Mfa1 specially recognized the amino acid residues 1167—1250 at the C-terminal region of the SspB protein, which was named SspB adherence region (BAR) (Park et al., 2005).

5.3 Host immune responses induced by fimbriae

Compared with wild type *P. gingivalis*, the mutant lacking FimA induced less severe bone loss in an animal periodontitis model (Malek et al., 1994). In addition to the fimbriae's roles in biofilm formation, their effects on host immune responses have also been well investigated (Enersen et al., 2013; Pathirana, O'Brien-Simpson, & Reynolds, 2010). The fimbriae can be recognized by the receptors on epithelial, endothelial, and immune cells, which leads to the activation of the cells and the production of cytokines and adhesion molecules (Amano, 2003).

Long fimbriae activated NF-κB via TLR2 and induce the production of pro-inflammatory cytokines, such as tumor necrosis factor-α (TNF-α), interleukin-1β (IL-1β), IL-8, and IL-6 (Hajishengallis et al., 2006). CD14 is an essential TLR2 co-receptor for *P. gingivalis* FimA (Eskan,

Hajishengallis, & Kinane, 2007). Gingival epithelial cells do not express CD14, therefore respond poorly to *P. gingivalis* FimA. In contrast, monocytes express CD14 and respond to FimA (Eskan et al., 2007). *P. gingivalis* fimbriae can induce differential immune responses in different host cells, which might be a strategy exploited by *P. gingivalis* to optimize its colonization and infection.

Long fimbriae of *P. gingivalis* are also able to exploit complement pathway to subvert the host immune clearance. Our research showed that fimbriae interacted with complement receptor 3 (CR3; CD11b/CD18) in monocytes/macrophages, resulting in extracellular signal-regulated kinase 1/2 (ERK1/2) phosphorylation and inhibition of LPS-induced IL-12 production. In vivo results proved that CR3 blockage enhanced IL-12 production and *P. gingivalis* clearance. Furthermore, CR3 antagonist protected the mice from *P. gingivalis*-induced periodontitis (Hajishengallis, Shakhatreh, Wang, & Liang, 2007). Since IL-12 is involved in oral pathogen clearance, fimbriae activated CR3 signaling cross talk with *P. gingivalis*-activated TLR pathways, and inhibit IL-12 production to promote the survival of *P. gingivalis*. Later, our research illustrated that the activation of CXCR4 with *P. gingivalis* interacted with TLR2 signaling. In this process, binding of long fimbriae to CXCR4 and TLR2 led to cAMP-dependent protein kinase A activation, which inhibited host immune clearance of bacteria (Hajishengallis, Wang, Liang, Triantafilou, & Triantafilou, 2008). Another crosstalk pathway is between CXCR4 and CR3, in which *P. gingivalis* fimbriae utilize CXCR4 to induce PI3K-dependent activation of CR3 (Hajishengallis, McIntosh, Nishiyama, & Yoshimura, 2012). Both CR3 and CXCR4 exploitation require FimA and FimCDE to assemble fully functional long fimbriae (Min Wang et al., 2007; Pierce et al., 2009).

Research on the effect of short fimbriae is relatively limited. Infection of FimA (long fimbriae) knockout mutant, mfa1 (short fimbriae) knockout, and double mutant *P. gingivalis* in an animal model showed that both long fimbriae and short fimbriae contributed to alveolar bone loss in periodontitis (Umemoto & Hamada, 2003). Short fimbriae, similar to long fimbriae, interacted with TLR2 and CD14 and induced the cytokine production in both human monocyte cell lines and mouse macrophages (Hamada, Watanabe, Arai, Hiramine, & Umemoto, 2002; Hiramine, Watanabe, Hamada, & Umemoto, 2003). In dendritic cells, Mfa1 interacts with DC-SIGN (Dendritic Cell-Specific Intercellular adhesion molecule-3-Grabbing Non-integrin), which stimulates the engulfment of *P. gingivalis* by dendritic cell but promotes the evasion of antibacterial autophagy and lysosome fusion, and leads to the intracellular persistence of *P. gingivalis* in myeloid dendritic cells (El-Awady et al., 2015).

6. Gingipain

Gingipain is a family of cysteine proteinase, which is also known as "trypsin-like" enzyme (Bostanci & Belibasakis, 2012). Gingipains include arginine-specific gingipains (Rgp, including RgpA and RgpB), which hydrolyze peptide bonds at the carbonyl groups of arginine residue, and lysine-specific gingipain (Kpg), which cleave polypeptides at the C-terminal after lysine residue. Structures of the gingipains are highly conserved among different *P. gingivalis* strains. Kgp is composed of a catalytic domain that specifically recognize Lys-Xaa peptide bond and an immunoglobulin-like domain, which are followed by a C-terminal that contains hemagglutinin-adhesin domain. RgpA's structure is similar to Kgp, with an Arg-Xaa recognizing catalytic domain, an immunoglobulin-like domain, and C-terminal hemagglutinin-adhesin domain. On the other hand, RgpB possesses the catalytic domain and immunoglobulin-like domain that similar to RgpA and Kgp, but lacks the COOH-terminal hemagglutinin domain (Guo, Nguyen, & Potempa, 2010). Newly translated products of gingipain genes are inactive zymogens to avoid the unwanted proteolytic activity inside the cell and require further post-translational processing to produce mature functional gingipains. For example, pro-RgpA and pro-Kgp need to go through proteolysis to liberate the hemagglutinin-adhesin domains. In different *P. gingivalis* strains, the mature gingipains can either be transported and docked on the bacterial outer membrane or released into extracellular environment.

Gingipains are accountable for 85% of the extracellular proteolytic activities of *P. gingivalis* and are important in the pathogenesis of periodontitis (de Diego et al., 2014). They cause dysregulated immune responses and inflammation through their involvement in the activation of the host matrix metalloproteinases, inactivation of immune inhibitors, degradation of immune factors, and cleavage of immune cell receptors (Guo et al., 2010; Imamura, Travis, & Potempa, 2003; Potempa, Banbula, & Travis, 2000). Inhibitors of gingipains can effectively suppress the pathogenicity of *P. gingivalis* and are studied for their potential therapeutic functions (Kadowaki et al., 2004).

6.1 Roles of gingipains in biofilm formation

Gingipains enhance the interactions of *P. gingivalis* with other periodontal pathogens, including *T. forsythia*, *T. denticola*, and *A. actinomycetemcomitans*, to facilitate their survival and biofilm formation (Bao et al., 2014; Haraguchi, Miura, Fujise, Hamachi, & Nishimura, 2014; Ito, Ishihara, Shoji, Nakayama, & Okuda, 2010). Gingipains can be directly involved in coaggregation with other bacteria or adhesion to host tissues and cells. *P. gingivalis* mutants without all three gingipains (*rgpA-rgpB- kpg-)* exhibit no coaggregative activity (Abe et al., 2004; Ito et al., 2010). Reports showed that the hemagglutinin-adhesin domains of RgpA and Kgp are responsible for the coaggregation (Abe et al., 2004; Kamaguchi et al., 2003), which was proved by the observation that *P. gingivalis* mutants deficient in hemagglutinin domain-bearing gingipains (i.e. RgpA and Kgp) and hemagglutinin hagA genes were deprived of their coaggregative capability (Abe et al., 2004). Furthermore, gingipains are involved in the fimbria-mediated adherence process through the modification and maturation of pro-fimbrilin into fimbriae. RgpB is vital for the processing and maturation of FimA (Kristoffersen, Solli, Nguyen, & Enersen, 2015), while both RgpA and RgpB are required for the proteolytic processing and polymerization of Mfa1 to produce short fimbriae (Lee et al., 2018). Besides their ability to coaggregate with other bacteria, gingipains participate in *P. gingivalis* adhesion to host cells like gingival epithelial cells and gingival fibroblasts, which also expedite their survival and infection (Andrian, Grenier, & Rouabhia, 2004; Grenier et al., 2003; Sakanaka, Takeuchi, Kuboniwa, & Amano, 2016).

Gingipains can also provide nutrients for the bacteria and help their proliferation. They degrade fibrinogen/fibrin, which contribute to the gingival tissue breakdown, inhibition of blood coagulation, and increased bleeding (Ally et al., 2003). Gingipain Kgp cleaves host heme proteins, such as hemoglobin, haptoglobin, and hemopexin, to provide *P. gingivalis* with heme for its growth (Sroka, Sztukowska, Potempa, Travis, & Genco, 2001).

6.2 Immune regulation by gingipains

Gingipains provide *P. gingivalis* the abilities to evade host immune responses and clearance. Some of the mechanisms involved the degradation of extracellular matrix components, such as collagens, cytokines, immunoglobulins, and complement factors (Curtis, Aduse-Opoku, & Rangarajan, 2001). Gingipains degrade the junctional adhesion molecules of the gingival epithelial cells to break down epithelial barrier and increase the LPS and proteoglycan penetration (Takeuchi et al., 2019). Gingipains are also involved in regulating the immune responses and the production of immune mediators in various cells. Defensins are antimicrobial peptides in innate immune processes that play an important role in pathogen clearance. In humans, defensins are classified into α-defensins (human neutrophil peptides) and ß-defensins (hBDs) (Dommisch & Jepsen, 2015). Gingipains are reported to degrade both neutrophil-derived α-defensins and ß-defensins (Carlisle, Srikantha, & Brogden, 2009; Maisetta, Brancatisano, Esin, Campa, & Batoni, 2011), which might protect the bacteria from being eradicated. Gingipains reduce the expression of innate immune receptors CD14 on macrophage cell surface, which results in decreased response to bacterial infection (Wilensky, Tzach-Nahman, Potempa, Shapira, & Nussbaum, 2015). In human monocytes, chemical inhibitor of gingipains reduced the IL-1β production, indicated its participation in the regulation of inflammatory (Hamedi et al., 2009).

Recently, complement pathways regained the interests of dental research in that they are critical for *P. gingivalis* to cause microbial dysbiosis in oral cavity (Hajishengallis et al., 2011). Complement system is composed of many proteins and protein fragments, which work through a cascade of reactions to (1) generate opsonization factors to promote phagocytosis of microbes and damaged host cells, (2) promote inflammation, and (3) lyse and destroy foreign cells by forming the membrane attack complex (MAC). The system can be activated through classical pathway, lectin pathway, or alternative pathway. All three pathways converge at the central stage of C3 activation. C3 is cleaved by C3 convertase to generate C3b and C3a. C3b can act as an opsonization factor to promote phagocytosis, while C3a is an important anaphylatoxin. More importantly, C3b is required for complement cascade to cleaves C5 into C5b and C5a. C5b works together with C6, C7, C8, and C9 to form membrane attack complex (MAC), which generates a pore on cell membrane and leads to the lysis of foreign cells. The "side products" of this process, anaphylatoxins C3a, C4a, C5a, are important pro-inflammatory factors. Gingipains disrupt the activation of complement system and host-microbe homeostasis in the periodontal tissue, which further cause periodontitis. Gingipains degrade components C3, C4, and C5 to inhibit complement activation and the formation of MAC, therefore inhibit the bacterial clearance (Popadiak, Potempa, Riesbeck, & Blom, 2007; Potempa et al., 2008). Degradation of C3 and C3b molecules by gingipains could also contribute to decreased bacterial opsonization to reduce antibody-mediated bacterial clearance (Wingrove et al., 1992). Rpgs are more potent in degrading C3, C4, and C5 than Kpg is (Popadiak et al., 2007; Potempa et al., 2008). Disruption of complement pathway by gingipains not only inhibits bacterial clearance, but also elevates inflammation. For example, Rgps cleave complement component C5 molecule and degrade C5b to block the formation of MAC. At the same time, anaphylatoxin C5a is released and result in higher inflammation (Liang et al., 2011; Schenkein, Fletcher, Bodnar, & Macrina, 1995; Wingrove et al., 1992). On the other hand, Kpg, but not Rpg, can cleave the C5a receptor (C5aR) on neutrophils, which might be a mechanism to curb the recruitment and anti-bacterial action of neutrophils (Jagels, Ember, et al., 1996; Jagels, Travis, Potempa, Pike, & Hugli, 1996).

Moreover, complement systems are involved in a network of cross-interactions with other signaling pathways (Hajishengallis, Abe, Maekawa, Hajishengallis, & Lambris, 2013; Ricklin, Hajishengallis, Yang, & Lambris, 2010). Indeed, *P. gingivalis* fimbriae exploit the crosstalks between complement receptor CR3, chemokine receptor CXCR4, and TLR signaling pathways to evade bacterial clearance (Maekawa et al., 2014). Additional complement-TLR crosstalk has been discovered to act through the C5a receptor (C5aR) and C3a receptor (C3aR) signaling, which involve mitogen-activated protein kinases (MAPKs), leading to extracellular signal-regulated kinase (ERK1/2) and the c-Jun N-terminal kinase (JNK) activation (Zhang et al., 2007). The production of C5a from C5 by Rgps facilitates the crosstalk (Schenkein et al., 1995; Wingrove et al., 1992). While the complement—TLR crosstalk might intend to serve as a protective mechanism against bacterial infection for the host, enhanced production of TNF-α, IL-1β and IL-6 could upregulate inflammation and cause a detrimental effect in inflammatory diseases (Zhang et al., 2007). C5aR and C3aR can be potential therapeutical targets since reports showed that blockade of C5a receptor (C5aR) and C3a receptor (C3aR) signaling decrease inflammation and ameliorate the periodontitis (Abe et al., 2012; Hajishengallis et al., 2011; Ricklin et al., 2010).

7. Capsules of *Porphyromonas gingivalis*

The bacterial capsule is an outer envelope structure that lies outside bacterial cell. Capsule usually consists of polysaccharides and contains water to keep the bacterial cells from desiccated. It is important for the bacteria to survive in less-than-optimal environment. Capsule also acts as a protective mechanism in that encapsulated bacteria are more resistant to phagocytosis and intracellular killing. There are many bacterial strains that possess capsules, which protect the bacteria from host immune clearance. One well-known example in textbooks is *Streptococcus pneumonia*, whose capsule was verified as important virulence factors in early 20th century, which showed that the encapsulated pneumococci were virulent while the non-encapsulated variants were avirulent (Griffith, 1928).

The capsule of *P. gingivalis* is also known as K-antigen (Bostanci & Belibasakis, 2012; Holt, Kesavalu, Walker, & Genco, 1999; Schifferle, Reddy, Zambon, Genco, & Levine, 1989). The gene loci of *P. gingivalis* capsule have been identified and characterized (Aduse-Opoku et al., 2006). The chemical composition of *P. gingivalis* capsules varies, by which different *P. gingivalis* strains can be categorized into different K-serotypes, while some *P. gingivalis* strains are not encapsulated at all (Laine, Appelmelk, & van Winkelhoff, 1997). For example, *P. gingivalis* strains ATCC 33277 and W83 have been extensively used in our and others' previous work (Chen et al., 2004; Hajishengallis, 2009; Hajishengallis et al., 2011; Hajishengallis et al., 2008; Liang et al., 2011; Nishikawa & Duncan, 2010; Min Wang et al., 2007; Pierce et al., 2009; Popadiak et al., 2007; Scheres & Crielaard, 2013; Wang et al., 2010). Although they both show strong capability to regulate host immune responses and induce periodontitis, ATCC 33277 is not encapsulated and W83 possesses K1 serotype polysaccharide capsules (Brunner et al., 2010; Sheets, Potempa, Travis, Casiano, & Fletcher, 2005; Sheets, Potempa, Travis, Fletcher, & Casiano, 2006). Similar to its functions in other bacteria, *P. gingivalis* capsule protects bacteria from the phagocytosis by host leukocytes, while non-encapsulated *P. gingivalis* strains are less resistant to phagocytosis and killing by macrophages and dentritic cells (Laine et al., 1997; Sundqvist, Figdor, Hanstrom, Sorlin, & Sandstrom, 1991). Encapsulated *P. gingivalis* strain showed much higher survival rate than the non-encapsulated strain (Singh et al., 2011). Report showed that different serotype of capsules exhibited differential adhesion capabilities. Overall, the adhesion capacity of encapsulated *P. gingivalis* to gingival epithelial cells was lower than the non-encapsulated ones (Dierickx et al., 2003). The roles of capsules in the decreased adhesion ability might contribute to its protective mechanism from phagocytosis by host cells. Capsules of *P. gingivalis* can also bind to other periodotopathogens to facilitate the co-aggregation of the bacteria and biofilm formation. The binding of *P. gingivalis* to other bacteria is K serotype specific. For example, K5 and K6, but not K1 capsules, can bind to *F. nucleatum* PK 1594 (Rosen & Sela, 2006).

Capsule was shown to induce host immune responses. Different *P. gingivalis* capsular serotypes induced differential cytokine expression in dendritic cells (Vernal et al., 2009) and chemokine production in macrophages (d'Empaire, Baer, & Gibson, 2006). In both cases, W83 strain of K1 serotype was most potent in stimulating chemokine and cytokine production (d'Empaire et al., 2006; Vernal et al., 2009). When the role of capsule in a specific strain was tested, a wild-type encapsulated strain W83 exhibited weaker ability to induce cytokine production in human gingival fibroblasts than its noncapsulated mutant (Brunner et al., 2010). It is possible that the *P. gingivalis* capsule is a weaker immunity inducer than other virulence factors. *P. gingivalis* capsule antagonizes the strong stimuli and promotes bacterial evasion of host immune responses, therefore promotes bacterial survival in hosts.

Another report demonstrated that encapsulated strain W50 was more resistant to bactericidal effect of antimicrobial peptide, such as defensin HBD3, than its nonencapsulated mutant (Igboin, Tordoff, Moeschberger, Griffen, & Leys, 2011).

In a mouse abscess model, the encapsulated *P. gingivalis* exhibited increased virulence than nonencapsulated strains in forming abscesses. The encapsulated strains caused invasive and spreading abscesses, while nonencapsulated strains caused non-invasive and localized abscesses (Katz, Ward, & Michalek, 1996; Laine & van Winkelhoff, 1998; Singh et al., 2011). Furthermore, when animals were subcutaneously injected with non-encapsulated *P. gingivalis* strain 33277 or strain A7A1-28 of K3 serotype, *P. gingivalis* 33277-challenged animals produced lower levels of serum immunoglobulin G (IgG) and salivary IgA than A7A1-28-challenged ones, implying that capsules may help to induce adaptive immunity and antibody production (Katz et al., 1996). In summary, various reports proved that capsule affected the virulence of *P. gingivalis*. These results showed that the observed differential virulence among the different *P. gingivalis* strains could be at least partially attributed to the inequality in capsular structure and function, while the detailed mechanism of capsule's involvement in the responses needs further investigation and clarification.

8. T cell immunity induced by *Porphyromonas gingivalis* and its virulence factors

Bacteria and their virulence factors play important roles in the pathogen interactions and inflammation. Dysregulated host factors like innate immune cells, cytokines, and other soluble factors, which are affected by the virulence factors, caused periodontitis progression. T cells are involved in various inflammatory diseases and are critical in host immune responses. Almost a half-century ago, Ivanyi and Lehner first reported that oral bacteria stimulated lymphocyte activation in mild periodontal disease patients, while this activation was suppressed in severe periodontitis patients (Ivanyi & Lehner, 1970). Later reports showed that T cells from periodontitis patients exhibited a reduced ability to proliferate, suggesting an impaired T cell response during periodontitis (Cole, Seymour, & Powell, 1987). The suppression of T cell activation can be induced by various periodontal pathogens (Shenker & Datar, 1995; Stashenko, Resmini, Haffajee, & Socransky, 1983). Gingipain Rpgs, which down-regulate IL-2 production and attenuate T-cell proliferation, might be used by *P. gingivalis* to evade host adaptive immune system (Khalaf & Bengtsson, 2012). Gingipains can also degrade T cell receptors, such as CD4 and CD8, and down-regulate T cell activation (Hajishengallis et al., 2013; Kitamura, Matono, Aida, Hirofuji, & Maeda, 2002). Interestingly, when tested for mixed lymphocyte reactions (MLRs), not all the periodontitis patients showed impaired T cell reactions. For the patients who showed lower MLR responses, the T cells were comprised of lower percentage of naïve T cells. After periodontal therapy in these patients, the MLR responses and naïve T cells went back to normality (Kimura, Fujimoto, & Okada, 1991). While these findings pointed to a protective function of T cells, recent research demonstrated that the roles of T cell immunity in periodontitis are complicated, partially due to the existence of divergent T cell subsets.

Baker and colleagues found that *P. gingivalis* infection induced bone loss at a lower level in the severe combined immunodeficient (SCID) mice, which lack B and T lymphocytes, than immunocompetent mice (Baker, Evans, & Roopenian, 1994). Later, Baker et al. further demonstrated that MHC-II deficient mice, which cannot induce $CD4^+$ T cell development and responses, were resistant to alveolar bone loss in a *P. gingivalis*-induced periodontitis model. On the other hand, MHC-I deficient mice that lack $CD8^+$ T cells showed no change in bone loss (Baker et al., 1999). In another report, adoptive transfer of human $CD4^+$ T cells, but not $CD8^+$ T cells, were essential for *A. actinomycetemcomitans* infection-induced alveolar bone destruction in NOD/SCID mice (Teng et al., 2000). After the initial infection of periodontal pathogens, the bacterial antigens will be processed and presented to $CD4^+$ helper T cells by professional antigen-presenting cells, such as dendritic cells, macrophages, and B cells. These findings imply that $CD4^+$ T cells are the major T cell contributors to bone destruction in periodontitis. Therefore, our following discussion has been focused on the $CD4^+$ T-helper cells (Th) and the effects of *P. gingivalis* and its virulence factors on Th cell responses. $CD4^+$ T cells are divided into different subsets Th1, Th2, Th17, Treg, and some other subsets, based on their distinct cytokines and specific transcription factors (Ahlers & Belyakov, 2010). Differentiation and activation of these $CD4^+$ T cell subsets from naïve cells are affected by T cell receptor activation, co-stimulatory factors, and cytokine milieu.

8.1 Th1/Th2 cells

From its establishment, Th1/Th2 paradigm classified the differentiated $CD4^+$ T cells into Th1 and Th2 cells and has prevailed for around two decades (Mosmann, Cherwinski, Bond, Giedlin, & Coffman, 1986). Th1 cells express transcription factors T-bet and STAT-4. They preferably produced cytokines IL-12 and IFN-γ and their differentiation is driven by IL-12. On the other hand, Th2 cells express GATA3 and STAT6, and produce cytokines IL-4, IL-5, IL-6, IL-10, and IL-13 (Jankovic & Feng, 2015). Th1 and Th2 cells respond to different stimulations and are associated with different outcomes in infectious and inflammatory diseases. Th1/Th2 paradigm has been investigated to identify the roles of both cell subsets in periodontitis. By testing the T cell subset cytokine profiles, some studies have shown that the Th1 cells are associated with stable lesions while Th2 cells with progressive lesions (Aoyagi, Sugawara-Aoyagi, Yamazaki, & Hara, 1995; Tokoro, Matsuki, Yamamoto, Suzuki, & Hara, 1997; Wassenaar, Reinhardus, Thepen, Abraham-Inpijn, & Kievits, 1995), suggesting that Th2 cells are associated with more severe disease. Meanwhile, other studies demonstrated that Th1 cells were predominant in the gingival tissues from periodontitis patients and were associated with inflammation and periodontitis severity (Takeichi et al., 2000; Ukai, Mori, Onoyama, & Hara, 2001). Interestingly, there are also papers showing that Th1 and Th2 responses were up-regulated to a similarly higher level during periodontitis (Berglundh, Liljenberg, & Lindhe, 2002).

Th cell differentiation during periodontitis can be at least partially attributed to periodontal pathogens and their virulence factors, with controversy on which specific Th subsets are preferably promoted. Tests on *P. gingivalis*-responsive T-cell lines from periodontitis and healthy or gingivitis subjects implied that *P. gingivalis* facilitated Th1 differentiation (Gemmell, Grieco, Cullinan, Westerman, & Seymour, 1999). *In vitro* test suggested that fimbriae might induce the IFN-γ production by dendritic cells and provide Th1-biased environment (Ravi Jotwani & Cutler, 2004). On the other hand, there are reports showing that *P. gingivalis* and its LPS induce Th2 responses (Aoyagi et al., 1995; Jotwani, Pulendran, Agrawal, & Cutler, 2003). *P. gingivalis* FimA acts through TLR2 and produce IL-10, which suppresses Th1 cell differentiation and function (Gaddis, Maynard, Weaver, Michalek, & Katz, 2013). The capability of *P. gingivalis* to induce Th cell was explored to develop potential vaccine.

In an animal model for periodontitis, IFN-γ- and IL-6-deficient mice were resistant to *P. gingivalis*-induced alveolar bone loss, implying that Th1 and Th2 cells play destructive effects during periodontitis (Baker et al., 1999). Later reports showed that transfer of Th1 or Th1 cytokine IFN-γ caused higher alveolar bone loss in the mice injected with *A. actinomycetemcomitans* or its virulence factors, which indicated the detrimental effects of Th1 cells in periodontitis (Kawai et al., 2000). Interestingly, deficiency in Th1/Th2 cytokines, such as IFN-γ, IL-12p40, IL-4, or IL-10, can lead to more severe alveolar bone loss in aged mice (Alayan, Ivanovski, & Farah, 2007). Since IFN-γ and IL-12 are important in bacterial clearance, and IL-4 and IL-10 are famous for their anti-inflammatory functions, the seemingly contradictory reports might be explained by that the balance between pathogens and different host immune components is required to maintain the periodontal health.

8.2 Th17 cells

Pathogenesis of periodontitis had not been fully explained through the Th1/Th2 paradigm. The discovery of more Th subsets in recent years would help us to further our understanding of the pathogenesis of periodontitis. Among these new subsets are Th17 cells, which express transcription factors RORγt and produce cytokines IL-17 and IL-21. Th17 development and maintenance require cytokines TGF-β, IL-6, IL-21, and IL-23. Th17 is important in protecting against extracellular pathogens and fungi as well as promoting inflammation (O'Connor, Zenewicz, & Flavell, 2010). Recent researches link Th17 to various autoimmune and inflammatory diseases (Singh et al., 2014; Stadhouders, Lubberts, & Hendriks, 2018; van Bruggen & Ouyang, 2014). Th17 specific cytokine IL-17 and other related cytokines have been found in the gingival tissues or gingival crevicular fluid from periodontitis patients (Cardoso et al., 2009; Dutzan et al., 2012; Honda et al., 2008; Lester, Bain, Johnson, & Serio, 2007; Moutsopoulos et al., 2012; Ohyama et al., 2009; Vernal et al., 2005). Increased infiltration of Th17 cells in periodontal lesions further proved the association of Th17 with periodontitis (Adibrad et al., 2012; Okui, Aoki, Ito, Honda, & Yamazaki, 2012).

It has been shown that *P. gingivalis* regulate Th17 development and function (Moutsopoulos et al., 2012; Oda, Yoshie, & Yamazaki, 2003). *P. gingivalis*, especially the encapsulated strain W83, induced Th17 in periodontal lesions (Moutsopoulos et al., 2012). Capsules of different serotypes seem to possess differential effects on Th subset polarization. After priming DCs with the *P. gingivalis* strains of different K-serotypes, strains W83 of serotype K1 and HG184 of K2 induced a Th1/Th17 pattern of immune response, while K3, K4 and K5 strains promoted a Th2 response (Vernal, Diaz-Guerra, Silva, Sanz, & Garcia-Sanz, 2014). Moreover, *P. gingivalis* LPS can promote an environment for Th17 development by inducing the production of IL-17 and IL-23, a cytokine critical for Th17 expansion and maintenance, in human periodontal ligament cells (Park et al., 2012). Besides producing pro-inflammatory cytokine IL-17, Th17 lymphocytes also promote bone loss by inducing the expression of RANKL, which is required for osteoclastogenesis (Sato et al., 2006; Stadhouders et al., 2018).

The role of IL-17 in periodontitis is not determined yet and under intense investigation. There are reports showing that IL-17 receptor signaling is protective in periodontitis (Yu, Ruddy, Conti, Boonanantanasarn, & Gaffen, 2008; Yu et al., 2007), and the ones showing that it is detrimental in the disease as well (Eskan et al., 2012). The reason of the inconsistency might be that different animal models were used.

8.3 Treg cells

Regulatory T cell (Treg) represents a special suppressive lymphocyte subset that down-regulates the activation, proliferation, and effector functions of a wide range of immune cells. Tregs specifically express transcription factor Foxp3 and secrete cytokines TGF-β and IL-10. Treg cells also express some cell surface molecules at higher level than naïve helper T cells, such as CD25, CTLA-4, GITR, and etc. Treg cells are critical in the maintenance of host immune homeostasis by controlling the direction and intensity of both adaptive and innate immunity and modulating various host immune responses, including inflammation (Chaudhry & Rudensky, 2013; Shevach, 2018).

Numerous publications on human patients demonstrated that Treg cells were enriched in periodontitis lesions (Cristina Ribeiro Cardoso et al., 2008; Dutzan, Gamonal, Silva, Sanz, & Vernal, 2009; Nakajima et al., 2005). Interestingly, there was also a finding showing reduced $Foxp3^{+}CD25^{+}$

cell number in periodontitis lesions (Ernst et al., 2007). It is still unclear what caused the inconsistency. In a series of reports with animal models for periodontitis, the association of Tregs with periodontitis and the role of Tregs in the disease have been further illustrated. Inhibition of Treg functions with blocking antibody anti-GITR treatment enhanced *A. actinomycetemcomitans*-induced inflammation and bone loss (Garlet et al., 2010), while recruitment of Foxp3^{+} Tregs into gingival tissue inhibited pro-inflammatory cytokine expression and *A. actinomycetemcomitans*-induced alveolar bone resorption in mice (Glowacki et al., 2013). Because all-trans retinoic acid (atRA) promotes the differentiation of Treg cells and sustains Treg stability (Liu, Wang, Ma, & Guo Zheng, 2015), atRA was orally administrated to induce an increase of Treg cells and decrease in Th17 cells in cervical lymph nodes, which were accompanied by the inhibition of *P. gingivalis*-induced bone loss (Wang, Wang, Jin, Gao, & Lin, 2014).

The roles of Treg-associated molecules have also been tested for their roles in periodontitis. TGF-β and IL-10 are famous anti-inflammatory cytokines and major Treg cytokines. TGF-β and IL-10 were expressed higher in periodontitis tissues (Cristina Ribeiro Cardoso et al., 2008; Nakajima et al., 2005; Yamazaki et al., 1997), while their expression levels were negatively correlated to the severity of the disease (Dutzan et al., 2009, 2012). *P. gingivalis* infection caused more severe alveolar bone loss in IL-10-deficient mice, (Sasaki et al., 2004), and the IL-10 administration suppressed alveolar bone loss, indicating the protective role of IL-10 (Zhang et al., 2014; Zhang & Teng, 2006). These findings all support the idea that Tregs play a key role in alleviating periodontitis, while the impairment of Treg function may cause exacerbated tissue damage. Indeed, some reported that during periodontitis, Foxp3^{+}CD4^{+} T cells harvested from periodontitis lesions lost their suppressive function (Okui et al., 2008). Studies have shown that Tregs can convert into IL-17-producing cells in a pro-inflammatory environment (Ayyoub et al., 2009; Koenen et al., 2008; Leveque et al., 2009). Tregs that express RORγt and IL-17 were recently found and believed to play a role in the phenotypic plasticity in T cell lineages, transiting between inflammatory and regulatory function responding to outside stimulation (Nyirenda et al., 2011; Valmori, Raffin, Raimbaud, & Ayyoub, 2010). On the other hand, there are other reports showed that IL-17^{+}Foxp3^{+} or RORγt^{+}Foxp3^{+} cells are an independent stable Treg lineage, and are potent in regulating Th17-related diseases (Kluger et al., 2016; Yang et al., 2016). IL-17^{+}Foxp3^{+} cells have been detected and seemed to be involved in periodontitis, whose roles are not fully elucidated either

(Okui et al., 2012). Since these cells possess both pro-inflammatory and suppressive properties, their roles, protective or destructive, in inflammatory diseases are still unclear (Blatner et al., 2012; Kryczek et al., 2011; Tartar et al., 2010; Voo et al., 2009).

Later *in vitro* and *in vivo* research confirmed the ability of *P. gingivalis* to induce Treg development (Kim et al., 2010; Kobayashi et al., 2011). Interestingly, the effect of *P. gingivalis* on Treg development varies under different conditions. For example, we found that *P. gingivalis* induced fewer Treg cells in pregnant mice than non-pregnant mice. Similarly, infection with *P. gingivalis*, especially the ones with type II FimA, reduced Tregs in atherosclerotic patients, implying that type II FimA might be associated with Treg dysregulation (Yang et al., 2014).

9. Discussion

In this chapter, we have introduced periodontal pathogens and their virulence factors. We focused on *P. gingivalis* because it has been widely investigated and regarded as the keystone pathogen for periodontitis. Despite of its role as a "keystone" pathogen in periodontitis, *P. gingivalis* is not a lone criminal in the progression of periodontitis. Other pathogens and their virulence factors also play critical roles in the pathogenesis of periodontitis, which have been reviewed in other articles (de Andrade, Almeida-da-Silva, & Coutinho-Silva, 2019). *P. gingivalis* uses many methods to cooperate with other pathogens and evade immune clearance in periodontitis. It develops strategies, including their virulence factors, to overcome and exploit host immune pathways to facilitate its own and other pathogens' colonization and survival. The virulence factors are required for the interactions between *P. gingivalis* and other bacteria. Fimbriae and capsules are able to adhere to the molecules on other bacteria and host tissues and cells to promote the formation of biofilm. The virulence factors can also act on the host cells to inhibit bacterial clearance, promote bacterial invasion, and cause inflammation and tissue damage (Fig. 1). They are critical in the manipulation and exploitation of host immune responses by *P. gingivalis* to lead to dysbiosis in oral cavity and periodontitis progression. T cells are major components of host immunity. Previous studies confirmed the importance of Th cells in periodontitis pathogenesis. *P. gingivalis* and its virulence factors induce biased Th differentiation and function, while the roles of the T cells in the periodontitis pathogenesis remains controversial.

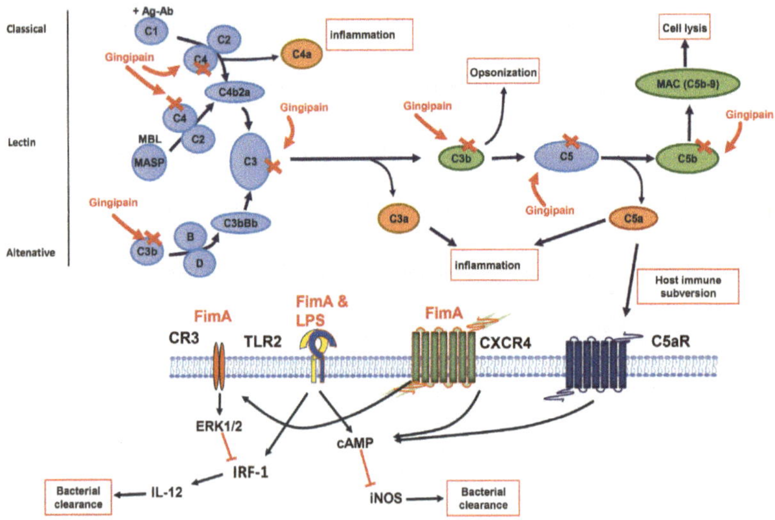

Fig. 1 *Porphyromonas gingivalis* virulence factors subvert immune responses to evade bacterial clearance and promote inflammation. *P. gingivalis* gingipains degrade C3, C4, and C5 to block complement cascade. Besides, anaphylatoxins C3a, C4a, and C5a are generated to promote inflammation. Degradation of C3b prevents opsonization and phagocytosis, and C5b cleavage prevents the formation of membrane attack complex (MAC). Furthermore, C5a crosstalks with TLR2/1 activation and enhances the cAMP responses, which inhibits iNOS production and bacterial killing. Fimbriae promote the crosstalk between CXCR4 and TLR and further enhance cAMP responses. Fimbriae also induce CXCR4-dependent activation of CR3. The interaction between CR3 and *P. gingivalis* requires fimbriae and induces ERK1/2 signaling to down-regulate IL-12 production, which results in impaired bacterial clearance.

Great progress has been achieved to understand the complex inter-microbial interactions and host-pathogen interactions during the disease. Previous studies demonstrate that oral bacteria and the host immune system need to maintain a delicate balance. Disruption of the balance would cause dysbiosis in oral cavity, unwanted tissue damage, and diseases. Further exploration on the immune responses induced by the periodontal pathogens and their virulence factors will help us to better understand the pathogenesis of periodontitis and pave the road for its future therapy.

Acknowledgments

S.L. is supported by grants KSEF-3283-RDE-018, KSEF-3835-RDE-020, and NIH DE025388. H.W. is supported by DE026727 from National Institute of Dental and Craniofacial Research, NIH, USA.
The authors report no conflicts of interest related to this work.

References

Aas, J. A., Paster, B. J., Stokes, L. N., Olsen, I., & Dewhirst, F. E. (2005). Defining the normal bacterial flora of the oral cavity. *Journal of Clinical Microbiology, 43*(11), 5721–5732. http://dx.doi.org/10.1128/JCM.43.11.5721-5732.2005.

Abe, N., Baba, A., Takii, R., Nakayama, K., Kamaguchi, A., Shibata, Y., et al. (2004). Roles of Arg- and Lys-gingipains in coaggregation of *Porphyromonas gingivalis*: Identification of its responsible molecules in translation products of rgpA, kgp, and hagA genes. *Biological Chemistry, 385*(11), 1041–1047. http://dx.doi.org/10.1515/BC.2004.135.

Abe, T., Hosur, K. B., Hajishengallis, E., Reis, E. S., Ricklin, D., Lambris, J. D., et al. (2012). Local complement-targeted intervention in periodontitis: Proof-of-concept using a C5a receptor (CD88) antagonist. *The Journal of Immunology, 189*(11), 5442–5448. http://dx.doi.org/10.4049/jimmunol.1202339.

Abusleme, L., Dupuy, A. K., Dutzan, N., Silva, N., Burleson, J. A., Strausbaugh, L. D., et al. (2013). The subgingival microbiome in health and periodontitis and its relationship with community biomass and inflammation. *The ISME Journal,* 7(5), 1016–1025. http://dx.doi.org/10.1038/ismej.2012.174.

Adibrad, M., Deyhimi, P., Ganjalikhani Hakemi, M., Behfarnia, P., Shahabuei, M., & Rafiee, L. (2012). Signs of the presence of Th17 cells in chronic periodontal disease. *Journal of Periodontal Research, 47*(4), 525–531. http://dx.doi.org/10.1111/j.1600-0765.2011.01464.x.

Aduse-Opoku, J., Slaney, J. M., Hashim, A., Gallagher, A., Gallagher, R. P., Rangarajan, M., et al. (2006). Identification and characterization of the capsular polysaccharide (K-antigen) locus of *Porphyromonas gingivalis*. *Infection and Immunity, 74*(1), 449–460. http://dx.doi.org/10.1128/IAI.74.1.449-460.2006.

Ahlers, J. D., & Belyakov, I. M. (2010). Molecular pathways regulating CD4($^{+}$) T cell differentiation, anergy and memory with implications for vaccines. *Trends in Molecular Medicine, 16*(10), 478–491. http://dx.doi.org/10.1016/j.molmed.2010.07.007.

Alayan, J., Ivanovski, S., & Farah, C. S. (2007). Alveolar bone loss in T helper 1/T helper 2 cytokine-deficient mice. *Journal of Periodontal Research, 42*(2), 97–103. http://dx.doi.org/10.1111/j.1600-0765.2006.00920.x.

Ally, N., Whisstock, J. C., Sieprawska-Lupa, M., Potempa, J., Le Bonniec, B. F., Travis, J., et al. (2003). Characterization of the specificity of arginine-specific gingipains from *Porphyromonas gingivalis* reveals active site differences between different forms of the enzymes. *Biochemistry, 42*(40), 11693–11700. http://dx.doi.org/10.1021/bi0349726.

Amano, A. (2003). Molecular interaction of *Porphyromonas gingivalis* with host cells: Implication for the microbial pathogenesis of periodontal disease. *Journal of Periodontology, 74*(1), 90–96.

Amano, A. (2010). Host-parasite interactions in periodontitis: Microbial pathogenicity and innate immunity. *Periodontology 2000, 54*(1), 9–14. http://dx.doi.org/10.1111/j.1600-0757.2010.00376.x.

de Andrade, K. Q., Almeida-da-Silva, C. L. C., & Coutinho-Silva, R. (2019). Immunological pathways triggered by *Porphyromonas gingivalis* and *Fusobacterium nucleatum*: Therapeutic possibilities? *Mediators of Inflammation*. http://dx.doi.org/10.1155/2019/7241312, 7241312.

Andrian, E., Grenier, D., & Rouabhia, M. (2004). In vitro models of tissue penetration and destruction by *Porphyromonas gingivalis*. *Infection and Immunity, 72*(8), 4689–4698. http://dx.doi.org/10.1128/IAI.72.8.4689-4698.2004.

Aoyagi, T., Sugawara-Aoyagi, M., Yamazaki, K., & Hara, K. (1995). Interleukin 4 (IL-4) and IL-6-producing memory T-cells in peripheral blood and gingival tissue in periodontitis patients with high serum antibody titers to *Porphyromonas gingivalis*. *Oral Microbiology and Immunology, 10*(5), 304–310.

Aruni, A. W., Dou, Y., Mishra, A., & Fletcher, H. M. (2015). The biofilm community-rebels with a cause. *Current Oral Health Reports, 2*(1), 48–56. http://dx.doi.org/10.1007/s40496-014-0044-5.

Ayyoub, M., Deknuydt, F., Raimbaud, I., Dousset, C., Leveque, L., Bioley, G., et al. (2009). Human memory FOXP3^{+} Tregs secrete IL-17 ex vivo and constitutively express the T(H)17 lineage-specific transcription factor RORgammat. *Proceedings of the National Academy of Sciences of the USA, 106*(21), 8635–8640. http://dx.doi.org/10.1073/pnas.0900621106.

Bainbridge, B. W., & Darveau, R. P. (2001). *Porphyromonas gingivalis* lipopolysaccharide: An unusual pattern recognition receptor ligand for the innate host defense system. *Acta Odontologica Scandinavica, 59*(3), 131–138.

Baker, P. J., Dixon, M., Evans, R. T., Dufour, L., Johnson, E., & Roopenian, D. C. (1999). CD4^{+} T cells and the proinflammatory cytokines gamma interferon and interleukin-6 contribute to alveolar bone loss in mice. *Infection and Immunity, 67*(6), 2804–2809.

Baker, P. J., Evans, R. T., & Roopenian, D. C. (1994). Oral infection with *Porphyromonas gingivalis* and induced alveolar bone loss in immunocompetent and severe combined immunodeficient mice. *Archives of Oral Biology, 39*(12), 1035–1040.

Bao, K., Belibasakis, G. N., Thurnheer, T., Aduse-Opoku, J., Curtis, M. A., & Bostanci, N. (2014). Role of *Porphyromonas gingivalis* gingipains in multi-species biofilm formation. *BMC Microbiology, 14*, 258. http://dx.doi.org/10.1186/s12866-014-0258-7.

Berglundh, T., Liljenberg, B., & Lindhe, J. (2002). Some cytokine profiles of T-helper cells in lesions of advanced periodontitis. *Journal of Clinical Periodontology, 29*(8), 705–709. http://dx.doi.org/10.1034/j.1600-051x.2002.290807.x.

Bergstrom, J. (2006). Periodontitis and smoking: An evidence-based appraisal. *Journal of Evidence-Based Dental Practice, 6*(1), 33–41. http://dx.doi.org/10.1016/j.jebdp.2005.12.018.

Blatner, N. R., Mulcahy, M. F., Dennis, K. L., Scholtens, D., Bentrem, D. J., Phillips, J. D., et al. (2012). Expression of RORgammat marks a pathogenic regulatory T cell subset in human colon cancer. *Science Translational Medicine, 4*(164), 164ra159. http://dx.doi.org/10.1126/scitranslmed.3004566.

Bostanci, N., & Belibasakis, G. N. (2012). *Porphyromonas gingivalis*: An invasive and evasive opportunistic oral pathogen. *FEMS Microbiology Letters, 333*(1), 1–9. http://dx.doi.org/10.1111/j.1574-6968.2012.02579.x.

Bozkurt, S. B., Hakki, S. S., Hakki, E. E., Durak, Y., & Kantarci, A. (2017). *Porphyromonas gingivalis* lipopolysaccharide induces a pro-inflammatory human gingival fibroblast phenotype. *Inflammation, 40*(1), 144–153. http://dx.doi.org/10.1007/s10753-016-0463-7.

Brown, L. J., Johns, B. A., & Wall, T. P. (2002). The economics of periodontal diseases. *Periodontology 2000, 29*, 223–234.

van Bruggen, N., & Ouyang, W. (2014). Th17 cells at the crossroads of autoimmunity, inflammation, and atherosclerosis. *Immunity, 40*(1), 10–12. http://dx.doi.org/10.1016/j.immuni.2013.12.006.

Brunner, J., Scheres, N., El Idrissi, N. B., Deng, D. M., Laine, M. L., van Winkelhoff, A. J., et al. (2010). The capsule of *Porphyromonas gingivalis* reduces the immune response of human gingival fibroblasts. *BMC Microbiology, 10*, 5. http://dx.doi.org/10.1186/1471-2180-10-5.

Cardoso, C. R., Garlet, G. P., Crippa, G. E., Rosa, A. L., Junior, W. M., Rossi, M. A., et al. (2009). Evidence of the presence of T helper type 17 cells in chronic lesions of human periodontal disease. *Oral Microbiology and Immunology, 24*(1), 1–6. http://dx.doi.org/10.1111/j.1399-302X.2008.00463.x.

Cardoso, C. R., Garlet, G. P., Moreira, A. P., Junior, W. M., Rossi, M. A., & Silva, J. S. (2008). Characterization of CD4^{+}CD25^{+} natural regulatory T cells in the inflammatory infiltrate of human chronic periodontitis. *Journal of Leukocyte Biology, 84*(1), 311–318. http://dx.doi.org/10.1189/jlb.0108014.

Carlisle, M. D., Srikantha, R. N., & Brogden, K. A. (2009). Degradation of human alpha- and beta-defensins by culture supernatants of *Porphyromonas gingivalis* strain 381. *Journal of Innate Immunity, 1*(2), 118–122. http://dx.doi.org/10.1159/000181015.

Chaudhry, A., & Rudensky, A. Y. (2013). Control of inflammation by integration of environmental cues by regulatory T cells. *Journal of Clinical Investigation, 123*(3), 939–944. http://dx.doi.org/10.1172/JCI57175.

Chen, T., Hosogi, Y., Nishikawa, K., Abbey, K., Fleischmann, R. D., Walling, J., et al. (2004). Comparative whole-genome analysis of virulent and avirulent strains of *Porphyromonas gingivalis*. *Journal of Bacteriology, 186*(16), 5473–5479.

Coats, S. R., Jones, J. W., Do, C. T., Braham, P. H., Bainbridge, B. W., To, T. T., et al. (2009). Human toll-like receptor 4 responses to *P. gingivalis* are regulated by lipid A 1- and 4'-phosphatase activities. *Cellular Microbiology, 11*(11), 1587–1599. http://dx.doi.org/10.1111/j.1462-5822.2009.01349.x.

Cole, K. L., Seymour, G. J., & Powell, R. N. (1987). Phenotypic and functional analysis of T cells extracted from chronically inflamed human periodontal tissues. *Journal of Periodontology, 58*(8), 569–573. http://dx.doi.org/10.1902/jop.1987.58.8.569.

Curtis, M. A., Aduse-Opoku, J., & Rangarajan, M. (2001). Cysteine proteases of *Porphyromonas gingivalis*. *Critical Reviews in Oral Biology & Medicine, 12*(3), 192–216.

Curtis, M. A., Percival, R. S., Devine, D., Darveau, R. P., Coats, S. R., Rangarajan, M., et al. (2011). Temperature-dependent modulation of *Porphyromonas gingivalis* lipid A structure and interaction with the innate host defenses. *Infection and Immunity, 79*(3), 1187–1193. http://dx.doi.org/10.1128/IAI.00900-10.

d'Empaire, G., Baer, M. T., & Gibson, F. C., 3rd (2006). The K1 serotype capsular polysaccharide of *Porphyromonas gingivalis* elicits chemokine production from murine macrophages that facilitates cell migration. *Infection and Immunity, 74*(11), 6236–6243. http://dx.doi.org/10.1128/IAI.00519-06.

Darveau, R. P., Arbabi, S., Garcia, I., Bainbridge, B., & Maier, R. V. (2002). *Porphyromonas gingivalis* lipopolysaccharide is both agonist and antagonist for p38 mitogen-activated protein kinase activation. *Infection and Immunity, 70*, 1867–1873.

Darveau, R. P., Pham, T. T., Lemley, K., Reife, R. A., Bainbridge, B. W., Coats, S. R., et al. (2004). *Porphyromonas gingivalis* lipopolysaccharide contains multiple lipid A species that functionally interact with both toll-like receptors 2 and 4. *Infection and Immunity, 72*(9), 5041–5051.

Dewhirst, F. E., Chen, T., Izard, J., Paster, B. J., Tanner, A. C., Yu, W. H., et al. (2010). The human oral microbiome. *Journal of Bacteriology, 192*(19), 5002–5017. http://dx.doi.org/10.1128/JB.00542-10.

Dickinson, D. P., Kubiniec, M. A., Yoshimura, F., & Genco, R. J. (1988). Molecular cloning and sequencing of the gene encoding the fimbrial subunit protein of *Bacteroides gingivalis*. *Journal of Bacteriology, 170*(4), 1658–1665.

de Diego, I., Veillard, F., Sztukowska, M. N., Guevara, T., Potempa, B., Pomowski, A., et al. (2014). Structure and mechanism of cysteine peptidase gingipain K (Kgp), a major virulence factor of *Porphyromonas gingivalis* in periodontitis. *Journal of Biological Chemistry, 289*(46), 32291–32302. http://dx.doi.org/10.1074/jbc.M114.602052.

Dierickx, K., Pauwels, M., Laine, M. L., Van Eldere, J., Cassiman, J. J., van Winkelhoff, A. J., et al. (2003). Adhesion of *Porphyromonas gingivalis* serotypes to pocket epithelium. *Journal of Periodontology, 74*(6), 844–848. http://dx.doi.org/10.1902/jop.2003.74.6.844.

Ding, P. H., Wang, C. Y., Darveau, R. P., & Jin, L. (2013). *Porphyromonas gingivalis* LPS stimulates the expression of LPS-binding protein in human oral keratinocytes in vitro. *Innate Immunity, 19*(1), 66–75. http://dx.doi.org/10.1177/1753425912450348.

Dixon, D. R., & Darveau, R. P. (2005). Lipopolysaccharide heterogeneity: Innate host responses to bacterial modification of lipid a structure. *Journal of Dental Research, 84*(7), 584–595.

Dommisch, H., & Jepsen, S. (2015). Diverse functions of defensins and other antimicrobial peptides in periodontal tissues. *Periodontology 2000, 69*(1), 96–110. http://dx.doi.org/10.1111/prd.12093.

Dutzan, N., Gamonal, J., Silva, A., Sanz, M., & Vernal, R. (2009). Over-expression of forkhead box P3 and its association with receptor activator of nuclear factor-kappa B ligand, interleukin (IL) -17, IL-10 and transforming growth factor-beta during the progression of chronic periodontitis. *Journal of Clinical Periodontology, 36*(5), 396–403.

Dutzan, N., Vernal, R., Vaque, J. P., Garcia-Sesnich, J., Hernandez, M., Abusleme, L., et al. (2012). Interleukin-21 expression and its association with proinflammatory cytokines in untreated chronic periodontitis patients. *Journal of Periodontology, 83*(7), 948–954. http://dx.doi.org/10.1902/jop.2011.110482.

Eke, P. I., Dye, B. A., Wei, L., Slade, G. D., Thornton-Evans, G. O., Borgnakke, W. S., et al. (2015). Update on prevalence of periodontitis in adults in the United States: NHANES 2009 to 2012. *Journal of Periodontology, 86*(5), 611–622. http://dx.doi.org/10.1902/jop.2015.140520.

Eke, P. I., Dye, B. A., Wei, L., Thornton-Evans, G. O., & Genco, R. J. (2012). Prevalence of periodontitis in adults in the United States: 2009 and 2010. *Journal of Dental Research, 91*(10), 914–920. http://dx.doi.org/10.1177/0022034512457373.

El-Awady, A. R., Miles, B., Scisci, E., Kurago, Z. B., Palani, C. D., Arce, R. M., et al. (2015). *Porphyromonas gingivalis* evasion of autophagy and intracellular killing by human myeloid dendritic cells involves DC-SIGN-TLR2 crosstalk. *PLoS Pathogens, 10*(2), e1004647. http://dx.doi.org/10.1371/journal.ppat.1004647.

Enersen, M., Nakano, K., & Amano, A. (2013). *Porphyromonas gingivalis* fimbriae. *Journal of Oral Microbiology, 5*. http://dx.doi.org/10.3402/jom.v5i0.20265.

Enersen, M., Olsen, I., Kvalheim, O., & Caugant, D. A. (2008). fimA genotypes and multilocus sequence types of *Porphyromonas gingivalis* from patients with periodontitis. *Journal of Clinical Microbiology, 46*(1), 31–42. http://dx.doi.org/10.1128/JCM.00986-07.

Ernst, C. W., Lee, J. E., Nakanishi, T., Karimbux, N. Y., Rezende, T. M., Stashenko, P., et al. (2007). Diminished forkhead box P3/CD25 double-positive T regulatory cells are associated with the increased nuclear factor-kappaB ligand (RANKL+) T cells in bone resorption lesion of periodontal disease. *Clinical and Experimental Immunology, 148*(2), 271–280. http://dx.doi.org/10.1111/j.1365-2249.2006.03318.x.

Eskan, M. A., Hajishengallis, G., & Kinane, D. F. (2007). Differential activation of human gingival epithelial cells and monocytes by *Porphyromonas gingivalis* fimbriae. *Infection and Immunity, 75*, 892–898.

Eskan, M. A., Jotwani, R., Abe, T., Chmelar, J., Lim, J. H., Liang, S., et al. (2012). The leukocyte integrin antagonist Del-1 inhibits IL-17-mediated inflammatory bone loss. *Nature Immunology, 13*(5), 465–473. http://dx.doi.org/10.1038/ni.2260.

Fujiwara, T., Morishima, S., Takahashi, I., & Hamada, S. (1993). Molecular cloning and sequencing of the fimbrilin gene of *Porphyromonas gingivalis* strains and characterization of recombinant proteins. *Biochemical and Biophysical Research Communications, 197*(1), 241–247. http://dx.doi.org/10.1006/bbrc.1993.2467.

Gaddis, D. E., Maynard, C. L., Weaver, C. T., Michalek, S. M., & Katz, J. (2013). Role of TLR2-dependent IL-10 production in the inhibition of the initial IFN-gamma T cell response to *Porphyromonas gingivalis*. *Journal of Leukocyte Biology, 93*(1), 21–31. http://dx.doi.org/10.1189/jlb.0512220.

Garlet, G. P., Cardoso, C. R., Mariano, F. S., Claudino, M., de Assis, G. F., Campanelli, A. P., et al. (2010). Regulatory T cells attenuate experimental periodontitis progression in mice. *Journal of Clinical Periodontology, 37*(7), 591–600. http://dx.doi.org/10.1111/j.1600-051X.2010.01586.x.

Gemmell, E., Grieco, D. A., Cullinan, M. P., Westerman, B., & Seymour, G. J. (1999). The proportion of interleukin-4, interferon-gamma and interleukin-10-positive cells in *Porphyromonas gingivalis*–specific T-cell lines established from *P. gingivalis*-positive subjects. *Oral Microbiology and Immunology, 14*(5), 267–274. http://dx.doi.org/10.1034/j.1399-302x.1999.140501.x.

Glowacki, A. J., Yoshizawa, S., Jhunjhunwala, S., Vieira, A. E., Garlet, G. P., Sfeir, C., et al. (2013). Prevention of inflammation-mediated bone loss in murine and canine periodontal disease via recruitment of regulatory lymphocytes. *Proceedings of the National Academy of Sciences of the USA, 110*(46), 18525–18530. http://dx.doi.org/10.1073/pnas.1302829110.

Goulbourne, P. A., & Ellen, R. P. (1991). Evidence that *Porphyromonas (Bacteroides) gingivalis* fimbriae function in adhesion to actinomyces-viscosus. *Journal of Bacteriology, 173*(17), 5266–5274. http://dx.doi.org/10.1128/jb.173.17.5266-5274.1991.

Grenier, D., Roy, S., Chandad, F., Plamondon, P., Yoshioka, M., Nakayama, K., et al. (2003). Effect of inactivation of the Arg- and/or Lys-gingipain gene on selected virulence and physiological properties of *Porphyromonas gingivalis*. *Infection and Immunity, 71*(8), 4742–4748. http://dx.doi.org/10.1128/iai.71.8.4742-4748.2003.

Griffith, F. (1928). The significance of pneumococcal types. *Journal of Hygiene, 27*(2), 113–159. http://dx.doi.org/10.1017/s0022172400031879.

Guo, Y., Nguyen, K. A., & Potempa, J. (2010). Dichotomy of gingipains action as virulence factors: From cleaving substrates with the precision of a surgeon's knife to a meat chopper-like brutal degradation of proteins. *Periodontology 2000, 54*(1), 15–44. http://dx.doi.org/10.1111/j.1600-0757.2010.00377.x.

Hajishengallis, G. (2009). *Porphyromonas gingivalis*-host interactions: Open war or intelligent guerilla tactics? *Microbes and Infection, 11*(6–7), 637–645.

Hajishengallis, G. (2014). Immunomicrobial pathogenesis of periodontitis: Keystones, pathobionts, and host response. *Trends in Immunology, 35*(1), 3–11. http://dx.doi.org/10.1016/j.it.2013.09.001.

Hajishengallis, G., Abe, T., Maekawa, T., Hajishengallis, E., & Lambris, J. D. (2013). Role of complement in host-microbe homeostasis of the periodontium. *Seminars in Immunology, 25*(1), 65–72. http://dx.doi.org/10.1016/j.smim.2013.04.004.

Hajishengallis, G., Darveau, R. P., & Curtis, M. A. (2012). The keystone-pathogen hypothesis. *Nature Reviews. Microbiology, 10*(10), 717–725. http://dx.doi.org/10.1038/nrmicro2873.

Hajishengallis, G., & Lambris, J. D. (2011). Microbial manipulation of receptor crosstalk in innate immunity. *Nature Reviews Immunology, 11*(3), 187–200. http://dx.doi.org/10.1038/nri2918.

Hajishengallis, G., & Lamont, R. J. (2012). Beyond the red complex and into more complexity: The polymicrobial synergy and dysbiosis (PSD) model of periodontal disease etiology. *Molecular oral microbiology, 27*(6), 409–419. http://dx.doi.org/10.1111/j.2041-1014.2012.00663.x.

Hajishengallis, G., Liang, S., Payne, M. A., Hashim, A., Jotwani, R., Eskan, M. A., et al. (2011). Low-abundance biofilm species orchestrates inflammatory periodontal disease through the commensal microbiota and complement. *Cell Host & Microbe, 10*(5), 497–506. http://dx.doi.org/10.1016/j.chom.2011.10.006.

Hajishengallis, G., McIntosh, M. L., Nishiyama, S. I., & Yoshimura, F. (2012). Mechanism and implications of CXCR4-mediated integrin activation by *Porphyromonas gingivalis*. *Molecular oral microbiology*. http://dx.doi.org/10.1111/omi.12021.

Hajishengallis, G., Shakhatreh, M.-A. K., Wang, M., & Liang, S. (2007). Complement receptor 3 blockade Promotes IL-12-mediated clearance of *Porphyromonas gingivalis* and negates its virulence in vivo. *The Journal of Immunology, 179*(4), 2359–2367.

Hajishengallis, G., Tapping, R. I., Harokopakis, E., Nishiyama, S.-I., Ratti, P., Schifferle, R. E., et al. (2006). Differential interactions of fimbriae and lipopolysaccharide from *Porphyromonas gingivalis* with the Toll-like receptor 2-centred pattern recognition apparatus. *Cellular Microbiology, 8*(10), 1557–1570.

Hajishengallis, G., Wang, M., Liang, S., Triantafilou, M., & Triantafilou, K. (2008). Pathogen induction of CXCR4/TLR2 cross-talk impairs host defense function. *Proceedings of the National Academy of Sciences USA, 105*(36), 13532–13537. http://dx.doi.org/10.1073/pnas.0803852105.

Hamada, N., Watanabe, K., Arai, M., Hiramine, H., & Umemoto, T. (2002). Cytokine production induced by a 67-kDa fimbrial protein from *Porphyromonas gingivalis*. *Oral Microbiology and Immunology, 17*(3), 197–200.

Hamedi, M., Belibasakis, G. N., Cruchley, A. T., Rangarajan, M., Curtis, M. A., & Bostanci, N. (2009). *Porphyromonas gingivalis* culture supernatants differentially regulate interleukin-1beta and interleukin-18 in human monocytic cells. *Cytokine, 45*(2), 99–104. http://dx.doi.org/10.1016/j.cyto.2008.11.005.

Handley, P. S., & Tipler, L. S. (1986). An electron microscope survey of the surface structures and hydrophobicity of oral and non-oral species of the bacterial genus Bacteroides. *Archives of Oral Biology, 31*(5), 325–335. http://dx.doi.org/10.1016/0003-9969(86)90047-6.

Haraguchi, A., Miura, M., Fujise, O., Hamachi, T., & Nishimura, F. (2014). *Porphyromonas gingivalis* gingipain is involved in the detachment and aggregation of *Aggregatibacter actinomycetemcomitans* biofilm. *Molecular Oral Microbiology, 29*(3), 131–143. http://dx.doi.org/10.1111/omi.12051.

Hasegawa, Y., Iijima, Y., Persson, K., Nagano, K., Yoshida, Y., Lamont, R. J., et al. (2016). Role of Mfa5 in expression of Mfa1 fimbriae in *Porphyromonas gingivalis*. *Journal of Dental Research, 95*(11), 1291–1297. http://dx.doi.org/10.1177/0022034516655083.

Hasegawa, Y., Iwami, J., Sato, K., Park, Y., Nishikawa, K., Atsumi, T., et al. (2009). Anchoring and length regulation of *Porphyromonas gingivalis* Mfa1 fimbriae by the downstream gene product Mfa2. *Microbiology, 155*(Pt 10), 3333–3347. http://dx.doi.org/10.1099/mic.0.028928-0.

Hasegawa, Y., Nagano, K., Ikai, R., Izumigawa, M., Yoshida, Y., Kitai, N., et al. (2013). Localization and function of the accessory protein Mfa3 in *Porphyromonas gingivalis* Mfa1 fimbriae. *Mol Oral Microbiol, 28*(6), 467–480. http://dx.doi.org/10.1111/omi.12040.

Hashimoto, M., Ogawa, S., Asai, Y., Takai, Y., & Ogawa, T. (2003). Binding of *Porphyromonas gingivalis* fimbriae to *Treponema denticola* dentilisin. *FEMS Microbiology Letters, 226*(2), 267–271. http://dx.doi.org/10.1016/S0378-1097(03)00615-3.

Herath, T. D., Wang, Y., Seneviratne, C. J., Darveau, R. P., Wang, C. Y., & Jin, L. (2013). The expression and regulation of matrix metalloproteinase-3 is critically modulated by *Porphyromonas gingivalis* lipopolysaccharide with heterogeneous lipid A structures in human gingival fibroblasts. *BMC Microbiology, 13*, 73. http://dx.doi.org/10.1186/1471-2180-13-73.

Herath, T. D., Wang, Y., Seneviratne, C. J., Lu, Q., Darveau, R. P., Wang, C. Y., et al. (2011). *Porphyromonas gingivalis* lipopolysaccharide lipid A heterogeneity differentially modulates the expression of IL-6 and IL-8 in human gingival fibroblasts. *Journal of Clinical Periodontology, 38*(8), 694–701. http://dx.doi.org/10.1111/j.1600-051X.2011.01741.x.

Hiramine, H., Watanabe, K., Hamada, N., & Umemoto, T. (2003). *Porphyromonas gingivalis* 67-kDa fimbriae induced cytokine production and osteoclast differentiation utilizing

TLR2. *FEMS Microbiology Letters, 229*(1), 49–55. http://dx.doi.org/10.1016/S0378-1097(03)00788-2.

Holden, J. A., Attard, T. J., Laughton, K. M., Mansell, A., O'Brien-Simpson, N. M., & Reynolds, E. C. (2014). *Porphyromonas gingivalis* lipopolysaccharide weakly activates M1 and M2 polarized mouse macrophages but induces inflammatory cytokines. *Infection and Immunity, 82*(10), 4190–4203. http://dx.doi.org/10.1128/IAI.02325-14.

Holt, S. C., Kesavalu, L., Walker, S., & Genco, C. A. (1999). Virulence factors of *Porphyromonas gingivalis*. *Periodontology 2000, 20*, 168–238.

Honda, T., Aoki, Y., Takahashi, N., Maekawa, T., Nakajima, T., Ito, H., et al. (2008). Elevated expression of IL-17 and IL-12 genes in chronic inflammatory periodontal disease. *Clinica Chimica Acta, 395*(1–2), 137–141. http://dx.doi.org/10.1016/j.cca.2008.06.003.

How, K. Y., Song, K. P., & Chan, K. G. (2016). *Porphyromonas gingivalis*: An overview of periodontopathic pathogen below the gum line. *Frontiers in Microbiology, 7*, 53. http://dx.doi.org/10.3389/fmicb.2016.00053.

Igboin, C. O., Tordoff, K. P., Moeschberger, M. L., Griffen, A. L., & Leys, E. J. (2011). *Porphyromonas gingivalis*-host interactions in a *Drosophila melanogaster* model. *Infection and Immunity, 79*(1), 449–458. http://dx.doi.org/10.1128/IAI.00785-10.

Imamura, T., Travis, J., & Potempa, J. (2003). The biphasic virulence activities of gingipains: Activation and inactivation of host proteins. *Current Protein & Peptide Science, 4*(6), 443–450. http://dx.doi.org/10.2174/1389203033487027.

Ito, R., Ishihara, K., Shoji, M., Nakayama, K., & Okuda, K. (2010). Hemagglutinin/Adhesin domains of *Porphyromonas gingivalis* play key roles in coaggregation with *Treponema denticola*. *FEMS Immunology and Medical Microbiology, 60*(3), 251–260. http://dx.doi.org/10.1111/j.1574-695X.2010.00737.x.

Ivanyi, L., & Lehner, T. (1970). Stimulation of lymphocyte transformation by bacterial antigens in patients with periodontal disease. *Archives of Oral Biology, 15*(11), 1089–1096. http://dx.doi.org/10.1016/0003-9969(70)90121-4.

Jagels, M. A., Ember, J. A., Travis, J., Potempa, J., Pike, R., & Hugli, T. E. (1996). Cleavage of the human C5A receptor by proteinases derived from *Porphyromonas gingivalis*: Cleavage of leukocyte C5a receptor. *Advances in Experimental Medicine & Biology, 389*, 155–164. http://dx.doi.org/10.1007/978-1-4613-0335-0_19.

Jagels, M. A., Travis, J., Potempa, J., Pike, R., & Hugli, T. E. (1996). Proteolytic inactivation of the leukocyte C5a receptor by proteinases derived from *Porphyromonas gingivalis*. *Infection and Immunity, 64*(6), 1984–1991.

Jankovic, D., & Feng, C. G. (2015). CD4($^+$) T cell differentiation in infection: Amendments to the Th1/Th2 axiom. *Frontiers in Immunology, 6*, 198. http://dx.doi.org/10.3389/fimmu.2015.00198.

Jotwani, R., & Cutler, C. W. (2004). Fimbriated *Porphyromonas gingivalis* is more efficient than fimbria-deficient *P. Gingivalis* in entering human dendritic cells in vitro and induces an inflammatory Th1 effector response. *Infection and Immunity, 72*(3), 1725–1732. http://dx.doi.org/10.1128/iai.72.3.1725-1732.2004.

Jotwani, R., Pulendran, B., Agrawal, S., & Cutler, C. W. (2003). Human dendritic cells respond to *Porphyromonas gingivalis* LPS by promoting a Th2 effector response in vitro. *European Journal of Immunology, 33*(11), 2980–2986. http://dx.doi.org/10.1002/eji.200324392.

Kadowaki, T., Baba, A., Abe, N., Takii, R., Hashimoto, M., Tsukuba, T., et al. (2004). Suppression of pathogenicity of *Porphyromonas gingivalis* by newly developed gingipain inhibitors. *Molecular Pharmacology, 66*(6), 1599–1606. http://dx.doi.org/10.1124/mol.104.004366.

Kamaguchi, A., Ohyama, T., Sakai, E., Nakamura, R., Watanabe, T., Baba, H., et al. (2003). Adhesins encoded by the gingipain genes of *Porphyromonas gingivalis* are responsible for co-aggregation with Prevotella intermedia. *Microbiology, 149*(Pt 5), 1257–1264.

Kassebaum, N. J., Smith, A. G. C., Bernabe, E., Fleming, T. D., Reynolds, A. E., Vos, T., et al. (2017). Global, regional, and national prevalence, incidence, and disability-adjusted life years for oral conditions for 195 countries, 1990-2015: A systematic analysis for the global burden of diseases, injuries, and risk factors. *Journal of Dental Research, 96*(4), 380–387. http://dx.doi.org/10.1177/0022034517693566.

Katz, J., Ward, D. C., & Michalek, S. M. (1996). Effect of host responses on the pathogenicity of strains of *Porphyromonas gingivalis. Oral Microbiology and Immunology, 11*(5), 309–318. http://dx.doi.org/10.1111/j.1399-302x.1996.tb00187.x.

Kawai, T., Eisen-Lev, R., Seki, M., Eastcott, J. W., Wilson, M. E., & Taubman, M. A. (2000). Requirement of B7 costimulation for Th1-mediated inflammatory bone resorption in experimental periodontal disease. *The Journal of Immunology, 164*(4), 2102–2109.

Kayagaki, N., Wong, M. T., Stowe, I. B., Ramani, S. R., Gonzalez, L. C., Akashi-Takamura, S., et al. (2013). Noncanonical inflammasome activation by intracellular LPS independent of TLR4. *Science, 341*(6151), 1246–1249. http://dx.doi.org/10.1126/science.1240248.

Khader, Y. S., Dauod, A. S., El-Qaderi, S. S., Alkafajei, A., & Batayha, W. Q. (2006). Periodontal status of diabetics compared with nondiabetics: A meta-analysis. *Journal of Diabetic Complications, 20*(1), 59–68. http://dx.doi.org/10.1016/j.jdiacomp.2005.05.006.

Khalaf, H., & Bengtsson, T. (2012). Altered T-cell responses by the periodontal pathogen *Porphyromonas gingivalis. PLoS One,* 7(9), e45192. http://dx.doi.org/10.1371/journal.pone.0045192.

Kim, Y. C., Ko, Y., Hong, S. D., Kim, K. Y., Lee, Y. H., Chae, C., et al. (2010). Presence of *Porphyromonas gingivalis* and plasma cell dominance in gingival tissues with periodontitis. *Oral Diseases, 16*(4), 375–381. http://dx.doi.org/10.1111/j.1601-0825.2009.01649.x.

Kimura, S., Fujimoto, N., & Okada, H. (1991). Impaired autologous mixed-lymphocyte reaction of peripheral blood lymphocytes in adult periodontitis. *Infection and Immunity, 59*(12), 4418–4424.

Kinane, D. F., Stathopoulou, P. G., & Papapanou, P. N. (2017). Periodontal diseases. *Nature Reviews Disease Primers, 3,* 17038. http://dx.doi.org/10.1038/nrdp.2017.38.

Kitamura, Y., Matono, S., Aida, Y., Hirofuji, T., & Maeda, K. (2002). Gingipains in the culture supernatant of *Porphyromonas gingivalis* cleave CD4 and CD8 on human T cells. *Journal of Periodontal Research, 37*(6), 464–468. http://dx.doi.org/10.1034/j.1600-0765.2002.01364.x.

Kluger, M. A., Meyer, M. C., Nosko, A., Goerke, B., Luig, M., Wegscheid, C., et al. (2016). RORgammat(+)Foxp3(+) cells are an independent bifunctional regulatory T cell lineage and mediate crescentic GN. *Journal of the American Society of Nephrology, 27*(2), 454–465. http://dx.doi.org/10.1681/ASN.2014090880.

Kobayashi, R., Kono, T., Bolerjack, B. A., Fukuyama, Y., Gilbert, R. S., Fujihashi, K., et al. (2011). Induction of IL-10-producing CD4+ T-cells in chronic periodontitis. *Journal of Dental Research, 90*(5), 653–658. http://dx.doi.org/10.1177/0022034510397838.

Koenen, H. J., Smeets, R. L., Vink, P. M., van Rijssen, E., Boots, A. M., & Joosten, I. (2008). Human CD25highFoxp3pos regulatory T cells differentiate into IL-17-producing cells. *Blood, 112*(6), 2340–2352. http://dx.doi.org/10.1182/blood-2008-01-133967.

Kristoffersen, A. K., Solli, S. J., Nguyen, T. D., & Enersen, M. (2015). Association of the rgpB gingipain genotype to the major fimbriae (fimA) genotype in clinical isolates of the periodontal pathogen *Porphyromonas gingivalis. Journal of Oral Microbiology,* 7, 29124. http://dx.doi.org/10.3402/jom.v7.29124.

Kryczek, I., Wu, K., Zhao, E., Wei, S., Vatan, L., Szeliga, W., et al. (2011). IL-17$^+$ regulatory T cells in the microenvironments of chronic inflammation and cancer. *The Journal of Immunology, 186*(7), 4388–4395. http://dx.doi.org/10.4049/jimmunol.1003251.

Kuboniwa, M., Houser, J. R., Hendrickson, E. L., Wang, Q., Alghamdi, S. A., Sakanaka, A., et al. (2017). Metabolic crosstalk regulates *Porphyromonas gingivalis* colonization and virulence during oral polymicrobial infection. *Nature Microbiology, 2*(11), 1493–1499. http://dx.doi.org/10.1038/s41564-017-0021-6.

Kuboniwa, M., & Lamont, R. J. (2010). Subgingival biofilm formation. *Periodontology 2000, 52*(1), 38–52. http://dx.doi.org/10.1111/j.1600-0757.2009.00311.x.

Laine, M. L., Appelmelk, B. J., & van Winkelhoff, A. J. (1997). Prevalence and distribution of six capsular serotypes of *Porphyromonas gingivalis* in periodontitis patients. *Journal of Dental Research, 76*(12), 1840–1844. http://dx.doi.org/10.1177/00220345970760120601.

Laine, M. L., & van Winkelhoff, A. J. (1998). Virulence of six capsular serotypes of *Porphyromonas gingivalis* in a mouse model. *Oral Microbiology and Immunology, 13*(5), 322–325. http://dx.doi.org/10.1111/j.1399-302x.1998.tb00714.x.

Lamont, R. J., Hsiao, G. W., & Gil, S. (1994). Identification of a molecule of *Porphyromonas gingivalis* that binds to Streptococcus gordonii. *Microbial Pathogenesis, 17*(5), 355–360. http://dx.doi.org/10.1006/mpat.1994.1081.

Lamont, R. J., & Jenkinson, H. F. (2000). Subgingival colonization by *Porphyromonas gingivalis*. *Oral Microbiology and Immunology, 15*(6), 341–349.

Lamont, R. J., Koo, H., & Hajishengallis, G. (2018). The oral microbiota: Dynamic communities and host interactions. *Nature Reviews Microbiology, 16*(12), 745–759. http://dx.doi.org/10.1038/s41579-018-0089-x.

Lee, J. Y., Miller, D. P., Wu, L., Casella, C. R., Hasegawa, Y., & Lamont, R. J. (2018). Maturation of the Mfa1 fimbriae in the oral pathogen *Porphyromonas gingivalis*. *Frontiers in Cellular and Infection Microbiology, 8*, 137. http://dx.doi.org/10.3389/fcimb.2018.00137.

Lee, J. Y., Sojar, H. T., Bedi, G. S., & Genco, R. J. (1991). *Porphyromonas (Bacteroides) gingivalis* fimbrillin: Size, amino-terminal sequence, and antigenic heterogeneity. *Infection and Immunity, 59*(1), 383–389.

Lester, S. R., Bain, J. L., Johnson, R. B., & Serio, F. G. (2007). Gingival concentrations of interleukin-23 and -17 at healthy sites and at sites of clinical attachment loss. *Journal of Periodontology, 78*(8), 1545–1550. http://dx.doi.org/10.1902/jop.2007.060458.

Leveque, L., Deknuydt, F., Bioley, G., Old, L. J., Matsuzaki, J., Odunsi, K., et al. (2009). Interleukin 2-mediated conversion of ovarian cancer-associated CD4$^+$ regulatory T cells into proinflammatory interleukin 17-producing helper T cells. *Journal of Immunotherapy, 32*(2), 101–108. http://dx.doi.org/10.1097/CJI.0b013e318195b59e.

Liang, S., Krauss, J. L., Domon, H., McIntosh, M. L., Hosur, K. B., Qu, H., et al. (2011). The C5a receptor impairs IL-12-dependent clearance of *Porphyromonas gingivalis* and is required for induction of periodontal bone loss. *The Journal of Immunology, 186*(2), 869–877. http://dx.doi.org/10.4049/jimmunol.1003252.

Lien, E., Means, T. K., Heine, H., Yoshimura, A., Kusumoto, S., Fukase, K., et al. (2000). Toll-like receptor 4 imparts ligand-specific recognition of bacterial lipopolysaccharide. *Journal of Clinical Investigation, 105*, 497–504.

Li, C., Li, B., Dong, Z., Gao, L., He, X., Liao, L., et al. (2014). Lipopolysaccharide differentially affects the osteogenic differentiation of periodontal ligament stem cells and bone marrow mesenchymal stem cells through Toll-like receptor 4 mediated nuclear factor kappaB pathway. *Stem Cell Research & Therapy, 5*(3), 67. http://dx.doi.org/10.1186/scrt456.

Liu, R., Desta, T., Raptis, M., Darveau, R. P., & Graves, D. T. (2008). *P. gingivalis* and *E. coli* lipopolysaccharides exhibit different systemic but similar local induction of inflammatory markers. *Journal of Periodontology, 79*(7), 1241–1247.

Liu, Z. M., Wang, K. P., Ma, J., & Guo Zheng, S. (2015). The role of all-trans retinoic acid in the biology of Foxp3$^+$ regulatory T cells. *Cellular and Molecular Immunology, 12*(5), 553–557. http://dx.doi.org/10.1038/cmi.2014.133.

Madianos, P. N., Bobetsis, Y. A., & Offenbacher, S. (2013). Adverse pregnancy outcomes (APOs) and periodontal disease: Pathogenic mechanisms. *Journal of Periodontology, 84*(4 Suppl. l), S170–S180. http://dx.doi.org/10.1902/jop.2013.1340015.

Maeda, K., Nagata, H., Nonaka, A., Kataoka, K., Tanaka, M., & Shizukuishi, S. (2004). Oral streptococcal glyceraldehyde-3-phosphate dehydrogenase mediates interaction with *Porphyromonas gingivalis* fimbriae. *Microbes and Infection, 6*(13), 1163–1170. http://dx.doi.org/10.1016/j.micinf.2004.06.005.

Maeda, K., Nagata, H., Yamamoto, Y., Tanaka, M., Tanaka, J., Minamino, N., et al. (2004). Glyceraldehyde-3-phosphate dehydrogenase of *Streptococcus oralis* functions as a coadhesin for *Porphyromonas gingivalis* major fimbriae. *Infection and Immunity, 72*(3), 1341–1348.

Maekawa, T., Abe, T., Hajishengallis, E., Hosur, K. B., Deangelis, R. A., Ricklin, D., et al. (2014). Genetic and intervention studies implicating complement C3 as a major target for the treatment of periodontitis. *The Journal of Immunology*. http://dx.doi.org/10.4049/jimmunol.1400569.

Maisetta, G., Brancatisano, F. L., Esin, S., Campa, M., & Batoni, G. (2011). Gingipains produced by *Porphyromonas gingivalis* ATCC49417 degrade human-beta-defensin 3 and affect peptide's antibacterial activity in vitro. *Peptides, 32*(5), 1073–1077. http://dx.doi.org/10.1016/j.peptides.2011.02.003.

Malek, R., Fisher, J. G., Caleca, A., Stinson, M., van Oss, C. J., Lee, J. Y., et al. (1994). Inactivation of *Porphyromonas gingivalis* fimA gene blocks periodontal damage in gnotobiotic rats. *Journal of Bacteriology, 176*(4), 1052–1059.

Medzhitov, R. (2001). Toll-like receptors and innate immunity. *Nature Reviews Immunology, 1*(2), 135–145.

Mosmann, T. R., Cherwinski, H., Bond, M. W., Giedlin, M. A., & Coffman, R. L. (1986). Two types of murine helper T cell clone. I. Definition according to profiles of lymphokine activities and secreted proteins. *The Journal of Immunology, 136*(7), 2348–2357.

Moutsopoulos, N. M., Kling, H. M., Angelov, N., Jin, W., Palmer, R. J., Nares, S., et al. (2012). *Porphyromonas gingivalis* promotes Th17 inducing pathways in chronic periodontitis. *Journal of Autoimmunity, 39*(4), 294–303. http://dx.doi.org/10.1016/j.jaut.2012.03.003.

Nagano, K., Hasegawa, Y., Iijima, Y., Kikuchi, T., & Mitani, A. (2018). Distribution of *Porphyromonas gingivalis* fimA and mfa1 fimbrial genotypes in subgingival plaques. *PeerJ, 6*, e5581. http://dx.doi.org/10.7717/peerj.5581.

Nagano, K., Hasegawa, Y., Murakami, Y., Nishiyama, S., & Yoshimura, F. (2010). FimB regulates FimA fimbriation in *Porphyromonas gingivalis*. *Journal of Dental Research, 89*(9), 903–908. http://dx.doi.org/10.1177/0022034510370089.

Naito, M., Hirakawa, H., Yamashita, A., Ohara, N., Shoji, M., Yukitake, H., et al. (2008). Determination of the genome sequence of *Porphyromonas gingivalis* strain ATCC 33277 and genomic comparison with strain W83 revealed extensive genome rearrangements in *P. gingivalis*. *DNA Research: an International Journal for Rapid Publication of Reports on Genes and Genomes, 15*(4), 215–225. http://dx.doi.org/10.1093/dnares/dsn013.

Nakajima, T., Ueki-Maruyama, K., Oda, T., Ohsawa, Y., Ito, H., Seymour, G. J., et al. (2005). Regulatory T-cells infiltrate periodontal disease tissues. *Journal of Dental Research, 84*(7), 639–643. http://dx.doi.org/10.1177/154405910508400711.

Nelson, K. E., Fleischmann, R. D., DeBoy, R. T., Paulsen, I. T., Fouts, D. E., Eisen, J. A., et al. (2003). Complete genome sequence of the oral pathogenic Bacterium *Porphyromonas gingivalis* strain W83. *Journal of Bacteriology, 185*(18), 5591–5601.

Nemoto, E., Darveau, R. P., Foster, B. L., Nogueira-Filho, G. R., & Somerman, M. J. (2006). Regulation of cementoblast function by *P. gingivalis* lipopolysaccharide via

TLR2. *Journal of Dental Research, 85*(8), 733–738. http://dx.doi.org/10.1177/154405910608500809.

Nishikawa, K., & Duncan, M. J. (2010). Histidine kinase-mediated production and autoassembly of *Porphyromonas gingivalis* fimbriae. *Journal of Bacteriology, 192*(7), 1975–1987. http://dx.doi.org/10.1128/JB.01474-09.

Nishiyama, S.-I., Murakami, Y., Nagata, H., Shizukuishi, S., Kawagishi, I., & Yoshimura, F. (2007). Involvement of minor components associated with the FimA fimbriae of *Porphyromonas gingivalis* in adhesive functions. *Microbiology, 153*, 1916–1925.

Nyirenda, M. H., Sanvito, L., Darlington, P. J., O'Brien, K., Zhang, G. X., Constantinescu, C. S., et al. (2011). TLR2 stimulation drives human naive and effector regulatory T cells into a Th17-like phenotype with reduced suppressive function. *The Journal of Immunology, 187*(5), 2278–2290. http://dx.doi.org/10.4049/jimmunol.1003715.

O'Connor, W., Jr., Zenewicz, L. A., & Flavell, R. A. (2010). The dual nature of T(H)17 cells: Shifting the focus to function. *Nature Immunology, 11*(6), 471–476. http://dx.doi.org/10.1038/ni.1882.

Oda, T., Yoshie, H., & Yamazaki, K. (2003). *Porphyromonas gingivalis* antigen preferentially stimulates T cells to express IL-17 but not receptor activator of NF-kappaB ligand in vitro. *Oral Microbiology and Immunology, 18*(1), 30–36. http://dx.doi.org/10.1034/j.1399-302x.2003.180105.x.

Ogawa, T. (1993). Chemical structure of lipid A from *Porphyromonas (Bacteroides) gingivalis* lipopolysaccharide. *FEBS Letters, 332*(1–2), 197–201. http://dx.doi.org/10.1016/0014-5793(93)80512-s.

Ohyama, H., Kato-Kogoe, N., Kuhara, A., Nishimura, F., Nakasho, K., Yamanegi, K., et al. (2009). The involvement of IL-23 and the Th17 pathway in periodontitis. *Journal of Dental Research, 88*(7), 633–638. http://dx.doi.org/10.1177/0022034509339889.

Okui, T., Aoki, Y., Ito, H., Honda, T., & Yamazaki, K. (2012). The presence of IL-17^{+}/FOXP3^{+} double-positive cells in periodontitis. *Journal of Dental Research, 91*(6), 574–579. http://dx.doi.org/10.1177/0022034512446341.

Okui, T., Ito, H., Honda, T., Amanuma, R., Yoshie, H., & Yamazaki, K. (2008). Characterization of CD4+ FOXP3+ T-cell clones established from chronic inflammatory lesions. *Oral Microbiology and Immunology, 23*(1), 49–54. http://dx.doi.org/10.1111/j.1399-302X.2007.00390.x.

Olsen, I., Lambris, J. D., & Hajishengallis, G. (2017). *Porphyromonas gingivalis* disturbs host-commensal homeostasis by changing complement function. *Journal of Oral Microbiology, 9*(1), 1340085. http://dx.doi.org/10.1080/20002297.2017.1340085.

Papadopoulos, G., Weinberg, E. O., Massari, P., Gibson, F. C., 3rd, Wetzler, L. M., Morgan, E. F., et al. (2013). Macrophage-specific TLR2 signaling mediates pathogen-induced TNF-dependent inflammatory oral bone loss. *The Journal of Immunology, 190*(3), 1148–1157. http://dx.doi.org/10.4049/jimmunol.1202511.

Papapanou, P. N. (2012). The prevalence of periodontitis in the US: Forget what you were told. *Journal of Dental Research, 91*(10), 907–908. http://dx.doi.org/10.1177/0022034512458692.

Park, Y. D., Kim, Y. S., Jung, Y. M., Lee, S. I., Lee, Y. M., Bang, J. B., et al. (2012). *Porphyromonas gingivalis* lipopolysaccharide regulates interleukin (IL)-17 and IL-23 expression via SIRT1 modulation in human periodontal ligament cells. *Cytokine, 60*(1), 284–293. http://dx.doi.org/10.1016/j.cyto.2012.05.021.

Park, Y., Simionato, M. R., Sekiya, K., Murakami, Y., James, D., Chen, W., et al. (2005). Short fimbriae of *Porphyromonas gingivalis* and their role in coadhesion with *Streptococcus gordonii*. *Infection and Immunity, 73*(7), 3983–3989.

Pathirana, R. D., O'Brien-Simpson, N. M., & Reynolds, E. C. (2010). Host immune responses to *Porphyromonas gingivalis* antigens. *Periodontology 2000, 52*(1), 218–237. http://dx.doi.org/10.1111/j.1600-0757.2009.00330.x.

Pierce, D. L., Nishiyama, S., Liang, S., Wang, M., Triantafilou, M., Triantafilou, K., et al. (2009). Host adhesive activities and virulence of novel fimbrial proteins of *Porphyromonas gingivalis*. *Infection and Immunity,* 77(8), 3294–3301. http://dx.doi.org/10.1128/IAI.00262-09.

Poltorak, A., He, X., Smirnova, I., Liu, M.-Y., Huffel, C. V., Du, X., et al. (1998). Defective LPS signaling in C3H/HeJ and C57BL/10ScCr mice: Mutations in Tlr4 gene. *Science, 282*(5396), 2085–2088. http://dx.doi.org/10.1126/science.282.5396.2085.

Popadiak, K., Potempa, J., Riesbeck, K., & Blom, A. M. (2007). Biphasic effect of gingipains from *Porphyromonas gingivalis* on the human complement system. *The Journal of Immunology, 178*(11), 7242–7250.

Potempa, J., Banbula, A., & Travis, J. (2000). Role of bacterial proteinases in matrix destruction and modulation of host responses. *Periodontology 2000, 24*, 153–192.

Potempa, M., Potempa, J., Okroj, M., Popadiak, K., Eick, S., Nguyen, K. A., et al. (2008). Binding of complement inhibitor C4b-binding protein contributes to serum resistance of *Porphyromonas gingivalis*. *The Journal of Immunology, 181*(8), 5537–5544. http://dx.doi.org/10.4049/jimmunol.181.8.5537.

Rangarajan, M., Aduse-Opoku, J., Paramonov, N. A., Hashim, A., & Curtis, M. A. (2017). Hemin binding by *Porphyromonas gingivalis* strains is dependent on the presence of A-LPS. *Mol Oral Microbiol, 32*(5), 365–374. http://dx.doi.org/10.1111/omi.12178.

Ricklin, D., Hajishengallis, G., Yang, K., & Lambris, J. D. (2010). Complement: A key system for immune surveillance and homeostasis. *Nature Immunology, 11*(9), 785–797. http://dx.doi.org/10.1038/ni.1923.

Rosen, G., & Sela, M. N. (2006). Coaggregation of *Porphyromonas gingivalis* and *Fusobacterium nucleatum* PK 1594 is mediated by capsular polysaccharide and lipopolysaccharide. *FEMS Microbiology Letters, 256*(2), 304–310. http://dx.doi.org/10.1111/j.1574-6968.2006.00131.x.

Sakanaka, A., Takeuchi, H., Kuboniwa, M., & Amano, A. (2016). Dual lifestyle of *Porphyromonas gingivalis* in biofilm and gingival cells. *Microbial Pathogenesis, 94*, 42–47. http://dx.doi.org/10.1016/j.micpath.2015.10.003.

Sasaki, H., Okamatsu, Y., Kawai, T., Kent, R., Taubman, M., & Stashenko, P. (2004). The interleukin-10 knockout mouse is highly susceptible to *Porphyromonas gingivalis*-induced alveolar bone loss. *Journal of Periodontal Research, 39*(6), 432–441.

Sato, K., Suematsu, A., Okamoto, K., Yamaguchi, A., Morishita, Y., Kadono, Y., et al. (2006). Th17 functions as an osteoclastogenic helper T cell subset that links T cell activation and bone destruction. *Journal of Experimental Medicine, 203*(12), 2673–2682. http://dx.doi.org/10.1084/jem.20061775.

Schenkein, H. A., Fletcher, H. M., Bodnar, M., & Macrina, F. L. (1995). Increased opsonization of a prtH-defective mutant of *Porphyromonas gingivalis* W83 is caused by reduced degradation of complement-derived opsonins. *The Journal of Immunology, 154*(10), 5331–5337.

Scheres, N., & Crielaard, W. (2013). Gingival fibroblast responsiveness is differentially affected by *Porphyromonas gingivalis*: Implications for the pathogenesis of periodontitis. *Molecular oral microbiology, 28*(3), 204–218. http://dx.doi.org/10.1111/Omi.12016.

Schifferle, R. E., Reddy, M. S., Zambon, J. J., Genco, R. J., & Levine, M. J. (1989). Characterization of a polysaccharide antigen from Bacteroides gingivalis. *The Journal of Immunology, 143*(9), 3035–3042.

Schromm, A. B., Brandenburg, K., Loppnow, H., Moran, A. P., Koch, M. H., Rietschel, E. T., et al. (2000). Biological activities of lipopolysaccharides are determined by the shape of their lipid A portion. *European Journal of Biochemistry, 267*(7), 2008–2013. http://dx.doi.org/10.1046/j.1432-1327.2000.01204.x.

Sheets, S. M., Potempa, J., Travis, J., Casiano, C. A., & Fletcher, H. M. (2005). Gingipains from *Porphyromonas gingivalis* W83 induce cell adhesion molecule cleavage and apoptosis in endothelial cells. *Infection and Immunity, 73*(3), 1543–1552. http://dx.doi.org/10.1128/IAI.73.3.1543-1552.2005.

Sheets, S. M., Potempa, J., Travis, J., Fletcher, H. M., & Casiano, C. A. (2006). Gingipains from *Porphyromonas gingivalis* W83 synergistically disrupt endothelial cell adhesion and can induce caspase-independent apoptosis. *Infection and Immunity, 74*(10), 5667–5678. http://dx.doi.org/10.1128/IAI.01140-05.

Shenker, B. J., & Datar, S. (1995). Fusobacterium nucleatum inhibits human T-cell activation by arresting cells in the mid-G1 phase of the cell cycle. *Infection and Immunity, 63*(12), 4830–4836.

Shevach, E. M. (2018). Foxp3($^+$) T regulatory cells: Still many unanswered questions-A perspective after 20 Years of study. *Frontiers in Immunology, 9*, 1048. http://dx.doi.org/10.3389/fimmu.2018.01048.

Shi, J., Zhao, Y., Wang, Y., Gao, W., Ding, J., Li, P., et al. (2014). Inflammatory caspases are innate immune receptors for intracellular LPS. *Nature, 514*(7521), 187–192. http://dx.doi.org/10.1038/nature13683.

Singh, R. P., Hasan, S., Sharma, S., Nagra, S., Yamaguchi, D. T., Wong, D. T., et al. (2014). Th17 cells in inflammation and autoimmunity. *Autoimmunity Reviews, 13*(12), 1174–1181. http://dx.doi.org/10.1016/j.autrev.2014.08.019.

Singh, A., Wyant, T., Anaya-Bergman, C., Aduse-Opoku, J., Brunner, J., Laine, M. L., et al. (2011). The capsule of *Porphyromonas gingivalis* leads to a reduction in the host inflammatory response, evasion of phagocytosis, and increase in virulence. *Infection and Immunity, 79*(11), 4533–4542. http://dx.doi.org/10.1128/IAI.05016-11.

Slocum, C., Coats, S. R., Hua, N., Kramer, C., Papadopoulos, G., Weinberg, E. O., et al. (2014). Distinct lipid a moieties contribute to pathogen-induced site-specific vascular inflammation. *PLoS Pathogens, 10*(7), e1004215. http://dx.doi.org/10.1371/journal.ppat.1004215.

Socransky, S. S., Haffajee, A. D., Cugini, M. A., Smith, C., & Kent, R. L., Jr. (1998). Microbial complexes in subgingival plaque. *Journal of Clinical Periodontology, 25*(2), 134–144.

Sojar, H. T., & Genco, R. J. (2005). Identification of glyceraldehyde-3-phosphate dehydrogenase of epithelial cells as a second molecule that binds to *Porphyromonas gingivalis* fimbriae. *FEMS Immunology and Medical Microbiology, 45*(1), 25–30. http://dx.doi.org/10.1016/j.femsim.2005.01.006.

Sroka, A., Sztukowska, M., Potempa, J., Travis, J., & Genco, C. A. (2001). Degradation of host heme proteins by lysine- and arginine-specific cysteine proteinases (gingipains) of *Porphyromonas gingivalis*. *Journal of Bacteriology, 183*(19), 5609–5616. http://dx.doi.org/10.1128/JB.183.19.5609-5616.2001.

Stadhouders, R., Lubberts, E., & Hendriks, R. W. (2018). A cellular and molecular view of T helper 17 cell plasticity in autoimmunity. *Journal of Autoimmunity, 87*, 1–15. http://dx.doi.org/10.1016/j.jaut.2017.12.007.

Stashenko, P., Resmini, L. M., Haffajee, A. D., & Socransky, S. S. (1983). T cell responses of periodontal disease patients and healthy subjects to oral microorganisms. *Journal of Periodontal Research, 18*(6), 587–600. http://dx.doi.org/10.1111/j.1600-0765.1983.tb00396.x.

Sundqvist, G., Figdor, D., Hanstrom, L., Sorlin, S., & Sandstrom, G. (1991). Phagocytosis and virulence of different strains of *Porphyromonas gingivalis*. *Scandinavian Journal of Dental Research, 99*(2), 117–129. http://dx.doi.org/10.1111/j.1600-0722.1991.tb01874.x.

Takeichi, O., Haber, J., Kawai, T., Smith, D. J., Moro, I., & Taubman, M. A. (2000). Cytokine profiles of T-lymphocytes from gingival tissues with pathological pocketing. *Journal of Dental Research, 79*(8), 1548–1555. http://dx.doi.org/10.1177/00220345000790080401.

Takeuchi, O., Hoshino, K., Kawai, T., Sanjo, H., Takada, H., Ogawa, T., et al. (1999). Differential roles of TLR2 and TLR4 in recognition of gram-negative and gram-positive bacterial cell wall components. *Immunity, 11*(4), 443–451.

Takeuchi, H., Sasaki, N., Yamaga, S., Kuboniwa, M., Matsusaki, M., & Amano, A. (2019). *Porphyromonas gingivalis* induces penetration of lipopolysaccharide and peptidoglycan through the gingival epithelium via degradation of junctional adhesion molecule 1. *PLoS Pathogens, 15*(11), e1008124. http://dx.doi.org/10.1371/journal.ppat.1008124.

Tartar, D. M., VanMorlan, A. M., Wan, X., Guloglu, F. B., Jain, R., Haymaker, C. L., et al. (2010). FoxP3$^+$RORgammat$^+$ T helper intermediates display suppressive function against autoimmune diabetes. *The Journal of Immunology, 184*(7), 3377–3385. http://dx.doi.org/10.4049/jimmunol.0903324.

Teng, Y. T., Nguyen, H., Gao, X., Kong, Y. Y., Gorczynski, R. M., Singh, B., et al. (2000). Functional human T-cell immunity and osteoprotegerin ligand control alveolar bone destruction in periodontal infection. *Journal of Clinical Investigation, 106*(6), R59–R67.

Tokoro, Y., Matsuki, Y., Yamamoto, T., Suzuki, T., & Hara, K. (1997). Relevance of local Th2-type cytokine mRNA expression in immunocompetent infiltrates in inflamed gingival tissue to periodontal diseases. *Clinical and Experimental Immunology, 107*(1), 166–174. http://dx.doi.org/10.1046/j.1365-2249.1997.d01-880.x.

Ukai, T., Mori, Y., Onoyama, M., & Hara, Y. (2001). Immunohistological study of interferon-gamma- and interleukin-4-bearing cells in human periodontitis gingiva. *Archives of Oral Biology, 46*(10), 901–908. http://dx.doi.org/10.1016/s0003-9969(01)00057-7.

Umemoto, T., & Hamada, N. (2003). Characterization of biologically active cell surface components of a periodontal pathogen. The roles of major and minor fimbriae of *Porphyromonas gingivalis*. *Journal of Periodontology, 74*(1), 119–122.

Valmori, D., Raffin, C., Raimbaud, I., & Ayyoub, M. (2010). Human RORgammat$^+$ TH17 cells preferentially differentiate from naive FOXP3$^+$Treg in the presence of lineage-specific polarizing factors. *Proceedings of the National Academy of Sciences of the USA, 107*(45), 19402–19407. http://dx.doi.org/10.1073/pnas.1008247107.

Vernal, R., Diaz Guerra, E., Silva, A., Sanz, M., & Garcia-Sanz, J. A. (2014). Distinct human T-lymphocyte responses triggered by *Porphyromonas gingivalis* capsular serotypes. *Journal of Clinical Periodontology, 41*(1), 19–30. http://dx.doi.org/10.1111/jcpe.12176.

Vernal, R., Dutzan, N., Chaparro, A., Puente, J., Antonieta Valenzuela, M., & Gamonal, J. (2005). Levels of interleukin-17 in gingival crevicular fluid and in supernatants of cellular cultures of gingival tissue from patients with chronic periodontitis. *Journal of Clinical Periodontology, 32*(4), 383–389. http://dx.doi.org/10.1111/j.1600-051X.2005.00684.x.

Vernal, R., Leon, R., Silva, A., van Winkelhoff, A. J., Garcia-Sanz, J. A., & Sanz, M. (2009). Differential cytokine expression by human dendritic cells in response to different *Porphyromonas gingivalis* capsular serotypes. *Journal of Clinical Periodontology, 36*(10), 823–829. http://dx.doi.org/10.1111/j.1600-051X.2009.01462.x.

Voo, K. S., Wang, Y. H., Santori, F. R., Boggiano, C., Arima, K., Bover, L., et al. (2009). Identification of IL-17-producing FOXP3+ regulatory T cells in humans. *Proceedings of the National Academy of Sciences of the USA, 106*(12), 4793–4798. http://dx.doi.org/10.1073/pnas.0900408106.

Wang, M., Krauss, J. L., Domon, H., Hosur, K. B., Liang, S., Magotti, P., et al. (2010). Microbial hijacking of complement-toll-like receptor crosstalk. *Science Signaling, 3*(109), ra11. http://dx.doi.org/10.1126/scisignal.2000697.

Wang, P.-L., & Ohura, K. (2002). *Porphyromonas gingivalis* lipopolysaccharide signaling in gingival fibroblasts- CD14 and Toll-like receptors. *Critical Reviews in Oral Biology & Medicine, 13*, 132–142.

Wang, M., Shakhatreh, M.-A. K., James, D., Liang, S., Nishiyama, S.-i., Yoshimura, F., et al. (2007). Fimbrial proteins of *Porphyromonas gingivalis* mediate in vivo virulence and exploit TLR2 and complement receptor 3 to persist in macrophages. *The Journal of Immunology, 179*(4), 2349–2358.

Wang, L., Wang, J., Jin, Y., Gao, H., & Lin, X. (2014). Oral administration of all-trans retinoic acid suppresses experimental periodontitis by modulating the Th17/Treg imbalance. *Journal of Periodontology, 85*(5), 740–750. http://dx.doi.org/10.1902/jop.2013.130132.

Wassenaar, A., Reinhardus, C., Thepen, T., Abraham-Inpijn, L., & Kievits, F. (1995). Cloning, characterization, and antigen specificity of T-lymphocyte subsets extracted from gingival tissue of chronic adult periodontitis patients. *Infection and Immunity, 63*(6), 2147–2153.

Watanabe, K., Onoe, T., Ozeki, M., Shimizu, Y., Sakayori, T., Nakamura, H., et al. (1996). Sequence and product analyses of the four genes downstream from the fimbrilin gene (fimA) of the oral anaerobe *Porphyromonas gingivalis*. *Microbiology and Immunology, 40*(10), 725–734.

Whittaker, C. J., Klier, C. M., & Kolenbrander, P. E. (1996). Mechanisms of adhesion by oral bacteria. *Annual Review of Microbiology, 50*, 513–552. http://dx.doi.org/10.1146/annurev.micro.50.1.513.

Wilensky, A., Tzach-Nahman, R., Potempa, J., Shapira, L., & Nussbaum, G. (2015). *Porphyromonas gingivalis* gingipains selectively reduce CD14 expression, leading to macrophage hyporesponsiveness to bacterial infection. *Journal of Innate Immunity, 7*(2), 127–135. http://dx.doi.org/10.1159/000365970.

Wingrove, J. A., DiScipio, R. G., Chen, Z., Potempa, J., Travis, J., & Hugli, T. E. (1992). Activation of complement components C3 and C5 by a cysteine proteinase (gingipain-1) from *Porphyromonas (Bacteroides) gingivalis*. *Journal of Biological Chemistry, 267*(26), 18902–18907.

Yamazaki, K., Nakajima, T., Kubota, Y., Gemmell, E., Seymour, G. J., & Hara, K. (1997). Cytokine messenger RNA expression in chronic inflammatory periodontal disease. *Oral Microbiology and Immunology, 12*(5), 281–287. http://dx.doi.org/10.1111/j.1399-302x.1997.tb00392.x.

Yang, B. H., Hagemann, S., Mamareli, P., Lauer, U., Hoffmann, U., Beckstette, M., et al. (2016). Foxp3($^+$) T cells expressing RORgammat represent a stable regulatory T-cell effector lineage with enhanced suppressive capacity during intestinal inflammation. *Mucosal Immunology, 9*(2), 444–457. http://dx.doi.org/10.1038/mi.2015.74.

Yang, J., Wu, J., Liu, Y., Huang, J., Lu, Z., Xie, L., et al. (2014). *Porphyromonas gingivalis* infection reduces regulatory T cells in infected atherosclerosis patients. *PLoS One, 9*(1), e86599. http://dx.doi.org/10.1371/journal.pone.0086599.

Yu, J. J., Ruddy, M. J., Conti, H. R., Boonanantanasarn, K., & Gaffen, S. L. (2008). The interleukin-17 receptor plays a gender-dependent role in host protection against *Porphyromonas gingivalis*-induced periodontal bone loss. *Infection and Immunity, 76*(9), 4206–4213. http://dx.doi.org/10.1128/iai.01209-07.

Yu, J. J., Ruddy, M. J., Wong, G. C., Sfintescu, C., Baker, P. J., Smith, J. B., et al. (2007). An essential role for IL-17 in preventing pathogen-initiated bone destruction: Recruitment of neutrophils to inflamed bone requires IL-17 receptor-dependent signals. *Blood, 109*(9), 3794–3802.

Zenobia, C., Hasturk, H., Nguyen, D., Van Dyke, T. E., Kantarci, A., & Darveau, R. P. (2014). *Porphyromonas gingivalis* lipid A phosphatase activity is critical for colonization and increasing the commensal load in the rabbit ligature model. *Infection and Immunity, 82*(2), 650–659. http://dx.doi.org/10.1128/IAI.01136-13.

Zhang, Q., Chen, B., Yan, F., Guo, J., Zhu, X., Ma, S., et al. (2014). Interleukin-10 inhibits bone resorption: A potential therapeutic strategy in periodontitis and other bone loss diseases. *BioMed Research International, 2014.* http://dx.doi.org/10.1155/2014/284836, 284836.

Zhang, X., Kimura, Y., Fang, C., Zhou, L., Sfyroera, G., Lambris, J. D., et al. (2007). Regulation of Toll-like receptor-mediated inflammatory response by complement in vivo. *Blood, 110*(1), 228–236.

Zhang, X., & Teng, Y. T. (2006). Interleukin-10 inhibits gram-negative-microbe-specific human receptor activator of NF-kappaB ligand-positive $CD4^{+}$-Th1-cell-associated alveolar bone loss in vivo. *Infection and Immunity, 74*(8), 4927–4931. http://dx.doi.org/10.1128/IAI.00491-06.

Zheng, C., Wu, J., & Xie, H. (2011). Differential expression and adherence of *Porphyromonas gingivalis* FimA genotypes. *Molecular Oral Microbiology, 26*(6), 388–395. http://dx.doi.org/10.1111/j.2041-1014.2011.00626.x.

CHAPTER THREE

Atherosclerosis: orchestrating cells and biomolecules involved in its activation and inhibition

Ashok Munjal and Rekha Khandia*
Department of Genetics, Barkatullah University, Bhopal, MP, India
*Corresponding author: E-mail: rekha.morchhale@gmail.com

Contents

Advances in Protein Chemistry and Structural Biology, Volume 120
ISSN 1876-1623
https://doi.org/10.1016/bs.apcsb.2019.11.002

Abstract

The term atherosclerosis refers to the condition of deposition of lipids and other substances in and on the artery walls, called as plaque that restricts the normal blood flow. The plaque may be stable or unstable in nature. Unstable plaque can burst and trigger clot formation adding further adversities. The process of plaque formation involves various stages including fatty streak, intermediate or fibro-fatty lesion and advanced lesion. The cells participating in the formation of atherosclerotic plaque include endothelial cells, vascular smooth muscle cells (VSMC), monocytes, monocytes derived macrophages, macrophages and dendritic cells and regulatory T cells (T_{REG}). The role of a variety of cytokines and chemokines have been studied which either help in progression of atherosclerotic plaque or *vice versa*. The cytokines involved in atherosclerotic plaque formation include IL-1, IL-2, IL-3, IL-4, IL-5, IL-6, IL-9, IL-10, IL-12, IL-13, IL-15, IL-17, IL-18, IL-20, IL-25, IL-27, IL-33, IL-37, TNF-α, TGF-β and IFN-γ; whereas amongst the chemokines (family of small cytokines) are CCL2, CCL3, CXCL4, CCL5, CXCL1, CX3CL1, CCL17, CXCL8, CXCL10, CCL20, CCL19 and CCL21 and macrophage migration-inhibitory factor. These are involved in the atherosclerosis advancements, whereas the chemokine CXCL12 is play atheroprotective roles. Apart this, contradictory functions have been documented for few other chemokines such as CXCL16. Since the cytokines and chemokines are amongst the key molecules involved in orchestrating the atherosclerosis advancements, targeting them might be an effective strategy to encumber the atherosclerotic progression. Blockage of cytokines and chemokines via the means of broad-spectrum inhibitors, neutralizing antibodies, usage of decoy receptors or RNA interference have been proved to be useful intervention against atherosclerosis.

1. Introduction

Atherosclerosis refers to a disease condition of the deposition of fatty or fibrous material; which is one among the leading cause of vascular disorders. Accumulated fibrous and fatty mass entangle circulating cells such as macrophages and monocytes in the sub-endothelial intimal layer of the arteries. The occlusion of vessels with such plaques results in the obstruction of blood flow and leads to hypoxia, which complicates into myocardial infarction and stroke (Wolf & Ley, 2019). These developed plaques may be of stable or unstable types. Unstable plaques rupture and further occlude the vessels and are more dangerous in comparison to the stable ones. Unstable atherosclerotic lesions are characterized by intra-lesion angiogenesis and inflammatory cell recruitment. A panel of cytokines and chemokines are released by different cells; those have various roles in atherosclerosis development including both the pro-atherogenic as well as anti-atherogenic roles. Few have promiscuous roles too and collectively all orchestrate the atherosclerosis fate.

2. Atherosclerosis stages

The atherosclerotic plaque brings changes in the wall of blood vessels resulting in the myocardial infarction, cerebral infarction or stroke and gangrene in the peripheral vasculature. The process of atherosclerosis may be subdivided into following four categories.

2.1 Fatty streak

This stage consists of accumulation of lipid-filled, monocyte-derived macrophages alternatively called the foam cells, which is the hallmark of atheromata. Along with the phagocytes, smooth muscle cells and endothelial cells are also laden with lipids. An injury to the endothelium, promotes adhesion of monocytes to the endothelium and loosening of cell junctions. Adhered monocytes later migrate beneath the endothelium and differentiate into the macrophages. The endothelium layer is also permeable to low density lipoproteins (LDL), which are engulfed by macrophages. Accumulated LDL engulfed macrophages are termed as fatty streak. Such lipid filled cells occupies 2–6 layers of intima of the artery, and gives a yellow appearance. These lesions may appear in aorta and coronary arteries of most human population aging 20 years and progress into atherosclerotic plaques or may regress (Ross, 1995).

2.2 The intermediate or fibrofatty lesions

T lymphocytes and macrophages are present between the layers of muscles and surrounded by poorly developed connective tissue encompassing collagen fibrils, elastic fibers and proteoglycans.

2.3 Advanced lesions

During the advanced stage, an interaction between circulating blood monocytes and T lymphocytes takes place and upon entering into the arterial wall, these cells secrete cytokines and inflammatory molecules resulting in the proliferation of smooth muscles and increase in thickness of connective tissue inside the artery and that is eventually responsible for atherosclerosis and clinical sequelae.

2.4 The fibrous plaque

It is the advanced lesion of atherosclerosis; fibrous plaque grows and is covered with the dense cap of connective tissue formed by collagen with some elastic fibers. When the lesion is thick, small microvascular tunnels called vasa vasora may be present to provide the nutrient and oxygen supply to the thickened lesion. These advanced lesions may be stable or unstable.

2.4.1 Stable plaque

The plaque gradually develops due to the accumulation of lipids in foam cells and proliferation of smooth muscles. When the fibrous cap is thick and even, it become stable and do not rupture.

2.4.2 Unstable plaque

These plaques grow rapidly by a fast lipid deposition. These plaques possess thin fibrous cap and edges as well as rich in T helper 1 (T_H1) cells, macrophages and few natural killer (NK) cells with a core of lipid and necrotic material. Rheological forces generated by flow of blood may lead to rupture the thin shoulders of lesion and hemorrhage. The hemorrhage may be from lumen of vessel itself or from vasa vasora. The ruptured plaques trigger the cascade of thrombosis that further may reduce/occlude the diameter of vessel lumen and produce serious clinical pathology.

3. Various cells involved in atherosclerosis

Various cells involved in the atherosclerosis are endothelial cells, vascular smooth muscle cells, dendritic cells, macrophages and monocytes; which participate in the atheroma formation.

3.1 The endothelium

The major functions of immune cells are to maintain the tissue homeostasis by eradicating the damaged cells. Endothelial cells not only line the wall of blood vessels, but also have functional roles in the progression of atherosclerosis. As such endothelial cells remain quiescent throughout adult life, but under various physiological and pathological stimuli, these may get activated. An abrupt flow of blood is one amongst the important factors that initiate the endothelial cell activation. Under high unidirectional laminar shear stress, the gene expression profile switches toward the expression of atheroprotective genes including the upregulation of transcription factor, Krüppel-like factor 2 (KLF2). The KLF2 is responsible for maintaining the integrity of vascular barrier and confer the endothelial cell quiescence resulting in anti-inflammatory, anti-thrombotic phenotype (Theodorou & Boon, 2018). The atherosclerotic areas develop in regions where shear stress (mechanical drag) is low and are exemplified with excessive EC proliferation and inflammation by the transcription factor, hypoxia-inducible factor 1α (HIF1α) in a hypoxic manner despite being in oxygenated microenvironment (Feng et al., 2017). Under the influence of proinflammatory cytokines, glucose uptake and glycolysis is increased.

3.1.1 *Intra-lesion angiogenesis*

Neoangiogenesis is the characteristics of unstable lesions. The histopathological examination of rapidly growing plaques exhibits increased number of microvessels in case of intra-plaque hemorrhage. Intraplaque hypoxia and hemoglobin:haptoglobin complexes tend to induce HIF1α signaling via inhibition of prolyl hydroxylases and higher quantities of vascular endothelial growth factor-A (VEGF-A) by CD163^{+} macrophages, which culminates into the enhanced vascular permeability, intra-plaque neo-angiogenesis and inflammatory cell recruitment (Guo et al., 2018).

3.1.2 *Endothelial cell metabolism*

Glucose uptake and glycolysis is increased in endothelial cells through proinflammatory cytokine induced NF-κB activation presumably through lactate (Végran, Boidot, Michiels, Sonveaux, & Feron, 2011). In endothelial cells glutaminolysis is also enhanced through pro-inflammatory YAP/TAZ pathway. The statins that lowered the level of cholesterol in blood are shown to increase KLF2 expression and eventually plays atheroprotective roles. The KLF2 expression reduces the NF-κB, HIF1α, and YAP/TAZ proinflammatory pathways, suggesting that statins may reprogram the endothelial cell metabolism to exhibit anti-atherogenic effects (Theodorou & Boon, 2018).

3.1.3 *Secretion of chemokines by endothelium*

Quiescent endothelial cells do not bind to the leukocytes; in addition, these inhibit the protein like P-selectin, E-selectin, vascular cell-adhesion molecule 1 (VCAM1) and intercellular adhesion molecule 1 (ICAM1) those are responsible for leukocyte interaction with endothelial cells. The quiescent state is also maintained by the shear stress caused by flowing blood that signals for the basis production of nitric oxide (NO). The NO suppresses expression of various inflammatory genes (De Caterina et al., 1995). Various cell types are prominent in an atherosclerotic plaque like macrophage, activated platelets, T_H1 cells, few NK cells, endothelial cells, and smooth muscle cells which are responsible for secreting inflammatory mediators. The tumor necrosis factor (TNF)-α and interleukin (IL)-1 are secreted by activated leukocytes. The binding of IL-1 to the type 1 IL-1 receptor (IL-1R1) present on the endothelial cells, induce a signaling cascade encompassing myeloid differentiation primary response gene 88 (MyD88), Toll/IL 1 receptor accessory protein (TIRAP; also known as MAL), IL 1R-associated kinase 1 (IRAK1), IRAK4, and TRAF6. The pro-inflammatory cytokines such as TNF-α and IL-1 induce the expression of IL-6, which is secreted by the cells of vascular walls. IL-6 may release in blood and may travel to liver. There it may reprogram the expression of genes to produce C-reactive protein, serum amyloid A, fibrinogen, and plasminogen activator inhibitor-1. In such inflammatory milieu, thrombosis is promoted (Libby, Ridker, & Hansson, 2009). TNF and IL-1 induce the reorganization of actin and tubulin cytoskeletons of the endothelial cells and vascular endothelial cadherin-based compromised cell-cell junction (Petrache, Birukova, Ramirez, Garcia, & Verin, 2003). Simultaneously, the leakage of plasma proteins is also promoted. Inflamed endothelial cells are involved in the synthesis of chemokine CXCL10, which binds to its receptor, CXC-chemokine receptor 3 (CXCR3) and mediate the expression of various adhesion molecules including E-selectin to which leukocytes are adhered. Atherogenesis linked enzymes such as 5-lipoxygenase and 12/15-lipoxygenase (5-LO, 12/15-LO) are also induced in macrophages during the development process of atherosclerosis.

The inflammatory cells mediated damage to endothelial cells also contributes to the inflammation. TNF and IL-1 together with interferon-γ (IFN-γ) may mediate endothelial cell death. IFN-γ induced cell death may be mediated through two potential mechanisms. In the first mechanism, the IFN-γ increased the expression of procaspase 8 in endothelial cells and TRADD (TNFR1-associated death domain protein) activation recruits FAS-associated death domain protein (FADD). In the other mechanism, IFN-γ induce the mobilization of cathepsin B from lysosome to cytosol. TNF signaling may activate cathepsin and trigger cytochrome dependent death pathways (Pober & Sessa, 2007).

3.2 Vascular smooth muscle cells

Clinical diagnosis of a rupture prone plaque identified a fibrous cap of smooth muscle α-actin that is supposed to be derived from vascular smooth muscle cells (VSMC), CD68 positive macrophages and necrotic core composed of lipid filled macrophages. Under normal physiological circumstances VSMC produce a range smooth muscle cell (SMC) markers including smooth muscle cell myosin heavy chain 11 (MYH11), SMC lineage-restricted protein (SM22α/tagln), ACTA2, smoothelin and few others. Under atherosclerotic conditions, VSMC reduce expressing these markers, and gains ability to proliferate, migrate and secrete various extracellular matrix (ECM) proteins and cytokines (Alexander & Owens, 2012). Phenotypic switching of VSMC toward acquiring macrophage markers and properties helps in generating a pro-atherogenic VSMC phenotype. The phenotypic switching toward the macrophage like phenotype is accompanied by lipid accumulation. In the lipid laden cultured VSMC, various proinflammatory genes are activated; and activated macrophage marker genes are upregulated. The phenotypic switching requires a reduction in integrin β3 (Itgβ3) to upregulate toll-like receptor 4 (TLR4), CD36, CD68 and Mac-2 leading to a macrophage like phenotype (Misra et al., 2018). The VSMC acquire phagocytic phenotype alike macrophages and such macrophage like VSMC may be having role in advancement of atherosclerosis due to reduced ability to clear lipid, dying cells and necrotic masses (Shankman et al., 2015). Such VSMC-derived macrophage-like cells have a poor phagocytic capacity to engulf apoptotic cells and hence consequence in the formation of necrotic core of the lesion. In the rupture prone atherosclerotic plaque, the stability depends upon the thickness of the cap and degree of cap inflammation. The death of VSMC leads to an increase in the cap thinning and breakdown of collagen and ECM, which eventually may lead to myocardial infarction or stroke. However, many times, the consequences of atherosclerotic plaque rupture are subclinical, because it is repaired by VSMC. In fact, vulnerable plaques are ruptured and repaired several times that further decreases the lumen of vessel (Bennett, Sinha, & Owens, 2016). Increased proliferation of VSMC has been observed in during early atherogenesis and on vascular injury.

Macrophages and VSMC undergo apoptosis during atherosclerotic plaque formation. Initially the rate of apoptosis is low, later as the lesion develops, the frequency of apoptosis is increased in both the necrotic core and fibrous cap. The unstable plaques are ruptured most commonly in the area of shoulder, where a greater number of macrophages and a smaller number of VSMC are present. It is suggestive of macrophage induced VSMC apoptosis. The VSMC apoptosis in atherosclerosis has association with inflammation. IL-1 is released by dying VSMC; IL-1β by apoptosis, whilst IL-1α is released by necrosis (Clarke, Talib, Figg, & Bennett, 2010).

3.3 Monocytes

Monocytes are recruited to atherosclerotic plaque throughout the different stages of the lesion progression. Monocyte infiltration in the plaque is increased when the diet is fat and cholesterol rich as shown in mice model

(Potteaux et al., 2011). Often the number of monocytes present in the blood determine the number of cells to be infiltrated into the plaque. Hence increase in the number of monocytes may be considered as an independent risk factor for atherosclerosis (Potteaux et al., 2011). Further, in the work of Potteaux et al. (2011), it has been shown that the blockage of influx of monocytes in plaque might inhibit the macrophage content in the plaque. Under atherosclerotic conditions, monocytes are recruited into the vessels' wall that is mediated by chemokines and monocyte chemoattractant protein 1 (MCP-1). Amongst mice, two distinguished subsets of monocytes have been observed; these are classical $CCR2^{hi}CX3CR1^{lo}Ly6C^{hi}$ monocytes and non-classical $CCR2^{lo}CX3CR1^{hi}Ly6C^{lo}$ monocytes. $Ly6C^{hi}$ subset is the progenitor of $Ly6C^{lo}$ monocytes. The $Ly6C^{hi}$ monocytes are comparatively more inflammatory and infiltrate into the arterial wall. In a study on mice, deficiency of Nur77, an orphan nuclear receptor, which is associated with the survival and differentiation of $Ly6C^{lo}$ monocytes, has been found linked with an enhanced atherosclerosis progression (Hanna et al., 2011).

3.4 Monocyte-derived macrophages

It is known that monocytosis is associated with both the accelerated atherosclerotic lesion progression and impairment in the lesion regression. A large proportion of macrophages are originated from the circulating monocytes infiltrated in plaque. Endothelial cells activation leads to the adherence of monocytes to the vessel wall and transmigration to arterial wall to mature into macrophage. Further, inhibition of monocyte migration to plaque site results in impaired plaque development (Flynn, Pernes, Lee, Nagareddy, & Murphy, 2019). The expression of acyl-CoA:cholesterol acyltransferase, which is present only in intestinal enterocytes, hepatocytes, and macrophages, is increased in monocytes after differentiation into macrophages (Sakashita et al., 2003). The atherosclerotic plaque not only contains the foam cell macrophages but also a good number of activated $CD4^+$ T helper cells and activated DC. Monocytes are recruited into atherosclerotic foci and where the monocyte-derived cells engulf modified proteins and secretes proinflammatory molecules. Lipid uptake and scavenger receptors-to be taken from.

3.5 Dendritic cells (DC) in atherosclerosis

Dendritic cells (DC) are the member of a family of professional antigen-presenting cells which serve as an interface between innate and adaptive immune response and capability to induce both the naïve as well as memory

T cells. The DC in vessels may be identified by immunohistochemical staining for S100, langerin, CD1a and fascin. DC markers BDCA-1, BDCA-2 CD11c, and CD123 are found in atherosclerotic plaques (Van Vré, Ilse Van Brussel, Bosmans, Vrints, & Bult, 2011). In the atherosclerosis-prone regions of blood vessels like the less curved area which face higher shear stress by blood flow, DC are present in higher numbers. DC in the normal aorta are normally present in intima and present in two distinct subsets $CD11c^{+}CD11b^{-}$ $CD103^{+}$ and $CD11c^{+}CD11b^{+}CD103^{-}$ DC. In atherosclerosis, sub-endothelial retention of modified apolipoprotein B-containing lipoproteins triggers an inflammatory response that eventually causes infiltration of monocytes into intima region to differentiate into macrophage or DC. These cells have capacity to engulf modified lipoproteins and lipid-laden apoptotic cells and convert into "foam cells". The mechanism of uptake of lipoproteins is possibly through macropinocytosis, receptor-mediated endocytosis, scavenger receptor-mediated endocytosis and direct intake through dendritic processes. In a cholesterol-rich diet fed mice model, foam cells were found positive for DC markers CD11c, 33D1, and major histocompatibility complex class II. Electron microscopy revealed intracellular presence of the lipid content. Selective depletion of DC by injecting diphtheria toxin, reduced the accumulation of lipids in early plaques in comparison to control animals, suggestive of imperative role of DC in lipid assimilation in plaque (Paulson et al., 2010). Matured and CCR7 expressing DC migrates into the draining lymph nodes and present antigen to naïve T cells. In context of atherosclerosis, MyD88 signaling in DC is essential to both the maturation of DC and protective regulatory T cells (T_{REG}) response (Subramanian, Thorp, Hansson, & Tabas, 2013). T_{REG} produce immunosuppressive cytokines transforming growth factor (TGF)-β, IL-10 and IL-35 (Pastrana et al., 2012).

3.5.1 Various cytokines produced by DC in maintaining chronic inflammatory conditions

In addition to antigen presentation DC may produce cytokines by engaging TLRs. Pro-atherogenic TNF, IL-6 and IL-12 cytokines are produced by DC. In a double knock-out mice model with $ApoE^{-/-}$-IL-$6^{-}/^{-}$ phenotype the atherosclerotic plaque formation was faster than $ApoE^{-}/^{-}$, wild-type (WT) or IL-$6^{-}/^{-}$ mice, suggested that a lifetime IL-6 deficiency enhances the atherosclerotic plaque formation in ApoE-deficient mice as a model for human atherosclerosis (Schieffer et al., 2004). To elucidate the effects of IL-12 on process of atherosclerosis, when mice were injected daily with recombinant murine IL-12, C57BL/6 mice didn't develop the lesions, indicating that IL-12 is alone is not solely responsible for lesion development. In ApoE deficient mice, however IL-12 treatment produced >100% larger plaque than control (Lee, Yen, Pan, & Chau, 1999). Mice deficient in both ApoE and TNF-α (ApoE$^{-/-}$/TNF-α $^{-/-}$) has high serum cholesterol and lesser atherosclerotic plaque area in the aortic luminal surface indicating TNF-α has atherogenic role. This pro-atherogenic role is due to enhancement in expressions of ICAM-1, VCAM-1 and MCP-1 in the vessel wall (Ohta et al., 2005).

3.5.2 Chemokine production by DC

DC via the secretion of chemokines recruit monocytes and lymphocytes into the atherosclerotic lesion area. In the shoulder of unstable atherosclerotic plaques, CD83$^+$ DC are present; those secrete CC-motif chemokines (CCL)19 and CCL21and are able to recruit naïve T lymphocytes at the site of lesion. Mature DC attracts immature DC at the site of plaque via secretion of CCL2 and CCL14 and recruited immature DC are the differentiated into mature antigen-presenting DC under local inflammatory microenvironment. Immature DC produces CCL18 which stimulates the tolerogenicity (Chistiakov, Sobenin, Orekhov, & Bobryshev, 2014).

3.5.3 DC and T_{REG}

T_{REG} are CD4$^+$ T cells which reduce inflammation by inhibiting activation of endothelial cells and macrophages, and prevent the recruitment of monocytes inside the plaque region via secreting anti-inflammatory cytokines TGF-β and IL-10 (Subramanian & Tabas, 2014). Pro-atherogenic T_H1 cells response is also suppressed. During early atherosclerosis phase, DC mediated T_{REG} activation occurs after its migration to draining lymph node, where levels of IFN-γ and IL-12 are low, a non-inflammatory microenvironment might help in shaping atheroprotective T_{REG}. Contrarily under the advanced lesion conditions, antigen presentation by DC is reduced under inflammatory conditions and hence T effector cells are generated. The CD103$^+$ DC in lesions are tolerogenic and enhance the number of T_{REG}. The tolerogenic function is dependent upon the interaction of hematopoietic growth factor Flt3L that is important for development of myeloid and plasmacytoid DC, with Flt3 and the administration of is Flt3L is known to induce proliferation of T_{REG} (Klein et al., 2013).

T_{REG} could inhibit DC proliferation by inhibiting cytokines IL-10 and TGF-β, cell surface molecule like CTLA-4, programmed death-1 (PD-1), and its ligands PD-L1/2 and lymphocyte activation gene-3 (LAG-3). T_{REG} suppress effector T cells in atherosclerosis through cytokines IL-10 and TGF-β. In amice model production of IL-10 decreases the inflammation and plaque size in $ApoE^{-/-}$ mice. An inefficient TGF-β signaling may speed up the atherosclerosis progression and produce unstable plaque phenotype in high cholesterol fed mice model (Mallat, Gojova, et al., 2001). Cytokine IL-35 mediate T_{REG} mediated T effector cell suppression. Epstein-Barr virus-induced gene 3 (Ebi3), belongs to the interleukin (IL)-12 heterodimeric cytokine family and it forms a heterodimer either with IL-27α or IL-12α to form IL-27 or IL-35 respectively (Kempe et al., 2009). IL-35 is expressed at minimal level and expressed under inflammatory conditions only. In $ApoE^{-/-}$ mice with establish atherosclerotic lesion, level of IL-35 was found to be lesser in comparison to age matched wild type C57BL/6 mice without atherosclerotic lesions (Wang et al., 2014).

4. Various cytokines involved in arthrosclerosis

Many pro-atherogenic as well as anti-atherogenic cytokines with various roles are active while atherogenesis. On one hand pro-atherogenic cytokines leads to the atherosclerosis progression, on the other hand anti-atherosclerotic cytokines have the therapeutic potential.

4.1 IL-1

The IL-1 cytokines are present in two isoforms, IL-1α and IL-1β. The vascular wall cells, VSMC, and endothelial cells produce both the isoforms of this pro-inflammatory cytokine. The role of IL-1α or IL-1β has been demonstrated to be stage dependent in mice model (Vromman et al., 2019). Blood monocytes are shifted toward less inflammatory state by IL-1β blockade that results in the reduction in the size of atheroma and early neutralization of IL-1α reduces the plaque's SMC content. Collectively, IL-1α has role in arterial remodeling during the early stages of experimental atherogenesis, whilst IL-1β is responsible for inflammation during the evolution of advanced atheroma (Vromman et al., 2019). Other members of the IL-1 cytokine family are IL-1 receptor antagonist (IL-1Ra), IL-18, IL-33 and others. IL-1Ra specifically inhibits the activity of both IL-1α and IL-1β, whilst IL-36Rα (formerly IL-1F5) and IL-1F7 nonspecifically inhibit inflammation and the innate immune response. Both the IL-18 and its receptors are up-regulated in the atherosclerotic plaques. A positive correlation has been found between the expression level of IL-18 and plaque instability. The study of Mallat, Corbaz, et al. (2001) is suggestive of imperative role of IL-18 in atherosclerotic plaque destabilization and eventually development of acute ischemic syndromes. In ApoE$^{-/-}$ mice, NaCl co-transporter (NCC; 12-transmembrane-domain ion transporter protein), was found to act as a receptor for IL-18 and absence of both the IL-18 receptor as well as NCC, protected mice from atherosclerosis. However, absence of one of the two does not exhibit the same result (Wang et al., 2015). It is pleiotropic proinflammatory cytokine and indicative of visceral fat deposition and cardiovascular death in patients of atherosclerotic cardiovascular disease (Yamaoka-Tojo et al., 2011). It is a pro-inflammatory cytokine that accelerates the atherosclerosis via expressing IFN-γ and CXCL16 in ApoE$^{-/-}$ mice model (Bhat & Dhavan, 2015).

IL-1F7 is also called as IL-37 and is produced by various cells such as monocytes, tissue macrophages, synovial cells, tonsillar B cells, plasma cells, dendritic cells (DC), epithelial cells and memory T cells. It differentiates the T helper (T_H) cells into the anti-inflammatory phenotype and improves plaque stability by decreasing matrix metalloproteinase (MMP)-dependent degradation of collagen and suppressing SMC apoptosis and hence possess the atheroprotective roles (Liu et al., 2018). It has shown to inhibit IL-18-induced inflammation through competitive binding with the IL-18 binding protein (IL-18bp) and the IL-18Rα chain. A complex of IL-37, IL-18Rα and the orphan receptor IL-1R8 is essential to inhibit inflammatory signaling. The complex reduces the phosphorylation of TAK1, Fyn, and the NF-κB pathways mediators IκBε, p65 and p105 and increased phosphorylation of the anti-inflammatory mediators Stat3, Mer, PTEN and p62 (Nold-Petry et al., 2015). It prevents the foam cell formation and subsequent necrosis in the atherosclerotic plaque (McCurdy, Liu, Yap, & Boisvert, 2017).

IL-33 is involved in mast cell functions, and allergic and T_H2 responses (Dinarello, 2009). ApoE$^{-/-}$ mice fed on a high-fat diet have less severe and inflamed atherosclerotic plaques in the aortic sinus in comparison to PBS control mice. Contrary to this, when the mice were treated with soluble ST2, a decoy receptor capable of neutralizing IL-33, developed larger sized atherosclerotic plaques (Miller et al., 2008). In ApoE$^{-/-}$ mice model IL-33 treatment reduced the formation and accumulation of macrophage foam cells. Such changes are mediated through the expression of genes involved in oxidized low-density lipoprotein (oxLDL) uptake including CD36 (McLaren et al., 2010). NF-κB, PI3Kγ and MAPK pathways are involved in IL-33 mediated inhibition of two imperative pro-atherogenic genes, MCP-1 and ICAM-1 (Buckley et al., 2019).

4.2 IL-2

$ApoE^{-/-}$ mice fed with high-lipid diet were when immunized intraperitoneally for 6 weeks, they developed antibodies against IL-2. Presence of antibodies enhanced the plaque size. Anti-IL-2 therapy had a significant anti atherosclerotic effect (Upadhya, Mooteri, Peckham, & Pai, 2004). IL-2/anti-IL-2 mAb administration in $ApoE^{-/-}$ mice increased T_{REG}, reduced inflammation and attenuated the development as well as progression of plaque (Dinh et al., 2012). IL-2 receptors are abundantly present on the surface of T_{REG} but sparsely on T effector cells therefore decreased IL-2 increases IFN producing T effector cells in lupus mouse model. Therefore, IL-2 may be used to balance the homeostasis between T_{REG} and T effector cells, which is disturbed during inflammatory disorders like atherosclerosis. Intra plaque administration of IL-2 results in the shrinkage of pre-established plaque size owing to the local expansion of T_{REG} (Dietrich et al., 2012).

4.3 IL-3

The CD4 subset of activated T lymphocytes which infiltrate into inflamed tissue, produces IL-3. It also activates endothelial cells to express adhesion molecules including E and P selectins; therefore, adhesion of T cells is promoted. Its pro-atherogenic activities are evidenced through the activation of VSMC migration and proliferation (Brizzi et al., 2001).

4.4 IL-4

IL-4 is released by the T_H2 cells and is considered as anti-inflammatory cytokine. However, there are ample evidences which suggest its pro-inflammatory and pro-atherogenic role. It helps in the secretion of inflammatory mediators from the endothelial cells. Bone marrow stem cell of IL-4 deficient mouse were when transplanted amongst LDL receptor deficient mice, it caused the reduction in the atherosclerotic lesion. The inflammation is mediated through the stimulation of reactive oxygen species generating pathways such as NADPH oxidase and 5-lipoxygenase (Lee, Kim, Lee, & Hirani, 2010). However, the works of King, Cassis, and Daugherty (2007) demonstrated that the IL-4 does not significantly influence the development of atherosclerotic lesions in $ApoE^{-/-}$ mice irrespective of the mode of atherosclerosis development. It induces the endothelial cell death by enhanced apoptosis and lead to endothelium cells dysfunction, a process linked with atherosclerosis development (Lee & Hirani, 2006).

4.5 IL-5

IL-5 is expressed by the T_H-2 lymphocytes and mast cells and can stimulate a specific subset of B lymphocytes to produce and secrete natural T15/EO6 IgM antibodies. These antibodies prevent the uptake of oxLDL by the macrophage scavenger receptor CD36 and thereby prevents foam cell formation. Targeted expression of IL-5 in Ldlr$^{-/-}$ mice macrophages, reduced the lesion size and area (Zhao et al., 2015). In a cohort of 3534 patients, serum IL-5 level was found to linked weakly with protection against atherosclerosis in early settings (Silveira et al., 2015). The protective efficacy against oxLDL is linked with the IL-5 and its deficiency may lead to atherogenesis (Binder et al., 2004). An inverse relationship between plasma IL-6 level and subclinical atherosclerosis was established by measuring carotid intima-media thickness (IMT) in a study group comprising of 1,011 participants and indicative of atheroprotective role of IL-5 (Sämpi et al., 2008).

4.6 IL-6

IL-6 is a unique cytokine that exhibits both pro- and anti-atherogenic activities. Pro-atherogenic action of IL-6 has been evidenced by its ability to attract neutrophils, monocytes and macrophages. Simultaneously, the endothelial cells are activated to induce chemokines and expression of adhesion molecules contributing to the development of atherosclerosis. Weekly administration of recombinant interleukin-6 (rIL-6) in ApoE$^{-/-}$ mice fed with high fat high-fat diet showed an increase in the lesion size in comparison to mice injected with saline buffer (Huber, Sakkinen, Conze, Hardin, & Tracy, 1999). On the other hand, atheroprotective effects have been demonstrated by experiment of Schieffer et al. (2004); in the experiment, ApoE$^{-/-}$-IL-6$^{-/-}$ double-knockout mice were generated and fed with normal chow. ApoE$^{-/-}$-IL-6$^{-/-}$ double-knockout mice exhibited higher serum cholesterol level and increased lesion formation in comparison to the controls. Atheroprotection also involves the upregulation of LDL receptor gene expression and thereby reducing plasma LDL level (Reiss, Siegart, & Leon, 2017).

4.7 IL-9

Various cells including T cells, natural killer T cells (NKTs) and mast cells synthesize IL-9 and T_H9 cells those secretes large amount of IL-9 have been found linked with autoimmune inflammatory disorders. Serum oxLDL concentrations are important contributor to the atherosclerosis initiation progression and plaque destabilization (Uchida et al., 2013); and have positive correlations with IL-9 levels (Li et al., 2017). Carotid plaques possess elevated mRNA levels of IL-9 and IL-9R in comparison to non-atherosclerotic vessels (Gregersen et al., 2013). IL-9 is able to induce VCAM-1 expression in aortic endothelial cells and increase the endothelium-monocyte adherence. In atherosclerotic mice IL-9 level is higher in plasma and aortic arches. In western style diet fed mice, exogenous rIL-9 administration in ApoE$^{-/-}$ mice significantly increased the plaque size accompanied with increased macrophage and T cell infiltration (Zhang et al., 2015).

4.8 IL-10

Macrophages present in atherosclerotic plaques are the major source of IL-10 production in advanced plaque. It inhibits the expression of metalloproteinase (MMP), pro-inflammatory cytokines and cyclooxygenase-2 in lipid engulfed activated macrophages. It also inhibits the monocytes to release pro-inflammatory cytokines including IL-1β, TNF-α, and IL-8 and promotes macrophage polarization toward the M2 phenotype (Han & Boisvert, 2015). In a study that comprised 33 chronic hemodialysis patients, exhibiting chronic renal failure; plaque positive hemodialysis patients, showed significant decrease in serum levels of anti-inflammatory cytokine IL-10 (Seyrek et al., 2005).

IL-19 is a member of an IL-10 sub-family and considered to be anti-inflammatory since it promotes T_H2 response. It is known to express in endothelial cells and VSMC of injured arteries. LDLR$^{-/-}$ mice fed on high fat diet, develop atherosclerotic lesions. Intra peritoneal injection of recombinant IL-19 revealed a marked reduction in the plaque area in comparison to the PBS treated control mice. At the same time, it inhibits the leukocyte-endothelial adhesion (Ellison et al., 2013). The IL-19 is unique in the context that it is the only reported cytokine with pro-angiogenic as well as anti-atherogenic activity (Jain, Gabunia, Kelemen, Panetti, & Autieri, 2011). Generally, the pro-angiogenic cytokines are pro-atherogenic too. oxLDL induced VCAM-1 expression is reduced by treatment with IL-19 in cultured endothelial cells and simultaneously it reduced the adhesion of monocytes to the oxLDL activated endothelial cells. Daily systemic administration of IL-19 inhibits plaque formation and at lower doses, it reduces the plaque size, which is suggestive of therapeutic potential of IL-19 (Ellison et al., 2013). The results of *in vitro* studies are indicative of STAT3,

Rac1, and MAPK p44/42 that leads to the endothelial cells proliferation spreading, and migration (England & Autieri, 2012).

4.9 IL-12

IL-12 is a heterodimeric (p70) cytokine, produced by numerous kinds of cells present in the atherosclerotic lesion including macrophages. It is composed of a 35 kDa light chain (p35) and a 40 kDa heavy-chain (p40). IL-12 p40$^{-/-}$/ApoE$^{-/-}$ double mutant mice exhibit less lesion in comparison to ApoE$^{-/-}$ mice in the aortic root. Oil-in-water emulsion intramuscular vaccination of IL-12, elicit anti-IL-12 antibodies resulted in 68.5% reduction in atherogenesis in LDLr$^{-/-}$ mice along with improving plaque stability, hence vaccination with IL-12 may be a therapeutic strategy for atherosclerosis (Hauer et al., 2005).

4.10 IL-13

IL-13 is produced by hematopoietic cells and nuocytes, those are involved in type 2 immune responses. It exhibits atheroprotective roles. IL-13 mainly binds to IL-13Rα1 and alternatively to IL-13Rα2, to induce the expression of TGF-β1 in macrophages through STAT6-independent pathway resulting in collagen deposition. Exogenous administration of IL-13 to cholesterol-fed LDLR$^{-/-}$ mice showed enhanced collagen formation and reduced macrophages in plaque. The expression of VCAM-1 molecules is reduced and subsequently VACM-1 dependent monocyte recruitment is also decreased. IL-13 activated macrophages have enhanced capabilities to uptake oxLDL and cholesterol efflux without increasing the number of total foam cells. LDLR$^{-/-}$ mice supplied with bone marrow from IL-13-deficient mice rapidly developed atherosclerosis (Cardilo-Reis et al., 2012).

4.11 IL-15

IL-15 is a proinflammatory cytokine that is expressed by immune cells including monocytes and macrophages. It plays an important role in the survival of T cells and hence in the memory cell formation. Apart activation of T cells, IL-15 has been involved in attracting monocytes and NK cells and enhances the T cell adhesion to endothelial cells. Production of anti-IL-15 antibodies through oral vaccination resulted in an increase in lesion size by 75% signifying the role of IL-15 in atherogenesis (van Es et al., 2011). Contrary to this, IL-15 inactivation did not lead to rapid onset of atherosclerosis in carotid arteries of ApoE knock out mice fed with high cholesteric mice (http://hdl.handle.net/11375/19142). Four IL-15 gene polymorphisms (rs4956403, rs3806798, rs1057972 and rs10833) have been tested in 397 individuals with subclinical atherosclerosis and it was observed that rs10833T allele is associated with the increased risk of atherosclerosis (Angeles-Martínez et al., 2017).

4.12 IL-17

IL-17 (also known as IL-17A) is produced by T_H17 $CD4^+$ T cells. The role of IL-17 is atherogenic or atheroprotective, is still controversial. Its pro-atherogenic roles are evidenced by induced production of pro-inflammatory factors such as IL-6, CXCL8, and CXCL10 by VSMC (Pietrowski et al., 2011). The cytokine is able to enhance the expression of VCAM-1 in aortic VSMC, an important event for immune cell adherence to endothelial cells (Zhang et al., 2013); while in the works of Madhur et al. (2011), IL-17 doesn't seem to contribute significantly in atherosclerosis progression.

4.13 IL-18

Atherosclerotic plaque over-expresses the IL-18 and its receptor (IL-18r) in macrophages, T cells, endothelial cells and VSMC. IL-18 is possessing the role in polarization of immune response toward T_H1 cells and release of pro-inflammatory cytokines, chemokines, and vascular adhesion molecules. Among $ApoE^{-/-}$ mice, deficiency of IL-18 or blockade of its signaling inhibits the process of atherosclerotic lesion development and stability. Intraperitoneal administration of IL-18 in $ApoE^{-/-}$ mice enhances atherosclerosis via release of gamma interferon and CXCL16 release (Tenger, Sundborger, Jawien, & Zhou, 2005). In $ApoE^{-/-}$ mice model, deletion of both the IL-18r and NCC, but not the deficiency of either of one, exhibit protection from atherosclerosis (Wang et al., 2015).

4.14 IL-20

IL-20 is a member of IL-10 family and is a pro-atherogenic cytokine. It is expressed in the monocytes and upon binding with the receptor complex (IL-20R1/IL-20R2 or IL-22R1/IL-20R2), it activates STAT3. Expression of IL-20 and its receptors are upregulated in atherosclerotic lesions of both in human and mice and promotes angiogenesis. Systemic delivery through intramuscular electroporation has shown to increase atherosclerotic plaque in $ApoE^{-/-}$ mice (Chen, Cheng, Jiang, Hsieh, & Chang, 2006).

4.15 IL-25

It is a member of the IL-17 cytokine family. Administration of IL-25 in early or late stages of athero-prone ApoE$^{-/-}$ mice to evaluate whether it has role in atherosclerosis initiation or progression; the effects revealed that IL-25 is able to inhibit both the initiation as well as progression of lesion. The effect of IL-25 on plaque size is not dependent upon the cholesterol levels since no difference in cholesterol level was observed in treated or control groups. Also, it had no effect on infiltration of macrophages or CD3^{+} T cells, however collagen content was seen increased in treated mice indicative of IL-25 role in plaque stabilization (Mantani et al., 2015). Complete IL-25 deficiency or its blockage has shown to aggravate the atherosclerosis in the aortic arch of ApoE$^{-/-}$ mice (Mantani et al., 2018).

4.16 IL-27

IL-27 is a member of the IL-12 family of cytokines. Its administration inhibits the uptake of modified LDL and release of proinflammatory cytokines in macrophages, but no effect on VCAM-1 expression by endothelial cells; which is suggestive of its function attributed through the macrophages but not endothelial cells. The deficiency of IL-27 or its receptor leads to accelerated atherosclerosis and the effect may be ameliorated by treatment with recombinant IL-27 (Hirase et al., 2013). Contrary to the atheroprotective role, Gregersen et al. (2017) outlined its pro-atherogenic roles. Increased plasma levels of IL-27 has been reported in 140 patients with carotid atherosclerosis and in the carotid lesion area, higher mRNA expression of IL-27 and IL-27 receptor as well as increased expression of IL-1β and NLRP3 were observed in 159 patients. Therefore, expression of both the IL-27 and inflammasome components is indicative of inflammatory roles of IL-27 in clinical settings of atherosclerosis (Gregersen et al., 2017).

4.17 IL-33

IL-33, which is a novel IL-1—like cytokine, is able to inhibit the atherosclerotic lesion development in high fat fed ApoE$^{-/-}$ mice. IL-33 is mainly released by the damaged cells. Exogenous administration of IL-33 for 6 weeks is able to reduce the plaque size and foam cell formation in ApoE$^{-/-}$ mice possibly through a polarization of the immune response toward the T_H2 phenotype and induction of antibodies against oxLDL (Aimo et al., 2018). It alters the expression of key genes in macrophages that are associated with total and esterified cholesterol content (McLaren et al.,

2010). Treatment of mice with soluble ST2, a decoy receptor that neutralizes IL-33, produced larger lesions in in the aortic sinus of the ApoE$^{-/-}$ mice. The protective role of IL-33 is through induction of IL-5 and oxLDL antibodies (Miller et al., 2008). Proteases like matrix metalloproteinases and members of ADAMTS (a disintegrin and metalloproteinase with thrombospondin motifs) family pose great effect on atherosclerotic plaque stability (Salter, Ashlin, Kwan, & Ramji, 2010). ADAMTS-1, -4 and -5 are expressed mainly in macrophages and to a lesser extent in the smooth muscle cells and endothelial cells of atherosclerotic lesion. IL-33 reduces the expression of ADAMTS-1, -4 and -5 (Ashlin et al., 2014). IL-33 promotes the expression of IL-10 in macrophages to control the expression of cholesterol transport-related genes contributing in positive effects against atherosclerosis (Zhang et al., 2017).

4.18 IL-37

It is a member of IL-1 family and possess the anti-inflammatory activities. IL-18Rα, and the orphan receptor IL-1R8 (also known as SIGIRR) are the receptors for the IL-37. A tripartite complex between IL-37, IL-18Rα and IL-1R8 is formed that leads to reduced phosphorylation of TAK1, Fyn, and the NF-κB pathways mediators and increased phosphorylation of inflammatory mediators Stat3, Mer, PTEN and p62. Ng to Il-18Rα and IL-1R8 is central to the activation of anti-inflammatory activities (Nold-Petry et al., 2015).

4.19 TNF-α

TNF-α is a pro-atherogenic cytokine secreted by macrophages, lymphocytes, NK cells, and VSMC. A double knockout (TNF-α$^{-/-}$, ApoE$^{-/-}$) mice exhibit reduced athero-progression, endothelial adhesion and inflammatory markers to control. Similar results were demonstrated by using soluble receptors acting as decoy for TNF-α (Tousoulis, Oikonomou, Economou, Crea, & Kaski, 2016). The VCAM-1 molecules are upregulated followed by TNF-α mediated interaction between circulating leukocytes and the endothelium. Simvastatin, a member of statin family, appears to reduce TNF-α-induced VCAM-1 expression and endothelial nitric oxide synthase. Another statin, Atorvastatin ameliorates the vascular nitric oxide bioavailability (Tousoulis et al., 2016).

4.20 TGF-β

It is produced by various inflammatory cells such as leukocytes, macrophages, smooth muscle cells, and platelets. The TGF-β1 exhibit anti-inflammatory properties by inhibiting TNF-α and IL-1β and interferon-γ that leads to the reduction in VCAM-1 expression, leukocyte adhesion and reduction in macrophage activity. The low TGF-β1 level is an indicator of risk factor for atherosclerosis in chronic kidney disease (CKD) and non-CKD patients (Hassan et al., 2018). Endothelial TGF-β has shown to drive the vascular inflammation. In hyperlipidemic mice, inhibition of endothelial TGF-β signaling, lead to the reduced inflammation and vascular permeability. It resulted in the plaque regression and inhibited the disease progression (Chen et al., 2019).

4.21 IFN-γ

It participates in the initiation and modulation of a various pro-atherogenic immune responses. It activates macrophages and thereby activate production of proinflammatory cytokines like TNF-α IL-6, Reactive oxygen species promoting LDL oxidation and metalloproteinases. IFN-γ enhances VCAM-1 expression and leukocyte recruitment to the vessel wall (Zhang et al., 2011). IFN-γ signaling mediates the up regulation of LDL scavenger receptors expression and therefore foam cell formation is enhanced. IFN-γ mediated release of matrix metalloproteases results in the weakened plaques prone to rupture and manifest clinical complications of atherosclerosis like myocardial infarction and stroke (Voloshyna et al., 2014).

5. Various chemokines involved in the atherosclerosis progression

Chemokines are the small cytokines; those are secreted by immune cells as signaling proteins to induce the migration of other cells.

5.1 CCL2

It is also termed as monocyte chemoattractant protein-1 (MCP-1); which is highly upregulated in human atherosclerotic lesions. Deletion of CCR2 inhibits the macrophage accumulation and hypercholesterolemic diet mediated plaque formation in ApoE$^{-/-}$ mice.

5.2 CCL3 (macrophage inflammatory protein-1α)

Atherosclerotic plaque may be progressed by the leukocyte-derived CCL3 mediated chemotaxis of neutrophils. Its contribution in atherosclerotic lesion formation and lipid metabolism has been demonstrated by an experiment into which low density lipoprotein receptor deficient ($LDLR^{-/-}$) mice were transplanted with bone marrow from $CCL3^{-/-}$ or C57BL/6 wild type donors. The mice were fed on western type diet and after 12 weeks of experiment, recipients of $CCL3^{-/-}$ marrow exhibited lower cholesterol and lesser atherosclerotic lesion size (Kennedy et al., 2012). The result is indicative of the role of CCL3 in the lipid metabolism. Low density lipoprotein $receptor^{-/-}$ ($LDLr^{-/-}$) and $CCL3^{-/-}$ mice fed with western type diet, exhibited attenuated adhesion and presence of neutrophils in plaques, whilst plaque macrophage, collagen, and VSMC content remained unchanged (de Jager et al., 2013).

5.3 CXCL4

The platelet chemokine CXCL4 was initially considered as a chemotactic factor for monocytes and neutrophils. The earlier events in the atherosclerotic plaque formation involves the endothelial cell death. CXCL4-activated monocytes are cytotoxic to the endothelial cells. The characteristics of CXCL4-induced macrophages are different from other subsets of macrophages demonstrated by transcriptome analysis and termed as M4 (Gleissner, Shaked, Little, & Ley, 2010). A combination of CD68, S100A8, and MMP7 has been found reliable markers for the identification of M4 macrophages in atherosclerotic lesions (Erbel et al., 2015). The imperative role of CXCL4 in atherosclerosis is demonstrated in CXCL4-deficient $PF4^{-/-}$ mice in $ApoE^{-/-}$ background, where significant reduction in aortic atherosclerosis was observed (Sachais et al., 2002). In the human atherosclerotic plaques, adventitia and the intima are the places where M4 macrophages, those polarization is irreversible, are principally present and could serve as a marker for inflammatory activities in the plaque (Domschke & Gleissner, 2017).

5.4 CCL5

The study of CCL5/RANTES and its receptors CCR1 and CCR5 in atherosclerosis has been taken up in various studies. In atherosclerotic plaques, the expression of CCL5 in monocyte/macrophages, lymphocytes and smooth muscle cells has been regulated by Rel proteins p50 and p65, the members of NF-κB transcription factor, (KLF13) and Y-box binding protein-1 (YB-1) respectively. The CCL5 is oligomerized and forms a bridge between the endothelial cells and circulating mononuclear cells (von Hundelshausen et al., 2001). A gene profiling studies of atherosclerotic plaques, revealed that during different stages of plaque development, chemokine profile is different and during initial stages CCL5 is expressed, while later on CX3CL1 is expressed possibly owing to the distinct receptor usage at various points of time by different monocyte subsets (Zernecke, Shagdarsuren, & Weber, 2008).

5.5 CXCL1

Blood pressure is having pulsatile nature, which causes cyclic mechanical stretch, posing pressure on blood vessels. This cyclic mechanical stretch has been shown to be responsible for VSMC death in rat aorta via JNK- and p38-dependent pathways. This mechanical stretch induces CXCL1 and CX3CL1 and these protect VSMC form death and prevent vascular remodeling of vessel protective roles. Acute use of CXCL1 and CX3CL might protect from aortic dissection during acute rise in blood pressure (Zhao et al., 2017). On the contrary, chronic appearance of these cytokines may promote vascular remodeling owing to the neutrophil chemoattractant activity and contribute to atherosclerosis (Zhou et al., 2011).

5.6 CX3CL1

It is a cytokine that is expressed both as a membrane anchored protein as well as in the soluble form. It is expressed by monocytes/macrophages, endothelial and smooth muscle cells within human atherosclerotic coronary arteries and expressed in all the stages of atherosclerotic plaque (Stolla et al., 2012). The higher shear stress is responsible for inflamed plaque. Though expression of CX3CL1 is found to confine within low shear stress region only, blocking of this cytokine reduce the plaque growth and the firm adherence of leukocytes with inflamed endothelial cells is reduced. CX3CL1 may act on endothelial cells to promote the expression of intercellular adhesion molecule-1 (ICAM-1) and exposure of P-selectin on adherent plants to initiate the leukocyte accumulation (Apostolakis & Spandidos, 2013). In a study, to generate long acting soluble form of CX3CL1, its chemokine domain was fused with Fc region of mice and this fusion product CX3CL1-Fc was given to Ldlr KO mice fed with atherogenic western diet. The administration resulted in decreased appearance of aortic lesion surface area (Riopel et al., 2019).

5.7 CCL17

The chemokine CCL17, is chiefly produced by DC and binds to the receptor CCR4 and facilitates the T cell-DC interactions. The CCL17 recruits the circulating T cells into the plaque region and in rat air pouch model, CCL17 mediated enhancement of recruitment of both the $CD4^{+}$ T cells and T_{REG} has been observed. It is also hypothesized that CCL17 interferes the Stat5 phosphorylation by inducing the MAPK pathway. The $ApoE^{-/-}$ mice treated with anti-CCL17 revealed significant reduction in atherosclerotic lesion formation and macrophage content; and therefore this may be targeted as therapeutic strategy against athrosclerosis (Weber et al., 2011).

5.8 CXCL8 (IL-8)

The CXCL8 chemokine was identified and described by various group of workers and named as neutrophil-activating factor, monocyte derived neutrophil-activating peptide and monocyte-derived neutrophil chemotactic factor. It is produced by multiple cell types and similarly, it acts on various types of cells. When monocytes come into contact with the oxidized low density lipoprotein, it enhances the CXCR2 expression on the surface of cells (Russo, Garcia, Teixeira, & Amaral, 2014). In an experimental model, inhibition of CXCL8 or its receptors show decreased atherosclerosis progression. It promotes the adhesion of circulating monocytes during the early phase of plaque formation. The macrophages, endothelial cells and VSMC present in lesion area produces high amounts of CXCL8 (Apostolakis, Amanatidou, & Spandidos, 2010).

5.9 CXCL10

It is constitutively expressed at lower level in stroma of thymus, spleen and lymph nodes, however the expression may be induced at higher level via interferons amongst endothelial cells, fibroblasts, keratinocytes, monocytes, T lymphocytes and macrophages. The signaling of CXCL10 occurs after its interaction with CXCR3, a G protein coupled receptor (GPCR); however, CXCR3 independent pathways are also present through receptor expressed by epithelial and endothelial cells (Soejima & Rollins, 2001). It modulates the local balance of the effector and regulatory arms of the immune system to develop atherosclerotic lesion (Heller et al., 2006).

5.10 CCL20

The cytokine CCL20 is produced by the cells related to the inflammation and autoimmune responses including endothelial cells, neutrophils, NK cells, T_H17 cells and B cells. The chemokine is having selectivity for its receptor CCR6. CCL20 is induced by LDL in VSMC in a dose- and time-dependent manner by kinase signaling and NF-κb pathways. Increased CCL20 level has been observed in hypercholesterolemic patients and moreover, it is significantly increased in atherosclerotic lesions. Circulating levels of CCL20 has been found positively correlated with levels of IL-6 (Calvayrac et al., 2011). It is a chemokine that is chemoattractant to the dendritic cells, effector/memory T lymphocytes, and naïve B cells. Chronic inflammatory condition that gives rise to ailments like atherosclerosis, induces the production of plasmin. Plasmin stimulates monocytes, macrophages, and DC to secrete inflammatory molecules. Production of CCL20 is elicited by plasmin at site of inflammation and plays an important role in the recruitment and activation of pathogenic $CCR6^+$ T cells; and subsequently augment the T_H17 cell migration (Li, Laumonnier, Syrovets, & Simmet, 2013).

5.11 CXCL19 and CCL21

Both these cytokines have a common receptor, CCR7. The level of both CCL19 and CCL21 have been found to be increased in the atherosclerotic lesions of the $ApoE^{-/-}$ mice, and human atherosclerotic carotid plaques. Inflammation, MMP activation and thrombus formation are promoted within the lesions followed by an increase in the levels of CCL19 and CCL21. Macrophages have also revealed the increased expression of CCL19 and CCL21 in inflamed atherosclerotic vessel (Damås et al., 2007). Under the influence of CCL19, DC produce inflammatory cytokines (Marsland et al., 2005). Besides inflammation, these cytokines are also involved in increased matrix degradation (Damås et al., 2007).

5.12 Macrophage migration-inhibitory factor (MIF)

MIF is a member of the CLF chemokine family and as the name implicates, it inhibits the migration of macrophages. Unlike other cytokines, it is constitutively expressed and stored amongst intracellular spaces. CXCR2, CXCR4 and CD74 are the receptors for MIF. In endothelial cells, VSMC, monocytes and T cells, expression of MIF has been greatly enhanced in hyperlipidemic conditions and the expression further enhances as the atherosclerotic lesions progress (van der Vorst, Döring, & Weber, 2015). Upon inhibition of MIF through small interfering RNA (siRNA) in HUVEC cell line, expression of VCAM-1 (responsible for adhesion of lymphocytes and monocytes to vascular endothelium), ICAM-1 (facilitating leukocyte endothelial transmigration), and MCP-1 (monocyte chemoattractant cytokine) were reduced. Also exogenous MIF increases the migration and TNF induced adhesion through increasing P-selectin expression (Cheng et al., 2010).

5.13 Atheroprotective chemokine CXCL12

The unstable plaque rupture is the causative of symptomatic coronary thrombosis and can be defined as an injury to the fibrous cap. This rupture further accelerates the atherosclerosis progression. A rupture prone lesion is characterized by the fibrous cap and enhances leukocytes infiltration. The apoptosis of VSMC further leads to the thinning of the cap. To stabilize the lesion, fibrous cap comprised of VSMC has been essential. Administration of smooth muscle progenitor cells (SPC), in ApoE$^{-/-}$RAG2$^{-/-}$ mice was found to limit plaque progression and generated a stable lesion phenotype (Zoll et al., 2008). Hence, interventions to increase the SPC recruitment in the atherosclerotic lesion area could be a therapeutic strategy. Akhtar, Gremse, Kiessling, Weber, and Schober (2013) has demonstrated that the repetitive treatment of ApoE$^{-/-}$ mice with CXCL12 helps in recruiting the bone marrow (BM)-derived SPC into the atherosclerotic lesions. Besides SPC, endothelial progenitor cells (EPC) also mobilized toward the vascular injury site under the influence of CCL12 and lead to lesion regression (Yao et al., 2012). CXCL12 is also named as stromal cell-derived factor 1 (SDF-1). CCL12 is ubiquitously expressed protein playing imperative role in homeostasis. CCL12 binds with the CXCR4 (GPCR). Another important receptor is CXCR7. In macrophages, the receptor ligand signaling between CXCL12 and CXCR4 participates in macropinocytosis of oxidized lipids and help in the accumulation of lipids in macrophages (van der Vorst et al., 2015). When rupture prone atherosclerotic plaques block an artery by growth or its rupture followed by the thrombus formation, ischemic conditions such as myocardial infarction appeared. Under hypoxic conditions, the CXCL12 and CXCR4 are upregulated and protective cells are recruited to the site of damage. Exogenous administration of CXCL12, reduce the infarct size that is apparently associated with the neoangiogenesis (Hu et al., 2007). Shear stress induced by arterial laminar flow ($>$15 dyn/cm^2/sec^{-1}) leads to atheroprotective gene response, whilst low laminar flow ($<$5 dyn/cm^2/sec^{-1}) predispose toward atherosclerotic situations. Laminar shear stress mediated anti-atherogenic impact is possibly mediated through at least in part, CXCR4 down-regulation (Melchionna et al., 2005).

5.14 Controversial chemokines in atherogenesis

CXCL16 is a transmembrane protein comprising of chemokine domain and trans-membrane mucin stalk. It combines inflammatory chemokine functions and scavenger receptor which facilitate uptake of oxLDL, phosphatidylserine, and bacteria. (Lehrke et al., 2007). Extracellular chemokine domain is cleaved by metalloproteinase ADAM10 that produces a soluble CXCL16 leading to activation of CXCR6-expressing T cells. Contradictory reports are available for the role of CXCL16. On one hand, its role in atherosclerosis is evidenced by its accumulation in atherosclerotic plaques (Wuttge et al., 2004). The soluble CXCL16 is reported to exhibit inflammatory functions and it activates T cells, one of the pioneer cells present in developing atherosclerotic lesions and these T cells are present co localized with CXCL16-expressing endothelial cells in atherosclerotic lesions (Lehrke et al., 2007). On the other hand contradictory results were obtained by Aslanian and Charo (2006), who reported accelerated plaque formation in CXCL16 gene deficient mice. Evidences of both pro-atherogenic as well as anti-atherogenic functions of CXCL16 are available, however in a nested case-control study of acute and chronic coronary artery disease, the data obtained form 699 subjects reveled that soluble CXCL16 level is positively correlated with inflammation and suggestive of pro-inflammatory role in human atherosclerosis (Lehrke et al., 2007). It also exhibits proangiogenic activities (Zhuge et al., 2005) and possibly supports plaque stabilization by nourishing intimal cells (Jovanović et al., 2015).

6. Therapeutic approaches

The statins are cholesterol lowering agents those inhibit the 3-hydroxy-3-methylglutaryl-CoA (HMG CoA) reductase, a key enzyme involved in the cholesterol biosynthesis. Therefore, targeting this enzyme may be helpful in reducing the serum cholesterol levels that eventually reduce the risk of individual's predisposal to cardiovascular consequences. Since some of the cytokines are proinflammatory and pro-atherosclerotic in nature, these could be the potential therapeutic candidates. An example is IL-17, where the IL-17 enhances the recruitment and activation of myeloid cells in the intima, associated with increased release of chemokines and proinflammatory cytokines. Blockade of IL-17A in $ApoE^{-/-}$ mice via adenovirus-produced decoy receptor IL-17RA, resulted in reduction of plaque in animals fed with high fat diet for 15 weeks. IL-6 reduction, limited

CXCL1 expression and macrophage content in aorta also have been observed in such treated mice (Smith et al., 2010).

An anti-mouse IL-6 receptor antibody (MR16-1) is able to protect atherosclerotic lesions in atherogenesis induced mice via dyslipidemia and/ or inflammation (Akita et al., 2017). Seeking the role of IFN-γ in atherosclerosis progression, strategies to target IFN-γ via its neutralization or inhibition of its' signally pathways have been implicated. IFN-γ neutralization through the use of decoy receptor IFN-γR, prevents the phosphorylation of STAT1 in the JAK-STAT pathway. The strategy was successfully used by Koga et al. (2007), who demonstrated that repeated gene transfers of IFN-γR, not only prevented plaque progression but also made advanced plaques static in ApoE-knockout mice fed on high-fat diet. Chemokines such as CCL2, CCL5 and CX3CL1 play crucial role during the initial monocyte recruitment phase of atherosclerosis. Inhibition of these through viral proteins as broad spectrum inhibitors, pharmacological inhibitors, neutralizing antibodies, decoy receptors or RNA interference might be a useful strategy to hamper atherosclerotic progression (Wilson Tang & Hazen, 2017). siRNA-loaded nanoparticles have been used to downregulate the CCR-2 expression participating in the infiltration of inflammatory monocytes to the atherosclerotic plaque (Kao et al., 2018). Maraviroc, an antiviral drug, is used to target CCR5 and it reduces both the plaque size as well as recruitment of macrophages to plaque along with reduction of VCAM-1, ICAM-1, CCL2 and TNF-α expression in the plaque (Cipriani et al., 2013).

7. Outlook

Atherosclerosis is considered as an inflammatory disorder leading to cardiovascular problems along with the related morbidity and mortality. It starts with the lipid deposition amongst the arteries. Accumulation of LDL and other apolipoproteins triggers endothelial cells activation. Activated endothelial cells releases cytokines and chemokines into the blood stream and express cell adhesion molecules to recruit inflammatory cells. Monocytes migrate into the arterial wall and differentiate into macrophages to uptake oxLDL to form foam cells. Unstable plaques grow rapidly by fast lipid deposition and have a thin fibrous cap and edges and rich in T_H1 cells, macrophages and few NK cells with a core of lipid and necrotic material. Ruptured plaques may trigger the cascade of thrombosis that further reduces the diameter of vessel lumen and produce sever clinical pathology. Various inflammatory molecules are involved in orchestrating the atherogenesis and IL-1, IL-2, IL-3, IL-4, IL-5, IL-6, IL-9, IL-10, IL-12, IL-13, IL-15, IL-17, IL-18, IL-20, IL-25, IL-27, IL-33, IL-37, TNF-α, TGF-β and IFN-γ are among them. CXCL16 has produced both atheroprotective as well as pro-atherogenic evidences and hence, its role is controversial. The CXCL12 is amongst atheroprotective roles. Many therapeutic interventions are developed to ameliorate this life-threatening inflammatory disorder and among them, use of decoy receptor, usage of monoclonal antibodies, usage of regulatory miRNA and siRNA, pharmaceutical broad-spectrum inhibitors are few. The complex interplay between fitness, genetics, lifestyle, and inflammation in the atherosclerotic pathogenesis is present; and by controlling the factors which influence the onset of atherosclerosis can be partially controlled by adapting a healthy life style.

Acknowledgments

Authors are thankful to the Prof. R.J. Rao, Hon'ble Vice-Chancellor, Barkatullah University for extending the all kind of support.

References

Aimo, A., Migliorini, P., Vergaro, G., Franzini, M., Passino, C., Maisel, A., et al. (2018). The IL-33/ST2 pathway, inflammation and atherosclerosis: Trigger and target? *International Journal of Cardiology, 267*, 188–192. http://dx.doi.org/10.1016/j.ijcard.2018.05.056.

Akhtar, S., Gremse, F., Kiessling, F., Weber, C., & Schober, A. (2013). CXCL12 promotes the stabilization of atherosclerotic lesions mediated by smooth muscle progenitor cells in ApoE-deficient mice. *Arteriosclerosis, Thrombosis, and Vascular Biology, 33*(4), 679–686. http://dx.doi.org/10.1161/atvbaha.112.301162.

Akita, K., Isoda, K., Sato-Okabayashi, Y., Kadoguchi, T., Kitamura, K., Ohtomo, F., et al. (2017). An interleukin-6 receptor antibody suppresses atherosclerosis in atherogenic mice. *Frontiers in Cardiovascular Medicine, 4*, 84. http://dx.doi.org/10.3389/fcvm.2017.00084.

Alexander, M. R., & Owens, G. K. (2012). Epigenetic control of smooth muscle cell differentiation and phenotypic switching in vascular development and disease. *Annual Review of Physiology, 74*, 13–40. http://dx.doi.org/10.1146/annurev-physiol-012110-142315.

Angeles-Martínez, J., Posadas-Sánchez, R., Pérez-Hernández, N., Rodríguez-Pérez, J. M., Fragoso, J. M., Bravo-Flores, E., et al. (2017). IL-15 polymorphisms are associated with subclinical atherosclerosis and cardiovascular risk factors. The genetics of atherosclerosis disease (GEA) Mexican Study. *Cytokine, 99*, 173–178. http://dx.doi.org/10.1016/j.cyto.2017.09.006.

Apostolakis, S., Amanatidou, V., & Spandidos, D. A. (2010). Therapeutic implications of chemokine-mediated pathways in atherosclerosis: Realistic perspectives and utopias. *Acta Pharmacologica Sinica, 31*(9), 1103–1110. http://dx.doi.org/10.1038/aps.2010.131.

Apostolakis, S., & Spandidos, D. (2013). Chemokines and atherosclerosis: Focus on the CX3CL1/CX3CR1 pathway. *Acta Pharmacologica Sinica, 34*(10), 1251–1256. http://dx.doi.org/10.1038/aps.2013.92.

Ashlin, T. G., Buckley, M. L., Salter, R. C., Johnson, J. L., Kwan, A. P., Ramji, D. P., et al. (2014). The anti-atherogenic cytokine interleukin-33 inhibits the expression of a disintegrin and metalloproteinase with thrombospondin motifs-1, -4 and -5 in human macrophages: Requirement of extracellular signal-regulated kinase, c-Jun N-terminal kinase and phosphoinositide 3-kinase signaling pathways. *The International Journal of Biochemistry and Cell Biology, 46*, 113–123. http://dx.doi.org/10.1016/j.biocel.2013.11.008.

Aslanian, A. M., & Charo, I. F. (2006). Targeted disruption of the scavenger receptor and chemokine CXCL16 accelerates atherosclerosis. *Circulation, 114*, 583–590. http://dx.doi.org/10.1161/CIRCULATIONAHA.105.540583.

Bennett, M. R., Sinha, S., & Owens, G. K. (2016). Vascular smooth muscle cells in atherosclerosis. *Circulation Research, 118*(4), 692–702. http://dx.doi.org/10.1161/circresaha.115.306361.

Bhat, O. M., & Dhavan, V. (2015). Role of IL-18 and its signaling in atherosclerosis. *Inflammation and Cell Signaling, 2*, e707. http://dx.doi.org/10.14800/ics.707.

Binder, C. J., Hartvigsen, K., Chang, M. K., Miller, M., Broide, D., Palinski, W., et al. (2004). IL-5 links adaptive and natural immunity specific for epitopes of oxidized LDL and protects from atherosclerosis. *Journal of Clinical Investigation, 114*(3), 427–437. http://dx.doi.org/10.1172/JCI20479.

Brizzi, M. F., Formato, L., Dentelli, P., Rosso, A., Pavan, M., Garbarino, G., et al. (2001). Interleukin-3 stimulates migration and proliferation of vascular smooth muscle cells: A potential role in atherogenesis. *Circulation, 103*(4), 549–554. http://dx.doi.org/10.1161/01.cir.103.4.549.

Buckley, M. L., Williams, J. O., Chan, Y. H., Laubertová, L., Gallagher, H., Moss, J. W. E., et al. (2019). The interleukin-33-mediated inhibition of expression of two key genes implicated in atherosclerosis in human macrophages requires MAP kinase, phosphoinositide 3-kinase and nuclear factor-κB signaling pathways. *Scientific Reports, 9*(1), 11317. http://dx.doi.org/10.1038/s41598-019-47620-8.

Calvayrac, O., Rodríguez-Calvo, R., Alonso, J., Orbe, J., Martín-Ventura, J. L., & Guadall, A. (2011). CCL20 is increased in hypercholesterolemic subjects and is upregulated by LDL in vascular smooth muscle cells: Role of NF-κB. *Arteriosclerosis, Thrombosis, and Vascular Biology, 31*(11), 2733–2741. http://dx.doi.org/10.1161/ATVBAHA.111.235721.

Cardilo-Reis, L., Gruber, S., Schreier, S. M., Drechsler, M., Papac-Milicevic, N., Weber, C., et al. (2012). Interleukin-13 protects from atherosclerosis and modulates plaque composition by skewing the macrophage phenotype. *EMBO Molecular Medicine, 4*(10), 1072–1086. http://dx.doi.org/10.1002/emmm.201201374.

Chen, W. Y., Cheng, B. C., Jiang, M. J., Hsieh, M. Y., & Chang, M. S. (2006). IL-20 is expressed in atherosclerosis plaques and promotes atherosclerosis in apolipoprotein E-deficient mice. *Arteriosclerosis, Thrombosis, and Vascular Biology, 26*(9), 2090–2095. http://dx.doi.org/10.1161/01.ATV.0000232502.88144.6f.

Cheng, Q., McKeown, S. J., Santos, L., Santiago, F. S., Khachigian, L. M., Morand, E. F., et al. (2010). Macrophage migration inhibitory factor increases leukocyte-endothelial interactions in human endothelial cells via promotion of expression of adhesion molecules. *The Journal of Immunology, 185*(2), 1238–1247. http://dx.doi.org/10.4049/jimmunol.0904104.

Chen, P. Y., Qin, L., Li, G., Wang, Z., Dahlman, J. E., Malagon-Lopez, J., et al. (2019). Endothelial TGF-β signalling drives vascular inflammation and atherosclerosis. *Nature Metabolism, 1*(9), 912–926. http://dx.doi.org/10.1038/s42255-019-0102-3.

Chistiakov, D. A., Sobenin, I. A., Orekhov, A. N., & Bobryshev, Y. V. (2014). Dendritic cells in atherosclerotic inflammation: The complexity of functions and the peculiarities of pathophysiological effects. *Frontiers in Physiology, 5*, 196. http://dx.doi.org/10.3389/fphys.2014.00196.

Cipriani, S., Francisci, D., Mencarelli, A., Renga, B., Schiaroli, E., D'Amore, C., et al. (2013). Efficacy of the CCR5 antagonist maraviroc in reducing early, ritonavir-induced atherogenesis and advanced plaque progression in mice. *Circulation, 127*(21), 2114–2124. http://dx.doi.org/10.1161/CIRCULATIONAHA.113.001278.

Clarke, M. C., Talib, S., Figg, N. L., & Bennett, M. R. (2010). Vascular smooth muscle cell apoptosis induces interleukin-1-directed inflammation: Effects of hyperlipidemia-mediated inhibition of phagocytosis. *Circulation Research, 106*, 363–372. http://dx.doi.org/10.1161/CIRCRESAHA.109.208389.

Damås, J. K., Smith, C., Øie, E., Fevang, B., Halvorsen, B., Waehre, T., et al. (2007). Enhanced expression of the homeostatic chemokines CCL19 and CCL21 in clinical and experimental atherosclerosis: Possible pathogenic role in plaque destabilization. *Arteriosclerosis, Thrombosis, and Vascular Biology, 27*(3), 614–620. http://dx.doi.org/10.1161/01.ATV.0000255581.38523.7c.

De Caterina, R., Libby, P., Peng, H. B., Thannickal, V. J., Rajavashisth, T. B., Gimbrone, M. A., et al. (1995). Nitric oxide decreases cytokine-induced endothelial activation. Nitric oxide selectively reduces endothelial expression of adhesion molecules and proinflammatory cytokines. *Journal of Clinical Investigation, 96*, 60–68. http://dx.doi.org/10.1172/JCI118074.

Dietrich, T., Hucko, T., Schneemann, C., Neumann, M., Menrad, A., & Willuda, J. (2012). Local delivery of IL-2 reduces atherosclerosis via expansion of regulatory T cells. *Atherosclerosis, 220*(2), 329–336. http://dx.doi.org/10.1016/j.atherosclerosis.2011.09.05.

Dinarello, C. A. (2009). Immunological and inflammatory functions of the interleukin-1 family. *Annual Review of Immunology, 27*(1), 519–550. http://dx.doi.org/10.1146/annurev.immunol.021908.132.

Dinh, T. N., Kyaw, T. S., Kanellakis, P., To, K., Tipping, P., Toh, B. H., et al. (2012). Cytokine therapy with interleukin-2/anti-interleukin-2 monoclonal antibody complexes expands $CD4^+CD25^+Foxp3^+$ regulatory T cells and attenuates development and progression of atherosclerosis. *Circulation, 126*(10), 1256–1266. http://dx.doi.org/10.1161/CIRCULATIONAHA.112.099044.

Domschke, G., & Gleissner, C. A. (2017). XCL4-induced macrophages in human atherosclerosis. *Cytokine, 122*, 154141. http://dx.doi.org/10.1016/j.cyto.2017.08.021.

Ellison, S., Gabunia, K., Kelemen, S. E., England, R. N., Scalia, R., Richards, J. M., et al. (2013). Attenuation of experimental atherosclerosis by interleukin-19. *Arteriosclerosis, Thrombosis, and Vascular Biology, 33*(10), 316–324. http://dx.doi.org/10.1161/ATVBAHA.113.301521.

England, R. N., & Autieri, M. V. (2012). Anti-inflammatory effects of interleukin-19 in vascular disease. *International Journal of Inflammation, 2012*, 253583. http://dx.doi.org/10.1155/2012/253583.

Erbel, C., Tyka, M., Helmes, C. M., Akhavanpoor, M., Rupp, G., Domschke, G., et al. (2015). CXCL4-induced plaque macrophages can be specifically identified by co-expression of $MMP7^+S100A8^+$ *in vitro* and *in vivo*. *Innate Immunity, 21*(3), 255–265. http://dx.doi.org/10.1177/1753425914526461.

van Es, T., van Puijvelde, G. H., Michon, I. N., van Wanrooij, E. J., de Vos, P., Peterse, N., et al. (2011). IL-15 aggravates atherosclerotic lesion development in LDL receptor deficient mice. *Vaccine, 29*(5), 976–983. http://dx.doi.org/10.1016/j.vaccine.2010.11.037.

Feng, S., Bowden, N., Fragiadaki, M., Souilhol, C., Hsiao, S., Mahmoud, M., et al. (2017). Mechanical activation of hypoxia-inducible factor 1α drives endothelial dysfunction at atheroprone sites. *Arteriosclerosis, Thrombosis, and Vascular Biology, 37*(11), 2087–2101. http://dx.doi.org/10.1161/ATVBAHA.117.309249.

Flynn, M. C., Pernes, G., Lee, M. K. S., Nagareddy, P. R., & Murphy, A. J. (2019). Monocytes, macrophages, and metabolic disease in atherosclerosis. *Frontiers in Pharmacology, 10*, 666. http://dx.doi.org/10.3389/fphar.2019.00666.

Gleissner, C. A., Shaked, I., Little, K. M., & Ley, K. (2010). CXC chemokine ligand 4 induces a unique transcriptome in monocyte-derived macrophages. *The Journal of Immunology, 184*(9), 4810–4818. http://dx.doi.org/10.4049/jimmunol.0901368.

Gregersen, I., Sandanger, Ø., Askevold, E. T., Sagen, E. L., Yang, K., Holm, S., et al. (2017). Interleukin 27 is increased in carotid atherosclerosis and promotes NLRP3 inflammasome activation. *PLoS One, 12*(11), e0188387. http://dx.doi.org/10.1371/journal.pone.0188387.

Gregersen, I., Skjelland, M., Holm, S., Holven, K. B., Krogh-Sørensen, K., Russell, D., et al. (2013). Increased systemic and local interleukin 9 levels in patients with carotid and coronary atherosclerosis. *PLoS One, 8*(8), e72769. http://dx.doi.org/10.1371/journal.pone.0072769.

Guo, L., Akahori, H., Harari, E., Smith, S. L., Polavarapu, R., Karmali, V., et al. (2018). $CD163^{+}$ macrophages promote angiogenesis and vascular permeability accompanied by inflammation in atherosclerosis. *Journal of Clinical Investigation, 128*, 1106–1124. http://dx.doi.org/10.1172/JCI93025.

Han, X., & Boisvert, W. A. (2015). Interleukin-10 protects against atherosclerosis by modulating multiple atherogenic macrophage function. *Thrombosis and Haemostasis, 113*(3), 505–512. http://dx.doi.org/10.1160/TH14-06-0509.

Hanna, R. N., Carlin, L. M., Hubbeling, H. G., Nackiewicz, D., Green, A. M., Punt, J. A., et al. (2011). The transcription factor NR4A1 (Nur77) controls bone marrow differentiation and the survival of Ly6C − monocytes. *Nature Immunology, 12*(8), 778–785. http://dx.doi.org/10.1038/ni.2063.

Hassan, M. O., Duarte, R., Dix-Peek, T., Dickens, C., Naidoo, S., Vachiat, A., et al. (2018). Transforming growth factor-β protects against inflammation-related atherosclerosis in South African CKD patients. *International Journal of Nephrology, 2018*, 8702372. http://dx.doi.org/10.1155/2018/8702372.

Hauer, A. D., Uyttenhove, C., de Vos, P., Stroobant, V., Renauld, J. C., van Berkel, T. J., et al. (2005). Blockade of interleukin-12 function by protein vaccination attenuates atherosclerosis. *Circulation, 112*(7), 1054–1062. http://dx.doi.org/10.1161/CIRCULATIONAHA.104.533463.

Heller, E. A., Liu, E., Tager, A. M., Yuan, Q., Lin, A. Y., Ahluwalia, N., et al. (2006). Chemokine CXCL10 promotes atherogenesis by modulating the local balance of effector and regulatory T cells. *Circulation, 113*(19), 2301–2312. http://dx.doi.org/10.1161/CIRCULATIONAHA.105.605121.

Hirase, T., Hara, H., Miyazaki, Y., Ide, N., Nishimoto-Hazuku, A., Fujimoto, H., et al. (2013). Interleukin 27 inhibits atherosclerosis via immunoregulation of macrophages in mice. *American Journal of Physiology – Heart and Circulatory Physiology, 305*(3), H420–H429. http://dx.doi.org/10.1152/ajpheart.00198.2013.

Huber, S. A., Sakkinen, P., Conze, D., Hardin, N., & Tracy, R. (1999). Interleukin-6 exacerbates early atherosclerosis in mice. *Arteriosclerosis, Thrombosis, and Vascular Biology, 19*(10), 2364–2367. http://dx.doi.org/10.1161/01.atv.19.10.2364.

Hu, X., Dai, S., Wu, W. J., Tan, W., Zhu, X., Mu, J., et al. (2007). Stromal cell derived factor-1 alpha confers protection against myocardial ischemia/reperfusion injury: Role of the cardiac stromal cell derived factor-1 alpha CXCR4 axis. *Circulation, 116*(6), 654–663. http://dx.doi.org/10.1161/CIRCULATIONAHA.106.672451.

von Hundelshausen, P., Weber, K. S., Huo, Y., Proudfoot, A. E., Nelson, P. J., Ley, K., et al. (2001). RANTES deposition by platelets triggers monocyte arrest on inflamed and atherosclerotic endothelium. *Circulation, 103*, 1772–1777. http://dx.doi.org/10.1161/01.cir.103.13.1772.

de Jager, S. C., Bot, I., Kraaijeveld, A. O., Korporaal, S. J., Bot, M., van Santbrink, P. J., et al. (2013). Leukocyte-specific CCL3 deficiency inhibits atherosclerotic lesion development by affecting neutrophil accumulation. *Arteriosclerosis, Thrombosis, and Vascular Biology, 33*(3), e75–e83. http://dx.doi.org/10.1161/atvbaha.112.300857.

Jain, S., Gabunia, K., Kelemen, S. E., Panetti, T. S., & Autieri, M. V. (2011). The anti-inflammatory cytokine interleukin 19 is expressed by and angiogenic for human endothelial cells. *Arteriosclerosis, Thrombosis, and Vascular Biology, 31*(1), 167–175. http://dx.doi.org/10.1161/ATVBAHA.110.214916.

Jovanović, I., Zivković, M., Djurić, T., Popović, M., Alavantić, D., & Stanković, A. (2015). CXCL16 in vascular pathology research: From macro effects to microRNAs. *Journal of Atherosclerosis and Thrombosis, 22*(10), 1012–1024. http://dx.doi.org/10.5551/jat.29942.

Kao, C. W., Wu, P. T., Liao, M. Y., Chung, I. J., Yang, K. C., Tseng, W. I., et al. (2018). Magnetic nanoparticles conjugated with peptides derived from monocyte chemoattractant protein-1 as a tool for targeting atherosclerosis. *Pharmaceutics, 10*(2), E62. http://dx.doi.org/10.3390/pharmaceutics10020062.

Kempe, S., Heinz, P., Kokai, E., Devergne, O., Marx, N., & Wirth, T. (2009). Epstein-barr virus-induced gene-3 is expressed in human atheroma plaques. *American Journal of Pathology, 175*(1), 440–447. http://dx.doi.org/10.2353/ajpath.2009.080752.

Kennedy, A., Gruen, M. L., Gutierrez, D. A., Surmi, B. K., Orr, J. S., Webb, C. D., et al. (2012). Impact of macrophage inflammatory protein-1α deficiency on atherosclerotic lesion formation, hepatic steatosis, and adipose tissue expansion. *PLoS One, 7*(2), e31508. http://dx.doi.org/10.1371/journal.pone.0031508.

King, V. L., Cassis, L. A., & Daugherty, A. (2007). Interleukin-4 does not influence development of hypercholesterolemia or angiotensin II-induced atherosclerotic lesions in mice. *American Journal of Pathology, 171*(6), 2040–2047. http://dx.doi.org/10.2353/ajpath.2007.060857.

Klein, O., Ebert, L. M., Zanker, D., Woods, K., Tan, B. S., Fucikova, J., et al. (2013). Flt3 ligand expands $CD4^+$ $FoxP3^+$ regulatory T cells in human subjects. *European Journal of Immunology, 43*(2), 533–539. http://dx.doi.org/10.1002/eji.201242603.

Koga, M., Kai, H., Yasukawa, H., Yamamoto, T., Kawai, Y., Kato, S., et al. (2007). Inhibition of progression and stabilization of plaques by postnatal interferon-gamma function blocking in ApoE-knockout mice. *Circulation Research, 101*(4), 348–356. http://dx.doi.org/10.1161/CIRCRESAHA.106.147256.

Lee, Y. W., & Hirani, A. A. (2006). Role of interleukin-4 in atherosclerosis. *Archives of Pharmacal Research, 29*(1), 1–15. http://dx.doi.org/10.1007/BF02977462.

Lee, Y. W., Kim, P. H., Lee, W. H., & Hirani, A. A. (2010). Interleukin-4, oxidative stress, vascular inflammation and atherosclerosis. *Biomolecules and Therapeutics (Seoul), 18*(2), 135–144. http://dx.doi.org/10.4062/biomolther.2010.18.2.135.

Lee, T. S., Yen, H. C., Pan, C. C., & Chau, L. Y. (1999). The role of interleukin 12 in the development of atherosclerosis in ApoE-deficient mice. *Arteriosclerosis, Thrombosis, and Vascular Biology, 19*(3), 734–742. http://dx.doi.org/10.1161/01.atv.19.3.734.

Lehrke, M., Millington, S. C., Lefterova, M., Cumaranatunge, R. G., Szapary, P., Wilensky, R., et al. (2007). CXCL16 is a marker of inflammation, atherosclerosis, and acute coronary syndromes in humans. *Journal of the American College of Cardiology, 49*(4), 442–449. http://dx.doi.org/10.1016/j.jacc.2006.09.034.

Libby, P., Ridker, P. M., & Hansson, G. K. (2009). Inflammation in atherosclerosis: From pathophysiology to practice. *Journal of the American College of Cardiology, 54*(23), 2129–2138. http://dx.doi.org/10.1016/j.jacc.2009.09.009.

Li, Q., Laumonnier, Y., Syrovets, T., & Simmet, T. (2013). Recruitment of CCR6-expressing Th17 cells by CCL20 secreted from plasmin-stimulated macrophages. *Acta Biochimica et Biophysica Sinica, 45*(7), 593–600. http://dx.doi.org/10.1093/abbs/gmt049.

Li, Q., Ming, T., Wang, Y., Ding, S., Hu, C., Zhang, C., et al. (2017). Increased Th9 cells and IL-9 levels accelerate disease progression in experimental atherosclerosis. *American Journal of Tourism Research, 9*(3), 1335–1343. PMCID: PMC5376024.

Liu, J., Lin, J., He, S., Wu, C., Wang, B., Liu, J., et al. (2018). Transgenic overexpression of IL-37 protects against atherosclerosis and strengthens plaque stability. *Cellular Physiology and Biochemistry, 45*(3), 1034–1050. http://dx.doi.org/10.1159/000487344.

Madhur, M. S., Funt, S. A., Li, L., Vinh, A., Chen, W., Lob, H. E., et al. (2011). Role of interleukin 17 in inflammation, atherosclerosis, and vascular function in apolipoprotein e-deficient mice. *Arteriosclerosis, Thrombosis, and Vascular Biology, 31*(7), 1565–1572. http://dx.doi.org/10.1161/ATVBAHA.111.227629.

Mallat, Z., Corbaz, A., Scoazec, A., Besnard, S., Lesèche, G., Chvatchko, Y., et al. (2001). Expression of interleukin-18 in human atherosclerotic plaques and relation to plaque instability. *Circulation, 104*(14), 1598–1603. http://dx.doi.org/10.1161/hc3901.096721.

Mallat, Z., Gojova, A., Marchiol-Fournigault, C., Esposito, B., Kamaté, C., Merval, R., et al. (2001). Inhibition of transforming growth factor-beta signaling accelerates atherosclerosis and induces an unstable plaque phenotype in mice. *Circulation Research, 89*(10), 930–934. http://dx.doi.org/10.1161/hh2201.099415.

Mantani, P. T., Dunér, P., Bengtsson, E., Alm, R., Ljungcrantz, I., Söderberg, I., et al. (2015). IL-25 inhibits atherosclerosis development in apolipoprotein E deficient mice. *PLoS One, 10*(1), e0117255. http://dx.doi.org/10.1371/journal.pone.0117255.

Mantani, P. T., Dunér, P., Bengtsson, E., Ljungcrantz, I., Sundius, L., & To, F. (2018). Interleukin-25 (IL-25) has a protective role in atherosclerosis development in the aortic arch in mice. *The Journal of Biological Chemistry, 293*(18), 6791–6801. http://dx.doi.org/10.1074/jbc.ra117.000292.

Marsland, B. J., Bättig, P., Bauer, M., Ruedl, C., Lässing, U., Beerli, R. R., et al. (2005). CCL19 and CCL21 induce a potent proinflammatory differentiation program in licensed dendritic cells. *Immunity, 22*(4), 493–505. http://dx.doi.org/10.1016/j.immuni.2005.02.010.

McCurdy, S., Liu, C. A., Yap, J., & Boisvert, W. A. (2017). Potential role of IL-37 in atherosclerosis. *Cytokine, 122*, 154169. http://dx.doi.org/10.1016/j.cyto.2017.09.025.

McLaren, J. E., Michael, D. R., Salter, R. C., Ashlin, T. G., Calder, C. J., Miller, A. M., et al. (2010). IL-33 reduces macrophage foam cell formation. *The Journal of Immunology, 185*(2), 1222–1229. http://dx.doi.org/10.4049/jimmunol.1000520.

Melchionna, R., Porcelli, D., Mangoni, A., Carlini, D., Liuzzo, G., Spinetti, G., et al. (2005). Laminar shear stress inhibits CXCR4 expression on endothelial cells: Functional consequences for atherogenesis. *The FASEB Journal, 19*(6), 629–631. http://dx.doi.org/10.1096/fj.04-2219fje.

Miller, A. M., Xu, D., Asquith, D. L., Denby, L., Li, Y., Sattar, N., et al. (2008). IL-33 reduces the development of atherosclerosis. *The Journal of Experimental Medicine, 205*(2), 339–346. http://dx.doi.org/10.1084/jem.20071868.

Misra, A., Feng, Z., Chandran, R. R., Kabir, I., Rotllan, N., Aryal, B., et al. (2018). Integrin beta3 regulates clonality and fate of smooth muscle-derived atherosclerotic plaque cells. *Nature Communications, 9*(1), 2073. http://dx.doi.org/10.1038/s41467-018-04447-7.

Nold-Petry, C. A., Lo, C. Y., Rudloff, I., Elgass, K. D., Li, S., Gantier, M. P., et al. (2015). IL-37 requires the receptors IL-18Rα and IL-1R8 (SIGIRR) to carry out its multifaceted anti-inflammatory program upon innate signal transduction. *Nature Immunology, 16*(4), 354–365. http://dx.doi.org/10.1038/ni.3103.

Ohta, H., Wada, H., Niwa, T., Kirii, H., Iwamoto, N., Fujii, H., et al. (2005). Disruption of tumor necrosis factor-alpha gene diminishes the development of atherosclerosis in ApoE-deficient mice. *Atherosclerosis, 180*(1), 11–17. http://dx.doi.org/10.1016/j.atherosclerosis.2004.11.016.

Pastrana, J. L., Sha, X., Virtue, A., Mai, J., Cueto, R., Lee, I. A., et al. (2012). Regulatory T cells and atherosclerosis. *Journal of Clinical and Experimental Cardiology, 2012*(Suppl. 12), 002. http://dx.doi.org/10.4172/2155-9880.S12-002.
Paulson, K. E., Zhu, S. N., Chen, M., Nurmohamed, S., Jongstra-Bilen, J., & Cybulsky, M. I. (2010). Resident intimal dendritic cells accumulate lipid and contribute to the initiation of atherosclerosis. *Circulation Research, 106*(2), 383–390. http://dx.doi.org/10.1161/CIRCRESAHA.109.210781.
Petrache, I., Birukova, A., Ramirez, S. I., Garcia, J. G., & Verin, A. D. (2003). The role of the microtubules in tumor necrosis factor-alpha-induced endothelial cell permeability. *American Journal of Respiratory Cell and Molecular Biology, 28*, 574–581. http://dx.doi.org/10.1165/rcmb.2002-0075OC.
Pietrowski, E., Bender, B., Huppert, J., White, R., Luhmann, H. J., & Kuhlmann, C. R. (2011). Pro-inflammatory effects of interleukin-17A on vascular smooth muscle cells involve NAD(P)H- oxidase derived reactive oxygen species. *Journal of Vascular Research, 48*(1), 52–58. http://dx.doi.org/10.1159/000317400.
Pober, J. S., & Sessa, W. C. (2007). Evolving functions of endothelial cells in inflammation. *Nature Reviews Immunology,* 7(10), 803–815. http://dx.doi.org/10.1038/nri2171.
Potteaux, S., Gautier, E. L., Hutchison, S. B., van Rooijen, N., Rader, D. J., & Thomas, M. J. (2011). Suppressed monocyte recruitment drives macrophage removal from atherosclerotic plaques of $ApoE^{-}/^{-}$ mice during disease regression. *The Journal of Clinical Investigation, 121*(5), 2025–2036. http://dx.doi.org/10.1172/JCI43802.
Reiss, A. B., Siegart, N. M., & Leon, J. D. (2017). Interleukin-6 in atherosclerosis: Atherogenic or atheroprotective? *Clinical Lipidology, 12*(1), 14–23. http://dx.doi.org/10.1080/17584299.2017.1319787.
Riopel, M., Vassallo, M., Ehinger, E., Pattison, J., Bowden, K., Winkels, H., et al. (2019). CX3CL1-Fc treatment prevents atherosclerosis in Ldlr KO mice. *Molecular Metabolism, 20*, 89–101. http://dx.doi.org/10.1016/j.molmet.2018.11.011.
Ross, R. (1995). Cell biology of atherosclerosis. *Annual Review of Physiology, 57*(1), 791–804. http://dx.doi.org/10.1146/annurev.ph.57.030195.004043.
Russo, R. C., Garcia, C. C., Teixeira, M. M., & Amaral, F. A. (2014). The CXCL8/IL-8 chemokine family and its receptors in inflammatory diseases. *Expert Review of Clinical Immunology, 10*(5), 593–619. http://dx.doi.org/10.1586/1744666X.2014.894886.
Sachais, B. S., Kuo, A., Nassar, T., Morgan, J., Kariko, K., Williams, K. J., et al. (2002). Platelet factor 4 binds to low-density lipoprotein receptors and disrupts the endocytic machinery, resulting in retention of low-density lipoprotein on the cell surface. *Blood, 99*(10), 3613–3622. http://dx.doi.org/10.1182/blood.v99.10.3613.
Sakashita, N., Miyazaki, A., Chang, C. C., Chang, T. Y., Kiyota, E., Satoh, M., et al. (2003). Acyl-coenzyme A: Cholesterol acyltransferase 2 (ACAT2) is induced in monocyte-derived macrophages: *In vivo* and *in vitro* studies. *Laboratory Investigation, 83*, 1569–1581. PMID: 14615411.
Salter, R. C., Ashlin, T. G., Kwan, A. P., & Ramji, D. P. (2010). ADAMTS proteases: Key roles in atherosclerosis? *Journal of Molecular Medicine, 88*(12), 1203–1211. http://dx.doi.org/10.1007/s00109-010-0654-x.
Sämpi, M., Ukkola, O., Päivänsalo, M., Kesäniemi, Y. A., Binder, C. J., & Hörkkö, S. (2008). Plasma interleukin-5 levels are related to antibodies binding to oxidized low-density lipoprotein and to decreased subclinical atherosclerosis. *Journal of the American College of Cardiology, 52*(17), 1370–1378. http://dx.doi.org/10.1016/j.jacc.2008.06.047.
Schieffer, B., Selle, T., Hilfiker, A., Hilfiker-Kleiner, D., Grote, K., Tietge, U. J., et al. (2004). Impact of interleukin-6 on plaque development and morphology in experimental atherosclerosis. *Circulation, 110*(22), 3493–3500. http://dx.doi.org/10.1161/01.CIR.0000148135.08582.97.

Seyrek, N., Karayaylali, I., Balal, M., Paydas, S., Aikimbaev, K., Cetiner, S., et al. (2005). Is there any relationship between serum levels of interleukin-10 and atherosclerosis in hemodialysis patients? *Scandinavian Journal of Urology and Nephrology, 39*(5), 405–409. http://dx.doi.org/10.1080/00365590500386734.

Shankman, L. S., Gomez, D., Cherepanova, O. A., Salmon, M., Alencar, G. F., & Haskins, R. M. (2015). KLF4-dependent phenotypic modulation of smooth muscle cells has a key role in atherosclerotic plaque pathogenesis. *Nature Medicine, 21*, 628–637. http://dx.doi.org/10.1038/nm.3866.

Silveira, A., McLeod, O., Strawbridge, R. J., Gertow, K., Sennblad, B., Baldassarre, D., et al. (2015). Plasma IL-5 concentration and subclinical carotid atherosclerosis. *Atherosclerosis, 239*(1), 125–130. http://dx.doi.org/10.1016/j.atherosclerosis.2014.12.046.

Smith, E., Prasad, K. M., Butcher, M., Dobrian, A., Kolls, J. K., Ley, K., et al. (2010). Blockade of interleukin-17A results in reduced atherosclerosis in apolipoprotein E-deficient mice. *Circulation, 121*(15), 1746–1755. http://dx.doi.org/10.1161/CIRCULATIONAHA.109.924886.

Soejima, K., & Rollins, B. J. (2001). A functional IFN-gamma-inducible protein-10/CXCL10-specific receptor expressed by epithelial and endothelial cells that is neither CXCR3 nor glycosaminoglycan. *The Journal of Immunology, 167*(11), 6576–6582. http://dx.doi.org/10.4049/jimmunol.167.11.6576.

Stolla, M., Pelisek, J., von Brühl, M. L., Schäfer, A., Barocke, V., Heider, P., et al. (2012). Fractalkine is expressed in early and advanced atherosclerotic lesions and supports monocyte recruitment via CX3CR1. *PLoS One, 7*(8), e43572. http://dx.doi.org/10.1371/journal.pone.0043572.

Subramanian, M., & Tabas, I. (2014). Dendritic cells in atherosclerosis. *Seminars in Immunopathology, 36*(1), 93–102. http://dx.doi.org/10.1007/s00281-013-0400-x.

Subramanian, M., Thorp, E., Hansson, G. K., & Tabas, I. (2013). Treg-mediated suppression of atherosclerosis requires MYD88 signaling in DCs. *The Journal of Clinical Investigation, 123*(1), 179–188. http://dx.doi.org/10.1172/JCI64617.

Tang, W. H., & Hazen, S. L. (2017). Atherosclerosis in 2016: Advances in new therapeutic targets for atherosclerosis. *Nature Reviews Cardiology, 14*(2), 71–72. http://dx.doi.org/10.1038/nrcardio.2016.216.

Tenger, C., Sundborger, A., Jawien, J., & Zhou, X. (2005). IL-18 accelerates atherosclerosis accompanied by elevation of IFN-gamma and CXCL16 expression independently of T cells. *Arteriosclerosis, Thrombosis, and Vascular Biology, 25*, 791–796. http://dx.doi.org/10.1161/01.ATV.0000153516.02782.65.

Theodorou, K., & Boon, R. A. (2018). Endothelial cell metabolism in atherosclerosis. *Frontiers in Cell and Developmental Biology, 6*, 82. http://dx.doi.org/10.3389/fcell.2018.00082.

Tousoulis, D., Oikonomou, E., Economou, E. K., Crea, F., & Kaski, J. C. (2016). Inflammatory cytokines in atherosclerosis: Current therapeutic approaches. *European Heart Journal, 37*(22), 1723–1732. http://dx.doi.org/10.1093/eurheartj/ehv759.

Uchida, Y., Maezawa, Y., Uchida, Y., Hiruta, N., Shimoyama, E., & Kawai, S. (2013). Localization of oxidized low-density lipoprotein and its relation to plaque morphology in human coronary artery. *PLoS One, 8*(2), e55188. http://dx.doi.org/10.1371/journal.pone.0055188.

Upadhya, S., Mooteri, S., Peckham, N., & Pai, R. G. (2004). Atherogenic effect of interleukin-2 and antiatherogenic effect of interleukin-2 antibody in apo-E-deficient mice. *Angiology, 55*(3), 289–294. http://dx.doi.org/10.1177/000331970405500308.

Van Vré, E. A., Ilse Van Brussel, I. V., Bosmans, J. M., Vrints, C. J., & Bult, H. (2011). Dendritic cells in human atherosclerosis: From circulation to atherosclerotic plaques. *Mediators of Inflammation, 2011*, 941396. http://dx.doi.org/10.1155/2011/941396.

Végran, F., Boidot, R., Michiels, C., Sonveaux, P., & Feron, O. (2011). Lactate influx through the endothelial cell monocarboxylate transporter MCT1 supports an NF-κB/IL-8 pathway that drives tumor angiogenesis. *Cancer Research, 71*, 2550–2560. http://dx.doi.org/10.1158/0008-5472.CAN-10-2828.

Voloshyna, I., Littlefield, M. J., & Reiss, A. B. (2014). Atherosclerosis and interferon-γ: New insights and therapeutic targets. *Trends in Cardiovascular Medicine, 24*(1), 45–51. http://dx.doi.org/10.1016/j.tcm.2013.06.003.

van der Vorst, E. P., Döring, Y., & Weber, C. (2015). MIF and CXCL12 in cardiovascular diseases: Functional differences and similarities. *Frontiers in Immunology, 6*, 373. http://dx.doi.org/10.3389/fimmu.2015.00373.

Vromman, A., Ruvkun, V., Shvartz, E., Wojtkiewicz, G., Santos Masson, G., Tesmenitsky, Y., et al. (2019). Stage-dependent differential effects of interleukin-1 isoforms on experimental atherosclerosis. *Europian Heart Journal, 40*(30), 2482–2491. http://dx.doi.org/10.1093/eurheartj/ehz008.

Wang, B., Dai, S., Dong, Z., Sun, Y., Song, X., Guo, C., et al. (2014). The modulation of endoplasmic reticulum stress by chemical chaperone upregulates immune negative cytokine IL-35 in apolipoprotein E-deficient mice. *PLoS One, 9*(1), e87787. http://dx.doi.org/10.1371/journal.pone.0087787.

Wang, J., Sun, C., Gerdes, N., Liu, C., Liao, M., Liu, J., et al. (2015). Interleukin 18 function in atherosclerosis is mediated by the interleukin 18 receptor and the Na-Cl co-transporter. *Nature Medicine, 21*(7), 820–826. http://dx.doi.org/10.1038/nm.3890.

Weber, C., Meiler, S., Döring, Y., Koch, M., Drechsler, M., Megens, R. T., et al. (2011). CCL17-expressing dendritic cells drive atherosclerosis by restraining regulatory T cell homeostasis in mice. *The Journal of Clinical Investigation, 121*(7), 2898–2910. http://dx.doi.org/10.1172/JCI44925.

Wolf, D., & Ley, K. (2019). Immunity and inflammation in atherosclerosis. *Circulation Research, 124*(2), 315–327. http://dx.doi.org/10.1161/CIRCRESAHA.118.313591.

Wuttge, D. M., Zhou, X., Sheikine, Y., Wågsäter, D., Stemme, V., Hedin, U., et al. (2004). CXCL16/SR-PSOX is an interferon-gamma-regulated chemokine and scavenger receptor expressed in atherosclerotic lesions. *Arteriosclerosis, Thrombosis, and Vascular Biology, 24*(4), 750–755. http://dx.doi.org/10.1161/01.ATV.0000124102.11472.36.

Yamaoka-Tojo, M., Tojo, T., Wakaume, K., Kameda, R., Nemoto, S., Takahira, N., et al. (2011). Circulating interleukin-18: A specific biomarker for atherosclerosis-prone patients with metabolic syndrome. *Nutrition and Metabolism, 8*, 3. http://dx.doi.org/10.1186/1743-7075-8-3.

Yao, L., Heuser-Baker, J., Herlea-Pana, O., Iida, R., Wang, Q., Zou, M. H., et al. (2012). Bone marrow endothelial progenitors augment atherosclerotic plaque regression in a mouse model of plasma lipid lowering. *Stem Cells, 30*(12), 2720–2731. http://dx.doi.org/10.1002/stem.1256.

Zernecke, A., Shagdarsuren, E., & Weber, C. (2008). Chemokines in atherosclerosis: An update. *Arteriosclerosis, Thrombosis, and Vascular Biology, 28*, 1897–1908. http://dx.doi.org/10.1161/ATVBAHA.107.161174.

Zhang, J., Alcaide, P., Liu, L., Sun, J., He, A., Luscinskas, F. W., et al. (2011). Regulation of endothelial cell adhesion molecule expression by mast cells, macrophages, and neutrophils. *PLoS One, 6*(1), e14525. http://dx.doi.org/10.1371/journal.pone.0014525.

Zhang, H., Chen, J., Liu, X., Awar, L., McMickle, A., Bai, F., et al. (2013). IL-17 induces expression of vascular cell adhesion molecule through signalling pathway of NF-κB, but not Akt1 and TAK1 in vascular smooth muscle cells. *Scandinavian Journal of Immunology, 77*(4), 230–237. http://dx.doi.org/10.1111/sji.12030.

Zhang, W., Tang, T., Nie, D., Wen, S., Jia, C., Zhu, Z., et al. (2015). IL-9 aggravates the development of atherosclerosis in ApoE^{-}/$^{-}$ mice. *Cardiovascular Research, 106*(3), 453–464. http://dx.doi.org/10.1093/cvr/cvv110.

Zhang, H. F., Wu, M. X., Lin, Y. Q., Xie, S. L., Huang, T. C., Liu, P. M., et al. (2017). IL-33 promotes IL-10 production in macrophages: A role for IL-33 in macrophage foam cell formation. *Experimental and Molecular Medicine, 49*(11), e388. http://dx.doi.org/10.1038/emm.2017.183.

Zhao, W., Lei, T., Li, H., Sun, D., Mo, X., Wang, Z., et al. (2015). Macrophage-specific overexpression of interleukin-5 attenuates atherosclerosis in LDL receptor-deficient mice. *Gene Therapy, 22*(8), 645–652. http://dx.doi.org/10.1038/gt.2015.33.

Zhao, J., Nishimura, Y., Kimura, A., Ozawa, K., Kondo, T., Tanaka, T., et al. (2017). Chemokines protect vascular smooth muscle cells from cell death induced by cyclic mechanical stretch. *Scientific Reports,* 7(1), 16128. http://dx.doi.org/10.1038/s41598-017-15867-8.

Zhou, Z., Subramanian, P., Sevilmis, G., Globke, B., Soehnlein, O., Karshovska, E., et al. (2011). Lipoprotein-derived lysophosphatidic acid promotes atherosclerosis by releasing CXCL1 from the endothelium. *Cell Metabolism, 13*(5), 592–600. http://dx.doi.org/10.1016/j.cmet.2011.02.016.

Zhuge, X., Murayama, T., Arai, H., Yamauchi, R., Tanaka, M., Shimaoka, T., et al. (2005). CXCL16 is a novel angiogenic factor for human umbilical vein endothelial cells. *Biochemical and Biophysical Research Communications, 331*(4), 1295–1300. http://dx.doi.org/10.1016/j.bbrc.2005.03.200.

Zoll, J., Fontaine, V., Gourdy, P., Barateau, V., Vilar, J., Leroyer, A., et al. (2008). Role of human smooth muscle cell progenitors in atherosclerotic plaque development and composition. *Cardiovascular Research,* 77(3), 471–480. http://dx.doi.org/10.1093/cvr/cvm034.

CHAPTER FOUR

Sphingolipids as mediators of inflammation and novel therapeutic target in inflammatory bowel disease

Olga A. Sukocheva[a], Elena Lukina[a], Eileen McGowan[b] and Anupam Bishayee[c,*]

[a]Discipline of Health Sciences, College of Nursing and Health Sciences, Flinders University, Bedford Park, SA, Australia
[b]School of Life Sciences, Faculty of Science, University of Technology Sydney, Sydney, NSW, Australia
[c]Lake Erie College of Osteopathic Medicine, Bradenton, FL, United States
*Corresponding author: E-mails: abishayee@lecom.edu; abishayee@gmail.com

Contents

Abstract

Morbidity of inflammatory gastrointestinal (GI) diseases continues to grow resulting in worsen quality of life and increased burden on public medical systems. Complex and heterogenous illnesses, inflammatory bowel diseases (IBDs) encompass several inflammation -associated pathologies including Crohn's disease and ulcerative colitis. IBD is often initiated by a complex interplay between host genetic and environmental factors, lifestyle and diet, and intestinal bacterial components. IBD inflammatory signature was linked to the pro-inflammatory cytokine tumor necrosis factor-α (TNF-α) signaling pathway that is currently targeted by IBD therapies. Sphingolipid signaling was identified as one of the key mediators and regulators of pro-inflammatory conditions, and, specifically, TNF-α related signaling. All GI tissues and circulating immune/blood cells

Advances in Protein Chemistry and Structural Biology, Volume 120
ISSN 1876-1623
https://doi.org/10.1016/bs.apcsb.2019.11.003

contain activated sphingolipid-metabolizing enzymes, including sphingosine kinases (SphK1 and SphK2) that generate sphingosine-1-phosphate (S1P), a bioactive lipid and ligand for five G-protein coupled membrane S1P receptors (S1PRs). Numerous normal and pathogenic inflammatory responses are mediated by SphK/S1P/S1PRs signaling axis including lymphocyte trafficking and activation of cytokine signaling machinery. SphK1/S1P/S1PRs axis has recently been defined as a target for the treatment of GI diseases including IBD/colitis. Several SphK1 inhibitors and S1PRs antagonists have been developed as novel anti-inflammatory agents. In this review, we discuss the mechanisms of SphK/S1P signaling in inflammation-linked GI disorders. The potential role of SphK/S1PRs inhibitors in the prevention and treatment of IBD/colitis is critically evaluated.

1. Introduction on inflammatory bowel disease (IBD) as group of inflammatory gastrointestinal diseases

IBD is an immune mediated illness associated with intestinal aberrant and inflammatory responses to dietary components and/or commensal bacteria antigens (Biasi, Leonarduzzi, Oteiza, & Poli, 2013). IBD is a complex multifactorial pathology that includes Crohn's Disease (CrD) and Ulcerative Colitis (UC), both chronic inflammatory intestinal disorders of unknown etiology (Baumgart & Sandborn, 2007). Complex intrinsic genetic and acquired immune system characteristics, environmental factors (i.e. exposure to external pathogenic microorganisms, toxins, diet), and adverse internal/luminal microbial flora impact the pathogenesis of IBD (Argollo, Fiorino, Hindryckx, Peyrin-Biroulet, & Danese, 2017; Danese, 2011). Notably, loss of immune tolerance is an IBD hallmark when healthy tissues are targeted by chronic inflammation that can progress towards the development of cancer and life-threatening conditions (Edwards, Zhang, Frauwirth, & Mosser, 2006; O'Connor et al., 2010).

Population-based studies indicate a worldwide increase in the overall incidence of IBD (Benchimol et al., 2019; Sýkora et al., 2018). As a major cause of lifetime morbidity, IBD severely impacts quality of life and up to 15% of patients undergo a surgical procedure of colectomy as the only option to treat the disease. Despite significant advances in treatment with anti-inflammatory agents, IBD remains a grievous and refractory pathology incurring significant discomfort, human and health care costs. Progress in the development of successful IBD-resolving medicine is slower than anticipated because molecular and immunologic mechanisms underlying the IBD-triggering inflammation are largely unclear. To improve current IBD treatment and prevention methods, it is necessary to advance understanding of preclinical internal human IBD mechanisms and the role of environmental/external triggers in triggering of IBD symptoms.

The etiology of both UC and CrD is not completely understood, however, recent investigations support a genetic predisposition along with development of autoimmune responses as consequences of previous infections. Combined as IBD, UC and CrD share some similarities, but differ in genetic predisposition, risk factors, and clinical presentation including some variations in endoscopic and histological features (Bandzar, Gupta, & Platt, 2013; Feuerstein et al., 2019). UC is more prevalent than CrD and marked by continuous and diffuse inflammation that can spread proximally up from the rectum and localize at the colonic mucosa, the primary barrier of defense that release antimicrobial peptides and supports immune tolerance through separation between host immune cells and luminal microbiome (Feuerstein et al., 2019). The degree of UC is stratified by the amount of colonic involvement starting from proctitis to left-sided colitis or extensive colitis (pancolitis) (Ordás, Eckmann, Talamini, Baumgart, & Sandborn, 2012). UC symptoms include bloody diarrhea, abdominal pain, fecal urgency, tenesmus, and several extraintestinal manifestations (EIMs). Arthritis was the most commonly observed EIM in patients diagnosed with UC (Karreman, Luime, Hazes, & Weel, 2017). The main tissue and cell pathology markers of UC are linked to mucosal layers and include Paneth cell metaplasia, mucin depletion (Goblet cell-associated pathology), distortion of crypt architecture (histological marker), crypt abscesses, and infiltrates of the mucosa with lymphocytes, plasma cells, and granulocytes (Danese et al., 2014; Feuerstein et al., 2019). For instance, the mucinous layer of epithelial cells in UC is less prominent and produces either insufficient amounts of colonic mucin subtype 2 (mucin 2), or the mucin is not properly folded and/or sulfated (Ordás et al., 2012; Van Klinken, Van der Wal, Einerhand, Büller, & Dekker, 1999). Accordingly, epithelial cell layers continue to deteriorate. Another prominent marker of UC, defective regulation of epithelial cell tight junctions results in the loss of barrier function and facilitates uptake of luminal antigens which further augment inflammation (Ordás et al., 2012).

Compared to UC, CrD is a chronic inflammatory condition of the gastrointestinal (GI) tract at any point from the mouth to the rectum. Notably, patients diagnosed with CrD are at significantly higher risk of premature death and shorter overall life expectancy (Bernstein, Nugent, Targownik, Singh, & Lix, 2015). CrD patients may also experience diarrhea, anemia, abdominal pain and masses, fever, and weight loss. Acute inflammatory phase indicators, such as C-reactive protein level and erythrocyte sedimentation rate, are often used in the diagnosis of CrD (Bandzar et al., 2013). EIMs of CrD are like UC and include inflammatory arthropathies (Bandzar et al., 2013; Cohen & Sachar, 2017). TNFα, a first line proinflammatory cytokine was shown to play a key-role in the development of inflammation in CrD and UC (Cohen & Sachar, 2017). Accordingly, several TNFα inhibitors are commonly prescribed as CrD treatment options, although none of the treatments provide effective recovery.

Investigation of the immunopathogenesis of IBD suggests that the main cause of intestinal tissues deterioration is the dysfunctional regulation of the immune system through unbalanced polarization of intestinal immune cells toward a Th1 (T-helper) response. Accordingly, the immunologic elements that mediate disturbed intestinal homeostasis represent the main target of IBD research. Aberrant invasion of mast cells, monocytes, macrophages and polymorphonuclear neutrophils into the colon epithelial layers, followed by amplification of multiple inflammation processes, is considered as a main diagnostic feature of IBD (Edwards et al., 2006). Infiltration of GI tissues by immune cells is directed by surrounding cytokine background. For instance, colonic tissue infiltration by the granulocytes was linked to increased production of several inflammatory cytokines including TNFα (Baumgart & Sandborn, 2007) supporting the use of anti-TNFα agents in IBD treatment for more than 15 years (Cohen & Sachar, 2017). Cytokines, their associated signaling pathways, and mediators of signaling are known as essential regulators of leukocyte activation, trafficking, and programmed cell death. During the last several decades cytokine signaling networks were explored as key immune mechanisms in the pathogenesis of IBD (Neuman, 2007). Currently, all moderately successful IBD therapies include agents that control various key-effectors of the immune system. Besides the TNFα network, gut-homing $\alpha_4\beta_7$ integrin, interleukin-12/interleukin-23, and Janus kinases were targeted by anti-IBD agents. However, many IBD patients do not show significant improvement after application of these agents which also cause several serious side effects (Giuffrida, Cococcia, Delliponti, Lenti, & Di Sabatino, 2019).

The sphingolipid signaling network was identified as a key regulator and mediator of pro-inflammatory gastrointestinal (GI) events, and, specifically, TNFα-related signaling (Hait & Maiti, 2017). Sphingolipid metabolism is activated during various immune responses (Hait & Maiti, 2017; Niwa et al., 2000; Rivera, Proia, & Olivera, 2008; Xia et al., 1998). A range of pro-inflammatory cytokines and coagulation-related substances stimulates sphingolipid metabolism associated with initiation and resolution of inflammation (Proia & Hla, 2015; Rivera et al., 2008). Sphingolipids were shown to intervene in several key immune regulatory processes both as positive and negative factors and multidirectional effectors. Recent advances in the understanding of the role of sphingolipids as regulators and mediators of cytokines and associated mucosal immunity and inflammation in IBD are discussed in the current Chapter. We also describe and critically evaluate major *in vitro* and *in vivo* human and animal studies related to understanding of sphingolipid impact on IBD development and treatment, currently available therapies, clinical trials targeting sphingolipid network, and potential therapy targets in IBD. Furthermore, the potential use of agonists and antagonists of sphingolipid signaling as cytokine-related biologic therapies and novel targets for the treatment of IBD are described.

2. Role of sphingolipid signaling during inflammation

The large and diverse class of sphingolipids includes more than 4000 distinct species with sphingoid amino alcohol bases. The most abundant structural sphingoid base is sphingosine ((2S,3R,4E)-2-aminooctadec-4-ene-1,3-diol) that is often used to assemble other naturally occurring sphingoid species with variations of chain lengths, saturation level, and number of hydroxyl groups (Pruett et al., 2008). The more complex group of sphingolipids contains ceramides as the structural backbone with longer amide-linked acyl chains ranging from 14 to 36 carbons in length (Hannun & Obeid, 2011). Sphingolipid metabolism, specifically ceramide and sphingosine synthesis are described elsewhere (Gault, Obeid, & Hannun, 2010). Sphingosine is produced by the hydrolysis of ceramide that is synthesized in the cytosolic leaflet of the endoplasmic reticulum (ER) (Gault et al., 2010). Sphingosine is quickly phosphorylated by sphingosine kinases (SphK1 and SphK2) to produce sphingosine-1-phosphate (S1P) described as one of the main regulators of inflammation (Fig. 1) (Maceyka & Spiegel, 2014).

Ceramide is a pro-apoptotic molecule that triggers both caspase-dependent and -independent mechanisms of cell death through interactions with several downstream targets including ceramide-activated protein phosphatase, protein kinase B/C, cathepsin D, and ROS (Wang, Lv, & Du, 2009). Like ceramide, sphingosine was also shown to activate programmed

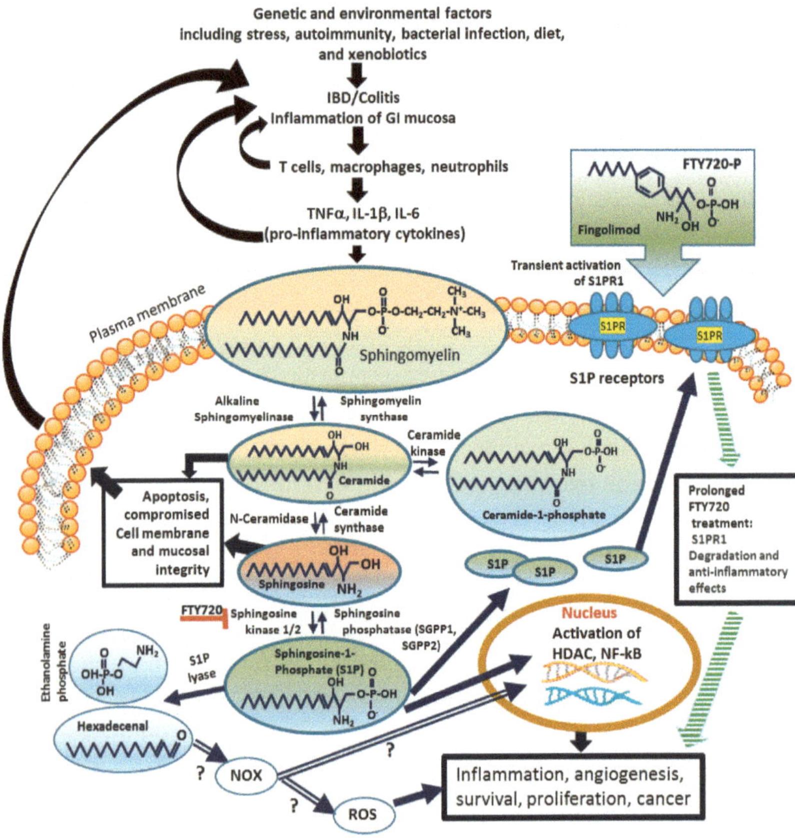

Fig. 1 Activation of sphingolipid metabolism by pro-inflammatory effectors in gastrointestinal tissues. Pro-inflammatory cytokines activate sphingomyelin hydrolysis by alkaline sphingomyelinase to produce ceramide in membranes of intestinal mucosal cells. Sphingomyelin synthase reverses this reaction. Neutral ceramidase (N-Ceramidase) hydrolyzes ceramide to sphingosine. The reaction can be reversed by ceramide synthase. Both ceramide and sphingosine are pro-apoptosis agents. Sphingosine can be further converted to S1P by SphKs (SphK1 in the cytosol; SphK2 in the nucleus). SphK1/2 produces S1P which promotes mitogenesis and concurrently depletes ceramide, thereby inhibiting apoptosis. SphK1 is also a critical mediator of the actions of pro-inflammatory cytokine TNFα. S1P can be converted back to sphingosine by sphingosine phosphatase (SGPP1 and SGPP2) or can be irreversibly broken down by sphingosine phosphate lyase (SPL) to ethanolamine phosphate and 2-trans hexadecenal. S1P can trigger epigenetic regulation of genes activation in the nuclear space (activation of histone deacetylases (HDACs) and nuclear transcription factors NF-kb). Hexadecenal can form adducts with proteins, lipids and DNA, stimulate production of reactive oxygen species (ROS) via activation of NOX proteins, and inhibit HDAC in the nucleus. Binding of S1P to the S1PRs results in activation signaling cascades, internalization and proteosomal degradation of S1PRs, or recycling of the receptors to the plasma membrane. FTY720/Fingolimod can activate S1PR1, S1PR3, S1PR4, and S1PR5. Binding of FTY720 to S1PR1 was shown to cause immune suppression during inflammation. However, some mechanisms of signaling were not tested in GI mucosal cells and questions marks (?) indicate that pathways remain to be confirmed.

cell death. Beyond apoptosis and associated cell signaling, changes in ceramide and sphingosine intracellular concentrations were linked to inflammation and associated metabolic diseases (Nixon, 2009). To avoid apoptosis, sphingosine is quickly phosphorylated by SphK1 and the resulting product S1P was shown to activate cell division, survival, and proliferation (Sukocheva, 2018). Accordingly, the pro-apoptotic substances (including cytokines) were shown to increase intracellular concentration of ceramide/sphingosine, while anti-apoptotic agents were shown to stimulate production of S1P via activation of SphK1; the mechanism is commonly called sphingolipid rheostat (Baumruker & Prieschl, 2002). S1P is a multifunctional second messenger involved in the regulation of various biological responses including cell growth, transformation, and migration. Intracellular S1P can be further degraded by S1P lyase or dephosphorylated back to sphingosine by phosphatases (Olivera, Allende, & Proia, 2013). When present in the extracellular space, S1P can bind and activate specific S1P receptors (S1PR) (S1Pn, n = 1 to 5) differentially expressed in various tissues including all GI and associated organs (Kawakita et al., 2017; Matula et al., 2015; Wang et al., 2014). Notably, the expression levels of different S1PRs were shown to change during inflammation (Kawakita et al., 2017; Peyrin-Biroulet, Christopher, Behan, & Lassen, 2017).

The SphK/S1P signaling network can be activated by a range of pro-inflammatory cytokines and coagulation-related substances during the initiation and progression of immune responses (Fig. 1) (Hait & Maiti, 2017; Niwa et al., 2000; Rivera et al., 2008; Xia et al., 1998). S1PRs are expressed in various immune cell subtypes including monocytes/macrophages, neutrophils (during inflammation), dendritic cells (DCs), eosinophils, natural killer (NK) cells, and mast cells (Hait & Maiti, 2017; Proia & Hla, 2015; Rivera et al., 2008). An important element of inflammation, lymphocyte trafficking, is also regulated by S1P concentration levels in blood, lymph, and/or interstitial fluids (Chiba, Matsuyuki, Maeda, & Sugahara, 2006; Rivera et al., 2008). Specifically, immune cell chemotaxis is coordinated by the strong S1P gradient between tissues and blood/lymph plasma compartments. While present at very low concentration in tissues, the S1P concentrations are much higher in the blood and the lymphatic system (Camerer et al., 2009; Sukocheva, 2018). This concentration gradient has been suggested to facilitate immune cell migration, including lymphocyte egress from primary lymphoid organs towards higher S1P concentrations present in blood/lymph (Chiba et al., 2006; Hait & Maiti, 2017; Rivera et al., 2008).

Vascular inflammation is marked by increased levels of S1P and activation of sphingosine metabolizing enzymes SphK1 and SphK2 that were shown to mediate pro-inflammatory responses in neutrophils (Niwa et al., 2000), monocytes, platelets, and endothelial cells (Knorr, Münzel, & Wenzel, 2014; Proia & Hla, 2015). Besides the designated kinases, extracellular S1P receptors (S1PR) are activated during inflammation and can bind and remove the excess of the sphingolipid from circulation through internalization (Sukocheva et al., 2018). However, the role of S1PRs in activation and resolution of inflammation remains largely unclear and often controversial. For instance, S1PRs promoted DCs-dependent inflammatory mechanisms including DCs migration to lymph nodes (Gollmann et al., 2008). Impediment in leukocytes rolling was associated with S1P receptor 3 (S1PR3) contribution to P-selectin-dependent signaling in vasculature during early stages of atherosclerosis (Nussbaum et al., 2015). Accordingly, S1PR3 can augment inflammation via stimulation of IL-1β and tissue factor production (Niessen et al., 2008). Vascular inflammation and associated vascular leakage result in increased expression and activation of signaling of endothelial S1PRs (specifically S1PR1) that bind S1P released into circulation from activated platelets, or erythrocytes, or endothelial cells (Camerer et al., 2009). The binding of S1P to S1PRs can either promote or reduce the inflammatory cascade confirming that S1PR-mediated signaling is orchestrated by other effectors of inflammation. For instance, during an acute phase inflammatory response, the S1PRs can reduce sepsis-associated mortality via protein C activation (Niessen et al., 2009). Multidirectional interactions between S1PRs and pro-inflammatory cascades in the vasculature were recently reviewed (Obinata & Hla, 2012; Rauch, 2014). Although many details of SphK/S1P axis involvement in vascular inflammation were uncovered during the last decade, the exact mechanism remains unclear and require further investigation, particularly considering its strong link to facilitation of angiogenesis and progression of cancer.

SphK/S1P axis was shown to mediate the inflammatory cascade initiated by various pro-inflammatory effectors including interleukin 1β (IL1β) (Pettus et al., 2003), TNFα (Billich et al., 2005; Xia et al., 1998), thrombin, and its receptors in the vasculature (Rivera et al., 2008). For instance, activation of sphingolipid signaling was noted during progression of the most acute and fast-developing inflammatory reactions in vasculature, sepsis. In sepsis, hyper-stimulated innate immune system effectors switched on the blood coagulation cascade that resulted in discharge of tissue factor and microvascular thrombosis (Riewald & Ruf, 2003), triggered by tissue factor, coagulation activates anticoagulatory processes by feedback mechanism via protease-activated receptor-1 (PAR-1) that binds thrombin and other coagulation-related proteases. Coagulation has been shown to deliver pro-inflammatory effects in sepsis. However, divergent responses are initiated to resolve inflammation. PAR-1 can switch on both protective (anti-inflammatory) and disruptive (pro-inflammatory) effects during development of sepsis (Kaneider et al., 2007). Thrombin was shown to induced cross talk with the S1PRs during activation of endothelial cell contraction and permeability (Niessen et al., 2009). Furthermore, S1P axis served as a downstream component of thrombin signaling in DCs during sepsis (Mahajan-Thakur et al., 2014; Niessen et al., 2008). The increased production of S1P is supposed to limit thrombin effects in the endothelium. However, S1P was also shown to promote thrombin signaling during development of atherosclerosis (Niessen et al., 2009). To explain the divergent effects of the thrombin/PAR-1/S1PRs signaling pathway, it was suggested that the balance is tipped by involvement of different vascular S1PRs (Niessen et al., 2009). Expression of specific S1PRs has a dual effect on the PAR-1 signaling pathway resulting in activation of several feedback systems and multifactorial networks (Obinata & Hla, 2012; Rauch, 2014).

Besides involvement in activation of coagulation, the SphK/S1P axis was linked to other diversified effects in the endothelium and surrounding tissues. It has been shown that the pro-inflammatory cytokine TNFα activates SphK1 and increases S1P production during induction of adhesion processes in endothelial cells (Xia et al., 1998). In vascular smooth muscle cells (VSMCs) activated S1PR3 induced expression of COX-2 via calcium-dependent PKC and Src-family tyrosine kinases (Nodai et al., 2007). Furthermore, the SphK/S1PR network triggers major inflammation-related transcription factors such as NF-κB (Billich, Urtz, Reuschel, & Baumruker, 2009) and forkhead-box-O (Mahajan-Thakur et al., 2014).

SphK/S1P signaling is activated during monocyte- and macrophage-associated immune reactions. All five S1PRs are expressed by human monocytes (Duong et al., 2004; Shatrov, Lehmann, & Chouaib, 1997) and were shown to mediate monocyte apoptosis and chemotaxis (Gude et al., 2008). Increased SphK1 activity was shown to direct macrophage recruitment and required for manifestation of anti-inflammatory macrophage characteristics (Duong et al., 2004; Furuya et al., 2017). For instance, generation of S1P by SphK1 protected LPS-activated macrophages from apoptosis and prevented oversupply of ceramide through inhibition of acid sphingomyelinase

(Gómez-Muñoz, Kong, Salh, & Steinbrecher, 2004). This observation was supported by experimental inhibition of S1P production. Down-regulation of SphK1 expression was associated with increased basal activity of interferon (INF)-induced genes and elevated pro-inflammatory responses during progression of viral infections (Clarke et al., 2016). S1P protective effects has been demonstrated during systemic infection and overproduction of pathogenic cytokines (Mahajan-Thakur et al., 2014). Accordingly, SphK and increased level of extracellular S1P were shown to regulate respiratory macrophage function and promote antibody-mediated phagocytosis in alveolar macrophages (Hait & Maiti, 2017).

The level of S1P production is important for monocyte differentiation into dendritic cells (DCs), DCs migration, and DCs relocation within the spleen (Czeloth et al., 2007). Heterogeneous expression pattern of S1PRs expression was suggested to indicate the maturation status and origin of DCs (Karuppuchamy et al., 2017). For instance, in bone marrow—derived DCs SphK1/S1PRs are considered key components involved in regulation of selection-induced cell death, DCs responses to specific pathogens or adverse conditions, and release of specific cytokines (Schaper et al., 2013). Pro-inflammatory behavior of monocytes controlled by the thrombin/PAR pathway, can be also be influenced by the SphK/S1PRs axis. Increased S1P production was linked to stimulated expression of PAR-1/PAR-4 in human monocytes, resulting in pro-inflammatory effects including induction of chemotaxis and elevated generation of COX-2 (Mahajan-Thakur et al., 2014). Furthermore, released from coagulation-induced platelets during vascular injury, S1P triggers various pro-inflammatory responses in circulating monocytes. The S1P-related pro-inflammatory and migration-directing process was also observed during metastasis and angiogenesis (Porta et al., 2009).

Besides the observed effects in vascular tissues, increased S1P concentrations were detected during spreading of inflammatory cascades in tissues and organs. Originating in blood plasma, sphingolipid-related signals penetrate the tissues along the vascular leakage and spread over the inflamed tissues. It has been shown that hematopoietically derived S1P induced S1PR1 signaling in hepatocytes and, speculatively, can initiate similar responses in other tissue-located parenchymal cells during inflammation (Ikeda et al., 2009; Proia & Hla, 2015). However, mechanisms of SphK/S1P signaling are very organ- and tissue-specific and generalization of the observed effects should be conducted very carefully. Similarly to vascular inflammation, controversial effects of SphK/S1P signaling during resolution of inflammation in tissues were detected. Activation of the SphK/S1P axis can both promote and prevent development of adverse effects of inflammation associated with various pathologies. For instance, activation of S1PR2/S1PR3 was noted in liver, lung, and kidney tissues during inflammation-linked fibrotic processes (Sato et al., 2016; Sobel et al., 2013; Wang et al., 2017; Zhang, Ritter, & Li, 2018). However, activation of the SphK/S1P/S1PRs axis signaling is not limited to promotion of inflammation, the pathway triggers reparative processes (Ikeda et al., 2009; Proia & Hla, 2015; Zhang et al., 2018).

The causative association between inflammation and gastro-intestinal (GI) malignancies has been demonstrated in multiple studies (Pennel, Park, McMillan, Roseweir, & Edwards, 2019; Porta et al., 2009). A diverse range of inflammation-activating effectors has been suggested and many of those effectors were shown to activate the sphingolipid network. Accordingly, the role of sphingolipids as mediators of pro-inflammatory responses and triggers of GI cancers is currently under vigorous investigation. The detailed understanding of mechanisms of sphingolipid signaling networking and downstream targets in health and disease will provide new strategies to influence immunity and prevent progression of inflammation-linked pathologies to GI cancer. Below, we analyzed the known mechanisms of SphK/S1P axis involvement in mediation and regulation of IBD and discuss its potential use as a treatment target within clinical settings.

3. Sphingolipid metabolism as anti-inflammatory target in IBD: pre-clinical studies

The strong proinflammatory cytokine TNFα has been shown to provoke and maintain the Th1 program responses including induction of nuclear factor kappa B (NFκB) signaling in the pathogenesis of IBD (Baumgart & Sandborn, 2007; Argollo et al., 2017). Among the downstream TNFα targets are the key proinflammatory enzymes such as nitric oxide synthase (NOS) and cyclooxygenase-2 (COX-2) (Mahadevan, Loftus, Tremaine, & Sandborn, 2002), and a net of enzymes involved in propagation of oxidative stress (Kruidenier, Kuiper, Lamers, & Verspaget, 2003). The SphK/S1P/S1PRs axis is one of the most prominent TNFα-induced downstream targets in various cells (Billich et al., 2005; Pettus et al., 2003; Xia et al., 1998). Switched by TNFα, SphK1 triggers COX-2 expression and production of prostaglandin E2 (PGE2) in L929 fibroblasts (Pettus et al., 2003) and colon cancer cells (Kawamori et al., 2006; Kohno et al., 2006). Accordingly, enhanced SphK1 protein expression exacerbated IBD conditions in pre-clinical colitis mouse model and in patients with UC (Crespo et al., 2017; Karuppuchamy et al., 2017; Snider et al., 2009). The oxazolone-induced colitis model was marked by mild DNA fragmentation and significantly increased SphK1 activity in colonic tissue (Abdin, 2013). Considering the pro-inflammatory effects of SphK1, an adverse role of elevated S1P in IBD was suggested (see Table 1). Supporting this theory, deletion of *sphk1* was associated with lower S1P concentrations and reduced severity of dextran sodium sulfate (DSS)—induced colitis (Snider et al., 2009) and loss of neutral ceramidase, an enzyme that breaks down ceramide to sphingosine, resulted in an increased S1P-associated inflammation (Snider et al., 2012). Notably, SphK2 seems to play a protective role in IBD as genetic *sphk2* deletion was shown to aggravate inflammatory conditions in both IBD (Liang et al., 2013) and rheumatoid arthritis (Baker, Eudaly, Smith, Obeid, & Gilkeson, 2013). The mechanism of the SphK2 effect in IBD requires further detailed investigation.

Tissue-specific localization of SphK1 was suggested to play a regulatory role in the direction of the enzyme responses. For instance, assessing local and systemic inflammatory responses, Snider et al. (2014) observed distinct roles for hematopoietic and extra-hemopoietic pools of SphK1/S1P in IBD. Hematopoietic cell derived SphK1 regulates S1P concentration in circulation that, in turn, can stimulate neutrophilia and systemic inflammation (Snider et al., 2014). Although S1P has been implicated in the regulation of lymphocyte egress from lymphoid organs (Chiba et al., 2006), the source of the regulatory S1P pool in IBD was identified only recently. Hematopoietic cell-located SphK1 was suggested as a regulator of lymphocyte egress from the spleen in DSS-induced colitis (Snider et al., 2014).

Located in extra-hematopoietic cells, such as colon epithelial cells, activated SphK1 was shown to stimulate COX2 expression but was not involved in triggering of STAT3 activation in pre-clinical model of acute colitis (Snider et al., 2014). Reported links between SphK1 and STAT are controversial as other groups reported SphK1-dependent STAT3 activation (Gurgui, Broere, Kalff, & van Echten-Deckert, 2010; Liang et al., 2013). Liang et al. (2013) registered that elevated S1P induces STAT3 and NF-κB activation resulting in production of cytokine IL-6. Simultaneously, upregulation of the S1PR1 was observed in UC and colitis-associated cancer models *in vivo*. Accordingly, it has been suggested that the SphK1/S1P/S1PR1 axis might serve as an interconnecting bridge between NF-κB and

Table 1 SphK/S1P/S1PRs effects in IBD/colitis associated pathologies: key-stone findings.

Mechanisms/roles of SphK/S1P/S1PRs	Type of study, pathology and # of patients; method/model	Reference
SphK1 facilitated progression of spontaneous intestinal adenoma *in vivo*; epithelial cell proliferation in the polyps was attenuated when SphK1 was downregulated.	Min mice model *in vivo*; Colitis transformation to colon cancer	Kohno et al., 2006
S1P/S1PRs (1-3) stimulated expression of IL-1 and COX-2 via Erg-1, and expression of IL-6 via activation of STAT3.	Rat intestinal smooth muscle cells; *in vitro* inflammation model in postoperative ileus	Gurgui et al., 2010
SphK1 expression was increased in IBD animal models and in human colon tissues/IBD patients. SphK1 mediated colonic damage during intestinal inflammation/Crohn's disease; SphK1 inhibitor, ABC294640 was used.	IBD, Crohn's disease was tested in animal models *in vivo*, human IBD/colon patient tissues, IHC	Maines et al., 2010
Increased S1P generation resulted in higher levels of inflammation and activation of COX-2. Otherwise, deletion of *sphk1* reduced severity of colitis.	*In vivo* neutral ceramidase KO mice models of IBD/DSS–induced colitis/UC; flow cytometry analysis	Snider et al., 2009; 2012; 2014
Colitis model was produced by a single immunomodulating helminth. The formation of new population of regulatory macrophages with high levels of SphK1.	*In vivo* colitis model in rodents; Flow cytometry, IHC	Ziegler et al., 2015
Intestinal symbiotic microbe *Bacteroides fragilis* can release sphingolipids that modify the homeostasis of host iNKT cells and decrease susceptibility to oxazolone-induced colitis.	*In vivo* colitis model in rodents	An et al., 2014

(*Continued*)

Table 1 SphK/S1P/S1PRs effects in IBD/colitis associated pathologies: key-stone findings.—cont'd

Mechanisms/roles of SphK/S1P/S1PRs	Type of study, pathology and # of patients; method/model	Reference
Elevated SphK1, S1P, S1PR1, and S1P phosphatase 2 were detected in mice IBD models; PCA was used to block S1P network and reverse IBD associated conditions	*In vivo* IBD model in rodents; IHC, WB, PCR	Crespo et al., 2017
SphK1/S1P/S1P1 axis might serve as an interconnecting bridge between NF-κB and STAT3 in colitis	Colitis model *in vivo*	Yuza et al., 2018
Deoxycholic acid can induce S1PR2 and trigger ERK1/2/cathepsin B signaling cascade that activates formation of NLRP3 inflammasomes in IBD	*In vivo* IBD model in rodents; Flow cytometry, IHC, WB, PCR	Zhao et al., 2018
Increased SphK1 activity in colonic tissue coincided with the mild DNA fragmentation	oxazolone-induced colitis model	Abdin, 2013
Elevated S1P can activate STAT3 and production of NF-κB-regulated cytokine IL-6. The signaling cascade led to up-regulation of the S1PR1 in colitis models	Colitis and cancer models *in vivo*	Liang et al., 2013
Enhanced levels of SphK1 and/or S1PR1 were found in inflamed intestinal mucosa	Patients with ulcerative colitis	Montrose et al., 2013
Deletion of *sgpp2* lessened DSS-induced colitis. Alternatively, deletion of *sgpp1* was shown to deteriorate IBD conditions and stimulate expression of IL-6, TNF-α, and IL-1β, and promote inflammation.	*In vivo* and *in vitro* models	Huang et al., 2016

SphK1 selective inhibitor, LCL351 decreased expression of pro-inflammatory markers and neutrophil infiltration.	Colitis model in mice *in vivo*	Pulkoski-Gross et al., 2017
S1P1 antagonist FTY720 cures colitis conditions and symptoms marked by abnormal T cell responses	Murine colitis models *in vivo*	Daniel et al., 2007
S1PR4 deficiency was linked to the changed composition of peritoneal B cell populations and reduced secretory IgA levels in inflammatory colitis	Colitis model in mice *in vivo*	Kleinwort et al., 2018
Selective immunomodulatory agonists for S1PRs blocked leukocyte migration into the inflamed mucosa	Modulation of S1PRs signaling in IBD clinical trials	Peyrin-Biroulet et al., 2017
Increased tissue S1P levels and expression of S1PR1 in several types of immune cells including a sub-set of T and B cells, GI-located endothelial and dendritic cells during chronic inflammation. Anti-inflammatory properties of FTY720 via S1PRs blokade: lymphopenic effects, regulation of DC migration, and vascular barrier function	DSS-model of IDB was used *in vivo* with S1PR1 transgenic mice; IHC, flow cytometry and imaging analysis with mouse tissues *in situ*	Karuppuchamy et al., 2017

STAT3, linking chronic inflammation with colitis-associated cancer (Fig. 1) (Liang et al., 2013; Yuza et al., 2018). Secretion of pro-inflammatory cytokine IL-6 was also activated by S1P/S1PRs (1-3) via STAT3 in rat intestinal smooth muscle cells used to pro-inflammatory colitis model in post-operative ileus. Besides, IL-6 in the same study, expression levels of IL-1β and COX-2 were also stimulated by S1P via Egr-1 (Gurgui et al., 2010).

It is considered established that SphK/S1P activation facilitate worsening of IBD symptoms. Accordingly, IBD-related beneficial effects of SphK/S1P signaling are controversial, despite reported protective effects of sphingolipids that were shown to defend intestinal mucosa against pathogenic bacteria and toxins (Greenspon et al., 2011). Although elevated S1P concentration was identified as an adverse factor in IBD, elimination of S1P via dephosphorylation back to sphingosine was not able to relieve disease severity (Huang et al., 2016). To reduce S1P level, it is necessary to degrade it or convert back to sphingosine. The conversion of S1P to sphingosine is catalyzed by S1P-specific phosphohydrolase isoforms (SGPP1 and SGPP2) (Le Stunff, Peterson, Liu, Milstien, & Spiegel, 2002). While deletion of *sgpp2* lessened DSS-induced colitis severity symptoms and improved mucosal barrier integrity, removal of *sgpp1* was shown to worsen IBD conditions and stimulate the expression of several pro-inflammatory effectors including IL-6, TNF-α, and IL-1β, thus, promoting immune cell infiltration into the colon (Huang et al., 2016). It is necessary to note that the adverse SGPP1 effect can be also associated with pro-apoptotic effects of increased levels of sphingosine (Le Stunff et al., 2007; Nagahashi, Abe, Sakimura, Takabe, & Wakai, 2018). Alternatively, S1P lyase degrades S1P and produces hexadecenal and phosphoethanolamine, the substrates of glycerophospholipid synthesis (Harrison, Dunn, & Campopiano, 2018; Matula et al., 2015; Olivera et al., 2013). The effects of increased expression and activation of S1P lyase requires further detailed testing in IBD-related models.

Inflamed mucosa of patients with UC expresses enhanced levels of SphK1 and S1PR1 that were shown to mediate inflammatory signaling in this tissue (Montrose et al., 2013). Furthermore, overexpressed SphK1 promotes proliferation in intestinal epithelial cells through enhanced expression of c-Myc (Jiang et al., 2013), a transcription factor with oncogenic potential (Hartl, 2016). A link between SphK1 and c-Myc indicates on direct carcinogenesis-promoting mechanism of this enzyme. Supporting this suggestion, phosphorylated SphK1 (pSphK1) was found significantly more prevalent in colitis-associated cancer patients (Yuza et al., 2018). The same study detected that colitis-associated cancers have a higher pSphK1 immunohistochemistry (IHC) score than sporadic colorectal cancers (CRCs) (Yuza et al., 2018).

Targeted *in vitro* and *in vivo* studies with specific SphK1/S1PRs subtype inhibitors support pro-inflammatory roles for the SphK/S1P axis. Application of the SphK1 selective inhibitor, LCL351, resulted in the decreased expression of pro-inflammatory markers (chemokines C-X-C motif ligand (CXCL) 1 and CXCL2 mRNAs), and a reduced neutrophil infiltration in colon tissue from colitis models in mice. Interestingly, the LCL351-dependent decrease in S1P tissue concentrations was suggested to prevent leukocyte recruitment and trap immune cells in the circulation (Pulkoski-Gross et al., 2017). Furthermore, it was shown that specific S1PR1 inhibitors, including FTY720 (fingolimod), reduce the severity of colitis and improve symptoms in murine models marked by abnormal T cell responses. Fingolimod-induced beneficial effects were associated with increased functions of $CD4^{+}CD25^{+}$Tregs cells and Th2-type responses, suggesting that the SphK/S1P axis can directly regulate specific T cell activation (Daniel et al., 2007) and T cell-dependent release of cytokines (Ottenlinger et al., 2016). Considering that TNF-α is one of the major inducers of the Th1 program, it is rational to conclude that the SphK1/S1P axis directly mediates TNFα signaling in IBD. However, this statement requires detailed verification.

The range of SphK/S1P-affected immune targets is not limited to the T cell pool. S1P signaling was also demonstrated to influence peritoneal B1 cell migration. In the recent study it was observed that S1PR4 deficiency results in the changed composition of peritoneal B cell populations and reduced secretory IgA concentrations. The data suggested S1P signaling as a potential therapeutic target to modulate B cell functions in inflammatory intestinal pathologies (Kleinwort, Lührs, Heidecke, Lipp, & Schulze, 2018). GI monocytes/macrophages might be involved in regulation of this process, however, a tripartite mechanism (SphK/S1P/S1PRsp(?), macrophages, B cells) has not been properly addressed yet. Supporting the complexity of interaction, a novel inflammation-suppressive/regulatory macrophage population was recently detected. The population was marked by high expression of SphK1 and managed to successfully protect against mucosal inflammation. Interestingly, this type of macrophages was primed and propagated by a single parasite immunomodulator secreted by an intestinal helminth (Ziegler et al., 2015).

The level of different S1PRs expression in GI cells and their role in regulation of immune responses in IBD remains largely unclear. Recent studies indicate the involvement of S1PR1 in regulation of vascular integrity during colitis. Inhibition of S1PR1 signaling by several new oral selective immunomodulatory antagonists of S1PRs blocked leukocyte migration into the inflamed mucosa (Peyrin-Biroulet et al., 2017). However, the specificity of the tested antagonists should be further explored to confirm or exclude the involvement of other S1PRs in this process. Exploring the DSS-induced tissue injury mouse model of colitis, it was shown that deletion of the *s1pr1* gene in the mouse endothelium resulted in colonic vascular leakage and enhanced bleeding. Interestingly, an alternative method of S1PR1 inhibition of signaling indicated different results. Treatment with fingolimod and the S1PR1-specific compound AUY954 did not increase bleeding in this model, suggesting that the ligand-induced S1PR1 degradation can be explored locally and have therapeutic potential (Montrose et al., 2013). Similarly to S1PR1 antagonists, SphK1 inhibitors (ABC747080 and ABC294640) were shown to direct cellular responses to TNFα in DSS mice models (Maines, Fitzpatrick, Green, Zhuang, & Smith, 2010). The drug treatment reduced expression of SphK1 and downregulated levels of S1P that was increased by DSS treatment. Further assessment indicated that the tested drugs can prevent TNFα-induced activation of NFκB, but stimulate expression of adhesion proteins, and enhance production of PGE2. Thus, the SphK1 inhibitors were able only partially prevent disease progression and reduce colonic levels of inflammatory cytokines (TNF-α, IL-1β, IFN-γ, IL-6), as well as S1P levels. Notably, the study demonstrated that SphK inhibitors did not eliminate all adverse signs of the colitis, particularly colon shortening and colonic neutrophil infiltration (Maines et al., 2010).

Sphingolipids are key structural components of all cell membranes. Accordingly, sphingolipids maintain the functional integrity of the intestinal mucosa and control intestinal absorption processes that are facilitated by membrane transport. However, the role of externally synthesized sphingolipids in the regulation of GI homeostatic processes is largely unclear. It was demonstrated that besides digestion, the GI tract plays a crucial role in the regulation of immune responses. The GI system is permanently exposed to a continuous flow of introduced and internal bacterial components. The interactions among bacteria (microbiome), diet, and host sphingolipids are potentially complex, mutual, and controversial (Hanada, 2005). The effect of essential food components on GI sphingolipid signaling requires detailed testing and should be considered within the interaction with symbiotic GI bacteria. Recent study indicated that fish oil, a common food supplement, down-regulates SphK1/S1P signaling in GI tissues (Li et al., 2012). The role of other diet components/lipids remains largely unclear (Merrill et al., 1995).

Guided by the immune system, GI cells adapt to tolerate a daily exposure to substances and bacteria from the external environment during food consumption. The toll-like receptor (TLR) signaling network, a crucial part of innate immunity system, was shown to participate in development of GI tolerance to the symbiotic bacteria that populate human intestines (Round, O'Connell, & Mazmanian, 2010). Although the role of sphingolipids in the regulation of innate immunity remains essentially unclear, TLRs were shown to mutually interact with sphingolipid metabolism (Arlt et al., 2014; Hamada et al., 2014). The innate immune system maintains normal GI homeostasis. However, during dysregulation of GI functions, innate immune responses were shown to intensify the pathogenesis of various inflammatory conditions including UC. Sphingolipids and their metabolites were suggested to influence microbial and viral pathogenesis, although this requires future investigation. For instance, first line guardians against invading pathogens, neutrophils were shown to engage sphingolipid signaling in UC (Espaillat, Kew, & Obeid, 2017). It is not entirely clear how the activation is initiated and how strong is the impact of intestinal bacteria/consumed food in this process. The complexity of GI tissue/bacteria and sphingolipid metabolism mutual interactions is aggravated by not only the large amount of bacterial population, but also by controversial findings collected so far. Both beneficial (symbiotic e.g. the microbiome) and harmful pathogenic bacteria were shown to impact the internal sphingolipid balance and host/microbial interactions (Harrison et al., 2018). Specific sphingolipids can be synthetized and released by bacteria. Originated from common human bacteria, phosphorylated dihydroceramides were shown to bind and activate TLR2 signaling resulting in progression of autoimmune diseases (Nichols et al., 2009). The intestinal symbiotic microbe *Bacteroides fragilis* can release sphingolipids that modify the homeostasis of host invariant natural killer T cells and impact microbial pathogenicity related to the oxazolone colitis susceptibility phenotype (An et al., 2014; Kato, Muto, Tanaka-Bandoh, Watanabe, & Ueno, 1995).

The potential impact of overall diet and related GI content on the sphingolipid network is supported by recent studies of IBD models in mice. Zhao et al. (2018) reported that deoxycholic acid, associated with a high fat diet, can activate S1PR2 and trigger the ERK1/2/cathepsin B signaling cascade, critical in the formation of the cytosolic innate immune signaling receptor

NLRP3 (NOD-, LRR- and pyrin domain-containing 3) inflammasomes. Previously, it was shown that SphK1 activation and membrane binding depends on plasma membrane composition that can be influenced by extracellular environment (Pulkoski-Gross et al., 2018), suggestively by bacterial and/or consumed sphingolipids. However, the mechanisms of SphK1/S1P network regulation by intestinal bacteria and consumed food are largely unclear. Further studies are warranted to clarify how the bacterial (or consumed) sphingolipids can impact the host sphingolipid protective barrier of intestinal mucosa against external pathogenic factors (Greenspon et al., 2011). Further description of the role of bacteria and diet in the regulation of sphingolipid metabolism and signaling is vital in the better understanding of the pathogenesis and development of GI inflammatory conditions and in the identification of novel therapeutic strategies.

4. Sphingosine kinase inhibitors are promising candidates for IBD treatment

As discussed in the previous section, the association of the well-characterized sphingosine kinase rheostat, SphK/S1P/S1PR1, with chronic inflammation and angiogenesis makes inhibitors of the SphK pathway strong candidates for treatment of acute and chronic IBDs. Activation of SphK1 is associated with IBDs (Yuza et al., 2018) and invasive gastric cancers (Hanyu et al., 2018). Persistence of IBDs, such as ulcerative colitis (UC), has been documented as high-risk factors for colorectal cancers (CRCs) and colitis-associates cancers (CACs) (Saleh & Trinchieri, 2011; Ullman & Itzkowitz, 2011). Given IBDs are linked to many types of inflammatory cancers, such as CRC and CAC, there is also enhanced interest in IBD-associated cancer prevention using SphK pathway inhibitors (Liang et al., 2013; Yuza et al., 2018).

Inflammatory cytokines, such as TNF-α, and growth factors are upstream activators of the SphK signaling, and play prominent roles in the inflammation associated with IBDs, both in initiation and persistence (Sandborn, 2003). Anti-TNF-α drugs were the first biologics to be developed for inflammatory diseases such as acute and chronic IBDs. Currently licensed anti-TNFα prescriptions are available for clinical use for IBDs, included Remicade® (infliximab), Humira® (adalimumab), and Cimzia® (certolizumab pegol) (Levin, Wildenberg, & van den Brink, 2016; Park et al., 2014). Side effects and contraindications of anti-TNFα make monitoring of patients critical during treatment as outlined in Gerriets and

Khaddour (2019). Therefore, targeting and blocking SphK/S1P activity, and thus blocking the inflammatory TNF-α response, is currently being actively research as an alternative biologic for inflammatory diseases.

The SphK isozymes, SphK1 and SphK2, are the only enzymes known to phosphorylate sphingosine to the active form S1P, making them crucial players in this rheostat, and key targets for anti-inflammatory therapies (Pyne, Adams, & Pyne, 2016). Due to their potential as anti-inflammatory and anti-cancer agents, there are a plethora of SphK inhibitors developed that target the specific ATP sphingosine-binding pocket of SphK, as well as the earlier more generic SphK inhibitors as detailed in Cao et al. (2018), Lynch, Thorpe, and Santos (2016), Pitman, Costabile, and Pitson (2016), Pitman and Pitson (2010). Although most of these compounds have had moderate success, the amino acid sequences of ATP binding domain of the SphKs are conserved across many of the diacyl-glycerol (DAG) kinase families, hence the earlier SphK inhibitors were problematic, due to non-discriminatory kinase inhibition. The first SphK inhibitors based on sphingosine derivatives, such as L-threo-dihydrosphingosine (DHS:Sfingol), N,N-dimethylsphingosine (DMS), and l-threo-dihydrosphingosine (Safingol) block the activity of SphK1 and 2 isozymes competing with the natural sphingosine substrate (Cao et al., 2018). Important off-target and low questionable specificity, included targeting ceramide synthase (CerS), a group of enzymes that catalyze the formation of ceramides in the sphingolipid pathway, and protein kinase C (PKC), modulators of major signaling pathways in the cell. Given the importance of SphK/S1P in inflammatory diseases, huge leaps have been made in the design, development and manufacturing and pre-clinical trials of more selective inhibitors directed at the individual SphK1 and SphK2 isozymes and S1PRs (Cao et al., 2018; Ding et al., 2019; Gupta et al., 2019; Haddadi, Lin, Simpson, Nassif, & McGowan, 2017; Hatoum, Haddadi, Lin, Nassif, & McGowan, 2017; Magli et al., 2019; Sun, Zhou, Shi, & Liu, 2019).

As mentioned, the TNFα pathway is a major player in the initiation of inflammatory diseases. The inflammatory TNFα response triggers the activation of ceramidase, an enzyme that catalyzes the conversion of ceramide to produce sphingosine, which is then converted to S1P by SphK and initiates proliferation. The formation of S1P is also involved in protection against apoptosis a key response inflaming the chronic IBDs state: macrophage and neutrophil apoptosis is blocked by S1P. As mentioned earlier, activation of S1P mimics many of TNFα downstream functions (such as induction of COX-2 and adhesion molecules) and these actions are blocked with the SphK inhibitor dimethyl sphingosine (DMS) (Pettus et al., 2003; Xia et al., 1998). Oral-bioavailable compounds, ABC747080 and ABC294640 also blocked the TNFα response (Maines et al., 2008). More recently, the use of a SphK1 selective inhibitor, PF543, was reported to improve the symptoms and pathological changes in DSS-induced colitis in pre-clinical *in vitro* and *in vivo* mouse models through inhibition of the pro-inflammatory cytokines, interleukin (IL)-1β and IL-6 (Schnute et al., 2012; Sun et al., 2019). These results show the use of SphK/S1P inhibitors to block/prevent the TNFα inflammatory response in IBD therapy as an alternative to anti-TNFα treatment, with potential to reduce some of the adverse side effects of current treatments, is enticing.

5. Specificity of SphK inhibitors as an indicator of clinical effectiveness

Although the two SphK isozymes, SphK1 and SphK2, both activate S1P they have very distinct, as well as similar, biochemical properties and functions, reviewed in Haddadi et al. (2017), Hatoum et al. (2017). Importantly they exhibit compensatory mechanisms when ablated in SphK-1 or SphK-2 mouse models. Hence, there is therapeutic advantages in targeting selective and/or non-selective SphK1 and SphK2 inhibitors and dual inhibitors depending on the different pathological conditions targeted. There is now a plethora of SphK inhibitors available to target one or both isoforms, reference Table 2 (Hatoum et al., 2017). Nonetheless most anti-SphK compounds have preferentially, not total, affinity to one isozyme (Vettorazzi et al., 2019): anti-SphK1 inhibitors usually have low affinity for SphK2 and vice versa. Most interest has focused on SphK1 inhibitors due to their strong association with inflammatory diseases and cancer and, as mentioned, the anti-SphK1, PF543 (Schnute et al., 2012; Sun et al., 2019), has shown some promise in improving IBDs in pre-clinical experiments. However, with the promising results of the immuno-suppressant drug, Fingolimod (2-amino-2[2-(4-octylphenyl)ethyl]- 1,3-propanediol), FTY720, which is phosphorylated to its active form FTY720-P primarily by SphK2, more focus has been directed on the design, molecular modeling and evaluation of SphK2 specific blockers as anti-inflammatory drugs (Vettorazzi et al., 2019). The selective inhibitor of SphK2, 7-bromo-2-(2-phenylethyl)-2,3,4,5-tetrahydro-1,4-epoxynaphtho[1,2-*b*]azepine (compound 17), developed by Vettorazzi and colleagues, also had promising results in their *in vitro* studies (Vettorazzi et al., 2019). This compound 17, designed through the molecular modeling of the SphK2 structural scaffolds, specifically targeting SphK2, did not exert any cytotoxic effects, exerted a potent anti-inflammatory and inhibited mononuclear cell adhesion in the dysfunctional endothelium with little impact on neutrophil-endothelial cell interaction. Mononuclear cell recruitment is associated with cardiometabolic disorders in the treatment of inflammatory disorders, making the anti-SphK2-Compound 17 a desirable compound to follow through in IBDs treatments to reduce adverse side effects.

Although there are now many patented SphK inhibitors, with 280 anti-SphK compounds (Lynch & Santos, 2018), and specific SphK1 inhibitors (Spiegel, Zipkin, and Adams 2019), recently filed with the US patent office, with many of these inhibitors considered to have clinical therapeutic potential, only a few compounds have entered clinical trials, and even fewer are used in the clinic. One successful anti-SphK candidate (with greater affinity for SphK2), currently in clinical use for multiple sclerosis (MS) patients, is the FTY720 drug. The use of FTY720 as a candidate therapy for IBDs is further discussed below.

6. Promising future for Fingolimod (FTY720) in IBD treatment

The use of FTY720 (Fingolimod) has been shown to be efficacious in several different experimental models of colitis (Daniel, Sartory, Zahn, Schmidt, et al., 2007; Daniel et al., 2007; Deguchi et al., 2006; Liang et al., 2013). Fingolimod, originally synthesized in 1995 by a Japanese chemist, Tetsuro Fujita (Cordi et al., 1995), is a potent immunomodulatory drug (Huwiler & Zangemeister-Wittke, 2018). Fingolimod is an analog of sphingosine, a prodrug phosphorylated *in vivo* by SphK2, and of all the sphingosine analogs it is the most studied. The similarity of Fingolimod to the chemical structure of the sphingosine molecule makes it an ideal substrate for both SphK1 and SpK2, and acts as a functional antagonist of S1P (Brinkmann et al., 2010). Albeit in principle both SphK1 and SphK2 can phosphorylate Fingolimod, the affinity of Fingolimod to SphK2 is much greater (>30 times) than to SphK1. Phosphorylated Fingolimod (P-FTY720) structural similarly to S1P allows binding of P-FTY720 with most of the S1PRs, with the exception of S1PR4 (Haass, Nassif, & McGowan, 2015). The major role identified for Fingolimod action was depletion of peripheral lymphocytes and therefore the drug was tested as a potential treatment for adverse immune response diseases. The results of administration of Fingolimod was found to be no better than conventional treatments for most immune-based diseases, with the exception of multiple sclerosis (MS) patients, and in 2010, under the trade name Gilenya^{R}, was approved for use for MS sufferers by the Food and Drug Administration (FDA) agency (Brinkmann et al., 2010). More recently in pre-clinical models, both *in vitro* and *in vivo,* Fingolimod has proven efficacious in many types of cancer, showing therapeutic anticancer potential (White, Alshaker, Cooper, Winkler, & Pchejetski, 2016). Although the anti-cancer mechanism of Fingolimod are not fully understood, Fingolimod has been shown to induces apoptosis, through a number of proposed mechanisms, the cleavage of caspase 9 and 3 followed by PARP cleavage and inhibition of Bcl-2, Bcl-xl and Mcl-1 and increase in Bax (Bai, Chiu, Chiu, Chu, & Weng, 2017). In the context of chronic colitis, upregulation of S1P by SphK1 links IBDs to the development of CAC through the NF-κB/IL-6/STAT3 amplification cascade (Liang et al., 2013). Although Fingolimod is preferentially associated with SphK2, in the context of chronic colitis it has been shown to be effective in the blocking the development of CAC preferentially through the SphK1/S1P pathway; Fingolimod is effective in blocking IDB-CAC in the absence of SphK2 (Liang et al., 2013). In essence, removal of SphK2 (in SphK2−/− mice) resulted in exacerbation colitis-associated tumorigenesis. Together the promising results of these Fingolimod-related experiments may have broader implications for the treatment of IBDs in preventing the development of CAC. Considering that Fingolimod/Gilenya^{R} has FDA approval for clinical use as MS treatment, this makes Fingolimod and/or analogs of FTY720 high candidates for fast track clinical use for colitis and associated complications.

7. Targeting S1P/S1PR in IBD

The targeting of the S1P/S1P/S1PR axis in the SphK/S1P/S1PR rheostat is also emerging as a viable alternative to TNFα biologics in IBDs treatment (Danese, Furfaro, & Vetrano, 2018; Neurath, 2017; Stepanovska & Huwiler, 2019). The bio-active lipid S1P signals mainly through its five S1PR(1-5), which belong to the class of transmembrane G-protein coupled receptors that transmit signals from the extracellular space to the intracellular space of the cell (Kihara, Maceyka, Spiegel, & Chun, 2014). S1PRs mediate a complexity of signaling roles playing key roles in physiological and pathophysiological responses, including autoimmune and inflammatory diseases

(Stepanovska & Huwiler, 2019). As described above, S1PR can selectively control lymphocyte migration, endothelial permeability, cell survival, apoptosis and differentiation (Danese et al., 2018). Specific and non-specific S1PR modulators (antagonists and agonists) have been developed to target one or more of the receptors (Hatoum et al., 2017). Most, if not all, these SIPR modulators target SIPR1, whilst many also target more than one SIPR. SIPR1 is the most widely expressed of the S1PRs, with high expression on endothelial cells and lymphocytes and is deregulated during the progression of IBDs (Karuppuchamy et al., 2017). The S1P/SIPR1 interaction has been recognized as crucial in the regulation of lymphocytes migration from the spleen to lymph nodes and into the systemic circulation, aberrant lymphocyte migration has been identified as key in inflammatory IBDs (Danese et al., 2018). Thus, SIPR1 modulators have been shown to have therapeutic effects in pre-clinical studies (Danese et al., 2018). Notably, these S1PR1 modulators have lower affinity for other S1PRs that indicates on both positive and unwanted side effects due to binding and activation of other S1PRs. An example being FTY720 which binds to all the S1PRs except S1PR2. Binding of S1P modulators to S1PR3 has been associated with cardiovascular complications, hence the observed negative side effects of FTY720 may be due to the binding of SIPR3.

Specificity of S1P/S1PRs is therefore a very important factor when promoting any S1P modulators to clinical trial. To-date thirteen S1PR modulators have gone to clinical trial for multiple diseases (Stepanovska & Huwiler, 2019), however only four S1P modulators have proceeded to clinical trials for IBDs. The suggested list of modulators includes Ozanimod (RPC-1063 - targeting S1PR1 and S1PR5), KRP-203 (similar to FTY720), Etrasimod (APD-334), and Amiselimod (MT-1303), the later three all targeting S1PR1, S1PR4-5 (Danese et al., 2018). *Note,* the more selective S1PRs modulator KRP-203 lacks binding specificity to S1PR3, thus reducing the adverse cardiovascular effects of FTY720-binding of S1PR3, making this a more attractive therapeutic agent for IBDs. Estrasimod, has been shown to regulate lymphocyte trafficking and reduces colitis in pre-clinical models (Al-Shamma et al., 2019), Crohn's disease (Biswas, Bryant, & Travis, 2019) and other IBDs (Pérez-Jeldres et al., 2019). Ozanimod has been tested in human ulcerative colitis and has a slight increase in clinical remission than placebo (Neurath, 2017). As the scope of more potent S1PRs modulators extend, with better selectivity, improved safety and tolerance, and independence of SphK activation, there are more promising translational clinical outcomes for IBDs.

8. Summary, challenges and future research directions

Pharmacological targeting of cancer-associated systemic inflammation and the inflammatory tumor microenvironment is considered as highly beneficial approach. The approach is strongly supported. Firstly, immune cells did not demonstrate similar mechanisms of drug resistance as solid tumors, and, secondly, anti-inflammatory therapies indicated promising data for cancer prevention at the earlier stages. Accordingly, development of substances that can target inflammation-related anti-cancer signaling pathways attract growing interest. Acute and chronic inflammatory processes are tightly intervened and mediated by sphingolipid signaling. Specifically, SphK/S1P/S1PRs axis was shown to regulate multiple immune responses and inflammation-linked pathologies in the GI tract. Considering the critical role of S1P in linking inflammation and cancer, the S1P signaling pathway has been suggested as an excellent therapeutic target for prevention and treatment of cancers triggered by inflammation (Nagahashi et al., 2018; Sukocheva et al., 2018). SphK/S1P/S1PR-targeting anti-inflammatory drugs are being developed and tested in pre-clinical and small clinical trial settings. The list of already successful or tested agents include, but not limited to, FTY720 (Tsai & Han, 2016), Etrasimod (APD-334) (Al-Shamma et al., 2019), ABC294640 (Chumanevich et al., 2010; Maines et al., 2010), JTE-013 (Liu, Li, Hylemon, & Zhou, 2018), Amiselimod (MT-1303) (Danese et al., 2018), and Ozanimod (RPC-1063) (Vetter &. Neurath, 2017). IBD mice models indicate that SphK1 inhibitors (ABC747080 and ABC294640) can modify TNF-α-linked inflammation and severity of colitis (Maines et al., 2010). Detected synergy between various anticancer drugs and SphK/S1PR inhibitors suggest more effective ways to abrogate tumorigenic activity in various GI cancers. For instance, combined inhibition of SphK1 by safignol and cisplatin promote synergistic cell death in esophageal cancer cells (Matula et al., 2015). However, combined treatment of IBDs with conventional drugs and novel SphK/S1PR-targeting substances remains largely unexplored.

Besides specifically designed and synthesized substances, several natural agents that influence the SphK/S1P axis have been explored as treatments against GI pathologies. Luteolin, present in celery, thyme, green peppers, watercress, and chamomile tea, inhibited S1P biosynthesis and ceramide traffic, suggesting its potential dietary introduction/supplementation as a new strategy to improve existing treatments in colorectal cancers (Abdel

Hadi, Di Vito, Marfia, Ferraretto, & Tringali, 2015). Another natural prenylflavonoid icaritin (a constituent of traditional Chinese herb belonging to the Epimedium genus) tested in different hepatocellular cancer cell models (HEPG2, KYN-2 and Huh-7 cell lines) and via oral administration in mice has credibly down-regulated SphK1 activity and JNK signaling associated with progression of carcinogenesis (Rosenberg et al., 2016). Vitamin A-like phytochemical peretinoin has been used to suppress SphK1 expression in human hepatocellular carcinoma cell lines, Huh-7 and in the SphK1-knockout mice (Funaki et al., 2017). Those non toxic phytocompounds promise to serve as potential dietary supplements or valuable additions to existing IBDs therapies. Future studies should address the role of natural modifiers of sphingolipid metabolism in IBDs.

The majority of IBD-associated symptoms were linked to bacterial or viral infection and consequent inflammation. However, the information about the role of sphingolipids as mediators of microbial pathogenesis has just recently emerged. We still do not understand how bacterial or viral sphingolipid metabolism can be modified to benefit the host GI health. Sphingolipids can be synthesized by bacteria and delivered with consumed food. Thus, those types of lipids can potentially impact the process of bacterial recognition by immune system of host, and, consequently, influence inflammation and specific disease resolution. Looking far ahead, it is tempting to speculate that specific dietary recommendations can be developed to facilitate resolution or prevention of GI pathologies and metabolic disorders. Currently, we do not know how pathogenic or healthy/symbiotic intestinal bacteria can impact the internal level of sphingolipids, or how the microbiome is affected by host level of sphingolipids. The deciphered regulation of sphingolipid metabolism by human microbiome may provide valuable alternatives to traditional approaches for prevention and treatment of IBDs and inflammation-linked GI cancers.

Declarations of Competing Interest

The authors declare that there are no conflicts of interest.

References

Abdel Hadi, L., Di Vito, C., Marfia, G., Ferraretto, A., Tringali, C., et al. (2015). Sphingosine kinase 2 and ceramide transport as key targets of the natural Flavonoid Luteolin to induce apoptosis in colon cancer cells. *PLoS One, 10*(11), e0143384.

Abdin, A. A. (2013). Targeting sphingosine kinase 1 (SphK1) and apoptosis by colon-specific delivery formula of resveratrol in treatment of experimental ulcerative colitis in rats. *European Journal of Pharmacology, 718*(1−3), 145−153.

Al-Shamma, H., Lehmann-Bruinsma, K., Carroll, C., Solomon, M., Komori, H. K., et al. (2019). The selective sphingosine 1-phosphate receptor modulator Etrasimod regulates lymphocyte trafficking and alleviates experimental colitis. *Journal of Pharmacology and Expimental Therapy, 369*(3), 311–317.

An, D., Oh, S. F., Olszak, T., Neves, J. F., Avci, F., et al. (2014). Sphingolipids from a symbiotic microbe regulate homeostasis of host intestinal natural killer T cells. *Cell, 156*(1–2), 123–133.

Argollo, M., Fiorino, G., Hindryckx, P., Peyrin-Biroulet, L., & Danese, S. (2017). Novel therapeutic targets for inflammatory bowel disease. *Journal of Autoimmunity, 85*, 103–116.

Arlt, O., Schwiebs, A., Japtok, L., Rüger, K., Katzy, E., et al. (2014). Sphingosine-1-phosphate modulates dendritic cell function: Focus on non-migratory effects in vitro and in vivo. *Cellular Physiology and Biochemistry, 34*(1), 27–44.

Bai, L. Y., Chiu, C. F., Chiu, S. J., Chu, P. C., & Weng, J. R. (2017). FTY720 induces autophagy-associated apoptosis in human oral squamous carcinoma cells, in part, through a reactive oxygen species/Mcl-1-dependent mechanism. *Scientific Reports, 7*, 5600.

Baker, D. A., Eudaly, J., Smith, C. D., Obeid, L. M., & Gilkeson, G. C. (2013). Impact of sphingosine kinase 2 deficiency on the development of TNF-alpha-induced inflammatory arthritis. *Rheumatology International, 33*(10), 2677–2681.

Bandzar, S., Gupta, S., & Platt, M. O. (2013). Crohn's disease: A review of treatment options and current research. *Cellular Immunology, 286*(1–2), 45–52.

Baumgart, D. C., & Sandborn, W. J. (2007). Inflammatory bowel disease: Clinical aspects and established and evolving therapies. *Lancet, 369*(9573), 1641–1657.

Baumruker, T., & Prieschl, E. E. (2002). Sphingolipids and the regulation of the immune response. *Seminars in Immunology, 14*(1), 57–63.

Benchimol, E. I., Bernstein, C. N., Bitton, A., Murthy, S. K., Nguyen, G. C., et al. (2019). The impact of inflammatory bowel disease in Canada 2018: A scientific report from the Canadian gastro-intestinal EpidemiologyConsortium to Crohn's and Colitis Canada. *Journal of Canadian Association for Gastroenterology, 2*(Suppl. 1), S1–S5.

Bernstein, C. N., Nugent, Z., Targownik, L. E., Singh, H., & Lix, L. M. (2015). Predictors and risks for death in a population based study of persons with IBD in Manitoba. *Gut, 64*(9), 1403–1411.

Biasi, F., Leonarduzzi, G., Oteiza, P. I., & Poli, G. (2013). Inflammatory bowel disease: Mechanisms, redox considerations, and therapeutic targets. *Antioxidants & Redox Signaling, 19*(14), 1711–1747.

Billich, A., Bornancin, F., Mechtcheriakova, D., Natt, F., Huesken, D., & Baumruker, T. (2005). Basal and induced sphingosine kinase 1 activity in A549 carcinoma cells: Function in cell survival and IL-1beta and TNF-alpha induced production of inflammatory mediators. *Cellular Signalling, 17*(10), 1203–1217.

Billich, A., Urtz, N., Reuschel, R., & Baumruker, T. (2009). Sphingosine kinase 1 is essential for proteinase-activated receptor-1 signalling in epithelial and endothelial cells. *The International Journal of Biochemistry & Cell Biology, 41*(7), 1547–1555.

Biswas, S., Bryant, R. V., & Travis, S. (2019). Interfering with leukocyte trafficking in Crohn's disease. *Best Practice & Research Clinical Gastroenterology, 38–39*, 101617.

Brinkmann, V., Billich, A., Baumruker, T., Heining, P., Schmouder, R., et al. (2010). Fingolimod (FTY720): Discovery and development of an oral drug to treat multiple sclerosis. *Nature Reviews Drug Discovery, 9*, 883–897.

Camerer, E., Regard, J. B., Cornelissen, I., Srinivasan, Y., Duong, D. N., & Coughlin, S. R. (2009). Sphingosine-1-phosphate in the plasma compartment regulates basal and inflammation-induced vascular leak in mice. *The Journal of Clinical Investigation, 119*(7), 1871–1879.

Cao, M., Ji, C., Zhou, Y., Huang, W., Ni, W., Tong, X., et al. (2018). Sphingosine kinase inhibitors: A patent review. *International Journal of Molecular Medicine, 41*, 2450–2460.

Chiba, K., Matsuyuki, H., Maeda, Y., & Sugahara, K. (2006). Role of sphingosine 1-phosphate receptor type 1 in lymphocyte egress from secondary lymphoid tissues and thymus. *Cellular & Molecular Immunology, 3*(1), 11–19.

Chumanevich, A. A., Poudyal, D., Cui, X., Davis, T., Wood, P. A., et al. (2010). Suppression of colitis-driven colon cancer in mice by a novel small molecule inhibitor of sphingosine kinase. *Carcinogenesis, 31*(10), 1787–1793.

Clarke, J. N., Davies, L. K., Calvert, J. K., Gliddon, W. B. L., Al Shujari, W. H., et al. (2016). Reduction in sphingosine kinase 1 influences the susceptibility to dengue virus infection by altering antiviral responses. *The Journal of General Virology, 97*(1), 95–109.

Cohen, B. L., & Sachar, D. B. (2017). Update on anti-tumor necrosis factor agents and other new drugs for inflammatory bowel disease. *British Medical Journal, 357*, j2505.

Cordi, A. A., Lacoste, J. M., Descombes, J. J., Courchay, C., Vanhoutte, P. M., et al. (1995). Design, synthesis, and structure-activity relationships of a new series of alpha-adrenergic agonists: spiro[(1,3-diazacyclopent-1-ene)-5,2'-(1',2',3',4'- tetrahydronaphthalene)]. *Journal of Medical Chemistry, 38*, 4056–4069.

Crespo, I., San-Miguel, B., Mauriz, J. L., Ortiz de Urbina, J. J., Almar, M., et al. (2017). Protective effect of protocatechuic acid on TNBS-induced colitis in mice is associated with modulation of the SphK/S1P signaling pathway. *Nutrients, 9*(3), E288.

Czeloth, N., Schippers, A., Wagner, N., Müller, W., Küster, B., et al. (2007). Sphingosine-1 phosphate signaling regulates positioning of dendritic cells within the spleen. *The Journal of Immunology, 179*(9), 5855–5863.

Danese, S. (2011). Immune and nonimmune components orchestrate the pathogenesis of inflammatory bowel disease. *American Journal of Physiology - Gastrointestinal and Liver Physiology, 300*(5), G716–G722.

Danese, S., Furfaro, F., & Vetrano, S. (2018). Targeting S1P in inflammatory bowel disease: New avenues for modulating intestinal leukocyte migration. *Journal of Crohn's and Colitis, 12*(Suppl. 2), S678–S686.

Daniel, C., Sartory, N., Zahn, N., Geisslinger, G., Radeke, H. H., & Stein, J. M. (2007). FTY720 ameliorates Th1-mediated colitis in mice by directly affecting the functional activity of $CD4^+CD25^+$ regulatory T cells. *The Journal of Immunology, 178*(4), 2458–2468.

Deguchi, Y., Andoh, A., Yagi, Y., Bamba, S., Inatomi, O., Tsujikawa, T., et al. (2006). The S1P receptor modulator FTY720 prevents the development of experimental colitis in mice. *Oncology Reports, 16*, 699–703.

Ding, X., Zhang, Y., Huang, T., Xu, G., Peng, C., et al. (2019). Targeting sphingosine kinase 2 suppresses cell growth and synergizes with BCL2/BCL-XL inhibitors through NOXA-mediated MCL1 degradation in cholangiocarcinoma. *American Journal of Cancer Research, 9*, 546–561.

Duong, C. Q., Bared, S. M., Abu-Khader, A., Buechler, C., Schmitz, A., & Schmitz, G. (2004). Expression of the lysophospholipid receptor family and investigation of lysophospholipid-mediated responses in human macrophages. *Biochimica et Biophysica Acta, 1682*(1–3), 112–119.

Edwards, J. P., Zhang, X., Frauwirth, K. A., & Mosser, D. M. (2006). Biochemical and functional characterization of three activated macrophage populations. *Journal of Leukocyte Biology, 80*(6), 1298–1307.

Espaillat, M. P., Kew, R. R., & Obeid, L. M. (2017). Sphingolipids in neutrophil function and inflammatory responses: Mechanisms and implications for intestinal immunity and inflammation in ulcerative colitis. *Advances in Biological Regulation, 63*, 140–155.

Feuerstein, J. D., Curran, T., Alvares, D., Alosilla, M., Lerner, A., & Poylin, V. (2019). Surgery for ulcerative colitis in geriatric patients is safe with similar risk to younger patients. *European Journal of Gastroenterology and Hepatology, 31*(11), 1356–1360.

Funaki, M., Kitabayashi, J., Shimakami, T., Nagata, N., Sakai, Y., et al. (2017). Peretinoin, an acyclic retinoid, inhibits hepatocarcinogenesis by suppressing sphingosine kinase 1 expression in vitro and in vivo. *Scientific Reports, 7*(1), 16978.

Furuya, H., Tamashiro, P. M., Shimizu, Y., Iino, K., Peres, R., et al. (2017). Sphingosine kinase 1 expression in peritoneal macrophages is required for colon carcinogenesis. *Carcinogenesis, 38*(12), 1218–1227.

Gault, C. R., Obeid, L. M., & Hannun, Y. A. (2010). An overview of sphingolipid metabolism: From synthesis to breakdown. *Advances in Experimental Medicine & Biology, 688*, 1–23.

Gerriets, V., & Khaddour, K. (2019). Tumor necrosis factor (TNF) inhibitors. *StatPearls (Treasure Island (FL)).*

Giuffrida, P., Cococcia, S., Delliponti, M., Lenti, M. V., & Di Sabatino, A. (2019). Controlling gut inflammation by restoring anti-inflammatory pathways in inflammatory bowel disease. *Cells, 8*(5), E397.

Gollmann, G., Neuwirt, H., Tripp, C. H., Mueller, H., Konwalinka, G., et al. (2008). Sphingosine-1-phosphate receptor type-1 agonism impairs blood dendritic cell chemotaxis and skin dendritic cell migration to lymph nodes under inflammatory conditions. *International Immunology, 20*(7), 911–923.

Gómez-Muñoz, A., Kong, J. Y., Salh, B., & Steinbrecher, U. P. (2004). Ceramide-1-phosphate blocks apoptosis through inhibition of acid sphingomyelinase in macrophages. *Journal of Lipid Research, 45*(1), 99–105.

Greenspon, J., Li, R., Xiao, L., Rao, J. N., Sun, R., et al. (2011). Sphingosine-1-phosphate regulates the expression of adherens junction protein E-cadherin and enhances intestinal epithelial cell barrier function. *Digestive Diseases and Sciences, 56*(5), 1342–1353.

Gude, D. R., Alvarez, S. E., Paugh, S. W., Mitra, P., Yu, J., et al. (2008). Apoptosis induces expression of sphingosine kinase 1 to release sphingosine-1-phosphate as a 'come-and-get-me' signal. *FASEB Journal, 22*(8), 2629–2638.

Gupta, P., Mohammad, T., Khan, P., Alajmi, M. F., Hussain, A., Rehman, M. T., et al. (2019). Evaluation of ellagic acid as an inhibitor of sphingosine kinase 1: A targeted approach towards anticancer therapy. *Biomedicine & Pharmacotherapy, 118*, 109245.

Gurgui, M., Broere, R., Kalff, J. C., & van Echten-Deckert, G. (2010). Dual action of sphingosine 1-phosphate in eliciting proinflammatory responses in primary cultured rat intestinal smooth muscle cells. *Cellular Signalling, 22*(11), 1727–1733.

Haass, N. K., Nassif, N., & McGowan, E. M. (2015). Switching the sphingolipid rheostat in the treatment of diabetes and cancer comorbidity from a problem to an advantage. *BioMed Research International, 2015*, 165105.

Haddadi, N., Lin, Y., Simpson, A. M., Nassif, N. T., & McGowan, E. M. (2017). "Dicing and splicing" sphingosine kinase and relevance to cancer. *International Journal of Molecular Sciences, 18*(9), E1891.

Hait, N. C., & Maiti, A. (2017). The role of sphingosine-1-phosphate and ceramide-1-phosphate in inflammation and cancer. *Mediators of Inflammation, 2017*, 4806541.

Hamada, Y., Nagasaki, H., Fujiya, A., Seino, Y., Shang, Q. L., et al. (2014). Involvement of de novo ceramide synthesis in pro-inflammatory adipokine secretion and adipocyte-macrophage interaction. *The Journal of Nutritional Biochemistry, 25*(12), 1309–1316.

Hanada, K. (2005). Sphingolipids in infectious diseases. *Japanese Journal of Infectious Diseases, 58*(3), 131–148.

Hannun, Y. A., & Obeid, L. M. (2011). Many ceramides. *Journal of Biological Chemistry, 286*(32), 27855–27862.

Hanyu, T., Nagahashi, M., Ichikawa, H., Ishikawa, T., Kobayashi, T., & Wakai, T. (2018). Expression of phosphorylated sphingosine kinase 1 is associated with diffuse type and lymphatic invasion in human gastric cancer. *Surgery, 163*, 1301–1306.

Harrison, P. J., Dunn, T. M., & Campopiano, D. J. (2018). Sphingolipid biosynthesis in man and microbes. *Natural Product Reports, 35*(9), 921–954.

Hartl, M. (2016). The quest for targets executing MYC-dependent cell transformation. *Frontiers in Oncology, 6*, 132.

Hatoum, D., Haddadi, N., Lin, Y., Nassif, N. T., & McGowan, E. M. (2017). Mammalian sphingosine kinase (SphK) isoenzymes and isoform expression: Challenges for SphK as an oncotarget. *Oncotarget, 8*, 36898–36929.

Huang, W.-C., Liang, J., Nagahashi, M., Avni, D., Yamada, A., et al. (2016). Sphingosine-1-phosphate phosphatase 2 promotes disruption of mucosal integrity and contributes to ulcerative colitis in mice and humans. *FASEB Journal, 30*(8), 2945–2958.

Huwiler, A., & Zangemeister-Wittke, U. (2018). The sphingosine 1-phosphate receptor modulator fingolimod as a therapeutic agent: Recent findings and new perspectives. *Pharmacology & Therapeutics, 185*, 34–49.

Ikeda, H., Watanabe, N., Ishii, I., Shimosawa, T., Kume, Y., et al. (2009). Sphingosine 1-phosphate regulates regeneration and fibrosis after liver injury via sphingosine 1-phosphate receptor 2. *Journal of Lipid Research, 50*(3), 556–564.

Jiang, P., Smith, A. D., Li, R., Rao, J. N., Liu, L., et al. (2013). Sphingosine kinase 1 over-expression stimulates intestinal epithelial cell proliferation through increased c-Myc translation. *American Journal of Physiology - Cell Physiology, 304*(12), C1187–C1197.

Kaneider, N. C., Leger, A. J., Agarwal, A., Nguyen, N., Perides, G., et al. (2007). 'Role reversal' for the receptor PAR1 in sepsis-induced vascular damage. *Nature Immunology, 8*(12), 1303–1312.

Karreman, M. C., Luime, J. J., Hazes, J. M. W., & Weel, A. E. A. M. (2017). The prevalence and incidence of axial and peripheral spondyloarthritis in inflammatory bowel disease: A systematic review and Meta-analysis. *Journal of Crohn's and Colitis, 11*(5), 631–642.

Karuppuchamy, T., Behrens, E. H., González-Cabrera, P., Sarkisyan, G., Gima, L., et al. (2017). Sphingosine-1-phosphate receptor-1 (S1P1) is expressed by lymphocytes, dendritic cells, and endothelium and modulated during inflammatory bowel disease. *Mucosal Immunology, 10*(1), 162–171.

Kato, M., Muto, Y., Tanaka-Bandoh, K., Watanabe, K., & Ueno, K. (1995). Sphingolipid composition in Bacteroides species. *Anaerobe, 1*(2), 135–139.

Kawakita, Y., Motoyama, S., Sato, Y., Koyota, S., Wakita, A., et al. (2017). Sphingosine-1-phosphate/sphingosine kinase 1-dependent lymph node metastasis in esophageal squamous cell carcinoma. *Surgery Today, 47*(11), 1312–1320.

Kawamori, T., Osta, W., Johnson, K. R., Pettus, B. J., Bielawski, J., et al. (2006). Sphingosine kinase 1 is up-regulated in colon carcinogenesis. *FASEB Journal, 20*(2), 386–388.

Kihara, Y., Maceyka, M., Spiegel, S., & Chun, J. (2014). Lysophospholipid receptor nomenclature review: IUPHAR review 8. *British Journal of Pharmacology, 171*(15), 3575–3594.

Kleinwort, A., Lührs, F., Heidecke, C. D., Lipp, M., & Schulze, T. (2018). S1P signalling differentially affects migration of peritoneal B cell populations in vitro and influences the production of intestinal IgA in vivo. *International Journal of Molecular Sciences, 19*(2), E391.

Knorr, M., Münzel, T., & Wenzel, P. (2014). Interplay of NK cells and monocytes in vascular inflammation and myocardial infarction. *Frontiers in Physiology, 5*, 595.

Kohno, M., Momoi, M., Oo, M. L., Paik, J. H., Lee, Y. M., et al. (2006). Intracellular role for sphingosine kinase 1 in intestinal adenoma cell proliferation. *Molecular and Cellular Biology, 26*(19), 7211–7223.

Kruidenier, L., Kuiper, I., Lamers, C. B., & Verspaget, H. W. (2003). Intestinal oxidative damage in inflammatory bowel disease: Semi-quantification, localization and association with mucosal antioxidants. *The Journal of Pathology, 201*(1), 28–36.

Le Stunff, H., Peterson, C., Liu, H., Milstien, S., & Spiegel, S. (2002). Sphingosine-1-phosphate and lipid phosphohydrolases. *Biochimica et Biophysica Acta, 1582*(1–3), 8–17.

Levin, A. D., Wildenberg, M. E., & van den Brink, G. R. (2016). Mechanism of action of anti-TNF therapy in inflammatory bowel disease. *Journal of Crohn's and Colitis, 10*, 989–997.

Li, Q., Wang, C., Zhang, Q., Tang, C., Li, N., & Li, J. (2012). The reduction of allograft arteriosclerosis in intestinal transplant is associated with sphingosine kinase 1/sphingosine-1-phosphate signaling after fish oil treatment. *Transplantation, 93*(10), 989–996.

Liang, J., Nagahashi, M., Kim, E. Y., Harikumar, K. B., Yamada, W. C., et al. (2013). Sphingosine-1-phosphate links persistent STAT3 activation, chronic intestinal inflammation, and development of colitis-associated cancer. *Cancer Cell, 23*(1), 107–120.

Liu, R., Li, X., Hylemon, P. B., & Zhou, H. (2018). Conjugated bile acids promote invasive growth of esophageal adenocarcinoma cells and cancer stem cell expansion via sphingosine 1-phosphate receptor 2 (S1PR2)-mediated YAP activation. *American Journal of Pathology, 188*(9), 2042–2058.

Lynch, K. R., & Santos, W. L. (2018). Sphingosine kinase inhibitors. In United States Patent, 141. United States: University of Virginia Patent Foundation, Charlotesvile, VA(US) (Ed.), *United States*. Blacksburg, VA(US): Virginia Tech Intelectual Properties, Inc.

Lynch, K. R., Thorpe, S. B., & Santos, W. L. (2016). Sphingosine kinase inhibitors: A review of patent literature (2006-2015). *Expert Opinion on Therapeutic Patents, 26*, 1409–1416.

Maceyka, M., & Spiegel, S. (2014). Sphingolipid metabolites in inflammatory disease. *Nature, 510*(7503), 58–67.

Magli, E., Corvino, A., Fiorino, F., Frecentese, F., Perissutti, E., et al. (2019). Design of sphingosine kinases inhibitors: Challenges and recent developments. *Current Pharmacological Design, 25*, 956–968.

Mahadevan, U., Loftus, E. V., Jr., Tremaine, W. J., & Sandborn, W. J. (2002). Safety of selective cyclooxygenase-2 inhibitors in inflammatory bowel disease. *The American Journal of Gastroenterology, 97*(4), 910–914.

Mahajan-Thakur, S., Sostmann, B. D., Fender, A. C., Behrendt, D., Felix, S. B., et al. (2014). Sphingosine-1-phosphate induces thrombin receptor PAR-4 expression to enhance cell migration and COX-2 formation in human monocytes. *Journal of Leukocyte Biology, 96*(4), 611–618.

Maines, L. W., Fitzpatrick, L. R., French, K. J., Zhuang, Y., Xia, Z., et al. (2008). Suppression of ulcerative colitis in mice by orally available inhibitors of sphingosine kinase. *Digestive Diseases and Sciences, 53*, 997–1012.

Maines, L. W., Fitzpatrick, L. R., Green, C. L., Zhuang, Y., & Smith, C. D. (2010). Efficacy of a novel sphingosine kinase inhibitor in experimental Crohn's disease. *Inflammopharmacology, 18*(2), 73–85.

Matula, K., Collie-Duguid, E., Murray, G., Parikh, K., Grabsch, H., et al. (2015). Regulation of cellular sphingosine-1-phosphate by sphingosine kinase 1 and sphingosine-1-phopshate lyase determines chemotherapy resistance in gastroesophageal cancer. *BMC Cancer, 15*, 762.

Merrill, A. H., Jr., Schmelz, E. M., Wang, E., Schroeder, J. J., Dillehay, D. L., & Riley, R. T. (1995). Role of dietary sphingolipids and inhibitors of sphingolipid metabolism in cancer and other diseases. *The Journal of Nutrition, 125*(6 Suppl. l), 1677S–1682S.

Montrose, D. C., Scherl, E. J., Bosworth, B. P., Zhou, X. K., Jung, B., et al. (2013). S1P1 localizes to the colonic vasculature in ulcerative colitis and maintains blood vessel integrity. *Journal of Lipid Research, 54*(3), 843–851.

Nagahashi, M., Abe, M., Sakimura, K., Takabe, K., & Wakai, T. (2018). The role of sphingosine-1-phosphate in inflammation and cancer progression. *Cancer Science, 109*(12), 3671–3678.

Neuman, M. G. (2007). Immune dysfunction in inflammatory bowel disease. *Translational Research, 149*(4), 173–186.
Neurath, M. (2017). Current and emerging therapeutic targets for IBD. *Nature Reviews Gastroenterology & Hepatology, 14*(11), 688.
Nichols, F. C., Housley, W. J., O'Conor, C. A., Manning, T., Wu, S., & Clark, R. B. (2009). Unique lipids from a common human bacterium represent a new class of Toll-like receptor 2 ligands capable of enhancing autoimmunity. *American Journal of Pathology, 175*(6), 2430–2438.
Niessen, F., Furlan-Freguia, C., Fernández, J. A., Mosnier, L. O., Castellino, F. J., et al. (2009). Endogenous EPCR/aPC-PAR1 signaling prevents inflammation-induced vascular leakage and lethality. *Blood, 113*(12), 2859–2866.
Niessen, F., Schaffner, F., Furlan-Freguia, C., Pawlinski, R., Bhattacharjee, G., et al. (2008). Dendritic cell PAR1-S1P3 signalling couples coagulation and inflammation. *Nature, 452*(7187), 654–658.
Niwa, M., Kozawa, O., Matsuno, H., Kanamori, Y., Hara, A., & Uematsu, T. (2000). Tumor necrosis factor-alpha-mediated signal transduction in human neutrophils: Involvement of sphingomyelin metabolites in the priming effect of TNF-alpha on the fMLP-stimulated superoxide production. *Life Sciences, 66*(3), 245–256.
Nixon, G. F. (2009). Sphingolipids in inflammation: Pathological implications and potential therapeutic targets. *British Journal of Pharmacology, 158*(4), 982–993.
Nodai, A., Machida, T., Izumi, S., Hamaya, Y., Kohno, T., et al. (2007). Sphingosine 1-phosphate induces cyclooxygenase-2 via Ca^{2+}-dependent, but MAPK-independent mechanism in rat vascular smooth muscle cells. *Life Sciences, 80*(19), 1768–1776.
Nussbaum, C., Bannenberg, S., Keul, P., Gräler, M. H., Gonçalves-de-Albuquerque, C. F., et al. (2015). Sphingosine-1-phosphate receptor 3 promotes leukocyte rolling by mobilizing endothelial P-selectin. *Nature Communications, 6*, 6416.
O'Connor, P. M., Lapointe, T. K., Beck, P. L., & Buret, A. G. (2010). Mechanisms by which inflammation may increase intestinal cancer risk in inflammatory bowel disease. *Inflammatory Bowel Diseases, 16*(8), 1411–1420.
Obinata, H., & Hla, T. (2012). Sphingosine 1-phosphate in coagulation and inflammation. *Seminars in Immunopathology, 34*(1), 73–91.
Olivera, A., Allende, M. L., & Proia, R. L. (2013). Shaping the landscape: Metabolic regulation of S1P gradients. *Biochimica et Biophysica Acta, 183*(1), 193–202.
Ordás, I., Eckmann, L., Talamini, M., Baumgart, D. C., & Sandborn, W. J. (2012). Ulcerative colitis. *Lancet, 380*(9853), 1606–1619.
Ottenlinger, F., Schwiebs, A., Pfarr, K., Wagner, A., Grüner, S., et al. (2016). Fingolimod targeting protein phosphatase 2A differently affects IL-33 induced IL-2 and IFN-γ production in CD8(+) lymphocytes. *European Journal of Immunology, 46*(4), 941–951.
Park, K. T., Crandall, W. V., Fridge, J., Leibowitz, I. H., Tsou, M., et al. (2014). Implementable strategies and exploratory considerations to reduce costs associated with anti-TNF therapy in inflammatory bowel disease. *Inflammatory Bowel Diseases, 20*, 946–951.
Pennel, K. A. F., Park, J. H., McMillan, D. C., Roseweir, A. K., & Edwards, J. (2019). Signal interaction between the tumor and inflammatory cells in patients with gastrointestinal cancer: Implications for treatment. *Cellular Signalling, 54*, 81–90.
Pérez-Jeldres, T., Tyler, C. J., Boyer, J. D., Karuppuchamy, T., Yarur, A., et al. (2019). Targeting cytokine signaling and lymphocyte traffic via small molecules in inflammatory bowel disease: JAK inhibitors and S1PR agonists. *Frontiers in Pharmacology, 10*, 212.
Pettus, B. J., Bielawski, J., Porcelli, A. M., Reames, D. L., Johnson, K. R., et al. (2003). The sphingosine kinase 1/sphingosine-1-phosphate pathway mediates COX-2 induction and PGE2 production in response to TNF-alpha. *FASEB Journal, 17*(11), 1411–1421.

Peyrin-Biroulet, L., Christopher, R., Behan, D., & Lassen, C. (2017). Modulation of sphingosine-1-phosphate in inflammatory bowel disease. *Autoimmunity Reviews, 16*(5), 495–503.
Pitman, M. R., Costabile, M., & Pitson, S. M. (2016). Recent advances in the development of sphingosine kinase inhibitors. *Cellular Signalling, 28*, 1349–1363.
Pitman, M. R., & Pitson, S. M. (2010). Inhibitors of the sphingosine kinase pathway as potential therapeutics. *Current Cancer Drug Targets, 10*(4), 354–367.
Porta, C., Larghi, P., Rimoldi, M., Totaro, M. G., Allavena, P., et al. (2009). Cellular and molecular pathways linking inflammation and cancer. *Immunobiology, 214*(9–10), 761–777.
Proia, R. L., & Hla, T. (2015). Emerging biology of sphingosine-1-phosphate: Its role in pathogenesis and therapy. *The Journal of Clinical Investigation, 125*(4), 1379–1387.
Pruett, S. T., Bushnev, A., Hagedorn, K., Adiga, M., Haynes, C. A., et al. (2008). Biodiversity of sphingoid bases ("sphingosines") and related amino alcohols. *Journal of Lipid Research, 49*(8), 1621–1639.
Pulkoski-Gross, M. J., Jenkins, M. L., Truman, J. P., Salama, M. F., Clarke, C. J., et al. (2018). An intrinsic lipid-binding interface controls sphingosine kinase 1 function. *Journal of Lipid Research, 59*(3), 462–474.
Pulkoski-Gross, M. J., Uys, J. D., Orr-Gandy, K. A., Coant, N., Bialkowska, A. B., et al. (2017). Novel sphingosine kinase-1 inhibitor, LCL351, reduces immune responses in murine DSS-induced colitis. *Prostaglandins & Other Lipid Mediators, 130*, 47–56.
Pyne, S., Adams, D. R., & Pyne, N. J. (2016). Sphingosine 1-phosphate and sphingosine kinases in health and disease: Recent advances. *Progress in Lipid Research, 62*, 93–106.
Rauch, B. H. (2014). Sphingosine 1-phosphate as a link between blood coagulation and inflammation. *Cellular Physiology and Biochemistry, 34*(1), 185–196.
Riewald, M., & Ruf, W. (2003). Science review: Role of coagulation protease cascades in sepsis. *Critical Care, 7*(2), 123–129.
Rivera, J., Proia, R. L., & Olivera, A. (2008). The alliance of sphingosine-1-phosphate and its receptors in immunity. *Nature Reviews Immunology, 8*(10), 753–763.
Rosenberg, Liu, H., Jin, H., Yue, X., Riley, S., et al. (2016). Design, synthesis, and in vitro and in vivo evaluation of an (18)F-Labeled sphingosine 1-phosphate receptor 1 (S1P1) PET tracer. *Journal of Medicinal Chemistry, 59*(13), 6201–6220.
Round, J. L., O'Connell, R. M., & Mazmanian, S. K. (2010). Coordination of tolerogenic immune responses by the commensal microbiota. *Journal of Autoimmunity, 34*(3), J220–J225.
Saleh, M., & Trinchieri, G. (2011). Innate immune mechanisms of colitis and colitis-associated colorectal cancer. *Nature Reviews Immunology, 11*, 9–20.
Sandborn, W. J. (2003). Strategies for targeting tumour necrosis factor in IBD. *Best Practice & Research Clinical Gastroenterology, 17*, 105–117.
Sato, M., Ikeda, H., Uranbileg, B., Kurano, M., Saigusa, D., et al. (2016). Sphingosine kinase-1, S1P transporter spinster homolog 2 and S1P2 mRNA expressions are increased in liver with advanced fibrosis in human. *Scientific Reports, 6*, 32119.
Schaper, K., Dickhaut, J., Japtok, L., Kietzmann, M., Mischke, R., et al. (2013). Sphingosine-1-phosphate exhibits anti-proliferative and anti-inflammatory effects in mouse models of psoriasis. *Journal of Dermatological Science, 71*(1), 29–36.
Schnute, M. E., McReynolds, M. D., Kasten, T., Yates, M., Jerome, G., et al. (2012). Modulation of cellular S1P levels with a novel, potent and specific inhibitor of sphingosine kinase-1. *Biochemical Journal, 444*, 79–88.
Shatrov, V. A., Lehmann, V., & Chouaib, S. (1997). Sphingosine-1-phosphate mobilizes intracellular calcium and activates transcription factor NF-κB in U937 cells. *Biochemical and Biophysical Research Communications, 234*(1), 121–124.

Snider, A. J., Ali, W. H., Sticca, J. A., Coant, N., Ghaleb, A. M., et al. (2014). Distinct roles for hematopoietic and extra-hematopoietic sphingosine kinase-1 in inflammatory bowel disease. *PLoS One, 9*(12), e113998.

Snider, A. J., Kawamori, T., Bradshaw, S. G., Orr, K. A., Gilkeson, G. S., et al. (2009). A role for sphingosine kinase 1 in dextran sulfate sodium-induced colitis. *FASEB Journal, 23*(1), 143–152.

Snider, A. J., Wu, B. X., Jenkins, R. W., Sticca, J. A., Kawamori, T., et al. (2012). Loss of neutral ceramidase increases inflammation in a mouse model of inflammatory bowel disease. *Prostaglandins and Other Lipid Mediators, 99*(3–4), 124–130.

Sobel, K., Menyhart, K., Killer, N., Renault, B., Bauer, Y., et al. (2013). Sphingosine 1-phosphate (S1P) receptor agonists mediate pro-fibrotic responses in normal human lung fibroblasts via S1P2 and S1P3 receptors and Smad-independent signaling. *Journal of Biological Chemistry, 288*(21), 14839–14851.

Spiegel, S., Wynnewood Zipkin, PA (US), R. E., & Adams, J. K. (2019). Sphingosine kinase inhibitors and uses thereof. In Enzo Therapeutics, Farmingdale, and Richmond NY(US);VirginiaCommonwealth University, VA (US), 20. United States (Ed.), *United States*. Farmingdale, NY(US): EnzoTherapeutics, Inc.

Stepanovska, B., & Huwiler, A. (2019). Targeting the S1P receptor signaling pathways as a promising approach for treatment of autoimmune and inflammatory diseases. *Pharmacological Research*. S1043-6618(18)32039-5.

Sukocheva, O. A. (2018). Expansion of sphingosine kinase and sphingosine-1-phosphate receptor function in normal and cancer cells: From membrane restructuring to mediation of estrogen signaling and stem cell programming. *International Journal of Molecular Sciences, 19*(2), E420.

Sun, M., Zhou, Y., Shi, Y., & Liu, B. (2019). Effect of the sphingosine kinase 1 selective inhibitor, PF543 on dextran sodium sulfate-induced colitis in mice. *DNA & Cell Biology*. http://dx.doi.org/10.1089/dna.2019.4737.

Sýkora, J., Pomahačová, R., Kreslová, M., Cvalínová, D., Štych, P., & Schwarz, J. (2018). Current global trends in the incidence of pediatric-onset inflammatory bowel disease. *World Journal of Gastroenterology, 24*(25), 2741–2763.

Tsai, H. C., & Han, M. H. (2016). Sphingosine-1-Phosphate (S1P) and S1P signaling pathway: Therapeutic targets in autoimmunity and inflammation. *Drugs, 76*(11), 1067–1079.

Ullman, T. A., & Itzkowitz, S. H. (2011). Intestinal inflammation and cancer. *Gastroenterology, 140*, 1807–1816.

Van Klinken, B. J., Van der Wal, J. W., Einerhand, A. W., Büller, H. A., & Dekker, J. (1999). Sulphation and secretion of the predominant secretory human colonic mucin MUC2 in ulcerative colitis. *Gut, 44*(3), 387–393.

Vetter, M., & Neurath, M. F. (2017). Emerging oral targeted therapies in inflammatory bowel diseases: Opportunities and challenges. *Therapeutic Advances in Gastroenterology, 10*(10), 773–790.

Vettorazzi, M., Vila, L., Lima, S., Acosta, L., Yepes, F., et al. (2019). Synthesis and biological evaluation of sphingosine kinase 2 inhibitors with anti-inflammatory activity. *Archiv der Pharmazie (Weinheim), 352*, e1800298.

Wang, C., Mao, J., Redfield, S., Mo, Y., Lage, J. M., & Zhou, X. (2014). Systemic distribution, subcellular localization and differential expression of sphingosine-1-phosphate receptors in benign and malignant human tissues. *Experimental and Molecular Pathology, 97*(2), 259–265.

Wang, J., Lv, X. W., & Du, Y. G. (2009). Potential mechanisms involved in ceramide-induced apoptosis in human colon cancer HT29 cells. *Biomedical and Environmental Sciences, 22*(1), 76–85.

Wang, Y., Aoki, H., Yang, J., Peng, K., Liu, R., et al. (2017). The role of sphingosine 1-phosphate receptor 2 in bile-acid-induced cholangiocyte proliferation and cholestasis-induced liver injury in mice. *Hepatology, 65*(6), 2005–2018.

White, C., Alshaker, H., Cooper, C., Winkler, M., & Pchejetski, D. (2016). The emerging role of FTY720 (Fingolimod) in cancer treatment. *Oncotarget, 7*, 23106–23127.

Xia, P., Gamble, J. R., Rye, K. A., Wang, L., Hii, C. S., et al. (1998). Tumor necrosis factor-alpha induces adhesion molecule expression through the sphingosine kinase pathway. *Proceedings of the National Academy of Sciences of the United States of America, 95*(24), 14196–14201.

Yuza, K., Nagahashi, M., Shimada, Y., Nakano, M., Tajima, Y., et al. (2018). Upregulation of phosphorylated sphingosine kinase 1 expression in colitis-associated cancer. *The Journal of Surgical Research, 231*, 323–330.

Zhang, X., Ritter, J. K., & Li, N. (2018). Sphingosine-1-phosphate pathway in renal fibrosis. *American Journal of Physiology - Renal Physiology, 315*(4), F752–F756.

Zhao, S., Gong, Z., Du, X., Tian, C., Wang, L., et al. (2018). Deoxycholic acid-mediated sphingosine-1-phosphate receptor 2 signaling exacerbates DSS-induced colitis through promoting cathepsin B release. *Journal of Immunology Research, 2018*, 2481418.

Ziegler, T., Rausch, S., Steinfelder, S., Klotz, C., Hepworth, M. R., et al. (2015). A novel regulatory macrophage induced by a helminth molecule instructs IL-10 in CD4+ T cells and protects against mucosal inflammation. *The Journal of Immunology, 194*(4), 1555–1564.

CHAPTER FIVE

Dietary plant flavonoids in prevention of obesity and diabetes

Biswanath Dinda[a,*], Manikarna Dinda[b], Arup Roy[c] and Subhajit Dinda[d]

[a]Department of Chemistry, Tripura University, Agartala, Tripura, India
[b]Department of Biochemistry and Molecular Genetics, University of Virginia, Charlottesville, VA, United States
[c]Chemical Science & Technology Division, CSIR - North East Institute of Science and Technology, Jorhat, Assam, India
[d]Department of Chemistry, Dasaratha Deb Memorial College, Khowai, Tripura, India
*Corresponding author: E-mail: dindabtu@gmail.com

Contents

Abstract

Obesity and diabetes are the most prevailing chronic metabolic diseases worldwide from mainly lipid and glucose metabolic dysfunctions and their incidence is increasing

Advances in Protein Chemistry and Structural Biology, Volume 120
ISSN 1876-1623
https://doi.org/10.1016/bs.apcsb.2019.08.006

at an alarming high rate. Obesity is characterized by excess fat accumulation in WAT and liver and is the central player of insulin resistance in the peripheral tissues from chronic inflammation, lipotoxicity and gut dysbiosis, and plays a key role for development of type 2 diabetes (T2DM) and vascular diseases. Diabetes mellitus, known as diabetes, is chiefly characterized by hyperglycaemia from impaired insulin secretion and insulin resistance. Several identified mutant genes in insulin secretion and resistance and various environmental factors are considered responsible for the onset of this disease. Currently available oral synthetic drugs, biguanides, incretin mimetic, GLP-1R and PPAR agonists and DPP-4 inhibitors for management of obesity and diabetes have several adverse effects in patients on long-term use. Emerging evidence supports the efficacy of dietary plant flavonoids in prevention and attenuation of obesity and diabetes by the protection and proliferation of pancreatic beta-cells and improvement of their insulin secretory function via activation of cAMP/PKA signaling pathway as well as in the improvement of insulin sensitivity in the peripheral metabolic tisssues for glucose uptake and utilization via inhibition of inflammation, lipotoxicity and oxidative stress. These flavonoids improve GLUT-4 expression and translocation to plasma membrane by activation of insulin-sensitive PI3K/Akt signaling and insulin-independent AMPK, SIRT-1 and MOR activation pathways for regulation of glucose homeostasis, and improve fat oxidation and reduce lipid synthesis by regulation of related genes for lipid homeostasis in the body of obese diabetic animals. In this chapter, we have highlighted all these beneficial anti-obesity and antidiabetic potentials of some dietary plant flavonoids along with their molecular actions, bioavailability and pharmacokinetics. In addition, the present understanding and management of obesity and diabetes are also focused.

Abbreviations

ABC	ATP-binding cassette
ACC	acetyl-coA carboxylase
ACOX	acyl coenzyme A oxidase
ACS	acetyl-CoA synthetase
Akt	protein kinase B
ALP	alkaline phosphatase
ALT	alanine transaminase
AMPK	adenosine monophosphate protein kinase
AP-1	activator protein-1
aP2	adipocyte-specific fatty acid-binding protein-P2
AST	aspartate transaminase
ATF6	activating transcription factor 6
ATGL	adipocyte triglyceride lipase
ATF-6	activating transcription factor-6
ATP	adenosine triphosphate
BAT	brown adipose tissue
Bax	Bcl-2-associated X protein
Bcl-2	B-cell lymphoma-2 protein
BDNF	brain-derived neurotrophic factor
BRB	blood-retinal barrier
BUN	blood urea nitrogen

bw	body weight
CaMKK	calcium/calmodulin-dependent protein kinase kinase
CAT	catalase
CD-36	platelet glycoprotein-4
CD-38	cluster of differentiation-38 protein
cdk	cyclin-dependent kinase
C/EBP	CCAAT-enhancer-binding protein
CHOP	C/EBP homologous protein
CK-MB	creatine kinase-myocardial brand
$CoCl_2$	cobalt(II) chloride
CPT-1	carnitine palmitoyltransferase-1
CREB	cyclic-AMP response element binding protein
CREBH	cyclic AMP response element-binding protein H
CRP	C-reactive protein
CYP2E1	cytochrome P450 2E1
dkn1a	cyclin dependent kinase inhibitor 1a
DPP-4	dipeptidyl peptidase-4
eNOS	endothelial nitric oxide synthase
EGCG	(−)-epigallocatechin gallate
EGC	(−)-epigallocatechin
ER	endoplasmic reticulum
ERK	extracellular signal-regulated kinase
FABP-1	fatty acid-binding protein-1
FAS	fatty acid synthase
FBPase	fructose 1,6-bisphosphatase
FFA	free fatty acid
FoxO1	forkhead box class protein O1
GK	glucokinase
GLP	glucagon-like peptide
GLUT	glucose transporter
G6Pase	glucose 6-phosphatase
G6PD	glucose 6-phosphate dehydrogenase
GPase	glycogen phosphorylase
GPAT	glycerol-3-phosphate acyltransferase
Gpr-146	G-protein coupled receptor-146
GS	glycogen synthase
GSH	glutathione
$GSH\text{-}P_X/GP_X$	glutathione peroxidase
GSK-3	glycogen synthase kinase-3
GSR	glutathione reductase
GST	glutathione S-transferase
HbA1c	glycosylated hemoglobin A1c
HDL-C	high-density lipoprotein cholesterol
HFD	high-fat diet
HG	high glucose
HK	hexokinase
HMGCR	3-hydroxy-3-methylglutaryl-coenzyme A reductase
HMGCS-1	hydroxymethylglutaryl-CoA-synthase-1
HP	high palmitate

HOMA-IR	homeostatic model assessment for insulin resistance
HSL	hormone-sensitive lipase
ICAM	intracellular adhesion molecule
IFN	interferon
IκB	inhibitor of NF-κB
IL	interleukin
iNOS	inducible nitric oxide synthase
INS-1	insulin-1
i.p.	intraperitoneal
IRS	insulin receptor substrate
i.v.	intravenous
JAK	Janus kinase
JNK	c-Jun N-terminal kinase
KLF-7	kruppel-like factor 7
LACD	long-chain acyl CoA-dehydrogenase
LC3B	microtubule-associated proteins 1A/1B light chain 3B
LDH	lactate dehydrogenase
LDL-C	low-density lipoprotein cholesterol
LKB1	liver kinase B1
LO	lipoxygenase
LPL	lipoprotein lipase
LPS	lipopolysaccharide
LXR-α	liver X receptor-alpha
MACD	medium-chain acyl coenzyme A-dehydrogenase
MafA	pancreatic β-cell activator transcription factor for insulin gene expression
MAPK	mitogen activated protein kinase
MCP-1	monocyte chemoattractant protein-1
MDA	malondialdehyde
MK	mevalonate kinase
MOR	μ-opioid receptor
m-TOR	mammalian target of rapamycin
NAD^+	nicotinamide adenine dinucleotide in oxidized form of coenzyme
NAFLD	non-alcoholic fatty liver disease
NGF	nerve growth factor
NLRP-3	nucleotide-binding domain, leucine-rich-repeat-containing family, pyrin domain containing 3
NO	nitric oxide
NOX	NADPH-oxidase
NRF-1	nuclear respiratory factor-1
OAT	organic anion transporter
OPN	osteopontin
PAI-1	plasminogen activator inhibitor-1
PAP	phosphatidic acid phosphohydrolase
PC	pyruvate carboxylase
PDK-4	pyruvate dehydrogenase kinase-4
PDX-1	pancreatic duodenal homeobox-1 protein
PEPCK	phosphoenolpyruvate carboxykinase
PFK	phosphofructokinase
PGC1α	peroxisome proliferator-activated receptor gamma coactivator 1-alpha
PI3K	phosphatidylinositide 3-kinase

PK	pyruvate kinase
PKC	protein kinase C
PKR	protein kinase RNA-activated
PON	paraoxonase
p.o.	post oral
PPAR	peroxisome proliferator-activated receptor
PSD-95	post-synaptic density protein-95
PSP	purple sweet potato
PTP1B	protein-tyrosine phosphatase 1B
PUFA	polyunsaturated fatty acid
RAGE	receptor of advanced glycation endproducts
RelA	v-rel avian reticuloendotheliosis viral oncogene homolog A, a subunit of NF-κB, also known as p65 protein
RGC	retinal ganglion cell
ROS	reactive oxygen species
RBP-4	retinol-binding protein-4
SCD	stearoyl-CoA desaturase
SCFA	short-chain fatty acid
SGLT	sodium-dependent glucose co-transporter
SIRT-1	silent-mating type information regulation 2 homolog 1, also known as sirtuin-1
SOCS	suppressor of cytokine signaling
SOD	superoxide dismutase
S6K	ribosomal protein S6 kinase
SQSTM1	sequestosome-1, also known as p62 protein
SREBP	sterol regulatory element-binding protein
STAT	signal-transducer and activator of transcription protein
SYP	synapsin
TBARS	thiobarbituric acid reactive substances
TC	total cholesterol
TCF7L2	transcription factor 7 like-2
T2DM	type-2 diabetes mellitus
TFAM	transcription factor A, mitochondrial
TG	triacylglyceride
TGF	transforming growth factor
TH	T helper cell
TLR	toll-like receptor
TNF	tumor-necrosis factor
Treg	regulatory T cell
TrKA	tropomyosin receptor kinase A
TXNIP	thioredoxin interacting protein
UCP	uncoupling protein
VAT	visceral adipose tissue
VCAM	vascular cell adhesion molecule
WAT	white adipose tissue
WHO	world health organization
wnt	wingless-type integration site protein
XBP-1	X-box binding protein-1

1. Introduction

Obesity and diabetes are closely interrelated global pandemic metabolic disorders and their incidence is increasing at an alarming rate with high rate of morbidity due to recent changes in food habits and lifestyle. Obesity is characterized by an excessive fat accumulation in white adipose tissues and liver from a positive energy balance due to imbalance between energy intake and energy expenditure and with a low grade systemic and islet inflammations. It is frequently observed among the people who lead a sedentary lifestyle and consume too much high calories fast foods and ethyl alcohol or suffer from genetic disorders. The incidence of obesity has risen dramatically in the last few decades, now affecting more than 650 million people worldwide. As per WHO report, people having a body mass index greater than 30 kg/m^2 are categorized as obese. The people of China, India, United States and United Kingdom are greatly affected to this disorder. Obesity-induced insulin resistance is the major driving force to the development of type 2 diabetes and its related vascular disorders particularly in kidney, heart and eyes. Indeed more than 80% of type 2 diabetic people in the United States are obese or overweight. It is expected that one of three children born in this century may develop obesity-related diabetes. Thus, it is an emerging area of research to confront and ameliorate this epidemic disease to save our future generation (Moller & Kaufman, 2005; World Health Organization, 2018).

Diabetes mellitus, simply known as diabetes, is a complex metabolic disorder, characterized by a chronic state of hyperglycemia resulting from impaired insulin secretion from pancreatic beta cells and insulin resistance to peripheral tissues and brain on insulin action. The disease diabetes was coined by Greek physician Apollonius of Memphis in 250 BC and the term, "mellitus" to this disease was introduced by British alchemist Thomas Willis in 1674–75 based on honey flavored urine of diabetic patients (Poretsky, 2010, p. 3). The major form of diabetes, known as type 2 diabetes mellitus (T2DM), formerly known as non-insulin dependent diabetes, NIDDM, accounts for 90%–95% of reported diabetic cases, while less common type 1 diabetes (T1DM), formerly known as insulin-dependent diabetes, IDDM, accounts only 5%–10% of reported cases. An extended form of T2DM, known as maturity-onset diabetes of the young, MODY, is reported among youngs of ages below 25 years (less than 5% of T2DM cases). Another form of diabetes, Gestational diabetes, GDM are reported among pregnant women (3%–9%) and this GDM is the risk factor for both offspring and mother. At present, more than 450 million people are affected to diabetes and this figure may be increased to 693 million in 2045 due to rapid changes in our lifestyle, food habits and urbanization (American Diabetes Association, 2014). Diabetic patients most often develop vascular complications, such as coronary artery disease, stroke, renal disease and blindness. Diabetes is the seventh leading cause of death in the United States (Murea, Ma, & Freedman, 2012). Currently available anti-diabetic drugs have several limitations on efficacy and safety to combat this pandemic disease. Therefore, the search of potent and safe drugs from medicinal foods for obesity and diabetes is a great challenge of clinical medicine.

Dietary plant flavonoids that are readily available in abundant quantities in dietary fruits, vegetables, tea and wine, have been found to exhibit potential benefits in prevention of obesity and diabetes. Some of these plant metabolites could be utilized in treatment of obesity and diabetes as safe and low cost drugs after extensive clinical trials. In this chapter, the etiology, pathogenesis, clinical diagnosis and current management of obesity and diabetes as well as the potential benefits of some dietary flavonoids in prevention of obesity and diabetes along with their molecular actions to the target cells and genes-responsible to these disorders are briefly discussed.

2. Etiology, pathogenesis, clinical diagnosis and management of obesity and diabetes

2.1 Etiology and pathogenesis of obesity and diabetes

Chronic consumption of high quantity of nutrients or high calorie fast food compared to their utilization in the body causes the storage of excessive fat/nutrients as energy in fat cells, adipocytes in obese individuals. Thus, long-term accumulation of excess fat in adipocytes of white adipose tissue is the main reason of obesity. The excessive storage fatty acids/lipid mainly as triacylglycerides in adipocytes creates an uneasy stress from adipocyte hypertrophic state and induces a chronic low-grade tissue inflammation, known as "metaflammation", mediated by several stress kinases such as PKC, JNK, IKK and PKR, molecular sensors/receptor TLR-2/TLR-4, and signaling proteins, CREBH and SOCS, is generated. This inflammation contributes a dysfunctional state of adipocytes and leads to the release of excessive fatty acids from enhanced lipolysis, stimulated by sympathetic sensory neurons, to the circulatory system for accumulation in distant organs, muscle, liver, brain and pancreas, which are not well-equipped for lipid storage. In addition to FFAs release, the hypertrophic adipocytes increase inflammatory M1 macrophage infiltration and secrete inflammatory adipokines, TNF-α, IL-6 and MCP-1 to the circulatory system by activation of infiltrate macrophages in distant organs. The increased lipotoxicity from excessive FFAs levels in liver, muscle, brain and pancreas activate the kinases, JNK, IKK and NF-κB, for induction of low-grade inflammation, which inhibits IRS-1 tyrosine phosphorylation and PI3K/Akt activation, decreases insulin-stimulated glucose utilization and leads to increase insulin resistance and metabolic dysfunction and to reduce insulin secretion from β-cells. Obesity-induced hypothalamic inflammation causes leptin resistance in the CNS, which contributes to systemic insulin resistance in liver via a brain-liver neuronal signal (Milanski et al., 2012). Moreover, chronic low-grade inflammation in intestine from obesity causes intestinal endothelial membrane leakage leading to the entry of increased levels of Gram-negative LPS and other harmful bacteria in the circulating system, and these LPS and other bacteria promote systemic inflammation in liver to harvest energy from the diet. Therefore, obesity-related lipotoxicity from excessive FFAs levels, inflammation and gastrointestinal microbiota (dysbiosis) are the major interacting components to induce insulin resistance in peripheral tissues and brain (Ertunc & Hotamisligil, 2016; Johnson & Olefsky, 2013; Redinger, 2007).

Accumulating evidence reveals that in type 1 diabetes, the autoimmune destruction of pancreatic β-cells occurs from chronic hyperglycemia induced secretion of high levels of pro-inflammatory cytokines, IL-1, TNF-α and IFN-γ in islet microenvironment, which resulted in the increased expression of iNOS gene and NO production to suppress the activity of immunoregulatory Th2 cells and their cytokines IL-4, IL-10 and IL-13, and to allow the Th1 cells to initiate the cell death cascade of β-cells for its destruction (Rabinovitch & Suarez- Pinzon, 1998). Genome-wide association studies (GWAS) identified some mutant genes, HLA-DR/DQ/DP, INS-VNTR, CTLA4, PTPN22, ERBB3, AIRE and IFIH1 that are the risk factor of T1DM because these genes act as negative regulators of Th2 cells (Nejentsev et al., 1997). Several viruses, namely coxsackievirus B, mumps virus, cytomegalovirus, rotavirus and rubella have been associated to the development of T1DM. Feeding of cow milk and cereals in early stages of childhood is the risk factor of this disease in children because these foods lack several growth factors and cytokines production that are essential for maturation of intestinal mucosa of children (Paschou, Papadopoulou-Marketou, Chrousos, & Kanaka-Gantenbein, 2018).

Several studies on the etiology of diabetes have focused that genetic factors related to impaired insulin secretion and insulin resistance and environmental factors namely obesity, overeating, too much drinking of ethyl alcohol, cigarette smoking and excessive stress in brain are the major risk factors for the development of type 2 diabetes. About 30% of reported T2DM cases are closely related to genetic factors and about 55% of reported T2DM cases are from obesity (Kaku, 2010). Recently GWAS have identified some mutant genes of glucokinase, mitochondrial functions and insulin receptor such as ADAMTS9, ADCY5, CAMK1D, CDKAL1, CDKN2A/B, CENTD2, DGKB, FOXO1, GCK, HHEX, IGF2BP2, KCNQ1, KCNJ11, MTNR1B, PROX1, SGK1, SLC30A8, TCF7L2, TSPAN8 and THADA that are directly or indirectly involved in impaired insulin secretion from pancreatic β-cells, and a few genes namely FTO, GCKR, IRS1, KIF14, PPARG and ADAMTS9 are associated with insulin resistance (Murea et al., 2012). In general, changes in food habits, particularly the increase in fat intake, decrease in starch and fiber intake and increase intake of simple sugars from soft drinks and sweets, contribute to the development of obesity and deteriorate oral glucose tolerance. Obesity from increased visceral fat mass is the high risk for the diabetes from insulin resistance. Recently attention has been paid in the regulation of inflammatory adipokines secretion from hypertrophic WAT of obese individuals for improvement of insulin resistance. Among the secreted adipokines, TNF-α, IL-6, MCP-1, TGF-β, leptin, resistin, CRP and FFAs act to increase insulin resistance, while adipokines, such as adiponectin, IL-10, and IL-4 reduce insulin resistance. A substantial part of released FFAs are accumulated in pancreas, skeletal muscle, liver and kidneys and induce lipotoxicity. A group of studies reveal that in obese adipose tissues of humans and mice, TNF-α gene is highly expressed and it activates stress related kinases JNK and IKK, and transcription factor NF-κB, which inhibit insulin receptor signaling. Furthermore, TNF-α activates the infiltrated macrophages in adipose tissues to secret more proinflammatory genes, some of them are leak out of the tissues to the circulating plasma and produce insulin resistance in pancreas, skeletal muscle, liver and kidneys (Osborn & Olefsky, 2012). High expression level of cytokine IL-6 in obese adipose tissues induces insulin resistance by reducing the expression levels of IRS-1 and GLUT-4 proteins in insulin sensitive tissues for glucose uptake (Rotter, Nagaev, & Smith, 2003). Obesity-induced hepatic inflammation produces chemokine MCP-1, which influences the migration of recruited hepatic macrophages in liver and activates Kupffer cells (resident macrophage in liver) to induce systemic insulin resistance in liver (Lanthier et al., 2010). Elevated FFAs levels in circulating system activate TLR4 sensor molecule for induction of JNK and IKK activation to inhibit insulin signaling via serine phosphorylation of IRS-1 (Glass & Olefsky, 2012). In many diabetic patients, impaired pancreatic α-cells increase fasting plasma glucagon level to induce hepatic glucose production. Therefore, suppression of glucagon secretion by increasing gut GLP-1 secretion could be a novel strategy for amelioration of fasting blood glucose level. Activation of nuclear hormone PPARγ has been found to stimulate its transcriptional program in adipocyte, liver, skeletal muscle and CNS to improve overall systemic insulin sensitivity by reduction of serum CRP, leptin, PAI-1, FFAs and TNF-α levels in obese diabetic humans and mice (Tontonoz & Spiegelman, 2008). Adipokine dipeptidyl-peptidase-4 (DPP-4) is highly expressed in obese diabetic patients and hence inhibition of DPP-4 activity could be an effective target to improve insulin sensitivity in diabetes (Rohrborn, Wronkowitz, & Eckel, 2015).

2.2 Clinical diagnosis of obesity and diabetes

As per guidelines of American Diabetes Association and American Association of Clinical Endocrinologists, an individual of 45 years of age or above having a body mass index > 25 kg/m^2, is considered as obese, while an individual of same age having fasting blood glucose level of 100–125 mg/dL and 2 h postprandial plasma glucose level of 140–199 mg/dL or glycated hemoglobin (HbA1c) level of 5.7%–6.4% is considered as diabetic patient. Individual persons having HbA1c levels of >6% are considered to have high risk of diabetes (Tabak, Herder, Rathmann, Brunner, & Kivimaki, 2012).

2.3 Current management of obesity and diabetes

Good glycemic control via protection of pancreatic β-cells mass and function, improvement of insulin action in peripheral metabolic tissues such as skeletal muscle, liver, kidney and adipose tissues for glucose utilization and reduced hepatic glucose production, amelioration of insulin sensitive neurotransmitter dysfunctions of brain, reduced renal glucose reabsorption and enhanced intestinal incretin GIP and GLP-1 secretions and incretin-stimulated insulin secretion from pancreatic β-cells and reduced elevated glucagon expression from impaired pancreatic α-cells, are the main foundations for treatment of obesity-induced diabetes. In addition, insulin-sensitive PPARα and PPARγ activation regulates lipolytic and lipogenic genes in liver and adipose tissues for reduction of lipotoxicity and lipid accumulation to improve insulin resistance in obese diabetic patients. In early stages of diabetes (HbA1c levels in the range >6.0%− < 6.5%), oral biguanide metformin and restricted food intake is recommended to diabetic patients for reduction of hyperglycemia, via improvement of insulin sensitivity and PPAR activity in peripheral tissues. The long-term use of metformin by elderly diabetic patients causes renal impairment and other adverse effects (Viollet et al., 2012). Incretin based drugs, GLP-1 receptor agonists such as liraglutide and exenatide, and expensive DPP-4 inhibitors such as sitagliptin and saxagliptin are also recommended for improvement of hyperglycemia, blood pressure and weight loss in obese diabetic patients. Long-term use of these drugs in high doses has high risk of gastrointestinal and respiratory tract infections and nasopharyngitis in patients (Stonehouse, Darsow, & Maggs, 2012). Increased insulin secreting second-generation antidiabetic drugs, sulfonylureas glipizide and glimepiride have risk factors of renal diseases and hypoglycaemia (Betonico, Titan, Correa-Giannella, Nery, & Queiroz, 2016). PPARγ agonists such as rosiglitazone and pioglitazone improved insulin sensitivity in adipose tissues, muscle and liver via increased adiponectin secretions and restoration of β-cell function and integrity, but have a risk factor of heart attack and stroke (Singh, Loke, & Furberg, 2007). Glucosuric antidiabetic drugs, sodium-glucose cotransporter-2 (SGLT-2) inhibitors namely canagliflozin and dapagliflozin are recommended in advanced stages of T2DM patients, whose pancreatic β-cell reserves are permanently lost, to reduce glucose re-absorption in renal proximal tubule in insulin-independent pathway, but its long-term use have high risk of urinary tract infections and related disorders such as urosepsis, pyelonephritis and genital mycosis (Hsia, Grove, & Cefalu, 2017). Antidiabetic potassium channels blockers glinides such as nateglinide and repaglinide improve insulin secretion from pancreatic β-cells via intracellular calcium influx for better control of postprandial high glucose levels, have a risk of bronchitis and muscoskeletal disorders including backpain in patients (Guardado-Mendoza, Prioletta, Jimenez-Ceja, Sosale, & Folli, 2013).

3. Dietary plant flavonoids in prevention of obesity and diabetes

Flavonoids, a large class of plant polyphenolic compounds, are present in high concentrations in dietary nut, colored fruits, vegetables, legumes, chocolate, tea and wines. A good number of chemical and clinical studies on flavonoids in last four decades demonstrated their strong antioxidant and anti-inflammatory potentials and efficacy in the treatment and prevention of various inflammatory-related diseases including viral-, neurodegenerative-, and cardiovascular diseases, cancer, obesity and diabetes (Kumar & Pandey, 2013; Hossain et al., 2016; Spagnuolo, Moccio & Russo, 2018).

3.1 Classification of flavonoids

Flavonoids (Latin word, flavan means yellow color) are abundantly found in plants as their glycosides or free aglycones. Chemically they have a flavan nucleus of 15 carbon atoms, that are arranged in three rings, C_6 (A)- C_3 (B)- C_6 (C), the middle ring C is a heterocyclic pyran ring, which is linked to aromatic rings A and B on either sides **1** (Fig. 1). These compounds are broadly categorized into six subclasses, flavones **2**, flavonols **3**, flavanones **4**, flavanols **5**, isoflavones **6**, and anthocyanins **7**, mainly based on which carbon of the C ring is connected to B ring and on the degree of unsaturation and oxidation of the C ring (Fig. 1). Within each group, individual compounds are categorized by the substitution pattern in the rings. Usually hydroxy, methoxy, *O*-glycosylated or C-glycosylated substituents are found in rings. Some minor flavonoids namely chalcones **8**, dihydrochalcones **8a** and aurones **9** (Fig. 1), based on C ring opening and rearrangement, have been isolated from several plants (Harborne & Mabry, 1982; Havsteen, 2002).

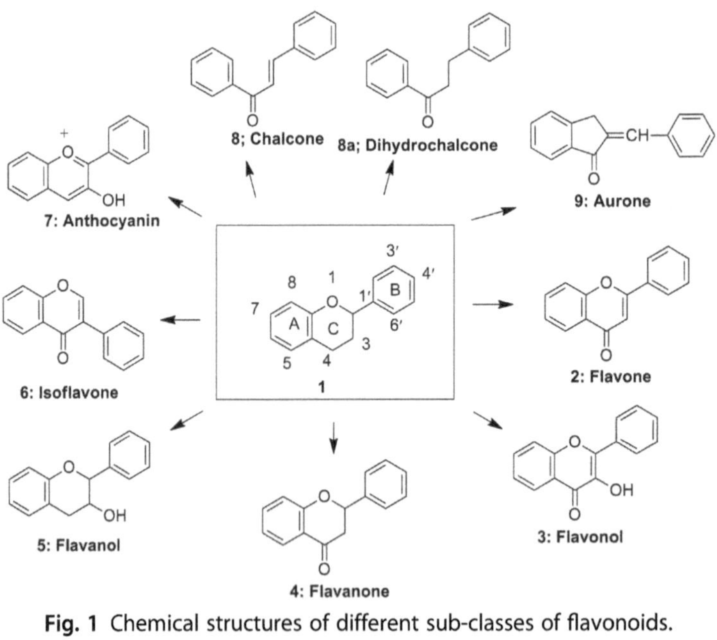

Fig. 1 Chemical structures of different sub-classes of flavonoids.

3.2 Plant flavonoids in prevention of obesity and diabetes

3.2.1 Flavonols in prevention of obesity and diabetes

Flavonols are the most abundant flavonoids in plants and are found in high concentrations in various dietary fruits and vegetables such as apples, grapes, oranges, berries, dates, onions, tomatoes, kale, cauliflowers, beans, broccoli, lettuce, peas, and tea. Among the isolated dietary flavonols, quercetin and its glycosides, kaempferol and its glycosides, isorhamnetin and its glycosides, myricetin, fisetin, and morin have been found significant anti-obesity and antidiabetic activities (Cao, Chen, Zhang, Zhang, & Zhao, 2010; Crozier, Jaganath, & Clifford, 2009; Hossain et al., 2016).

Quercetin **10**, and its glycosides, quercetin 3-*O*-glucoside **11**, and quercetin 3-*O*-galactoside **12** (Fig. 2), isolated from antidiabetic cowberry (*Vaccinium vitis-idaea*) fruits, exhibited significant glucose uptake in skeletal muscle C2C12 cells via AMPK activation and their activity was very similar to that of anti-diabetic drug metformin (Eid et al., 2010). In L6 myotubes, quercetin showed glucose uptake via GLUT-4 protein expression and its translocation to the cell membrane, and up-regulated the expression of the mRNA levels of CaMKK, AMPK and p38MAPK kinases about five fold and of IRS and PI3K only one fold compared to the control. Co-treatment of dorsomorphin (AMPK inhibitor) or STO-609 (CaMKK inhibitor) significantly reduced the glucose uptake effect of quercetin, while co-treatment of wortmannin (PI3K inhibitor) partially reduced the glucose uptake effect of quercetin. These findings reveal that quercetin improves glucose utilization in skeletal muscle in both AMPK and PI3K activation, but the former pathway being the dominant one (Dhanya, Arya, Nisha, & Jayamurthy, 2017). Blue berry peel extract rich in quercetin and its glycosides significantly inhibited the adipogenesis in 3T3-L1 adipocytes. Accumulating evidence indicates that inhibition of adipogenesis is one of the targets for reduction of lipid mass in obese individuals. Therefore, quercetin and its glycosides could be effective in prevention of obesity (Song et al., 2013). The study of molecular mechanism revealed that quercetin in 3T3-L1 adipocytes, significantly reduced the adipogenesis by decreasing the expression levels of adipogenic/lipogenic enzymes, FAS, SREBP-1, CEBP-α, and PPAR-γ via activation of AMPK and its substrate ACC. Furthermore, quercetin induced apoptosis of mature adipocytes by up-regulating the expression levels of pro-apoptotic genes caspase-3 and 9 and down-regulating the expression level of cell survival Bcl-2 proteins via suppression of stress kinases, ERK1/2 and JNK phosphorylation (Ahn et al., 2008). In

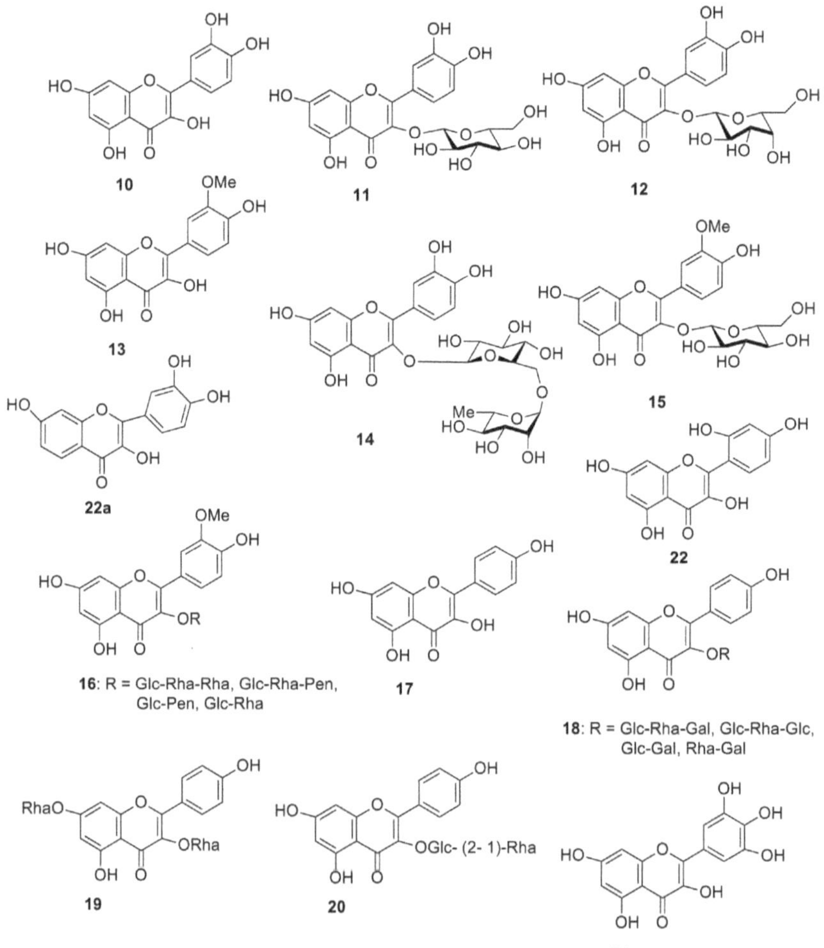

Fig. 2 Chemical structures of flavonols having anti-obesity and anti-diabetic effects.

animal model, quercetin on supplementation in high-fat diet (HFD) (0.03% of diet) to obese rats for 4 weeks significantly reduced the mass of WAT via decreasing the PPARγ mRNA levels in WAT, liver and quadriceps muscle and increasing plasma adiponectin level and reducing plasma 8-iso-prostaglandin-$F_{2\alpha}$ level in obese rats. Furthermore, quercetin and its major metabolite isorhamnetin **13** (Fig. 2) did not exhibit PPAR-γ binding activity. All these findings suggest that the anti-obesity effect of quercetin is related to the improvement of antioxidant status and adiponectin secretion in a PPAR-γ independent pathway, different from that of PPAR-γ binding thiazolidinediones (Wein, Behm, Petersen, Kristiansen, & Wolfram, 2010).

Administration of quercetin (15 mg/kg/d, *i.p.*) to STZ-induced diabetic rats for 4 weeks significantly improved hyperglycemia in rats by reducing plasma glucose and insulin levels and protecting pancreatic β-cell integrity and insulin secretion. Histochemical study of pancreatic homogenates of rats reveals that quercetin improves the activity of antioxidant enzymes SOD, GSH and CAT and reduces the levels of toxic markers MDA and NO for protection of pancreatic functional β-cells (Coskun et al., 2005). Quercetin on diet supplementation in diabetic mice (0.5% of diet) significantly reduced the elevated expression of cyclin-dependent kinase inhibitor 1a (dkn1a) to improve the expression of sex determining region Y (SRY)- box-4 (SOX-4) gene for survival and replication of β-cells in pancreas of diabetic mice. Furthermore, it reduced the elevated levels of heat-shock proteins ubiquitin C and G-protein coupled receptor 146 (Gpr-146) and TBARS proteins and improved the reduced SOD level in liver of mice for restoration of normal functioning of hepatic genes in glucose utilization (Kobori et al., 2009). The same group also reported that long-term diet supplementation of quercetin to HFD-fed obese mice for 20 weeks significantly reduced visceral and liver fat accumulation and weight gain in mice by suppression of plasma inflammatory cytokine TNF-α and FFAs levels and improvement of hepatic PPAR-α expression for β-oxidation of fatty acids and reduction of the expression of hepatic lipogenic enzymes, PPAR-γ, SREBP-1c, and FAS for reduction of lipid synthesis and of CD36 enzyme for reduction of lipid uptake in mice. Furthermore, quercetin improved the expression of hepatic GK gene for glucose utilization in glycolysis and reduced the expression of hepatic PEPCK enzyme to reduce glucose production in gluconeogenesis through AMPK activation (Kobori, Masumoto, Akimoto, & Oike, 2011). Quercetin on diet supplementation also reduced the level of iNOS level in WAT of obese Zucker rats (Rivera, Moron, Sanchez, Zarzuels, & Galisteo, 2008). Quercetin on diet supplementation to alloxan-induced diabetic mice significantly improved the renal and liver complications by reducing the elevated levels of serum urea, creatinine, ALT, AST and ALP levels (Alam, Meerza, & Naseem, 2014). Long-term diet supplementation of quercetin in HFD (0.05% of diet) to NAFLD associated obese mice for 16 weeks significantly reduced hepatic lipid accumulation through the regulation of lipid metabolism and improvement of insulin resistance and gut microbiota dysbiosis. Quercetin improved the gut microbiota by reducing the elevated plasma endotoxic LPS, inflammatory cytokines TNF-α and IL-6 levels to reduce the activation of NF-κB, and by increasing the expression levels of short-chain fatty acids (SCFAs) producing bacteria,

Bacteroidia, Akkermansia, Proteobacteria and decreasing the expression levels of Gram-negative bacteria, *Firmicules* and *Helicobactor* in intestinal gut for improvement of gut barrier function via mucin synthesis. The mucin layer on the intestinal epithelial cell membrane prevents the entry of harmful bacteria through the membrane into systemic circulation. Furthermore, quercetin reduced the hepatic lipid peroxidation level by reducing the elevated expression of CYP2E1 gene and NLRP3 inflammasome activation and improved the expression of fatty acid trafficking genes FABP1, and FOXA1 and reduced the expression of CD-36 to reduce lipotoxicity and lipid accumulation. Moreover, it reduced the expression of lipogenic genes, C/EBPα, SREBP-1c and LXRα to reduce hepatic lipid synthesis and TG level in the liver of obese mice. Hence, quercetin could be a potential drug in prevention of HFD-induced dysbiosis-related NAFLD development in diabetic patients (Porras et al., 2017).

Rutin (quercetin-3- O-rutinoside) **14** (Fig. 2), a major flavonoid constituent of oranges, lemons, grapes and berries, on supplementation in HFD (25 and 50 mg/kg diet) to HFD-fed obese mice for 4 weeks, significantly reduced the body weight and lipid mass in liver of mice by reducing the expression of the mRNA levels of lipogenic genes, PPAR-γ and CEBP-α in liver tissues. Treatment of rutin in 3T3-L1 adipocytes inhibited adipogenesis by down-regulation of PPAR-γ and CEBP-α proteins (Choi, Park, Choi, & Lee, 2006). Chronic diet supplementation of rutin at higher dose (100 mg/kg of diet) to HFD-fed obese rats for 8 weeks significantly reduced the weights of body, liver and adipose tissues and decreased serum glucose, insulin, TG, TC, leptin, AST and ALT levels in obese rats compared to the control group. Furthermore, rutin reduced the hepatic oxidative stress by decreasing lipid peroxidation and glutathione disulfide content and improved the levels of antioxidant enzymes, GSH, GP_X, GSR, and GST for amelioration of obesity in obese rats (Hsu, Wu, Huang, & Yen, 2009). Treatment of rutin at the same dose to STZ-induced diabetic rats for 45 days significantly exhibited anti-diabetic effect by improving pancreatic β-cells mass and insulin secretion and plasma insulin and C-peptide levels as well as restoring glycogen content in liver and muscles of diabetic rats. Furthermore, rutin increased the antioxidant status in pancreas by increasing the activity of antioxidant enzymes SOD, CAT and GP_X for protection of pancreatic β-cells and increased the activity of glycolytic hexokinase gene in liver and muscle for glucose utilization and reduced the expression of hepatic gluconeogenic genes G6Pase and FBPase to decrease glucose production in liver for maintenance of glucose homeostasis in the body of diabetic rats

(Kamalakkannan & Prince, 2006; Stanley Mainzen Prince & Kamalakkannan, 2006). Rutin in rat soleus muscle cells culture in presence of 2-deoxy-D-glucose, significantly stimulated glucose uptake by up-regulation of glucose transporter GLUT-4 proteins and its translocation from intracellular matrix to plasma membrane through the activation of PI3K, PKC and MAPK kinase, similar to that of insulin. Thus, rutin improves glucose utilization in skeletal muscle of diabetic animals by its insulin-mimetic effect (Kappel et al., 2013). Rutin treatment at low doses (8 mg/kg) to STZ-induced diabetic rats protected the cardiomyocytes from STZ-toxicated injury by reducing the elevated levels of myocardial-injury marker enzymes, creatine kinase-myocardial brand (CK-MB) and LDH, pro-apoptotic proteins, Bax and caspase-3 and improving the level of anti-apoptotic Bcl-2 proteins in cardiac tissues of diabetic rats (Wang et al., 2015). While rutin treatment at high dose (100 mg/kg bw) to STZ-induced diabetic rats for 5 weeks protected the rats from retinal dysfunction by improving the expression levels of BDNF and NGF and decreasing the expression level of caspase-3 proteins in retinal tissue homogenate of diabetic rats. Accumulating evidence reveals that elevated BDNF expression in retinal tissue improves retinal ganglion cells (RGCs) dysfunction and death from high glucose-induced glaucoma and restores vision function of eyes, while NGF expression improves axonal regeneration of RGC and its survival via binding to its receptor tyrosine kinase, TrKA (Ola et al., 2015). Rutin on dietary supplementation to healthy volunteers for 6 weeks resulted in the increased concentrations of quercetin, kaempferol and isorhamnetin in the tissues of the volunteers and no rutin was detected. Hence, rutin is biotransformed in to its metabolites in mammals and its metabolites are the actual bioactive components of rutin in vivo models (Boyle et al., 2000).

Isorhamnetin 3-*O*-glucoside **15** (Fig. 2), on oral administration (25 mg/kg bw) to STZ-induced diabetic rats showed significant improvement in diabetes and diabetic retinal complications by reducing serum glucose and insulin levels and sorbital accumulation in RBC, eye lenses and sciatic nerves in diabetic rats. In cellular model, compound **15** in rat lens homogenate strongly inhibited the activity of aldose reductase with an IC_{50} value of 1.4 μM. Several studies have shown that aldose reductase is highly expressed in eye lenses of diabetic patients and animals and it catalyzes the reduction of glucose into sorbitol under high glucose concentrations. This generated sorbital is accumulated in eye lenses and increases osmotic stress in eye lenses resulting in the degeneration of lens epithelial cells and leads to cataract formation on eye lenses (Lee et al., 2005). *Opuntia ficus-indica* (OFI) fruits are traditionally consumed by Mexican people to get relief from diabetes and hyperlipidemic disorders. A methanolic extract fraction of OFI, rich in isorhamnetin 3-*O*- glycosides mixture **16**, on diet supplementation in HFD (0.3 and 0.6 g/kg diet) to HFD-fed obese mice for 12 weeks significantly ameliorated hyperglycemia and hyperlipidemia in obese mice by reducing serum glucose, insulin, TC, TG and LDL-C levels. Furthermore, OFI extract significantly increased hepatic and muscle glucose utilization by increasing tyrosine phosphorylation of IRS-1 and activation of S6K (p70S6K), PI3K and Akt proteins and reduced hepatic lipid content by decreasing mRNA expression levels of lipogenic genes, XBP-1, ATF6, CHOP, SREBP1, FAS and SCD-1 and increasing mRNA expression levels of mitochondrial biogenetic genes, CPT-α and ACOX in obese mice. OFI extract in rat pancreatic islets culture, remarkably increased insulin secretion and GLUT-1 protein expression in β-cells. Thus, OFI extract increased insulin sensitivity and insulin secretion in obese mice to regulate glucose and lipid metabolism for amelioration of hyperglycemia and hyperlipidemia (Rodriguez- Rodriguez et al., 2015). Possibly, isorhamnetin glycosides are converted into isorhamnetin **13** and ultimately into quercetin **10** by intestinal flora, *E. coli* and *Enterococcus* spp., etc. in animals and thus quercetin would be the bioactive metabolite in vivo models of isorhamnetin glycosides (Du et al., 2017).

Kaempferol **17** (Fig. 2), abundant in several fruits such as apples, grapes and berries and in vegetables such as tomato, potato and broccoli, on treatment in rodent insulin secreting INS-1 β-cells and isolated human pancreatic islets under chronic HG concentrations, increased glucose-induced insulin secretion by improving intracellular cAMP, Akt proteins and ATP production. It also increased the synthesis of insulin in β-cells and protected β-cells by increasing the expression of anti-apoptotic Bcl-2 and reducing the expression of pro-apoptotic caspase 3 proteins (Zhang & Liu, 2011). In another cellular study, kaempferol in HP-exposed hyperlipidemic condition of INS-1E β-cells and human islets, increased insulin secretion and synthesis and protected the β-cells from lipotoxic-induced apoptosis by increasing the expression of cell survival PDX-1 proteins via cAMP production, PKA activation, CREB phosphorylation and its transcriptional activity in β-cells. Thus, kaempferol could be a prospective drug in treatment of obesity (Zhang et al., 2013). In animal model, kaempferol on diet supplementation in HFD (0.05% of diet) to HFD-fed obese mice for 5 months significantly reduced obesity-related complications in mice by decreasing plasma glucose, TG and LDL-C levels, and increasing glucose utilization of muscle and adipose tissues through GLUT-4 protein expression and AMPK activation. Kaempferol in skeletal C2C12 cells culture under HG and lipid (oleic acid and palmitate) concentrations, significantly increased lipolysis, glucose uptake and glycogen synthesis via AMPK activation and expression of GLUT-4 and its translocation from intracellular matrix to plasma membrane. In addition, kaempferol on dietary supplementation (0.05% of diet) to STZ plus HFD-fed induced obese diabetic mice significantly improved pancreatic β-cells mass and insulin secretion to reduce hyperglycemia, hyperlipidemia and lipid mass in adipose tissues of mice by insulin sensitive glucose uptake and AMPK activation. These findings indicate that kaempferol improves lipid and glucose metabolism in diabetic mice via increased insulin sensitivity and insulin secretion and AMPK activation (Alkhalidy et al., 2015). Kaempferol on diet supplementation in obese diabetic rats reduced hepatic inflammation by decreasing the levels of plasma inflammatory cytokines TNF-α and IL-6 via inhibition of the activation of IKKB and NF-κB kinases to improve insulin resistance in liver and other peripheral tissues (Luo et al., 2015). Kaempferol treatment to STZ-induced diabetic mice increased glycogen contents in liver and muscle of the diabetic mice by up-regulation of muscle hexokinase and hepatic glucokinase (GK) expression and reduced hepatic glucose production by down-regulation of gluconeogenic gene, pyruvate carboxylase (PC) (Alkhalidy et al., 2018). A methanolic extract fraction from soybean leaves rich in kaempferol 3-*O*-glycosides mixture **18** (Fig. 2), on oral supplementation in HFD (0.15 g/kg diet) to obese diabetic mice for 92 days significantly reduced the weights of body, adipose tissues and liver, and improved plasma glucose, HbA1c and TG levels in diabetic mice. Furthermore, the extract reduced the expression of hepatic lipogenic enzymes, PPAR-γ, SREBP-1c and FAS to reduce lipid synthesis, increased the expression of CPT-1 gene for fatty acid oxidation and reduced the hepatic glucose production by lowering the activity of gluconeogenic enzymes through improvement of AMPK activation. Thus, dietary consumption of soybean leaves and fruits, rich in kaempferol glycosides, could be effective in prevention of obesity (Zang, Zhang, Igarashi, & Yu, 2015). Kaempferol glycosides, kaempferitrin **19** (Fig. 2) and kaempferol 3-*O*-neohesperidoside **20** (Fig. 2), on treatment in rat soleus muscle cells in presence of 2-deoxy-D-glucose increased glucose uptake by increasing the expression of GLUT-4 proteins and its translocation from cytoplasm to cell membrane through activation of IR, PI3K and PKC proteins and increased glycogen content via up-regulation of glycogen synthase and other proteins related to glycogen synthesis in muscle cells (Cazarolli et al., 2013). In vivo model, compound **20**, on treatment (100 mg/kg bw, a single dose) in diabetic rats increased the glycogen content in the muscles of diabetic rats (Zanatta et al., 2008).

Myricetin **21** (Fig. 2), a flavonoid constituent of berries and other fruits and vegetables, on administration (1 mg/kg bw/thrice daily, *i.v.*) to high fructose plus high-fat fed obese diabetic rats for 14 days significantly reduced hyperglycemia and hyperlipidemia by decreasing plasma glucose, insulin and TG levels in diabetic rats through increased the expression levels of GLUT-4 proteins and phosphorylated IR, IRS-1 and Akt proteins in soleus muscle of diabetic rats (Liu, Tzeng, Liou, & Lan, 2007a). Myricetin in C2C12 myotubes under HG condition, increased glucose uptake via up-regulation of GLUT-4 proteins expression and increased activation of Akt and AMPK kinases. Thus, myricetin exhibits its anti-diabetic effect via both insulin-stimulated Akt activation and insulin-independent AMPK activation (Ding, Dai, Zhang, & Li, 2012). In another study, myricetin improved opioid peptide, β-endorphin secretion from adrenal gland for activation of μ-opioid receptor (MOR) in soleus muscle of fructose-fed obese rats to improve the expression level of GLUT-4 proteins and both expression and phosphorylated levels of IRS-1, p-85 (PI3K), Akt and AS160 proteins for glucose intake in insulin-stimulated IRS- PI3K signaling pathway (Tzeng, Liou, & Liu, 2011). Long-term treatment of myricetin to STZ plus cadmium-induced diabetic nephrotoxic rats significantly ameliorated the kidney injury by reducing the elevated levels of serum urea, uric acid, creatinine, and blood urea nitrogen (BUN) levels toward normal in nephrotoxic rats. Possibly, myricetin removes the nephrotoxic cadmium (Cd) from infected kidney tissues and other Cd sensitive organs of diabetic rats by chelate formation with Cd (Kandasamy & Ashokkumar, 2014). Myricetin on diet supplementation in starch diet to high starch fed diabetic rats significantly reduced serum fasting glucose and HbA1c levels via reduction of the activity of maltase/α-glucosidase in the intestine of diabetic rats (Liu, Tzeng, Liou, & Lan, 2007b).

Morin **22** (Fig. 2), a flavonoid constituent of guava, orange, apple, onion and beverages such as tea, coffee and wine, inhibited the activity of PTP1B, a key enzyme that acts as negative regulator of insulin action in peripheral tissues, in a non-competitive manner with an IC_{50} value of 15.0 ± 0.8 μM and Ki value of 5.9 ± 0.4 μM. Further, morin in HepG2 hepatoma cells, increased the activation of Akt via phosphorylation of IRβ and inhibition of the activity of transcription factor FoxO1 gene through its phosphorylation and translocation from nucleus to cytoplasm and reduced the expression of gluconeogenic PEPCK gene in the cells. Moreover, morin in HG-treated HepG2 cells, increased glycogen synthesis and this effect is increased in presence of insulin (Paoli et al., 2013). In an animal model, morin on oral administration (25 and 50 mg/kg bw/d) to STZ-induced diabetic rats for 30 days significantly reduced serum glucose level and improved serum insulin level by restoring pancreatic β-cells mass and insulin secretion. Moreover, morin regulated glucose metabolism in the liver of the diabetic rats by increasing glucose utilization via up-regulation of hexokinase and G6PD genes and reducing glucose production via down-regulation of gluconeogenic G6Pase and FBPase genes for maintenance of glucose homeostasis in the body of rats (Vanitha et al., 2014).

Fisetin **22a** (Fig. 2), an important bioactive flavonoid constituent of strawberries, onion and persimmon, on oral administration (10 mg/kg bw/d) to STZ-induced diabetic rats for 30 days significantly reduced plasma glucose and HbA1c levels and improved plasma insulin level in diabetic rats. Fisetin maintained glucose homeostasis in the blood of diabetic rats by reducing hepatic and renal glucose production by down-regulation of the expression of gluconeogenic genes G6Pase, FBPase and PEPCK, and increasing glucose utilization by up-regulation of the expression of glycolytic genes hexokinase (HK), pyruvate kinase (PK) and G6PD. Furthermore, fisetin restored hepatic glycogen content by up-regulation of glycogen synthase (GS) and down-regulation of glycogen phosphorylase (GPase) genes in liver and reduced the elevated expression of LDH to improve insulin secretion in pancreatic β-cells (Prasath & Subramanian, 2011; Prasath, Pillai, & Subramanian, 2014). In another study, fisetin on treatment to STZ-induced diabetic rats at the same dose and period significantly protected the pancreatic β-cells of diabetic rats from oxidative damage by reducing the levels of inflammatory cytokine IL-1β and NO production and of NF-κB activation in pancreatic tissues (Prasath, Sundaram, & Subramanian, 2013). In vitro, fisetin in primary human umbilical vein endothelial cells (HUVECs) under HG condition, prevented glucose-induced vascular inflammation processes by reducing the expression of inflammatory markers, MCP-1, IL-8, ICAM-1 and VCAM-1, and decreasing the production of ROS and activation of NF-κB kinase as well as inhibiting the adhesion of monocytes, MCP-1 and THP-1 cells to endothelial cells. Hence, fisetin could be effective in treatment of diabetic vascular inflammatory disorders including atherosclerosis (Kwak, Ku, & Bae, 2014). Fisetin, on long-term diet supplementation (0.05% of diet) to type 1 diabetic Akita mice for 18 weeks significantly reduced hyperglycemia and improved kidney hypertrophy and urinary albuminuria level in diabetic mice by lowering the elevated levels of methylglyoxal-associated proteins RAGE, amyloid A, CRP, TBARS and OPN in the kidney tissues. Thus, fisetin could be effective in amelioration of diabetic nephropathy (Maher et al., 2011).

3.2.2 Flavanones in prevention of obesity and diabetes

Citrus fruits such as oranges, tangerines, tangors, bergamots, grapes, lemons and tomatoes are rich sources of bioactive flavanones. Flavanones naringin and its aglycone naringenin, hesperidin and eriodictyol have been found to possess significant antioxidant, anti-inflammatory and antidiabetic activities (Havsteen, 2002; Peterson et al., 2006).

Naringin **23** (Fig. 3), a major flavanone glycoside constituent of grapes, on long-term diet supplementation in HFD (0.2 g/kg diet) to HFD-fed obese diabetic mice for 10 weeks significantly reduced hyperlipidemia and hyperglycemia through improvement of insulin resistance in diabetic mice. Naringin improved insulin resistance by reducing hepatic inflammation through decreasing the level of inflammatory cytokine TNF-α by inhibition of the activation of stress-related kinases, ERK1/2, JNK1/2 and p38MAPK, transcription factor NF-κB and AP-1 proteins, and reducing plasma lipotoxic FFAs level by increasing the expression levels of FA oxidation-related PPAR-α, CPT-1α, and ACOX genes in liver and adipose tissues of mice via AMPK activation. AMPK activation also reduced hepatic lipid accumulation by reducing the expression levels of hepatic lipogenic enzymes, SERBP-1c, FAS, ACC1α, PPAR-γ and SCD-1 for improvement of insulin sensitivity in liver as evidenced from increased IRS-1 tyrosine phosphorylation and reduced HOMA-IR index and decreased expression levels of gluconeogenic PEPCK and G6Pase genes. Moreover, this flavanone glycoside increased the expression of antioxidant enzymes SOD, GSH, GSH-P_X and CAT to restore the normal functions of hepatic genes in regulation of glucose and lipid metabolism (Pu et al., 2012a,b). Naringin on diet supplementation at higher doses (50 and 100 mg/kg diet) to HFD-fed plus STZ-induced obese type 2 diabetic rats for 28 days significantly improved pancreatic β-cell dysfunction, hepatic steatosis and kidney damage in rats by reducing the levels of inflammatory markers TNF-α, IL-6 and CRP proteins, and increasing the expression of stress-related heat shock proteins HSP-27 and HSP-72 in pancreas, liver and kidney and reducing the expression of LXR-α in liver for restoration of the normal functions of hepatic genes. Further, naringin improved PPAR-γ activation and adiponectin

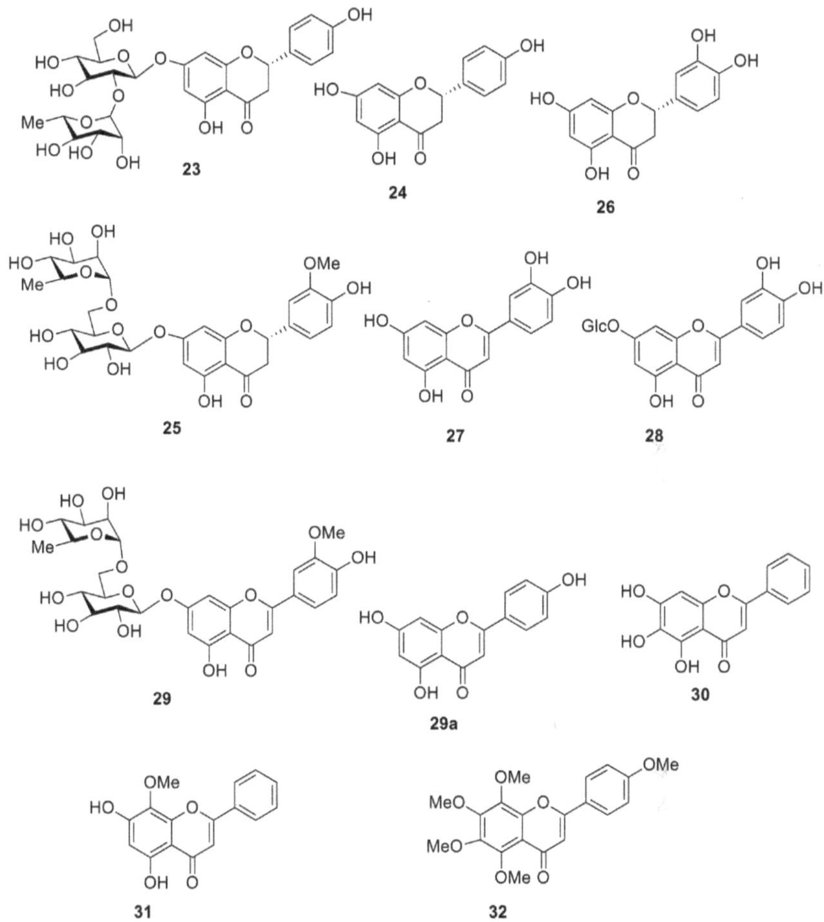

Fig. 3 Chemical structures of flavanones and flavones having anti-obesity and anti-diabetic effects.

secretion, similar to that of anti-diabetic drug, rosiglitazone to increase insulin sensitivity in peripheral tissues (Sharma et al., 2011). Another study reported that naringin, on diet supplementation in diabetic mice reduced the expression levels of hepatic cholesterol synthesis genes, HMG-CR and acyl coenzyme A: cholesterol acyltransferase (ACAT) to decrease hepatic TC level in diabetic mice (Jung, Lee, Park, Kang, & Choi, 2006). Treatment of naringin and its aglycone naringenin **24** (Fig. 3) (100 mg/kg bw) to STZ and nicotinamide-induced type 2 diabetic rats for 4 weeks showed significant anti-diabetic effect by increasing insulin secretion from pancreatic β-cells and improving insulin sensitive effects in adipose tissues and liver in diabetic rats. Both these flavanones increased hepatic glycogen content by reducing the expression of glycogen breaking enzyme, GPase and increased insulin-stimulated glucose utilization in adipose tissues by increased mRNA expression levels of insulin receptor (IR), GLUT-4 and adiponectin (Ahmed, Hassan, Abdel-Twab, & Abdel Azeem, 2007). Naringenin in skeletal L6 myotubes in presence of high glucose, increased glucose uptake by increasing GLUT-4 proteins expression and its translocation into plasma membrane through AMPK activation. But it did not show any significant effect in basal and insulin-stimulated Akt phosphorylation in L6 myotubes. Further, treatment of nicotinamide and EX527 (SIRT-1 inhibitors) did not decrease naringenin-induced AMPK phosphorylation in the myotubes. Thus, naringenin increases glucose uptake in skeletal muscle in an insulin-independent AMPK activation without SIRT-1 activation (Zygmunt, Faubert, MacNeil, & Tsiani, 2010). Naringenin in differentiated 3T3-L1 adipocytes decreased the expression of toll-like receptor-2 (TLR-2) gene by suppression of the levels of inflammatory adipokines TNF-α and MCP-1 via inhibition of the activation of JNK and NF-κB kinases. Accumulating evidence reveals that an over-expression of TLR-2 in adipose tissue contributes a significant role in the development of insulin resistance and diabetes in obese individuals (Yoshida et al., 2013). Treatment of naringenin in human white adipocytes for 7—14 days significantly increased mRNA levels of fatty acid oxidation-related genes, CPT1β, PGC-1α, PGC-1β and UCP-1 and mRNA levels of insulin-sensitive genes, adiponectin and GLUT-4 and increased the phosphorylation of AMPK. These findings suggest that anti-obesity effect of naringenin is regulated by up-regulation of thermogenic genes in AMPK pathway in diabetic animals (Rebello et al., 2019).

Hesperidin **25** (Fig. 3), a major flavonoid constituent of lemon, on supplementation in diet (0.2 g/kg diet) to *db/db* type 2 diabetic mice for 5 weeks significantly improved diabetes by reducing plasma glucose, TC, TG and FFAs levels through reduction of insulin resistance in liver and adipose tissues, and improvement of lipid and glucose metabolism in diabetic mice. Hesperidin improved hepatic and adipocyte PPAR-γ expression for reduction of lipid accumulation in liver and adipose tissues by suppression of the expression of lipogenic enzymes FAS, PAP, HMGCR and ACAT, related to TG and cholesterol synthesis and increased the expression of PPAR-α and CPT-1α for improvement of fatty acid oxidation in liver and adipose tissues. Further, it improved the expression of hepatic GK gene for glucose utilization and decreased the expression of hepatic GLUT-2, PEPCK and G6Pase proteins to reduce glucose production in liver. In addition, it increased the level of plasma antioxidant enzyme paraoxonase (PON) to reduce lipid peroxidation level in liver (Jung et al., 2006). Dietary supplementation of hesperidin to STZ-induced marginal type 1 diabetic rats improved serum glucose and lipid profile in rats by increasing adiponectin secretion and hepatic GK enzyme up-regulation and hepatic G6Pase gene down-regulation (Akiyama et al., 2010). Hesperidin in high glucose-toxicated retinal ganglion cells (RGCs) culture, protected the cells from oxidative stress-induced apoptosis by improving the levels of antioxidant enzymes SOD, CAT and GP_X, and decreasing the expression and activation of stress kinases JNK and p38MAPK as well as inhibiting the expression of pro-apoptotic proteins caspase-9, caspase-3 and Bax in the cells. Furthermore, hesperidin restores the mitochondrial functions in RGCs by improving the reduced level of mitochondrial membrane potential and inhibiting the release of cytochrome C from mitochondria to cytoplasm. Thus, dietary supplementation of hesperidin in humans could be effective in treatment of diabetic retinal disorders (Liu, Liou, Hong, & Liu, 2017). Hesperidin on diet supplementation to experimentally induced cardiac ischemia and reperfusion (I/R) injury in STZ-induced diabetic rats, protected the myocardial tissues from I/R injury in diabetic rats by reducing the oxidative stress and plasma CK-MB, LDH and TNF-α levels toward normal values through increased activation of PPAR-γ gene for amelioration of cardiac dysfunction (Agrawal et al., 2014).

Eriodictyol **26** (Fig. 3), a major flavonoid constituent of lemon fruits, on treatment in differentiated 3T3-L1 adipocytes under HG concentration, enhanced the expression of PPAR-γ2 and aP2 mRNA levels, and expression of GLUT-1 and GLUT-4 proteins, compared to control group for improvement of insulin resistance and glucose uptake by reduction of hyper triglyceridemia level in adipocytes. Moreover, eriodictyol in HG-toxicated HepG2 cells significantly increased insulin-stimulated Akt phosphorylation, which was inhibited in presence of LY294002, a PI3K inhibitor. The activity of eriodictyol in PPAR-γ activation, GLUT-4 expression and Akt activation/phosphorylation, was very similar to that of anti-diabetic PPAR-γ agonist rosiglitazone, used as positive control. These findings indicate that eriodictyol improves insulin sensitivity and glucose utilization in adipocytes and hepatocytes in hyperglycaemic condition via GLUT-4 and GLUT-1 expression in insulin-sensitive PI3K/Akt and PPAR-γ activation pathway (Zhang et al., 2012). Eriodictyol in isolated mice islets and insulin secreting MIN6 cells under HG concentrations, significantly increased insulin secretion through increased cAMP proteins expression and PKA activation. Pre-treatment of H-89 (PKA inhibitor) or verapamil (Ca^{2+}-channel blocker), almost completely inhibited the activity of eriodictyol in insulin secretion, while co-treatment of SQ22536 (adenylate cyclase inhibitor) partially inhibited this insulin secreting effect. These findings indicate that eriodictyol stimulates insulin secretion in both mice islets and MIN6 cells under hyperglycaemic condition in glucose-stimulated cAMP/PKA signaling pathway with an intracellular Ca^{2+}-dependent interactions. In animal model, oral treatment of eriodictyol (10 and 20 mg/kg bw/d) to STZ plus nicotinamide-induced type 2 diabetic rats for 28 days significantly reduced plasma glucose level and improved plasma insulin level for amelioration of diabetes in diabetic rats (Hameed et al., 2018). Eriodictyol, on long-term diet supplementation in HFD (0.005% of diet) to HFD-fed obese mice for 16 weeks significantly improved adiposity, hepatic steatosis, dyslipidemia and hyperglycemia in obese mice through improvement of insulin resistance and insulin action. Eriodictyol improved insulin resistance in liver and adipose tissues by reducing the levels of plasma inflammatory adipokines PAI-1, IFN-γ, IL-1β and leptin, and increasing the level of plasma anti-inflammatory cytokine IL-10. Eriodictyol improved dyslipidemia by reducing the elevated levels of plasma TC, TG, FFAs and ApoB (LDL-C) and decreasing the expression levels of lipogenic genes SREBP-1, ACC, PAP, FAS and SCD-1 and of lipid uptake genes CD36 and LPL, and increasing the expression of fatty acid oxidation related genes CPT and PPAR-α, in the liver of obese mice. Furthermore, eriodictyol improved insulin secretion and insulin activity by increasing the expression of GLP-1 and IRS-2 genes in pancreas and increased hepatic insulin-stimulated glucose utilization by up-regulation of GK gene and decreased hepatic glucose production by down-regulation of PEPCK and G6Pase genes. Eriodictyol also enhanced fecal excretion of cholesterol and fatty acids for suppression of exogenous fat absorption in intestine (Kwon & Choi, 2019). Eriodictyol on diet supplementation (10 mg/kg diet) to STZ-induced diabetic rats for 10 days significantly suppressed retinal inflammation by reducing the levels of pro-inflammatory cytokines TNF-α, ICAM-1, VEGF and eNOS, and the level of lipid peroxidation and protected the blood-retinal barrier (BRB) from inflammation-induced breakdown of BRB membrane (reduced 40% breakdown) in diabetic rats. Accumulating evidence indicates that BRB in retina of human eyes maintains a specialized environment of neural retina for normal vision function. Hyperglycemia-induced toxicity contributes to the development of retinal inflammation and causes breakdown of the membranes of BRB resulting in the loss of vision. Thus, eriodictyol on diet supplementation in diabetic patients could be effective in prevention of diabetic retinopathy (Bucola et al., 2012).

3.2.3 Flavones in prevention of obesity and diabetes

Flavones are another class of flavonoids and are mainly found in various dietary vegetables and fruits such as onions, cabbage, broccoli, carrots, green peppers, beans, citrus fruits and apples. Some of the dietary flavones such as luteolin and its glycosides, luteolin 7-O-glucoside and diosmin, apigenin, baicalein, wogonin and tangeretin have significant anti-obesity and antidiabetic effects (Hossain et al., 2016; Vinayagam & Xu, 2015).

Luteolin **27** (Fig. 3), a major flavonoid constituent of celery, broccoli, tansy leaf, green pepper and spinach, on treatment in 3T3-L1 adipocytes significantly increased glucose uptake, GLUT-4 expression, Akt-2 phosphorylation, adiponectin and leptin secretion, PPAR-γ activation and decreased the expression of inflammatory cytokines, TNF-α, IL-6 and MCP-1. Co-treatment of GW9662, an antagonist of PPAR-γ, reduced the expression of adiponectin and leptin in adipocytes. Luteolin also increased PPAR-γ activation, adiponectin and leptin secretion in primary adipose cells. These findings indicate that luteolin increased insulin action in glucose uptake and production of insulin sensitive adipokines, adiponectin and leptin by activation of PPAR-γ pathway in adipocytes and could be effective for improvement of insulin resistance in obese individuals (Ding, Jin, & Chen, 2010). In an animal model, dietary supplementation of luteolin and its 7-O-glucoside **28** (Fig. 3)in HFD (0.0437 and 0.0687% of diet) to HFD-fed obese diabetic KK-Ay mice for 28 days significantly improved obesity and diabetes by reducing plasma fasting glucose, HbA1c, insulin, TG and TNF-α levels in obese diabetic mice. Furthermore, these flavonoids significantly reduced hepatic TBARS level and decreased the activity of hepatic lipogenic genes SREBP-1c and FAS to reduce lipid synthesis in liver. These findings indicate that luteolin and its glycoside **28** exhibit anti-obesity and anti-diabetic effects in mice through improvement of insulin resistance via inhibition of hepatic lipid synthesis, oxidative stress and inflammation (Zang, Igarashi, & Li, 2016). Treatment of luteolin in uric acid-toxicated hyperglycemic MIN6 cells, significantly inhibited the expression of iNOS enzyme and increased glucose-stimulated insulin secretion and the expression of insulin secretory MafA gene in MIN6 cells. Available evidence reveals that an elevated level of serum uric acid in diabetic patients develops pancreatic β-cell injury and dysfunction by up-regulation of iNOS gene expression and NO production in pancreas from NF-κB activation. Therefore, luteolin improved insulin secretion in diabetic animals through increased expression of MafA gene and reduced expression of iNOS gene in pancreas (Ding et al., 2014). In another cellular model, luteolin and quercetin, major flavonoid constituents of Malaysian Gelam honey, in hamster pancreatic HIT-T15 cells under HG (50 μM) concentration, decreased the production of ROS and LPO contents to reduce the toxicity in cells and increased the insulin content and viability of the cells (Batumalaie, Qvist, Yusof, Ismail, & Sekaran, 2014). Luteolin on diet supplementation in HFD (10 mg/kg diet) to HFD-fed obese diabetic mice for 16 weeks significantly ameliorated obesity-induced cognitive impairment in obese diabetic mice through reduction of body weight, food intake, plasma glucose level, neuro-inflammation and oxidative stress, and improvement of synaptic dysfunction by increasing the expression of BDNF, synapsin-1 (SYP-1), and post-synaptic density protein-95 (PSD-95) levels in cortex and hippocampus of mice brain (Liu, Fu et al., 2014; Liu, Li et al., 2014).

Diosmin (3′- *O*-methyl-luteolin-7-*O*-rutinoside) **29** (Fig. 3), a major flavonoid constituent of some citrus fruits, on diet supplementation (100 mg/kg bw) to STZ plus nicotinamide-induced type 2 diabetic rats for 45 days exhibited significant anti-diabetic effect by reduction of serum glucose and HbA1c levels and increased insulin secretion from pancreatic β-cells. Further, diosmin increased insulin sensitivity in liver for hepatic glucose utilization via up-regulation of hexokinase and G6PD genes and decreased hepatic glucose production via down-regulation of gluconeogenic G6Pase and FBPase genes in liver of diabetic rats (Puri & Srinivasan, 2010). Diosmin on treatment (160 mg/kg, *i.p.*) to STZ-induced diabetic rats for one week reduced plasma glucose level and increased plasma β-endorphin level in rats. It also increased the expression of GLUT-4 protein in soleus muscle and reduced the expression of gluconeogenic enzyme PEPCK in liver of diabetic rats. These effects of diosmin were dramatically reversed on treatment of naloxone, an antagonist of MOR, in diabetic rats. These findings indicate that β-endorphin-induced MOR activation in soleus muscle and liver of diabetic rats, is involved, at least in part, for improvement of insulin sensitivity in glucose utilization in the muscle by GLUT-4 protein expression and its translocation into plasma membrane and in reduced glucose production in liver by decreasing the expression of gluconeogenic enzyme PEPCK (Hsu et al., 2017 a). Diosmin in isolated adrenal glands increased the expression of β-endorphin and this effect was completely blocked on treatment of BU-224, an antagonist of imidazoline receptor, I-2R. In chicken hamster ovary, CHO-K1 cells, diosmin increased the intracellular Ca^{2+} ions. This effect of diosmin was blocked on transfection of the cells with IR gene (NISCH). In an animal model, the hypoglycaemic and hypolipidemic effects of diosmin in type 1 diabetic rats were reversed on treatment of BU-224. These findings indicate that diosmin acts as I-2R agonist to increase MOR activation by improving intracellular Ca^{2+} influx in skeletal muscle and liver of diabetic rats to increase insulin sensitivity for amelioration of hyperglycemia and hyperlipidemia (Hsu, Lin, Cheng, & Wu, 2017b).

Apigenin **29a** (Fig. 3), a major flavonoid constituent of chamomile (*Matricaria recutita*) flowers, in HepG2 cells under HG concentration (30 μM/L) increased the levels of phosphorylated AMPK and its target gene ACC, and reduced the intracellular lipid accumulation in the cells (Zang et al., 2006). In an animal model, apigenin on treatment (0.78 mg/kg bw, *i.p.*) to alloxan-induced diabetic mice for 10 days significantly reduced serum glucose and TC levels, and increased serum insulin and thyroid hormones (T_3 and T_4) levels in diabetic mice. Further, apigenin reduced hepatic glucose production by lowering the expression of gluconeogenic enzyme G6Pase and reduced hepatic lipid peroxidation level by increasing the activity of anti-oxidant enzymes SOD, CAT and GSH in the liver of mice. Thus, apigenin could be effective in treatment of diabetes and diabetes-related thyroid dysfunction (Panda & Kar, 2007). A group of studies have shown that multifunctional CD38 enzyme is a key regulator in the degradation of cellular NAD^+ levels and these NAD^+ coenzymes regulate insulin secretion, energy metabolism and other functions in mammal tissues. The over-expression of CD38 enzyme reduces the NAD^+ levels in tissues and hence, inhibition of CD38 enzyme expression and activity in liver and other tissues of mammals, increases NAD^+ levels via SIRT-1 activation for improvement of obesity, type 2 diabetes and other aspects of metabolic disorders of mammals. In a cellular model, apigenin in human lung cancer A549 cells transfected with human CD38 small interfering (si)-RNA, reduced the expression of CD38 proteins and increased the intracellular NAD^+ level and SIRT-1 activity, and decreased the acetylation of p53 and RelA/p65 proteins. It inhibited the activity of CD38 proteins with an IC_{50} value of 14.8 ± 2.2 μM/L in the cells. In a vivo model, apigenin on treatment (100 mg/kg bw, *i.p.*) to HFD-fed obese mice for 7 days significantly improved hyperglycemia and dyslipidemia in obese mice. Apigenin reduced hepatic lipid accumulation, increased hepatic NAD^+ levels and reduced CD38 activity and acetylation of global proteins.

It also increased the expression levels of fatty acid oxidation-related LACD, MACD and CPT1α proteins in the liver of mice. Moreover, apigenin in HepG2 cells under hyperglycaemic and hyperlipidemic conditions, decreased the acetylation of p53 and p65 proteins and lipid accumulation in the cells, and this effect of apigenin was completely blocked on treatment with EX527, a SIRT-1 inhibitor. These findings indicate that apigenin ameliorates obesity-induced hyperlipidemia and hyperglycemia in obese mice by inhibition of CD38 protein activity and increased activation of SIRT-1 gene in the liver of mice (Escande et al., 2013). In another animal model, apigenin on oral administration (20 mg/kg bw, *p.o.*) to STZ-induced diabetic rats significantly reduced hyperglycemia and improved renal dysfunction in diabetic rats. Further, apigenin reduced serum creatinine, BUN, and urinary albumin levels and decreased collagen and fibronectin deposition in glomerular mesangial matrix via suppression of TGF-β1 expression, and reduced mesangial cell apoptosis via suppression of Bax and caspase 3 proteins expression in renal tissues of diabetic rats to improve renal injury and glomerular filtration rate. It reduced the levels of inflammatory cytokines TNF-α and IL-6 by suppression of the activation of stress kinases, ERK1/2, p38MAPK, JNK and NF-κB in renal tissues. These effects of apigenin were similar to that of ramipril, a well known drug for treatment of heart and kidney injuries. Thus, apigenin could be effective in treatment of diabetic nephropathy (Malik et al., 2017). Apigenin and luteolin in HepG2 cells, increased Akt activation and phosphorylation of FoxO1 gene for its removal from nucleus to cytoplasm and inhibited the mRNA expression levels of gluconeogenic enzymes PEPCK and G6Pase, and of lipogenic enzymes FAS and ACC in HepG2 cells via inhibition of FoxO1 activity. Thus, these flavonoids improved glucose and lipid homeostasis in diabetic animals via activation of Akt and inhibition of FoxO1 activity (Bumke-Vogt et al., 2014).

Baicalein **30** (Fig. 3), a major flavonoid constituent of dietary herbs, *Oroxylum inducum* and *Scutellaria baicalensis*, on diet supplementation in HFD (0.5 g/kg diet) to HFD-fed plus STZ-induced obese diabetic mice for 4 weeks significantly reduced plasma glucose level and improved plasma insulin level and pancreatic β-cell mass in obese diabetic mice. In vitro, baicalein in INS382/13 beta cells and isolated human islets under HG concentrations, increased glucose-stimulated insulin secretion. Moreover in INS382/13 cells under hyperlipidemic HP concentration, baicalein increased β-cell viability and its proliferation. Accumulating evidence reveals that 12-lipoxygenase (12-LO) is over-expressed in pancreas of obese rodents and humans and it produces 12-(*S*)-hydroxyeicosatetraenoic acid (12-*S*-HETE), which increases NO production in pancreas via JNK and p38MAPK activation to induce impairment of insulin secretion and apoptosis of β-cells. Baicalein is a potent inhibitor of 12-LO. Possibly baicalein protected the pancreatic β-cells from death by inhibition of the activity of 12-LO in obese mice (Fu et al., 2014). Baicalein, on long-term diet supplementation at high dose in HFD (400 mg/kg bw/d) to HFD-fed obese mice for 29 weeks significantly reduced fatty liver mass, hyperlipidemia and hyperglycemia in obese mice through improvement of insulin resistance by AMPKα_2 activation. The increased AMPK activation reduces hepatic inflammation by decreasing the expression levels of inflammatory cytokines TNF-α and MCP-1 (reduced about 50%) via inhibition of the activation of ERK1/2, JNK and NF-κB kinases and increases insulin activity by phosphorylation of IRS-1 and Akt. This AMPK activation also reduces fatty liver mass by decreasing the expression of hepatic genes SREBP-1c, FAS, PPAR-γ and aP2, related to lipid synthesis, and increased the expression of PPAR-α, CPT-1 and UCP-2 genes, related to mitochondrial fatty acid oxidation (Pu et al., 2012a,b). Baicalein on diet supplementation in HFD to HFD-fed plus STZ-induced type 2 diabetic mice improved hepatic and muscle glycogen content and decreased the expression of SOCS-3 and phosphorylation of GSK-3β for improvement of insulin sensitivity and insulin-stimulated glucose utilization via Akt activation and GLUT-4 protein expression and its translocation to cell membrane. Co-supplementation of acarbose (α-glucosidase inhibitor) with baicalein to diabetic mice showed synergistic effect in reduction blood glucose and lipid levels in diabetic mice (Sun et al., 2017). In another study, baicalein on diet supplementation in HFD (150 mg/kg diet/d) to HFD-fed plus STZ-induced obese diabetic rats significantly ameliorated obesity-induced hyperglycemia, hyperlipidemia and gut microbiota by decreasing plasma endotoxin LPS level and inflammatory cytokines and receptor TNF-α, IL-6 and TLR-4 levels in obese diabetic rats. Baicalein treatment improved the contents of short-chain fatty acids (SCFAs) producing bacteria, *Bacteroides*, *Bacteroidales* S24-7 and *Akkermannia muciniphila,* compared to *Firmicutes* in intestinal gut of obese diabetic rats to reduce the permeability of LPS and other Gram-negative bacteria in the circulating plasma via gut epithelial membrane. Growing evidence reveals that SCFAs producing bacteria increase the thickness of gut epithelial membrane having strong barrier function by increasing mucin synthesis and protect the entry of harmful LPS and other Gram-negative bacteria, which induce inflammation in liver for an access of nutrients as food for their growth. Thus, baicalein could be a potential pre-biotic drug for amelioration of obesity induced gut microbiota (dysbiosis) disorder in diabetic rodents and humans (Zhang et al., 2018).

Wogonin **31** (Fig. 3), a major bioactive flavonoid constituent of Chinese herb, *S. baicalensis*, on dietary supplementation to diabetic mice significantly reduced body weight and food intake, and decreased serum glucose, insulin and TC levels. Furthermore, wogonin improved glycogen content in liver and WAT, and increased PPAR-α expression in liver, and adiponectin secretion in WAT. In vitro, wogonin in 3T3-L1 adipocytes, increased AMPK activation. These findings indicate that wogonin regulates glucose and lipid metabolism by increasing adiponectin and PPAR-α expression via AMPK activation for amelioration of hyperglycemia and dyslipidemia in diabetic mice (Bak et al., 2014). Wogonin (10 mg/kg bw/once in 3 d) on treatment to STZ-induced type 1 diabetic mice for 8 weeks significantly reduced blood glucose by reducing the level of plasma adipokine, osteopontin (OPN) and its expression in adipose tissue of diabetic mice. In vitro, wogonin in 3T3-L1 adipocytes under HG concentration, increased PPAR-α expression and reduced its phosphorylation, and decreased the expression of OPN and phosphorylation of p38MAPK. Accumulating evidence demonstrates that OPN, an inflammatory phosphoprotein is highly expressed in adipose tissue of obese animals and humans via phosphorylation of p38MAPK kinase and contributes to the development of insulin resistance and increased hepatic glucose production in gluconeogenetic pathway. These findings indicate that wogonin exhibits its hypoglycaemic effect in diabetic mice by improving insulin resistance through suppression of OPN expression and increased PPAR-α expression and activity and inhibition of phosphorylation of PPAR-α and p38MAPK in adipose tissue (Zhang, Niu, Sun, Xu, & Yue, 2015).

Tangeretin **32** (Fig. 3), a major flavonoid constituent of tangerine and other citrus fruits such as oranges, lemons and lime, in differentiated 3T3-F442A adipocytes under high deoxyglucose and insulin concentrations, increased glucose uptake by increasing GLUT-4 and GLUT-1 proteins expression and increased PI3K, Akt1/2, PKA and CREB activation. Co-treatment of wortmannin (PI3K inhibitor) significantly reduced the effect of glucose uptake, while co-treatment of H89 (PKA inhibitor) reduced the effect of glucose uptake to some extent. Available evidence reveals that glucose transport by GLUT-4 expression occurs via PI3K and Akt1 activation, while glucose transport by GLUT-1 expression occurs via PKA and CREB activation. These facts indicate that tangeretin induces insulin-stimulated glucose uptake in the adipocytes via both GLUT-4 and GLUT-1 expression, the former being the major pathway (Onda, Horike, Suzuki, & Hirano, 2013). Tangeretin in C2C12 myotubes in presence of 2-deoxy-D-glucose, increased glucose uptake by expression of GLUT-4 protein and its translocation from cytosol to plasma membrane via activation of AMPK and AS160 proteins. In an animal model, tangeretin on dietary supplementation in HFD (200 mg/kg diet) to HFD-fed obese mice for 8 weeks showed significant anti-obesity and antidiabetic effects by reducing plasma glucose, TC and TG levels via improvement of insulin resistance in AMPK pathway. Tangeretin improved insulin resistance by reducing the levels of plasma inflammatory cytokines IL-6 and MCP-1, and increasing the secretion of adiponectin, leptin and resistin in adipocytes and down-regulated the expression of lipogenic ACC gene in adipose tissues and liver via AMPK activation for amelioration of obesity in obese mice (Kim, Hur, Kwon, & Hwang, 2012).

3.2.4 Isoflavones in prevention of obesity and diabetes

Several legumes such as soybean, peas, beans, lentils and peanut are the rich sources of isoflavones, proteins and fibers. Some bioactive isoflavones such as genistein, daidzein, daidzin, prunetin, biochanin A, formononetin and puerarin have potential health benefits in prevention of various diseases including cancer, diabetes, heart disease and bone loss (Messina, 1999).

Genistein **33** and daidzein **34** (Fig. 4), major isoflavone constituents of dietary soybean (*Glycine max*), on supplementation in diet (0.2 g/kg diet) to type 2 diabetic (*db/db*) mice for 6 weeks showed significant antidiabetic effect by reducing serum glucose, HbA1c, TC, TG and FFAs levels in diabetic mice. Furthermore, these isoflavones improved glucose metabolism in mice by increasing hepatic glycogen content by up-regulation of GK enzyme and reduced gluconeogenetic glucose production by down-regulation of PEPCK and G6Pase enzymes. Moreover, these isoflavones reduced hepatic lipid accumulation by reducing the expression of FAS gene for reduction of lipid synthesis and increasing the expression of CPT gene and other related genes to improve β-oxidation of fatty acids (Park et al., 2006). Another study reported that soy isoflavone mixture rich in genistein and daidzein on dietary supplementation in HFD (5.8 mg/kg diet) to HFD-fed obese Zucker rats significantly reduced lipid mass in adipose tissues and liver by increased PPAR-α and PPAR-γ activation for reduction of FFAs and TG levels in obese rats (Mezei et al., 2003). Genistein in pancreatic islets increased glucose-stimulated insulin secretion (5—10 folds) by inhibiting the

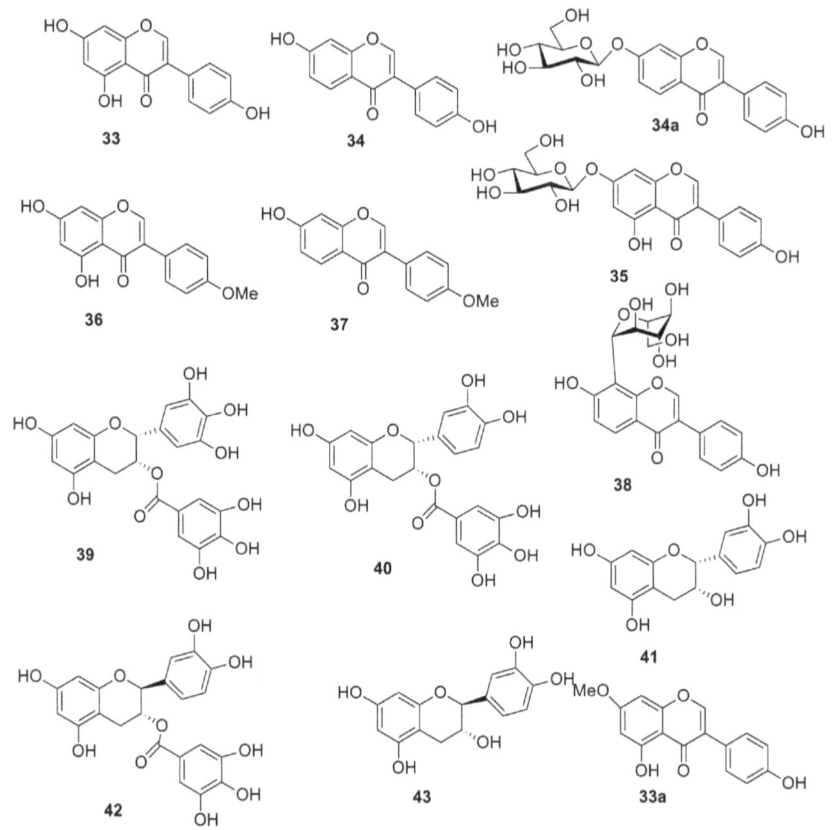

Fig. 4 Chemical structures of isoflavones and flavanols having anti-obesity and anti-diabetic effects.

activity of protein tyrosine kinase enzyme (a negative regulator of insulin secretion Sorenson, Brelje, & Roth, 1994). Moreover, genistein protected insulimia RINm5F cells and isolated rat islets from IL-1β and IFN-γ induced toxicity by inhibition of the expression levels of iNOS mRNA and protein via suppression of NF-κB, ERK1/2 and JAK/STAT activation. Available evidence demonstrates that NO production from increased iNOS expression by inflammatory cytokines IL-1β and IFN-γ, is one of the major factors for impaired insulin secretion and β-cell damage in pancreas leading to the development of type 1 diabetes. Therefore, genistein improves glucose metabolism by increased insulin secretion and viability of pancreatic β-cells in diabetic animals (Kim et al., 2007). Furthermore, genistein increased insulin secretion in insulimia INS-1 and MIN6 cells, and isolated mouse pancreatic islets via increased intracellular accumulation of cAMP and activation of PKA. In addition, genistein increased IRS-2 protein expression and activation of pancreatic duodenal homeobox-1 (PDX-1) proteins in pancreas for survival and growth of β-cells. Fu et al. reported that PKA activation improved MEK1/2 phosphorylation, which in turn, increased the expression of cell cycle activator, cyclin D1 for proliferation of β-cells. Genistein glycoside genistin **35** and biochanin A **36** (Fig. 4), a metabolite of genistein also exhibited β-cell proliferative effect (24%–27% over control) (Liu et al., 2006; Fu et al., 2010). In an animal model, genistein on diet supplementation to HFD-fed plus STZ-induced obese diabetic mice, increased β-cell mass and insulin secretion for amelioration of obesity and diabetes in mice (Fu et al., 2012). In vitro, genistein in HG-toxicated human aortic endothelial cells (HAECs) significantly inhibited the production of monocyte MCP-1 and the binding of monocytes to endothelial cells (ECs) via cAMP expression and PKA activation. In an animal model, genistein supplementation to obese type 2 diabetic mice significantly improved vascular inflammation in mice by reducing the levels of serum inflammatory cytokines MCP-1, IL-8, VCAM-1 and ICAM-1 and increasing the level of serum anti-inflammatory cytokine IL-10, and decreasing the binding of MCP-1 to EC by 87% (Babu, Si, Fu, Zhen, & Liu, 2012).

Prunetin, 7-O-methylated genistein **33a** (Fig. 4), a soy isoflavone and present in several *Prunus* species, on administration (20 μg/kg bw/d, *i.p.*) to HFD-fed obese mice for 3 weeks significantly reduced body weight and visceral fat pad weight and plasma glucose level in obese mice. Prunetin significantly reduced the elevated mRNA expression levels of adipogenic genes, PPARγ, C/EBPα, SREBP, aP2, LPL, adiponectin and its receptors R1/R2, and leptin in adipose tissues, and lipid metabolic genes, SREBP, PPARγ, LXR and HMGCR in liver, and increased AMPK activation for reduction of lipid mass in adipose tissues and liver of obese mice. Hence, prunetin could be effective in treatment of obesity (Ahn et al., 2013).

Daidzein **34** (Fig. 4), on diet supplementation (0.1% of diet) to diabetic *db/db* mice for 5 weeks significantly improved hyperglycemia and kidney function in diabetic mice by increasing glucose utilization in skeletal muscle and liver through GLUT-4 expression and reducing hepatic glucose production through down-regulation of gluconeogenic genes PEPCK and G6Pase, and suppressing urinary glucose excretion through increased AMPK activation (Cheong et al., 2014). Daidzin **34a** (Fig. 4), a major isoflavone glycoside of *Puerariae Radix*, on oral administration (400 mg/kg bw/d) to STZ-induced diabetic mice for 14 days significantly reduced serum glucose, TC, TG and LDL-C levels and improved oral glucose tolerance in diabetic mice. Daidzin improved insulin resistance by reducing the levels of inflammatory cytokines, IL-6, TNF-α and MCP-1 and increasing the level of antioxidant enzyme SOD in plasma of diabetic mice. In vitro, daidzin stimulated glucose uptake in C2C12 myotubes by increased expression and translocation of GLUT-4 proteins into plasma membrane under HG condition. It also inhibited the activity of α-glucosidase. Possibly, daidzin ameliorated hyperglycemia in diabetic mice by inhibition of α-glucosidase activity and utilization of glucose in muscle (Tian et al., 2016).

Biochanin A **36**, a major isoflavone constituent of red clover (*Trifolium pratense*) and other clover species and also abundant in soy, peanut and chickpeas, on oral administration in 0.5% DMSO (10 mg/kg bw/d) to STZ-induced diabetic rats for 45 days showed significant antidiabetic effect by reducing serum glucose and HbA1c levels and increasing insulin secretion and insulin activity in skeletal muscle and liver of diabetic rats. Biochanin A improved glycogen content in liver and muscle by increasing the activity of glycolytic enzymes GK and G6PD, and reduced hepatic glucose production by decreasing the activity of gluconeogenic enzymes G6Pase and FBPase (Harini, Ezhumalai, & Pugalendi, 2012). In another animal model, biochanin A, on diet supplementation (40 mg/kg diet) to HFD-fed plus STZ-induced obese diabetic rats for 28 days showed significant anti-obesity and antidiabetic effects by increasing SIRT-1 expression in pancreatic tissue to increase glucose-stimulated insulin secretion from β-cells and increasing glucose uptake and reducing hepatic lipid mass via AMPK activation in diabetic rats. Further, biochanin A repressed UCP-2 proteins in β-cells to stimulate insulin secretion and increased AMPK activation in liver to improve insulin activity in glucose utilization by glycolysis and to reduce hepatic glucose production in gluconeogenetic pathway and to reduce hepatic TG lipid mass by up-regulation of PPAR-γ and its target genes (Oza & Kulkarni, 2018a). In another animal study, biochanin A on diet supplementation in HFD (0.05%) to HFD-fed obese mice for 12 weeks significantly improved hepatic steatosis and insulin resistance in obese mice by regulating hepatic lipid and glucose metabolism through increased expression levels of PPAR-α and its regulatory proteins and suppression of gluconeogenic enzymes, G6Pase and PK in the liver of obese mice (Park et al., 2016).

Formononetin **37** (Fig. 4), another major isoflavone constituent of red clover (about 1%), on oral administration (20 and 40 mg/kg bw) to STZ-induced type 2 diabetic rats for 28 days significantly improved hyperglycemia and dyslipidemia by decreasing plasma glucose, HbA1c, TG and LDL-C levels, and reducing hepatic TG and lipid mass in diabetic rats. Further, formononetin increased SIRT-1 expression in pancreas to increase insulin secretion from β-cells and increased AMPK activation in liver for improvement of insulin sensitivity in glycogen synthesis and reduction of gluconeogenesis and fatty liver mass by activation of PPAR-α and PPAR-γ transcription factors (Oza & Kulkarni, 2018b). In another animal model, formononetin on oral administration (10 and 20 mg/kg bw) to alloxan-induced type 1 diabetic mice for 28 days significantly improved hyperglycemia in diabetic mice by promoting pancreatic β-cell proliferation, insulin secretion, hepatic glycogen synthesis and glycolysis. Formononetin up-regulated the expression levels of PDX-1, IRS-2, GK and GLUT-2, mRNA and proteins, and down-regulated the expression levels of death receptor Fas (CD-95) and caspase-3 mRNA and proteins in the pancreas tissue of diabetic mice for proliferation of β-cells and increased insulin secretion. Moreover, this isoflavone increased GK mRNA and protein levels in the liver of mice to improve glycogen synthesis and glycolysis (Qui et al., 2017).

Puerarin (daidzein-8-*C*-glucoside) **38** (Fig. 4), a major isoflavone glucoside of Asian antidiabetic herb, kudzu (*Pueraria montana var. lobata*), on diet supplementation in HFD (150 mg/kg diet) to HFD-fed *db/db* mice for 55 days significantly reduced body weight gain and hyperglycemia of diabetic mice through improvement of pancreatic β-cell mass and insulin secretion. In cellular model, puerarin in MIN6 cells and isolated mouse pancreatic islets under HG concentrations, significantly increased the β-cell mass, secretion of insulin and expression of the mRNA levels of GLP-1R, TCF7L2, PDX-1 and INS-1 (insulin 1) genes and reduced the expression of FoxO1 and caspase-3 proteins, compared to control. The protective effect of puerarin on β-cell mass was remarkably suppressed on co-treatment of exendin (9- 39), an antagonist of GLP-1R, while co-treatment of exendin-4, an agonist of GLP-1R, significantly enhanced the effect of puerarin on proliferation of β-cell and insulin secretion. Accumulating evidence demonstrates that transcription factor 7-like-2 (TCF7L2) gene is a master regulator of β-cell mass and function via pro-insulin expression and subsequent processing through several genes including MafA, INS-1 and PDX-1 genes for matured insulin, β-cell proliferation and survival in rodent and human islets. GLP-1R and its agonist increases TCF7L2 activation via activation of Akt and inhibition of GSK-3β to promote Wnt signaling for proliferation of pancreatic β-cell and improvement of its function. Thus, puerarin acting as GLP-1R agonist improves insulin secretion and β-cell mass in diabetic mice for amelioration of hyperglycemia (Yang et al., 2016; Liu & Habener, 2008). In another study, puerarin in

$CoCl_2$-induced toxicated pancreatic islets and MIN6 β-cells protected the cells from $CoCl_2$-induced apoptosis via activation of Akt and up-regulation of anti-apoptotic Bcl-2 proteins and antioxidant enzymes SOD and GP_X, while decreasing the expression of pro-apoptotic Bax proteins (Li et al., 2014). In another animal model, dietary supplementation of an aqueous extract of *Radix Pueraria lobata* (RP) rich in puerarin (300 mg/kg diet) to HFD-fed obese mice for 16 weeks significantly reduced body weight, muscle atrophy, lipid accumulation in liver and adipose tissues as well as decreased fasting plasma glucose and TC levels in obese mice. Moreover, RP extract increased PGC-1α expression and phosphorylated-AMPK level in skeletal muscle tissues of obese mice. Both RP extract and puerarin in C2C12 cells increased the expression levels of PGC-1α, nuclear respiratory factor-1 (NRF-1) and transcription factor A, mitochondrial (TFAM) proteins and activation of AMPK and ACC. These findings demonstrate that RP extract improves dyslipidemia in obese mice by regulating lipid metabolism through increased mitochondrial biogenesis in fatty acid oxidation and AMPK activation (Jung, Kang, Kang, Park, & Song, 2017). In vitro, puerarin in HG-toxicated murine podocytes increased both mRNA and protein levels of SIRT-1 gene and reduced superoxide production and NOX gene expression. In an animal model, puerarin, on oral administration (20 mg/kg bw/d) to STZ-induced diabetic eNOS null mice, a mimic model of human diabetic nephropathy, for 8 weeks significantly improved kidney injury by reducing NOX expression in kidney tissues and decreasing albumin-creatinine ratio in urine of diabetic kidney-injury mice. These findings indicate that increased expression of SIRT-1 gene by puerarin reduces NF-κB activated NOX expression in podocytes to reduce the deposition of fibronectin for improvement of 'kidney injury and glomerular filtration for amelioration of kidney dysfunction in mice (Li et al., 2017). Puerarin on treatment (100 mg/kg bw, *i.p.*) to STZ-induced diabetic rats for 4 weeks significantly reduced plasma glucose and HbA1c levels and improved plasma insulin level and oral glucose tolerance in diabetic rats. Furthermore, puerarin increased the expression and phosphorylation of MOR gene in soleus muscle of diabetic rats. But these effects of puerarin were reversed on pretreatment of naloxone (MOR inhibitor) in diabetic rats. This fact indicates that MOR activation by puerarin might be involved in its antidiabetic effect. In vitro, puerarin in palmitate-induced mimic hyperlipidemic L6 myotubes, increased insulin-stimulated glucose uptake by increasing phosphorylation of MOR, IRS-1, Akt and AS-160 proteins, and expression of GLUT-4 protein and its translocation into plasma membrane. These effects were reduced by pretreatment of naloxone, MOR antagonist. These findings suggest that the antidiabetic effect of puerarin, at least in part, involves MOR activation in skeletal muscle of diabetic rats to increase insulin sensitivity for insulin-stimulated glucose uptake by Akt activation and GLUT-4 expression and translocation (Chen et al., 2018).

3.2.5 Flavanols in prevention of obesity and diabetes

Naturally occurring flavan-3-ols, commonly known as flavanols or catechins, are present in high amounts in various fruits (e.g. apricots, apples, cherries, grapes, pears, etc), teas, cocoa beans and other beans. The most common plant-derived catechins are (+)-catechin, (−)-catechin, (+)-gallocatechin, (−)-epicatechin, (−)-epigallocatechin, (−)-catechin gallate, and (−)-epigallocatechin gallate. These dietary catechins have significant efficacy against cancer, obesity, diabetes, heart disease and other inflammatory-related disorders (Arts, van de Putte, & Hollman, 2000).

(−)-Epigallocatechin 3-*O*-gallate (EGCG) **39** (Fig. 4), a major catechin constituent of green tea (*Camellia sinensis*) (about 100 mg/g), on long-term diet supplementation in HFD (3.2 g/kg diet) to HFD-fed obese mice for 16 weeks significantly showed anti-obesity effect by reducing body weight and lipid mass in liver, and lowering plasma glucose, insulin, TC, TG, MCP-1 and ALT levels toward normal values in obese mice. Furthermore, EGCG improved fecal lipid excretion in mice. Possibly EGCG reduced hepatic lipid mass by reducing inflammation and toxicity in liver for restoration of normal functioning of hepatic genes in lipid and glucose metabolism in obese mice (Bose et al., 2008). In another study, EGCG on dietary supplementation (1.5 g/kg diet) to diabetic mice for 7 days significantly reduced hyperglycemia by decreasing the expression of gluconeogenic enzymes PEPCK and G6Pase in the liver of mice (Koyama et al., 2004). An oral administration of EGCG (75 mg/kg bw, a single dose) to diabetic rats promoted GLUT-4 translocation in skeletal muscle of diabetic rats. EGCG in insulin resistant L6 myotubes under HG and HP concentrations, increased glucose uptake by increasing the expression and translocation of GLUT-4 protein to plasma membrane via PI3K activation. Co-treatment of genistein (tyrosine kinase inhibitor) did not suppress the EGCG-induced glucose uptake, while co-treatment of wortmannin (a PI3K inhibitor), suppressed both insulin- and EGCG-induced glucose uptake. These findings suggest that the mechanism of glucose uptake by EGCG is different from that of insulin in diabetic rats (Ueda, Nishiumi, Nagayasu, Fukuda, & Yoshida, 2008). In another study, EGCG in dexamethasone-induced insulin resistant rat L6 cells, increased insulin-sensitive GLUT-4 protein translocation to plasma membrane through activation of AMPK and Akt. These results indicate that EGCG increases glucose uptake in insulin resistant L6 myotubes through AMPK activation in PI3K/Akt pathway (Zhang et al., 2010). EGCG in human HepG2 cells in a hyperglycemic condition, reduced IRS-1 serine phosphorylation and increased Akt activation and glycogen synthesis through increased AMPK activation. Co-treatment of compound C (AMPK inhibitor) significantly blocked EGCG-stimulated Akt phosphorylation and glycogen synthesis. These findings indicate that EGCG increases glucose utilization and glycogen synthesis in HepG2 cells through AMPK activation (Lin & Lin, 2008). EGCG in hyperlipidemic 3T3-L1 adipocytes, significantly reduced intracellular lipid accumulation by stimulation of lipolysis through increased expression of hormone-sensitive lipase (HSL) gene (Lee et al., 2009a,b). In an animal model, EGCG on diet supplementation in HFD (0.2% and 0.5% of diet) to HFD-fed obese mice for 8 weeks significantly reduced body weight and mass of adipose tissues and liver of obese mice. In the WAT of obese mice, it significantly reduced the mRNA levels of adipogenic genes, PPAR-γ, C/EBP-α, SREBP-1c, aP2, LPL and FAS, and increased the mRNA levels of fatty acid oxidation related genes, CPT-1 and UCP-2 as well as of lipolytic genes, HSL and ATGL. Possibly, EGCG reduced lipid accumulation in adipose tissues and liver of obese mice by regulation of genes related to lipogenesis, lipolysis and fat oxidation through AMPK activation (Lee, Kim, & Kim, 2009). In another study, EGCG on diet supplementation (0.5% and 1.0% of diet) to *db/db* mice for 7 weeks significantly reduced plasma glucose and TG levels and increased plasma insulin level in mice. Moreover, EGCG increased hepatic glucose utilization by increasing the expression of GK gene and reduced hepatic glucose production in gluconeogenic pathway. Furthermore, it increased the expression of acyl coenzyme A oxidase-1 (ACOX-1) and CPT-1 in both liver and adipose tissues for fatty acid oxidation to reduce the levels of FFAs. EGCG in rat H411E hepatoma cells, reduced the expression of gluconeogenic genes G6Pase and FBPase, and increased the expression of glycolytic genes, phosphofructokinase (PFK) and GLUT-1, and reduced the expression of lipogenic genes FAS, SCD-1/2, GPAT, HMGCS-1 and mevalonate kinase (MK). Thus, EGCG exhibits its antidiabetic effect through regulation of glucose and lipid metabolism via increased insulin sensitivity in peripheral liver and adipose tissues in mice (Wolfram et al., 2006). EGCG on dietary supplementation (10 mg/kg bw) to intralipid plus heparin induced- FFAs toxicated rats, significantly prevented FFAs-induced peripheral insulin resistance by increasing AMPK activation and GLUT-4 translocation in skeletal muscle and adipose tissues of rats (Li et al., 2011). EGCG on diet supplementation in HFD to HFD-fed obese rats significantly reduced plasma FFAs level and plasma fasting glucose and insulin levels in rats. Further, EGCG reduced the expression of TLR-4 and inhibited the activation of its target genes IκBβ and NF-κB for reduction of TNF-α, IL-6 and IRS-1 serine phosphorylation levels for improvement of insulin resistance in adipose tissues of rats (Bao et al., 2014). EGCG in IL-1β and IFN-γ-toxicated human islets and RINm5F cells protected the β-cells from damage by reducing the elevated expression levels of iNOS mRNA and proteins and caspase-3 proteins. Moreover, it restored glucose-stimulated insulin secretion by improving mitochondrial membrane potential and functions in the cells. In an animal model, EGCG on diet supplementation in drinking water (0.05%) to spontaneous non-obese T1D mice for 27 weeks significantly reduced plasma glucose and HbA1c levels and increased plasma insulin level in obese mice. These findings indicate that EGCG promotes pancreatic β-cell survival and function in insulin secretion and delays the development of T1D in non-obese mice and could reduce the risk of T1D (Han, 2003; Fu, Zhen, Yuskavage, & Liu, 2011; Zhang, Ding, Dai, Wang, & Li, 2011). EGCG and epicatechin gallate (EGC) **40** (Fig. 4), another major catechin constituent of green tea extract, inhibited the activity of sodium-dependent glucose co-transporter-1 (SGLT-1) proteins in isolated rabbit small intestine in a competitive manner. These galloylcatechins could be effective as SGLT-1 antagonists in reduction of glucose absorption (uptake) by the intestinal tract of diabetic patients (Kobayashi et al., 2000). EGC and (−)-epicatechin (EC) **41** (Fig. 4), another major catechin constituent of green tea extract in 3T3-L1 adipocytes under HG concentrations, increased glucose uptake by promoting GLUT-4 proteins expression and translocation to the plasma membrane through increased phosphorylation of PI3K and PKC in an insulin-independent pathway without phosphorylation of IRS-1 and Akt. While other catechin constituents of green tea namely (−)-catechin gallate (CG) **42** (Fig. 4) and EGCG, increased glucose uptake by increased phosphorylation of Akt (Ueda et al., 2010). In an animal model, (−)-epicatechin (EC), on diet supplementation in HFD (20 mg/kg diet) to HFD-fed obese mice for 2 weeks significantly reduced body weight and plasma glucose, TC, TG and FFAs levels in obese mice. Further, EC increased the activation of IRS-1 and Akt for glucose uptake and decreased the activation of JNK, IKK, NF-kB and PTP1B to improve insulin sensitivity in adipose tissues and liver of obese mice (Cremonini, Bettaieb, Haj, Fraga, & Oteiza, 2016).

(−)-Catechin **43** (Fig. 4), another major bioactive flavonoid constituent of green tea, in 3T3-L1 adipocytes, enhanced the expression and secretion of adiponectin, and markedly decreased the expression of kruppel-like factor-7 (KLF-7) gene and improved insulin sensitivity to increase insulin-stimulated glucose uptake in adipocytes. Available evidence demonstrates that KLF-7 gene is over-expressed in adipocytes of human diabetic patients and plays a significant role in the pathogenesis of T2DM via regulation of adipokines for the inhibition of glucose-induced insulin secretion in pancreatic β-cells and reduction of insulin sensitivity in peripheral tissues. Antidiabetic drug, pioglitazone, a PPAR-γ agonist, increases adiponectin secretion in adipocytes through activation of PPAR-γ and its target genes, but not through reduction of KLF-7 expression, while (−)-catechin does not alter the expression levels of PPAR-γ and its target genes. Thus, (−)-catechin of green tea could be a new therapeutic target in treatment of T2DM (Cho et al., 2007; Kawamura, Tanaka, Kawamori, & Maeda, 2006).

3.2.6 Anthocyanins in prevention of obesity and diabetes

Colored plant pigments anthocyanins and their aglycones anthocyanidins, another class of flavonoids, are found abundantly in various edible fruits such as berries, grapes, apples, etc. and vegetables such as sweet potato, carrot, radish, black beans, etc. The most common anthocyanidins are cyanidin **44**, peonidin **45**, delphinidin **46**, malvidin **47**, pelargonidin **48**, and petunidin **49** (Fig. 5). More than 600 anthocyanins have been identified and among them, cyanidin 3-*O*-glucoside is the most abundant anthocyanin in dietary fruits and vegetables, followed by pelargonidin glycosides and delphinidin glycosides. Anthocyanins are water soluble pigments and have potential health benefits in prevention of various inflammatory-related disorders including obesity, diabetes, cancer and cardiovascular diseases (Ghosh & Konishi, 2007).

Purple corn (*Zea mays*, commonly known as maize) color (PCC), rich in anthocyanins having cyanidin-3-*O*-glucoside **50** (Fig. 5) as major constituent, on diet supplementation in HFD (2 g/kg diet) to HFD-fed obese mice for 12 weeks significantly reduced body weight and lipid mass of liver and adipose tissues, and decreased plasma glucose, insulin, leptin and TG levels in obese mice. PCC reduced the lipid mass in liver and white adipose tissues by decreasing the expression of lipogenic enzymes FAS, ACS-1, GPAT and SREBP-1 mRNA levels and reduced the inflammation by lowering the

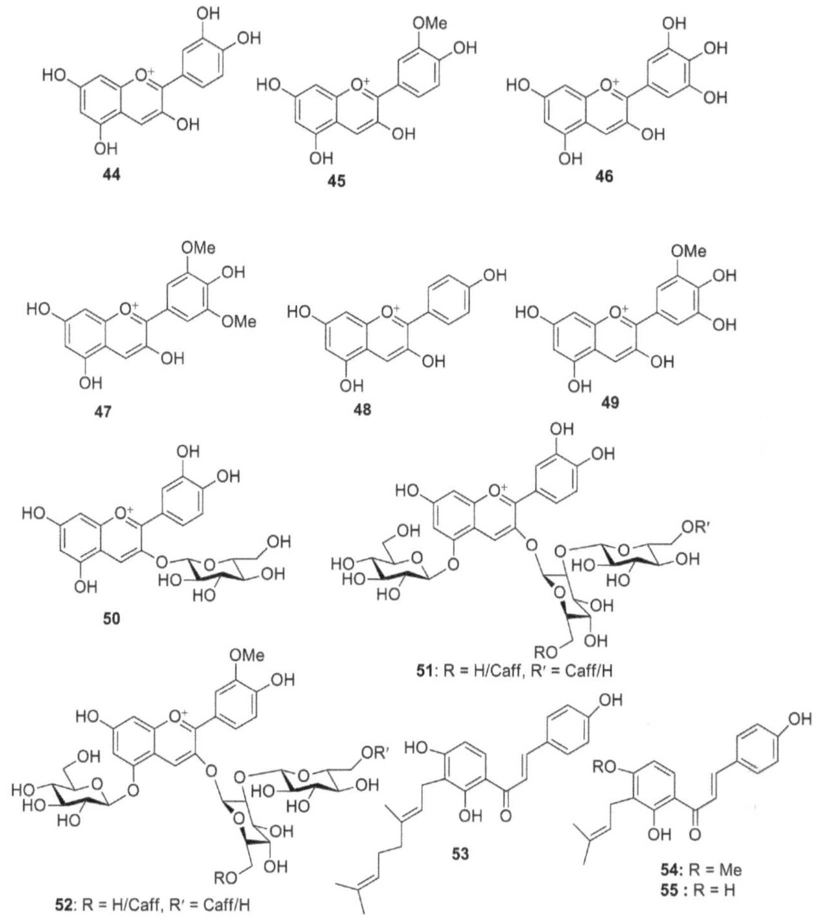

Fig. 5 Chemical structures of anthocyanins and chalcones having anti-obesity and anti-diabetic effects.

level of cytokine TNF-α through increased AMPK activation in liver and adipose tissues of obese mice (Tsuda, Horio, Uchida, Aoki, & Osawa, 2003). In vitro, cyanidin 3-*O*-glucoside (C3G) in isolated rat adipocytes, enhanced the secretion of adiponectin and leptin, upregulated the expression of lipolytic enzymes LPL, aP2, and UCP-2, and increased AMPK activation, and decreased AMP/ATP ratio in adipocytes. These findings indicate that PCC exhibits anti-obesity effect in obese mice by improving insulin sensitivity in liver and WAT via increased AMPK activation and secretion of adiponectin, leptin and other adipokines in adipocytes in WAT (Tsuda et al., 2004). C3G on supplementation in diet (2 g/kg diet) to diabetic KK-Ay mice for 5 weeks showed significant antidiabetic effect by decreasing serum glucose, insulin and TG levels, and improving insulin sensitivity in peripheral tissues of diabetic mice. C3G significantly increased glucose uptake in skeletal muscle and adipose tissues by increased AMPK activation and up-regulation of GLUT-4 expression and its translocation to plasma membrane and down-regulation of the expression of inflammatory adipokines, retinol binding protein 4 (RBP-4), MCP-1 and TNF-α in WAT. Furthermore, C3G reduced hepatic glucose production by decreasing the expression of gluconeogenic enzyme, G6Pase. Accumulating evidence reveals that an increased expression of RBP-4 gene in adipocytes is strongly associated to the impairment of insulin signaling in skeletal muscle and increased hepatic glucose production in rodents. These findings suggest that C3G improves glucose metabolism in diabetic mice by increasing insulin sensitivity in peripheral tissues through AMPK activation, reduction of RBP-4 and other inflammation levels in adipose tissue, up-regulation of GLUT-4 expression and reduction of gluconeogenic G6Pase gene (Sasaki et al., 2007). A study on the mechanism of C3G in increased insulin sensitivity in obese diabetic mice for reduction of hyperglycemia and hyperlipidemia reveals that under HG condition, C3G increases insulin sensitive Akt activation for improvement of phosphorylation of nuclear FoxO1 gene and its exclusion from nucleus to cytoplasm in both hepatocytes and adipocytes. This exclusion of FoxO1 from nucleus reduces hepatic glucose production by down-regulation of gluconeogenic gene, G6Pase and reduces TG synthesis in liver by binding to PPAR-γ, and increases FFAs oxidation in liver and adipose tissues (Guo et al., 2012). C3G and delphinidin 3-glucoside in rodent pancreatic beta INS-1 832/13 cells under HG concentrations, increased glucose-stimulated insulin secretion in the cells (Jayaprakasam, Vareed, Olson, & Nair, 2005). In an animal model, C3G on dietary supplementation in diabetic mice activates cAMP-PKA-eNOS signaling pathway in aortic endothelial cells to increase endothelial nitric oxide bioavailability for improvement of impaired vasorelaxation process and to increase blood flow in aorta of diabetic mice. Thus, C3G could be effective in prevention of endothelial dysfunction in diabetic patients (Liu, Fu et al., 2014; Liu, Li et al., 2014).

Anthocyanin fraction (AF) from purple sweet potato (PSP) (*Ipomoea batatas*) extract, on diet supplementation in HFD (200 mg/kg diet) to HFD-fed obese mice for 4 weeks significantly reduced weight gain and hepatic lipid content in obese mice. The AF of PSP significantly increased the phosphorylation of AMPK and ACC in the liver of mice. Further, the AF of PSP in

HepG2 cells, significantly reduced the expression of lipogenic enzyme SREBP-1 and its target genes ACC and FAS, and increased the expression level of phosphorylated AMPK. Treatment of compound C (AMPK inhibitor) reduced the effect of AF of PSP on lipogenic enzymes. These findings indicate that the AF of PSP reduced weight gain and hepatic lipid mass in obese mice by down-regulation of lipogenic enzymes in liver through AMPK activation (Hwang et al., 2011). Purple sweet potato (PSP)- anthocyanins extract rich in acylated cyanidin and peonidin (about 1% and 11% of total anthocyanins), in 3T3-L1 adipocytes decreased leptin secretion, suppressed the expression of the mRNAs levels of lipogenic enzymes SREBP-1c, ACS, FAS, and LPL, and increased the expression of mRNAs levels of lipolytic and fatty oxidation genes HSL, perilipin, CPT-1 and ACOX, and reduced the expression levels of inflammatory cytokines MCP-1, IL-6 and COX-2. These results suggest that PSP- anthocyanins extract exhibits anti-obesity effect through reduction of inflammation and lipid synthesis and increased lipolysis and fatty acid β-oxidation in obese mice (Ju et al., 2011). Purple sweet potato color (PSPC) is rich in acylated anthocyanins, containing cyanidin and peonidin acyl glycosides **51** and **52** (Fig. 5) as major anthocyanin constituents (about 90%). PSPC on supplementation in HFD (700 mg/kg diet) to HFD-fed obese mice for 20 weeks significantly improved obesity by reducing hepatic lipid accumulation, plasma glucose, AST and ALT levels, and hepatic TG, TC and FFAs levels through reduction of oxidative and ER stress in the liver of obese mice. PSPC improved insulin-stimulated PI3K activation and phosphorylation of Akt to reduce serine phosphorylated IRS-1 level through inhibition of JNK-1 and IKKβ/NF-κB activation and increased the activity of antioxidant enzymes GSH, GP_X, and SOD in the liver of mice. Thus, PSPC could be a potential drug in the treatment of obesity-induced non-alcoholic fatty liver disease (NAFLD) and diabetes (Zhang et al., 2013). In another study, PSPC on diet supplementation (400 mg/kg diet) to HFD-fed obese rats for 6 weeks significantly increased leptin secretion in hypothalamus to inhibit hypothalamic AMPK activation via β-endorphin secretion to reduce food intake and body weight of obese rats. PSPC decreased hepatic fatty acid synthesis and glucose production and increased β-oxidation of fatty acids in liver of obese rats. Further, PSPC increased m-TOR activity by reduction of oxidative stress-related DNA-toxic markers, ROS, RAGE and TXNIP levels in hypothalamus of obese rats (Zhang et al., 2015).

Anthocyanin rich bilberry (*Vaccinium myrtillus*) extract (BBE) containing a mixture of delphinidin-, cyanidin-, petunidin-, peonidin- and malvidin-glycosides, on supplementation in diet (27 g BBE equivalent to 10 g anthocyanin/kg diet) to diabetic KK-Ay mice for 5 weeks significantly improved hyperglycemia and insulin sensitivity by reduction of blood glucose and hepatic lipid content in diabetic mice. BBE increased AMPK activation in WAT and skeletal muscle to increase glucose uptake by GLUT-4 expression, and increased AMPK activation in liver to reduce glucose production by decreasing the expression of gluconeogenic G6Pase and PEPCK genes and to reduce hepatic lipid content by lowering the activity of lipogenic ACC gene and increasing the activity of fatty acid oxidation related genes, PPAR-α, ACOX and CPT-1 for maintenance of glucose and lipid homeostasis in the body of diabetic mice (Takikawa, Inoue, Horio, & Tsuda, 2010).

Blueberry powder (BB), a mixture of 1: 1 *Vaccinium ashei* and *Vaccinium corymbosa*, rich in anthocyanins, peonidin and malvidin glycosides, on diet supplementation in HFD (4% of diet) to HFD-fed obese mice for 8 weeks significantly ameliorated obesity-induced insulin resistance and protected the adipocytes from death by reduction of inflammation and plasma lipo- and gluco-toxicities in obese mice. BB powder significantly down-regulated the expression levels of inflammatory cytokines TNF-α, MCP-1, IL-6 and iNOS in the adipose tissues and liver of mice (DeFuria et al., 2009). In another in vivo study, BB powder on supplementation in HFD (2% of diet) to obese Zucker rats significantly reduced abdominal fat mass and reduced plasma glucose, insulin, leptin, TC and TG levels in obese rats. BB powder increased the activity of PPAR-α and PPAR-γ and their target genes in adipose tissues and skeletal muscle to increase fat oxidation and glycolysis by up-regulation of PPAR-α, UCP-3, CPT-1α, PFK, and PDK-4 genes, and to reduce lipid synthesis by down-regulation of FAS gene for reduction of lipid mass. Moreover, BB powder increased insulin stimulated glucose uptake in skeletal muscle and adipose tissues by PI3K activation and GLUT-4 protein expression. Thus, anthocyanins of BB could be useful in prevention of obesity and diabetes (Seymour et al., 2011).

Mulberry (*Morus alba* L.) anthocyanins extract (MAE), rich in cyanidin 3-glucoside, cyanidin 3-rutinoside and pelargonidin 3-glucoside, in HepG2 cells under HG and HP conditions, increased insulin sensitive glucose uptake and glycogen content via increased phosphorylation of Akt and expression of glycogen synthase-2 (GS-2). Further, it reduced the activity of gluconeogenic enzymes PEPCK and G6Pase via decreasing the activity of PGC-1α and FoxO1 genes. These effects of MAE on HepG2 cells were blocked on treatment of LY294002, an inhibitor of PI3K. Possibly, PGC-1α on binding to FoxO1 stimulates gluconeogenesis in liver during starvation period. In vivo, MAE on oral administration (50 and 125 mg/kg bw/d) to diabetic *db/db* mice for a week significantly decreased fasting blood glucose, serum insulin, leptin, TG and TC levels, and increased serum adiponectin level in diabetic mice. These metabolic improvements are related to PI3K/Akt signaling pathway via Akt activation in liver, skeletal muscle and adipose tissues in diabetic mice (Yan, Dai, & Zheng, 2016).

Strawberry (*Fragaria ananassa*) powder rich in anthocyanins of cyanidin and pelargonidin glycosides, on dietary supplementation to HFD-fed obese mice for 4 weeks significantly improved obesity and hyperglycemia by reducing serum glucose and lipid profile levels and increasing insulin sensitive adiponectin secretion in obese mice (Andersen, Fossen, Torskangerpoll, Fosen, & Hauge, 2004). Pelargonidin **48**, on treatment (3 mg/kg bw, *i.p.*) to STZ-induced diabetic rats showed significant antidiabetic effect by reducing serum glucose and HbA1c levels and improving serum insulin level in diabetic rats. Pelargonidin reduced glycated-hemoglobin- induced iron toxicity in liver by improving the levels of antioxidant enzymes SOD and CAT and reducing the level of MDA for maintenance of glucose homeostasis in diabetic rats (Roy, Sen, & Chakraborti, 2008).

3.2.7 Chalcones in prevention of obesity and diabetes

Xanthoangelol (XA) **53** (Fig. 5)and 4-hydroxyderricin (4- HD) **54** (Fig. 5), two major chalcone constituents of Japanese edible herb, "Ashitaba" (*Angelica keiskei*), on diet supplementation to genetically diabetic KK-Ay mice showed significant antidiabetic effect by lowering plasma glucose and TG levels in diabetic mice. These chalcones showed insulin-like activity in muscle and liver of diabetic mice by acting as PPARγ agonists. Further, 4-HD showed more GLUT-4 expression in skeletal muscle of diabetic mice. In 3T3-L1 cells, these chalcones reduced intracellular lipid accumulation by down-regulation of lipogenic genes expression (Enoki et al., 2007). In a cellular model, both XA and 4-HD in 3T3-L1 adipocytes under high 2-deoxyglucose concentration, increased glucose uptake and GLUT-4 translocation to the plasma membrane, increased phosphorylation of AMPK and its target gene, ACC, and its upstream kinase, LKB1. Treatment of si-RNA of LKB1 reduced the AMPK activation (phosphorylation) and glucose uptake effects of these chalcones. These findings demonstrate that these chalcones improve GLUT-4 dependent glucose uptake via activation of LKB1/AMPK signaling pathway in the adipocytes (Ohta et al., 2015). In another study, 4-HD in 3T3-L1 adipocytes, promoted adipogenesis by increasing the expression of adipogenic genes, PPARγ, C/EBPα and aP2 and adiponectin mRNA and protein levels, and increased insulin-dependent glucose uptake via increased Akt phosphorylation and GLUT-4 mRNA expression.

While XA did not promote TG accumulation in the adipocytes (Li, Ikutani et al., 2016; Li, Yamakuni et al., 2016). Both XA and 4-HD, on treatment in LPS-stimulated co-culture of RAW264.7 macrophage and 3T3-L1 adipocytes, inhibited the expression levels of inflammatory cytokines, TNF-α, MCP-1 and iNOS via suppression of the activation of JNK, NF-κB and AP-1 proteins.

While in a co-culture of RAW264.7 macrophage and C3H10T1/2 adipocytes, increased the expression of thermogenic UCP-1 gene mRNA level. In an animal model, XA on dietary supplementation in HFD (0.15% of diet) to HFD-fed obese mice for 14 weeks significantly reduced the body weight, adipose tissue mass and liver TG content of obese mice. Further, XA significantly decreased the expression levels of TNF-α and MCP-1, and increased the mRNA expression level of UCP-1 proteins in WAT of obese mice. While, a short-term (2 weeks) supplementation of XA to HFD-fed obese mice significantly increased the expression level of PGC-1α and reduced the phosphorylation of JNK in WAT of obese mice. These findings demonstrate that these chalcones could be useful in suppression of obesity-induced inflammation and adipocyte dysfunction (Li, Ikutani et al., 2016; Li, Yamakuni et al., 2016). Both XA and 4-HD in HepG2 cells under hyperlipidemic condition, decreased lipid accumulation by decreasing the expression level of lipogenic SREBP1 gene and increasing the expression level of PPARα through increased phosphorylation of LKB1 and AMPK kinases (Zhang, Yamamoto, & Ashida, 2014). Another chalcone, isobavachalcone **55** (Fig. 5), isolated from *A. Keiskei*, in 3T3-L1 adipocytes, inhibited adipogenesis (proliferation and differentiation of adipocytes) by suppression of mitotic clonal expansion (MCE) stage at G0/G1 phase through inhibition of the expression levels of cyclin D1, cyclin B1, cdk-4 and cdk-6 proteins, as well as reduction of the mRNA levels of adipogenic genes, PPARγ, C/EBPα, and of lipogenic genes, SREBP1α, ACC1 and FAS in adipocytes. Furthermore, it inhibited the autophagy flux at the MCE stage by inducing intracellular accumulation of autophagasome-related proteins, LC3BI, LC3BII and SQSTM1/p62 in the 3T3-L1 cells. In a Zebrafish model, isobavachalcone on dietary supplementation in HFD (100 nM) to HFD-fed Zebrafish larvae inhibited hepatic lipid accumulation via suppression of the mRNA levels of C/EBPα and PPARγ genes in liver, compared to the control group (Lee et al., 2018).

The major effects and the molecular actions of the representative flavonoids to the target cells, genes and enzymes against the clinical symptoms of obesity and diabetes in both cellular and animal models for prevention of these metabolic disorders are summarized in Table 1. The major activities in respect to reduce inflammation, oxidative stress, and lipid accumulation and to increase insulin secretion and protection of pancreatic β-cell mass and to induce glucose uptake and utilization as well as the mechanistic pathways in regulation of glucose and lipid metabolism by the flavonoids in different metabolic tissues of obese diabetic animals are summarized in three figures (Figs. 6–8). Fig. 6 summarizes the major anti-obesity and antidiabetic activities of the flavonoids in the mammal insulin-sensitive tissues, and Fig. 7 summarizes the major molecular actions to target genes and proteins by the flavonoids against the clinical symptoms of obesity and diabetes for prevention and treatment of these metabolic disorders. While, Fig. 8 shows the major metabolic pathways involved in protection of pancreatic β-cells, insulin secretion from β-cells and insulin sensitivity in regulation of glucose and lipid metabolism in the main metabolic tissues, muscle, liver, adipose tissues and adrenal glands of obese diabetic animals for reduction of hyperglycemia and hyperlipidemia by the flavonoids.

4. Bioavailability and bioefficacy of flavonoids in prevention of obesity and diabetes

Most of the bioactive polyphenolic flavonoids have been found to possess poor water solubility, low bioavailability to the target cells and short elimination half-life ($T_{1/2}$) in mammals and restricts to low bioefficacy on oral administration. Recent approach in encapsulation of flavonoids with polymer-based and lipid-based nanocarriers and complex formation with nanosize phytosomes and cyclodextrins improves stability in plasma, bioavailability and bioefficacy of flavonoids to the target cells. These nanoformulations increase surface area of flavonoids for better hydrophobic interactions with the cell membrane of intestine and other metabolic organs to increase higher absorption and to decrease their metabolic degradation by gastrointestinal microbiota compared to free flavonoids. Possibly, these nanoemulsions and nanoparticles absorb mucin from intestinal epithelial cell walls for protection from microbiota (Bilia, Isacchi, Righeschi, Guccione, & Bergonzi, 2014). For instance, anthocyanidin pelargonidin (PG) on encapsulation (NPG) with poly-lactide-co-glycolide (PLGA) increased about 10 fold efficacy than free PG in prevention of alloxan-

Table 1 A summary of major effects and molecular targets of flavonoids in prevention of obesity and diabetes.

Flavonoids	Experimental model	Effects on cells/tissues (molecular targets)	Reference
Quercetin **10**	3T3-L1 adipocytes	Lipogenesis↓(PPARγ↓,SREBP- 1↓,FAS↓, CEBPα↓, AMPK↑)	Ahn et al., 2008
	Obese rats	Insulin sensitivity↑ (adiponectin↑,PPARγ↓, 8-isoPGF$_{2\alpha}$↓)	Wein et al., 2010
	Rat L6 myotubes	Glucose uptake↑(GLUT-4↑, CaMKK↑, AMPK↑)	Dhanya et al., 2017
	Diabetic rats	Insulin secretion & β-cell mass↑(NO↓, dkn1a↓, SOD↑, GSH↑,CAT↑)	Coskun et al., 2005; Kobori et al., 2009
	Obese mice	Lipid mass↓(PPARα↑, PPARγ↓, SREBP1c↓, FAS↓, CD-36↓)	Kobori et al., 2011
	Obese mice	Gluconeogenesis↓(PEPCK↓) Inflammation↓(TNF-α↓, IL-6↓, LPS↓, NF-κB↓, NLRP-3↓, FoxA1↑)	Porras et al., 2017
Rutin **14**	Obese mice	Lipogenesis↓(PPARγ↓, CEBPα↓)	Choi et al., 2006
	Obese mice	Hepatotoxicity↓(AST↓, ALT↓, TG↓, GSH↑, GP$_X$↑, leptin↓)	Hsu et al., 2009
	Diabetic rats	Insulin secretion & β-cell mass↑ Glucose utilization↑(HK↑) Gluconeogenesis↓(G6Pase↓, FBPase↓)	Stanley Mainzen Prince & Kamalakkannan, 2006
	Rat soleus muscle cells	Glucose uptake↑(GLUT-4↑, PI3K↑, PKC↑)	Kappel et al., 2013
	Diabetic rats	Cardiac tissue protection↑(CK-MB↓, LDH↓)	Wang et al., 2015
	Diabetic rats	Retinal function↑(BDNF↑, NGF↑)	Ola et al., 2015
Isorhamnetin 3- glucoside **15**	Diabetic rats	Eye lens protection↑(sorbitol deposition↓, aldose reductase↓)	Lee et al., 2005

Isorhamnetin 3-glycosides **16** fraction	Obese mice	Insulin-sensitive glucose uptake↑(IRS-1↑, PI3K↑, Akt↑) Hepatic lipogenesis↓(SREBP1↓, FAS↓, SCD1↓, XBP1↓, CHOP↓) Hepatic fat oxidation↑(CPT-α↑, ACOX↑)	Rodriguez-Rodriguez et al., 2015
	Rat pancreatic islets	Insulin secretion↑(GLUT-2↑)	Rodriguez-Rodriguez et al., 2015
Kaempferol **17**	Rodent INS-1 β-cells and human pancreatic islets	Glucose-induced insulin secretion↑ (cAMP↑, Akt↑, Bcl-2↑)	Zhang & Liu, 2011
	Rodent INS-1E β- cells	β-cell survival↑(PDX-1↑, PKA↑, CREB↑)	Zhang et al., 2013
	Obese diabetic mice	Glucose uptake↑(GLUT-4↑, AMPK↑) Insulin secretion↑	Alkhalidy et al., 2015
	Diabetic mice	Glycolysis↑(HK↑, GK↑) Gluconeogenesis↓(PEPCK↓)	Alkhalidy et al., 2018
Kaempferol 3-glycosides **18**	Obese diabetic mice	Lipogenesis↓(PPARγ↓, SREBP-1c↓, FAS↓) Fat oxidation↑(CPT-1↑, AMPK↑)	Zang et al., 2015
Kaempferitrin **19** & kaempferol 3-neohesperidoside **20**	Rat soleus muscle cells	Glucose uptake↑(IR↑, PI3K↑, PKC↑, GLUT-4↑)	Cazarolli et al., 2013
Myricetin **21**	Obese diabetic rats	Insulin-dependent glucose uptake↑ (IRS-1↑, Akt↑, GLUT-4↑)	Liu et al., 2007a
	Obese rats	MOR-stimulated glucose uptake ↑(β-endorphin↑, PI3K↑, Akt↑, AS-160↑,GLUT-4↑)	Tzeng et al., 2011
	STZ +Cd-induceddiabetic nephrotoxic rats	Kidney damage↓(serum urea↓, creatinine↓, BUN↓, uric acid↓)	Kandasamy & Ashokkumar, 2014
	Starch-fed diabetic rats	Insulin sensitivity↑(α-glucosidase↓)	Liu et al., 2007b

(Continued)

Table 1 A summary of major effects and molecular targets of flavonoids in prevention of obesity and diabetes.—cont'd

Flavonoids	Experimental model	Effects on cells/tissues (molecular targets)	Reference
Morin **22**	HepG2 cells	Gluconeogenesis↓(PEPCK↓, Akt↑, FoxO1↓)	Paoli et al., 2013
	Diabetic rats	Insulin secretion↑Glycolysis↑(HK↑, G6PD↑) Gluconeogenesis↓(G6Pase↓, FBPase↓)	Vanitha et al., 2014
Fisetin **22a**	Diabetic rats	Gluconeogenesis↓(G6Pase↓, FBPase↓, PEPCK↓) Glycolysis↑(HK↑, G6PD↑, PK↑)	Prasath & Subramanian, 2011
		Glycogen synthesis↑(GP↑, GPase↓)	
	HUVECs	Inflammation↓(MCP-1↓, IL-8↓, ICAM↓, VCAM↓, ROS↓, NF-κB↓)	Kwak, Ku & Bae, 2014
Naringin **23**	Obese diabetic mice	Inflammation↓(TNF-α↓, ERK1/2↓, JNK1/2↓, p38MAPK↓, NF-κB↓, AP-1↓) Lipogenesis↓(SREBP-1c↓, FAS↓, ACC- 1α↓, PPARγ↓, SCD-1↓, AMPK↑) Gluconeogenesis↓(PEPCK↓, G6Pase↓)	Pu et al., 2012a,b
	Type 2 diabetic rats	Antioxidant status↑(GSH-P_X↑, CAT↑, SOD↑) Inflammation/Oxidative stress↓(TNF- α↓, IL-6↓, CRP↓, HSP-27↑, HSP-72↑)	Sharma et al., 2011
	Obese diabetic mice	Insulin sensitivity↑(Adiponectin↑)	Jung et al., 2006
		Lipid synthesis↓(HMGCR↓, ACAT↓)	
Naringenin **24**	Rat muscle L6 myotubes	Glucose uptake↑(GLUT-4↑, AMPK↑)	Zygmunt et al., 2010
	3T3-L1 adipocytes	Inflammation↓(TLR-2↓, TNF-α↓, MCP-1↓, JNK↓, NF-κB↓)	Yoshida et al., 2013

	Human white adipocytes	Insulin sensitive glucose uptake↑ (adiponectin↑, GLUT-4↑, AMPK↑) Fat oxidation↑(CPT-1β↑, PGC-1α/β↑, UCP-1↑)	Rebello et al., 2019
	STZ + NA-type 2 diabetic rats	Insulin secretion↑Insulin sensitive glucose uptake↑ (adiponectin↑, GLUT-4↑) Gluconeogenesis↓(GPase↓)	Ahmed et al., 2017
Hesperidin **25**	Diabetic *db/db* mice	Oxidative stress↓(PON↑) Lipogenesis↓(FAS↓, PAP↓, HMGCR↓, ACAT↓) Fat oxidation↑(CPT-1α↑)	Jung et al., 2006
	Type 1 diabetic rats	Gluconeogenesis↓(PEPCK↓, G6Pase↓) Insulin sensitivity↑(adiponectin↑) Glycolysis↑(GK↑)	Akiyama et al., 2010
	Diabetic rats	Gluconeogenesis↓(G6Pase↓) Cardioprotection↑(CK-MB↓, LDH↓, TNF-α↓)	Agrawal et al., 2014
	Retinal ganglion cells	Oxidative damage↓(JNK↓, p38MAPK↓, Bax↓, caspase-3↓, SOD↑, CAT↑, GP_X↑)	Liu et al., 2017
Eriodictyol **26**	3T3-L1 adipocytes under HG condition	Insulin sensitive glucose uptake↑ (PPARγ↑, aP2↑, GLUT-4↑, GLUT-1↑)	Zhang et al., 2012
	HepG2 cells under HG condition	Glucose uptake↑(Akt↑, GLUT-4↑)	Zhang et al., 2012
	Diabetic rats	Vascular inflammation↓(TNF-α↓, ICAM↓, VEGF↓, eNOS↓)	Bucola et al., 2012

(Continued)

Table 1 A summary of major effects and molecular targets of flavonoids in prevention of obesity and diabetes.—cont'd

Flavonoids	Experimental model	Effects on cells/tissues (molecular targets)	Reference
	Obese mice	Inflammation↓(IL-1β↓, IFN-γ↓, PAI-1↓, leptin↓, IL-10↑) Lipid mass↓(SREBP-1↓, ACC↓, PAP↓, FAS↓, SCD-1↓, CD-36↓, LPL↓, CPT↑, PPARα↑) Insulin secretion↑(GLP-1↑, IRS-2↑) Gluconeogenesis↓(PEPCK↓, G6Pase↓) Glycolysis↑(GK↑)	Kwon & Choi, 2019
Luteolin **27**	3T3-L1 adipocytes	Inflammation & insulin resistance↓ (TNF-α↓, IL-6↓, MCP-1↓, leptin↑, adiponectin↑)	Ding, Jin & Chen, 2010
	Obese diabetic KKAy mice	Glucose uptake↑(Akt↑,GLUT-4↑) Lipogenesis↓(SREBP-1c↓, FAS↓)	Zang, Igarashi & Li, 2016
	Hamster pancreatic HIT-T15 β-cells	Insulin secretion & β-cell protection↑ (ROS↓, LPO↓)	Batumalaie et al., 2014
	Uric acid toxicated mouse insulinoma MIN6 cells	Insulin secretion & cell survival↑ (iNOS↓, MafA↑)	Ding et al., 2014
	Obese diabetic mice	Brain dysfunction↓(BDNF↑, SYP↑, PSD-95↑)	Liu, Fu et al., 2014; Liu, Li et al., 2014
Diosmin **29**	STZ + NA-type 2 diabetic rats	Insulin secretion↑ Glycolysis↑(HK↑, G6PD↑) Gluconeogenesis↓(G6Pase↓, FBPase↓)	Puri & Srinivasan, 2010
	Type 1 diabetic rats	Insulin-stimulated glucose uptake↑ (β-endorphin↑, MOR↑, I-2R↑, GLUT-4↑) Gluconeogenesis↓(PEPCK↓)	Hsu et al., 2017a &b
Apigenin **29a**	Alloxan-diabetic mice	Insulin resistance & thyroid dysfunction↓(SOD↑, GSH↑, CAT↑, T_3 & T_4↑)	Panda & Kar, 2007

		Gluconeogenesis↓(G6Pase↓)	
	Diabetic rats	Renal tissues injury↓(TGF-β1↓, Bax↓, caspase-3↓, ERK1/2↓, JNK↓, NF-κB↓)	Malik et al., 2017
	HepG2 cells	Glucose uptake↑(Akt↑, p-FoxO1↑) Gluconeogenesis↓(PEPCK↓, G6Pase↓)	Bumke-Vogt et al., 2014
	Lung A549 cells	Lipogenesis↓(FAS↓, ACC↓)	Escande et al., 2013
		Insulin action↑(CD-38↓, NAD^{+}↑, SIRT- 1↑)	
	Obese mice	Lipid mass↓(CD-38↓, NAD+↑, SIRT-1↑, LACD↑, MACD↑, CPT-1α↑)	Escande et al., 2013
Baicalein **30**	Rodent INS 382/13 β-cells & human pancreatic islets under HP concentration	Insulin secretion & β-cell mass↑	Fu et al., 2014
	Obese mice	Inflammation↓(TNF-α↓, MCP-1↓)	Pu et al., 2012a,b
		Insulin-stimulated glucose uptake↑ (IRS-1↑, Akt↑, AMPK↑) Lipid synthesis↓ (SREBP-1c↓, FAS↓, PPARγ↓, aP2↓) Fatty acid oxidation↑(PPARα↑, CPT-1↑, UCP-2↑)	
	Obese type 2 diabetic mice	Inflammation↓(TNF-α↓, IL-6↓, TLR- 4↓, LPS↓, SOCS↓) Insulin-stimulated glucose uptake↑ (GSK-3β↓, Akt↑, GLUT-4↑)	Zhang et al., 2018
Wogonin **31**	Diabetic mice	Insulin sensitivity↑(adiponectin↑, PPAR-α↑, AMPK↑)	Bak et al., 2014
	Type 1 diabetic mice	Inflammation↓(OPN↓, p38MAPK↓, PPAR-α↑)	Zhang et al., 2015
Tangeretin **32**	Rat muscle C2C12 myotubes	Glucose uptake↑(GLUT-4↑, AMPK↑, AS-160↑)	Kim et al., 2012

(Continued)

Table 1 A summary of major effects and molecular targets of flavonoids in prevention of obesity and diabetes.—cont'd

Flavonoids	Experimental model	Effects on cells/tissues (molecular targets)	Reference
	3T3-F442A adipocytes	Glucose uptake↑(PI3K↑, Akt↑, GLUT-4↑, PKA↑, CREB↑, GLUT-1↑)	Onda et al., 2013
	Obese mice	Insulin sensitivity↑(IL-6↓, MCP-1↓, adiponectin↑, leptin↑, resistin↑, ACC↓, AMPK↑)	Kim et al., 2012
Genistein **33**	Mouse pancreatic islets	Insulin secretion↑	Sorenson, Brelje & Roth, 1994
	IL-1β + IFNγ-toxicated rat insulinoma RINm5F cells & rat pancreatic islets	Cell apoptosis↓(iNOS↓, ERK1/2↓, JAK/STAT↓, NF-κB↓)	Kim et al., 2007
	Rodent INS-1 β-cells & mouse insulinoma MIN6 cells	Insulin secretion & cell survival↑ (cAMP↑, PKA↑, IRS-2↑, PDX-1↑, cyclin D1↑)	Liu et al., 2006; Fu et al., 2010
	Human aortic endothelial cells	Vascular inflammation↓(MCP-1↓, cAMP↑, PKA↑)	Babu et al., 2012
	Obese diabetic mice	Vascular inflammation↓(MCP-1↓, IL-8↓, VCAM-1↓, ICAM-1↓, IL-10↑)	Babu et al., 2012
Genistein & Daidzein **34**	Diabetic *db/db* mice	Serum glucose & lipid profile↓ Lipid mass↓(FAS↓, CPT↑) Glycolysis↑(GK↑) Gluconeogenesis↓(PEPCK↓, G6Pase↓)	Park et al., 2006
Daidzein	Diabetic *db/db* mice	Glucose uptake↑(GLUT-4↑, AMPK↑)	Cheong et al., 2014
Daidzin **34a**	Diabetic mice	Inflammation↓(TNF-α↓, IL-6↓, MCP-1↓) Plasma glucose & lipid↓ Antioxidant status↑(SOD↑)	Tian et al., 2016

Prunetin **33a**	Obese mice	Adipogenesis↓(PPARγ↓, C/EBPα↓, SREBP↓, aP2↓, LPL↓, adiponectin↓,leptin↓) Hepatic lipogenesis↓(SREBP↓,PPARγ↓, LXR↓, HMGCR↓, adipoR1/R2↑, AMPK↑)	Ahn et al., 2013
Biochanin A **36**	Diabetic rats	Serum glucose & HbA1c↓ Glycolysis↑(GK↑, G6PD↑) Gluconeogenesis↓(G6Pase↓,FBPase↓)	Harini et al., 2012
	Obese diabetic rats	Fatty acid oxidation↑(PPARα↑) Insulin secretion & insulin sensitivity↑ (SIRT-1↑, AMPK↑)	Oza & Kulkarni, 2018a
Formononetin **37**	Type 2 diabetic rats	Plasma glucose, HbA1c, TG, LDL-C↓ Insulin secretion & insulin sensitivity↑ (SIRT-1↑, AMPK↑) Lipid mass↓(PPARγ↓, PPARα↑)Glycogen synthesis↑Gluconeogenesis↓	Oza & Kulkarni, 2018b
	Alloxan-type 1 diabetic mice	Insulin secretion & β-cell mass↑(GK↑, GLUT-2↑, IRS-2↑, PDX-1↑, Fas↓, caspase-3↓)	Qui et al., 2017
Puerarin **38**	Mouse insulinoma MIN6 cells & pancreatic islets	Insulin secretion & cell proliferation↑ (GLP-1R↑, TCF7L2↑, PDX-1↑, FoxO1↓, caspase-3↓, Akt↑)	Yang et al., 2016
	CoCl2-induced injured pancreatic islets & MIN6 cells	Cell apoptosis↓(Akt↑, Bcl-2↑, SOD↑, GP_X↑, Bax↓)	Liu, Fu et al., 2014; Liu, Li et al., 2014
	Obese mice & rodent muscle C2C12 cells	Lipid accumulation↓(ACC↓, PGC-1α↑, NRF-1↑, TFAM↑, AMPK↑)	Jung et al., 2017
	HG-induced podocytes	Oxidative stress↓(SIRT-1↑, NOX↓)	Li et al., 2017

(*Continued*)

Table 1 A summary of major effects and molecular targets of flavonoids in prevention of obesity and diabetes.—cont'd

Flavonoids	Experimental model	Effects on cells/tissues (molecular targets)	Reference
	Diabetic rats & high palmitate-induced L6 myotubes	Glucose uptake↑(MOR↑, IRS-1↑, Akt↑, AS-160↑, GLUT-4↑)	Chen et al., 2018
EGCG **39**	Diabetic mice	Gluconeogenesis↓(PEPCK↓, G6Pase↓)	Koyama et al., 2004
	HG & HP-induced rat L6 myotubes	Glucose uptake↑(PI3K↑, GLUT-4↑)	Ueda et al., 2008
	Dexamethasoneinduced rat L6 myotubes	Glucose uptake↑(Akt↑, AMPK↑, GLUT-4↑)	Zhang et al., 2010
	Obese mice	Lipogenesis↓(PPARγ↓, C/EBPα↓, SREBP-1c↓, aP2↓, LPL↓, FAS↓)	Lee, Kim & Kim, 2009
		Lipolysis↑(HSL↑, ATGL↑)	
		Fatty acid oxidation↑(CPT-1↑, UCP- 2↑)	
	Diabetic *db/db* mice	Glycolysis↑(GK↑)	Wolfram et al., 2006
		Gluconeogenesis↓(G6Pase↓, FBPase↓)Fatty acid oxidation↑(ACOX↑, CPT- 1↑)	
		Lipogenesis↓(FAS↓, SCD1/2↓, GPAT↓, HMGCS-1↓, MK↓)	
EGCG & EGC **40**	Rabbit small intestine	Glucose absorption↓(SGLT-1↓)	Kobayashi et al., 2000
EGC & epicatechin **41**	3T3-L1 adipocytes	Glucose uptake↑(GLUT-4↑, PI3K↑, PKC↑)	Ueda et al., 2010
Catechin gallate **42**	3T3-L1 adipocytes	Glucose uptake↑(GLUT-4↑, Akt↑)	Ueda et al., 2010
Epicatechin	Obese mice	Glucose uptake↑(IRS-1↑, Akt↑)	Cremonini et al., 2016
		Insulin sensitivity↑(JNK↓, IKK↓, NF- κB↓, PTP1B↓)	
(−)-Catechin **43**	3T3-L1 adipocytes	Insulin sensitivity↑(KLF-7↓)	Kawamura et al., 2006
Cyanidin 3- glucoside **50**	Obese mice	Lipogenesis↓(FAS↓, ACS-1↓, SREBP- 1↓, GPAT↓, AMPK↑)	Tsuda et al., 2003
		Inflammation↓(TNF-α↓)	

	Rat adipocytes	Insulin sensitivity↑(adiponectin↑, leptin↑, AMPK↑) Lipolysis & oxidation of FFAs↑(LPL↑, aP2↑, UCP-2↑)	Tsuda et al., 2004
	Diabetic KK-Ay mice	Glucose uptake↑(GLUT-4↑, AMPK↑) Insulin resistance↓(RBP-4↓, MCP-1↓, TNF-α↓)	Sasaki et al., 2007
	Rodent INS-1 β-cells under HG condition	Glucose-stimulated insulin secretion↑	Jayaprakasam et al., 2005
	Obese diabetic mice	Insulin sensitive gluconeogenesis↓ (G6Pase↓, p-FoxO1↑) Lipid mass↓(PPARα↑)	Guo et al., 2012
Anthocyanins of PSP (**51** & **52**)	3T3-L1 adipocytes	Lipogenesis↓(SREBP-1c↓, ACS↓, FAS↓,LPL↓) Lipolysis & FFAs oxidation↑(HSL↑, perilipin↑, CPT-1↑, ACOX↑) Inflammation↓(MCP-1↓, IL-6↓, COX- 2↓)	Ju et al., 2011
	HepG2 cells	Lipid accumulation↓(SREBP-1↓, ACC↓, FAS↓, AMPK↑)	Hwang et al., 2011
	Obese mice	Insulin action↑(PI3K↑, Akt↑) Antioxidant status↑(GSH↑, GP_X↑, SOD↑)	Zhang et al., 2013
	Obese rats	Insulin action↑(β-endorphin↑, AMPK↑) Oxidative stress↓(ROS↓, RAGE↓, TXNIP↓)	Zhang et al.,2015
Anthocyanins of bilberry	Diabetic KK-Ay mice	Glucose uptake↑(GLUT-4↑, AMPK↑) Gluconeogenesis↓(G6Pase↓, PEPCK↓)	Takikawa et al., 2010

(Continued)

Table 1 A summary of major effects and molecular targets of flavonoids in prevention of obesity and diabetes.—cont'd

Flavonoids	Experimental model	Effects on cells/tissues (molecular targets)	Reference
	Obese Zucker rats	Lipid mass↓(ACC↓, PPARα↑, ACOX↑, CPT-1↑) Fat oxidation↑(PPARα↑, CPT-1↑, UCP-3↑) Glycolysis↑(PFK↑, PDK-4↑) Lipogenesis↓(FAS↓)	Seymour et al., 2011
	Obese mice	Glucose uptake↑(GLUT-4↑, PI3K↑) Inflammation↓(TNF-α↓, MCP-1↓, iNOS↓, IL-6↓)	DeFuria et al., 2009
Anthocyanins of mulberry extract	HepG2 cells under HG & HP conditions	Glucose uptake↑(Akt↑) Glycogen content↑(GS-2↑) Gluconeogenesis↓(PEPCK↓, G6Pase↓, PGC-1α↓, FoxO1↓)	Yan, Dai & Zheng, 2016
	Diabetic *db/db* mice	Serum glucose, insulin, TG, TC & leptin↓(Akt↑) Insulin resistance↓(adiponectin↑)	Yan, Dai & Zheng, 2016
Xanthoangelol **53** & 4-hydroxyderricin **54**	3T3-L1 adipocytes under HG condition	Glucose uptake↑(GLUT-4↑, LKB1↑, AMPK↑)	Ohta et al., 2015
	HepG2 cells under HP condition	Lipid accumulation↓(SREBP1↓, PPARα↑, LKB1↑, AMPK↑)	Zhang, Yamamoto & Ashida, 2014; Enoki et al., 2007
	Diabetic KK-Ay mice	Plasma glucose & TG↓(GLUT-4↑, PPARγ↓)	Li, Ikutani et al., 2016; Li, Yamakuni et al., 2016

	RAW264.7 macrophage plus 3T3-L1 adipocytes	Inflammation↓(TNF-α↓, MCP–1↓, iNOS↓)	
Xanthoangelol	Obese mice	Lipid mass in WAT & liver↓(UCP–1↑, PGC1α↑, p-JNK↓)	Li, Ikutani et al., 2016; Li, Yamakuni et al., 2016
Isobavachalcone **55**	3T3-L1 adipocytes	Adipogenesis↓(PPARγ↓, C/EBPα↓, SREBP1α↓, ACC1↓, FAS↓, cyclin D1/B1↓, LC3BI/II↑, p62↑)	Lee et al., 2018

↑: Increase; ↓: Decrease. In animal models of obesity/diabetes, streptozotocin (STZ) was used to induce diabetes, STZ + HFD to induce obese diabetes, and HFD to induce obesity or otherwise stated. NA: Nicotinamide.

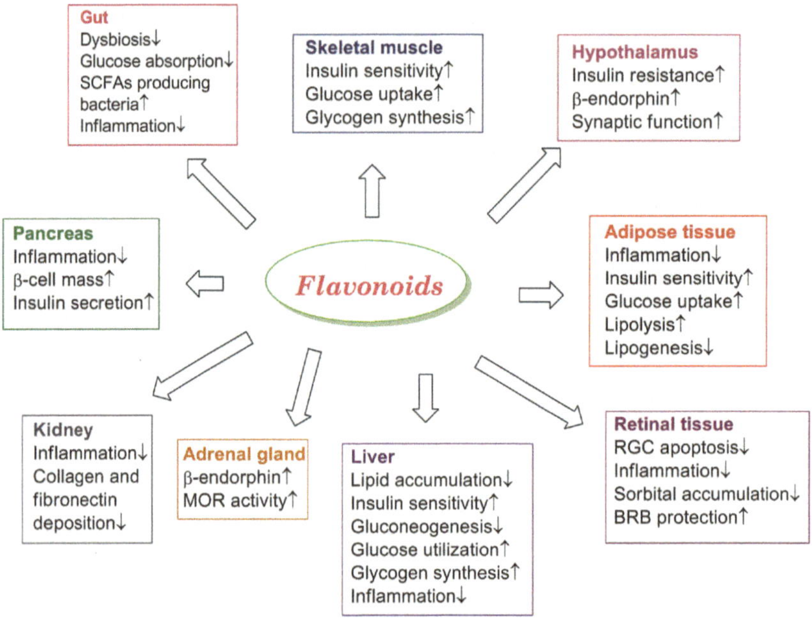

Fig. 6 Major anti-obesity and anti-diabetic effects of the flavonoids in metabolically-related tissues in obese diabetic animals (↑: Increase of expression/activation; ↓: Decrease of expression/activation).

induced oxidative stress and DNA damage in L6 cells in vitro. Further, NPG increased glucose uptake and utilization efficiency in up-regulation of the expression levels of GLUT-4, IRS1/2, GK and PK proteins and down-regulation of the expression of caspase-3 proteins via activation of PI3K, Akt and PPAR in L6 cells for protection of cells from hyperglycemia-induced oxidative damage and mitochondrial dysfunction (Samadder et al., 2016, 2017). Similarly, bioavailability and bioefficacy of quercetin (QUE) was improved by nanoparticle complex formation (QUE/P) with poly-(ethylene glycol)-*block*-(poly (ethylenediamine 1-glutamate)-*graft*-poly (ε-benzyloxycarbonyl-1-lysine (PEG-*b*-(PELG-*g*-PZLL) in STZ-induced diabetic rats for reduction of hyperglycemia and renal injury in diabetic rats. The nanocomplex QUE/P increased the stability of quercetin in plasma and improved the efficacy of quercetin in reduction of ICAM-1 expression and neutrophil-EC adhesion formation compared to free quercetin in diabetic rats for protection of renal tissues from vascular inflammation (Tong et al., 2017). Oral bioavailability of baicalein was increased to 200.7% by its microemulsion formation with cremophor RH40, transcutol P and caprylic capric triglyceride (Liu et al., 2012).

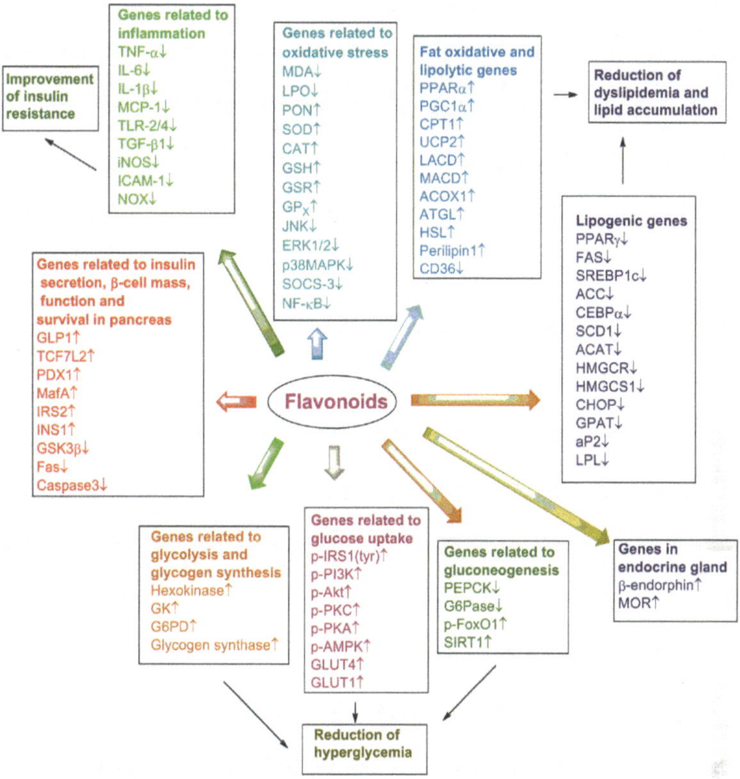

Fig. 7 Major therapeutic targets of the flavonoids to the genes in metabolically-related tissues of obese diabetic animals for prevention of obesity and diabetes (↑: Increase of expression/activation; ↓: Decrease of expression/activation).

5. Pharmacokinetics of flavonoids in prevention of obesity and diabetes

Flavonoids undergo extensive metabolism in small intestine and liver by microbiota and enzymes present in enterocytes and hepatocytes of rodents and humans prior to their entry into systemic circulation and to reach the target tissues/cells for bioactivity. Therefore, extensive study on pharmacokinetics and metabolism is necessary to understand the bioefficacy of the chemical forms of flavonoids at the site of action in the target cells for prevention of obesity and diabetes in mammals. Several enzymes and microbiota present in small intestine convert flavonoid *O*-glycosides to free aglycones and a part of free aglycones are converted into conjugate glucuronides,

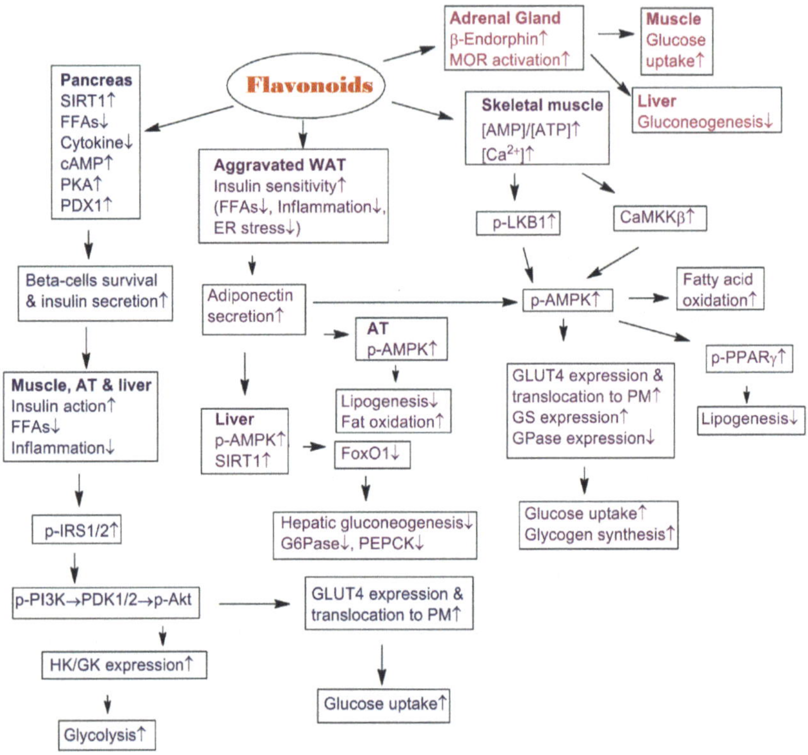

Fig. 8 Major metabolic pathways in metabolically-related tissues by which flavonoids ameliorate hyperglycemia and hyperlipidemia in obese diabetic animals. *AT*, Adipose tissue; *PM*, Plasma membrane. Please incorporate a suppl. captions for Figs. 6–8 as provided as a file (↑: Increase of expression/activation; ↓: Decrease of expression/activation).

sulfates and *O*-methylated forms in presence of glucuronyltransferase, sulphotransferase and *O*-methyl transferase enzymes present in intestinal epithelial membrane. Another part of free aglycones in liver are conjugated into these glucuronate, sulfate and methyl groups as well as into other metabolites by cleavage of *C*-ring of flavonoids. The distribution of these metabolites to the target tissues depends on their interactions with the ABC and OAT transporters present in the enterocytes and hepatocytes. All these metabolites enter into plasma and reach the target cells for their bioaction, but a part of these metabolites are found to be active against the target cells. Pharmacokinetic study of all these bioactive metabolites, mainly their maximum concentration in plasma (C_{max}), time to reach the maximum concentration in plasma (T_{max}), area under plasma concentration versus time curve (AUC), elimination half-life ($T_{1/2}$) and percent of dose excreted in urine and feces are necessary to evaluate the efficacy of these bioactive flavonoids. The AUC is the better indicator for evaluation of bioavailability of flavonoids. The emerging pharmacokinetic studies of flavonoids in humans and animals reveal that anthocyanins have very low concentrations in plasma (C_{max}), of range 0.01−0.02 μM/L, while flavanones have relatively better concentrations in plasma (C_{max}), in the range 0.46−5.99 μM/L and most of the flavonoids have C_{max} in the range 2 nM/L - 8 μM/l (Kay, 2006; Manach & Donovan, 2004; Williamson, Kay, & Crozier, 2018). In the most of these dietary bioactive flavonoids, several metabolites have been identified in animals and humans. For instance, in case of quercetin, 95% of applied dose was found as methylated, glucuronated and sulfated conjugates in human plasma. The major forms were quercetin 3′-*O*-sulfate (50% of total quercetin), quercetin 3-*O*-glucuronide, isorhamnetin 3-*O*-glucuronide and quercetin 3′-*O*-glucuronide (Mullen, Edwards, & Crozier, 2006). While in case of (−)-epigallocatechin gallate (EGCG), its oral administration in humans, an enzymatic hydrolysis of galloyl moiety occurred resulting into the formation of (−)-epigallocatechin (EGC) in oral cavity and both EGCG and EGC were found in saliva of humans (Yang, Lee, & Chen, 1999). The study of the metabolism of (−)-epicatechin in humans revealed the presence of 12 metabolites in plasma and among them, (−)-epicatechin-3′-*O*-glucuronide and (−)-epicatechin 3′-sulfate were found in high concentration in plasma (Ottaviani et al., 2016). Administration of puerarin and daidzin in diabetic mice, *C*-glucoside puerarin was rapidly absorbed in intestine without metabolism, while *O*-glucoside daidzin was hydrolyzed to its aglycone daidzein and both puerarin and daidzein were detected in plasma of diabetic mice (Meezan et al., 2005). Cyanidin-3-glucoside on oral ingestion by humans are converted in to several phenolic acids and related compounds, which are found in plasma (Czank et a., 2013). Accumulating evidence demonstrates that the composition and relative abundance of the metabolites of dietary plant flavonoids in the plasma of rats and humans are not same.

6. Summary

Based on the reported studies, dietary plant flavonoids could be a potential source for development of safe and effective anti-obesity and antidiabetic drugs. Most of the flavonoids discussed in this chapter, have significant insulin secretory activity from pancreatic beta-cells and insulin sensitive activities in both cellular and animal models for improvement of insulin resistance and lipid and glucose metabolism in metabolic tissues in maintenance lipid and glucose homeostasis in the body of mammals. Some of them exhibit potent activity in pancreatic β-cell function in glucose-stimulated insulin secretion and proliferation of β-cells. While most of these flavonoids improve insulin resistance by reduction of lipo- and gluco-toxicities, inflammation and oxidative stress in insulin sensitive peripheral adipose tissues, skeletal muscle, liver and kidney to reduce hyperglycemia and hyperlipidemia in obese diabetic animals. These flavonoids could be useful in treatment of diabetic vascular complications including cardiovascular disease, retinopathy, nephropathy and other inflammatory disorders including cancer and neurodegenerative disorders. Future studies on these flavonoids in respect to their effective dosages, improved efficacy and bioavailability to the target cells via nanoformulations and derivative formations, long-term toxicities, pharmacokinetics and detailed molecular actions to the target cells in animal models are needed before their extensive clinical trials in humans for commercial applications in drug industry.

References

Agrawal, Y. O., Sharma, P. K., Shrivastava, B., Ojha, S., Upadhya, H. M., Arya, D. S., et al. (2014). Hesperidin produces cardioprotective activity via PPAR-γ pathway in ischemic heart disease model in diabetic rats. *PLoS One, 9*(11), e111212.

Ahmed, O. M., Hassan, M. A., Abdel-Twab, S. M., & Abdel Azeem, M. N. (2007). Navel orange peel hydroethanolic extract, naringin and naringenin have anti-diabetic potentials in type 2 diabetic rats. *Biomedicine & Pharmacotherapy, 94*, 197–205.

Ahn, J., Lee, H., Kim, S., Park, J., & Ha, T. (2008). The anti-obesity effect of quercetin is mediated by the AMPK and MAPK signaling pathways. *Biochemical and Biophysical Research Communications, 373*(4), 545–549.

Ahn, T. G., Yang, G., Lee, H. M., Kim, M. D., Choi, H. Y., Park, K. S., et al. (2013). Molecular mechanisms underlying the anti-obesity potential of prunetin, an O-methylated isoflavone. *Biochemical Pharmacology, 85*, 1525–1533.

Akiyama, S., Katsumata, S. I., Suzuki, K., Ishimi, Y., Wu, J., & Uehara, M. (2010). Dietary hesperidin exerts hypoglycaemic and hypolipidemic effects in streptozotocin- induced marginal type 1 diabetic rats. *Journal of Clinical Biochemistry & Nutrition, 46*(1), 87–92.

Alam, M. M., Meerza, D., & Naseem, I. (2014). Protective effect of quercetin on hyperglycemia, oxidative stress and DNA damage in alloxan induced type 2 diabetic mice. *Life Sciences, 109*(1), 8–14.

Alkhalidy, H., Moore, W., Wang, Y., Luo, J., McMillan, R. P., Zhen, W., et al. (2018). The flavonoid kaempferol ameliorates streptozotocin- induced diabetes by suppressing hepatic glucose production. *Molecules, 23*(9), 2338.

Alkhalidy, H., Moore, W., Zhang, Y., McMillan, R., Wang, A., Ali, M., et al. (2015). Small molecule kaempferol promotes insulin sensitivity and preserved pancreatic β-cell mass in middle-aged obese diabetic mice. *Journal of Diabetes Research, 2015*, 532984.

American Diabetes Association on diabetes care. (2014). Diagnosis and classification of diabetes mellitus. *Diabetes Care, 37*(Suppl. 1), S81−S90.
Andersen, M., Fossen, T., Torskangerpoll, K., Fosen, A., & Hauge, U. (2004). Anthocyanin from strawberry (*Fragaria ananassa*) with novel aglycone, 5-carboxypyranpelargonidin. *Phytochemistry, 65*(4), 405−410.
Arts, I. C. W., van de Putte, B., & Hollman, P. C. H. (2000). Catechin contents of foods commonly consumed in The Netherlands. 1. Fruits, vegetables, staplefoods and processed foods. *Journal of Agricultural and Food Chemistry, 48*(5), 1746−1751.
Babu, P. V. A., Si, H., Fu, Z., Zhen, W., & Liu, D. (2012). Genistein prevents hypoglycaemia- induced monocyte adhesion to human aortic endothelial cells through preservation of cAMP signalling pathway and ameliorates vascular inflammation in obese diabetic mice. *The Journal of Nutrition, 142*(4), 724−730.
Bak, E. J., Kim, J., Choi, Y. H., Kim, J. H., Lee, D. E., Woo, G. H., et al. (2014). Wogonin ameliorates hyperglycemia and dyslipidemia via PPARα activation in db/db mice. *Clinical Nutrition, 33*(1), 156−163.
Bao, S., Cao, Y., Fan, C., Fan, Y., Bai, S., Teng, W., et al. (2014). Epigallocatechin gallate improves insulin signalling by decreasing toll-like receptor 4 (TLR4) activity in adipose tissues of high-fat diet rats. *Molecular Nutrition & Food Research, 58*(4), 677−686.
Batumalaie, K., Qvist, R., Yusof, K. M., Ismail, I. S., & Sekaran, S. D. (2014). The antioxidant effect of the Malaysian Gelam honey on pancreatic hamster cells cultured under hyperglycemic conditions. *Clinical and Experimental Medicine, 14*(2), 185−195.
Betonico, C. C., Titan, S. M., Correa-Giannella, M. L., Nery, M., & Queiroz, M. (2016). Management of diabetes mellitus in individuals with chronic kidney disease: Therapeutic perspectives and glycemic control. *Clinics (Sao Paulo)., 71*(1), 47−53.
Bilia, A. R., Isacchi, B., Righeschi, C., Guccione, C., & Bergonzi, M. C. (2014). Flavonoids loaded with nanocarriers: An opportunity to increase oral bioavailability and bioefficacy. *Food and Nutrition Sciences, 5*(13), 1212−1227.
Bose, M., Lambert, J. D., Ju, J., Reuhl, K. R., Shapes, S. A., & Yang, C. S. (2008). The major green tea polyphenol, (−)-epigallocatechin-3-O-gallate, inhibits obesity, metabolic syndrome, and fatty liver disease in high-fat-fed mice. *The Journal of Nutrition, 138*(9), 1677−1683.
Boyle, S. P., Dobson, V. L., Duthie, S. J., Hinselwood, D. C., Kyle, J. A., & Collins, A. R. (2000). Bioavailability and efficiency of rutin as an antioxidant: A human supplementation study. *European Journal of Clinical Nutrition, 54*(10), 774−782.
Bucola, C., Leggio, G. M., Drago, E., & Salomone, S. (2012). Eriodictyol prevents early retinal and plasma abnormalities in streptozotocin- induced diabetic rats. *Biochemical Pharmacology, 84*(1), 88−92.
Bumke-Vogt, C., Osterhoff, M. A., Borchert, A., Guzman-Perez, V., Sarem, Z., Birkenfeld, A. L., et al. (2014). The flavones apigenin and luteolin induce FOXO1 translocation but inhibit gluconeogenic and lipogenic gene expression in human cells. *PLoS One, 9*(8), e104321.
Cao, J., Chen, W., Zhang, Y., Zhang, Y., & Zhao, X. (2010). Content of selected flavonoids in 100 edible vegetables and fruits. *Food Science and Technology Research, 16*(5), 395−402.
Cazarolli, L. H., Pereira, D. F., Kappel, V. D., Folador, P., Fiqueiredo, M. S. R. B., Pizzolatti, M. G., et al. (2013). Insulin signalling: A potential signalling pathway for the stimulatory effect of kaempferitrin on glucose uptake in skeletal muscle. *European Journal of Pharmacology, 712*(1−3), 1−7.
Chen, X., Wang, L., Fan, S., Song, S., Min, H., Wu, Y., et al. (2018). Puerarin acts on the skeletal muscle to improve insulin sensitivity in diabetic rats involving μ-opioid receptor. *European Journal of Pharmacology, 818*, 115−123.
Cheong, S. H., Furuhashi, K., Ito, K., Nagaoka, M., Yonezawa, T., Miura, Y., et al. (2014). Daidzein promotes glucose uptake through glucose transporter-4 translocation to plasma

membrane in L6 myotubes and improves glucose homeostasis in type 2 diabetic model mice. *The Journal of Nutritional Biochemistry, 25*(2), 136–143.

Choi, I., Park, Y., Choi, H., & Lee, E. H. (2006). Anti-adipogenic activity of rutin in 3T3-L1 cells and mice fed with high-fat diet. *BioFactors, 26*(4), 273–281.

Cho, S. Y., Park, P. J., Shin, H. J., Kim, Y. K., Shin, D. W., Shin, E. S., et al. (2007). (−)-Catechin suppresses expression of kruppel-like factor 7 and increases expression and secretion of adiponectin protein in 3T3-L1 cells. *American Journal of Physiology - Endocrinology And Metabolism, 292*(4), E1166–E1172.

Coskun, O., Kanter, M., Korkmaz, A., & Oter, S. (2005). Quercetin, a flavonoid antioxidant, prevents and protects streptozotocin- induced oxidative stress and β-cell damage in rat pancreas. *Pharmacological Research, 51*(2), 117–123.

Cremonini, E., Bettaieb, A., Haj, F. G., Fraga, C. G., & Oteiza, P. I. (2016). (−)-Epicatechin improves insulin sensitivity in high fat diet-fed mice. *Archives of Biochemistry and Biophysics, 599*, 13–21.

Crozier, A., Jaganath, I. B., & Clifford, M. N. (2009). Dietary phenolics: Chemistry, bioavailability and effects on health. *Natural Product Reports, 26*(8), 1001–1043.

Czank, C., Cassidy, A., Zhang, Q., Morrison, D. J., Preston, T., Kroon, P. A., et al. (2013). Human metabolism and elimination of anthocyanin, cyanidin-3-glucoside: A ^{13}C-tracer study. *American Journal of Clinical Nutrition, 97*, 995–1003.

DeFuria, J., Bennett, G., Strissel, K. J., Perfield, J. W., Milbury, P. E., Greenberg, A. S., et al. (2009). Dietary blue berry attenuates whole-body insulin resistance in high-fat-fed mice by reducing adipocyte death and its inflammatory sequelae. *The Journal of Nutrition, 139*(8), 1510–1516.

Dhanya, R., Arya, A. D., Nisha, P., & Jayamurthy, P. (2017). Quercetin, a lead compound against type 2 diabetes, ameliorates glucose uptake via AMPK pathway in skeletal muscle cell line. *Frontiers in Pharmacology, 8*, 336.

Ding, L., Jin, D., & Chen, X. (2010). Luteolin enhances insulin sensitivity via activation of PPARγ transcriptional activity in adipocytes. *The Journal of Nutritional Biochemistry, 21*(10), 941–947.

Ding, Y., Dai, X. Q., Zhang, Z. F., & Li, Y. (2012). Myricetin attenuates hyperinsulinemia-induced insulin resistance in skeletal muscle cells. *European Food Research and Technology, 234*(5), 873–881.

Ding, Y., Shi, X., Shuai, X., Xu, Y., Liu, Y., Liang, X., et al. (2014). Luteolin prevents uric acid- induced pancreatic β-cell dysfunction. *The Journal of Biomedical Research, 28*(4), 292–298.

Du, L. Y., Zhao, M., Tao, J. H., Qian, D. W., Jiang, S., Shang, E. X., et al. (2017). The metabolic profiling of isorhamnetin-3-*O*-neohesperidoside produced by human intestinal flora employing UPLC-Q-TOF/MS. *Journal of Chromatographic Science, 55*(3), 243–250.

Eid, H. M., Martineau, L. C., Saleem, A., Muhammad, A., Vallerand, D., Benhaddou-Andaloussi, A., et al. (2010). Stimulation of AMP- activated protein kinase and enhancement of basal glucose intake in muscle cells by quercetin and quercetin glycosides, active principles of the anti-diabetic medicinal plant *Vaccinium vitis-idaea*. *Molecular Nutrition & Food Research, 54*(7), 991–1003.

Enoki, T., Ohnogi, H., Nagamine, K., Kudo, Y., Sugiyama, K., Tanabe, M., et al. (2007). Antidiabetic activities of chalcones isolated from a Japanese herb, *Angelica keilskei*. *Journal of Agricultural and Food Chemistry, 55*(15), 6013–6017.

Ertunc, M. E., & Hotamisligil, G. S. (2016). Lipid signaling and lipotoxicity in metaflammation: Indications for metabolic disease pathogenesis and treatment. *Journal of Lipid Research, 57*(12), 2099–2114.

Escande, C., Nin, V., Price, N. L., Capellini, V., Gomes, A. P., Barbosa, M. T., et al. (2013). Flavonoid apigenin is a inhibitor of the NAD^{+} ase CD38. *Diabetes, 62*(4), 1084–1093.

Fu, Y., Luo, J., Jia, Z., Zhen, W., Zhou, K., Gilbert, E., et al. (2014). Baicalein protects against type 2 diabetes via promoting islet β-cell function in obese diabetic mice. *International Journal of Endocrinology*, 846742.

Fu, Z., Gilbert, E. R., Pfeiffer, L., Zhang, Y., Fu, Y., & Liu, D. (2012). Genistein ameliorates hyperglycemia in a mouse model of nongenetic type 2 diabetes. *Applied Physiology Nutrition and Metabolism, 37*(3), 480–488.

Fu, Z., Zhang, W., Zhen, W., Lum, H., Nadler, J., Bassaganya-Riera, J., et al. (2010). Genistein induces pancreatic β-cell proliferation through activation of multiple signalling pathways and prevents insulin-deficient diabetes in mice. *Endocrinology, 151*(7), 3026–3037.

Fu, Z., Zhen, W., Yuskavage, J., & Liu, D. (2011). Epigallocatechin gallate delays the onset of type 1 diabetes in spontaneous non-obese diabetic mice. *British Journal of Nutrition, 105*(8), 1218–1225.

Ghosh, D., & Konishi, T. (2007). Anthocyanins and anthocyanin- rich extracts: Role in diabetes and eye function. *Asia Pacific Journal of Clinical Nutrition, 16*(2), 200–208.

Glass, C. K., & Olefsky, J. M. (2012). Inflammation and lipid signaling in the etiology of insulin resistance. *Cell Metabolism, 15*(5), 635–645.

Guardado-Mendoza, R., Prioletta, A., Jimenez-Ceja, L. M., Sosale, A., & Folli, F. (2013). The role of nateglinide and repaglinide, derivatives of meglitinide,in the treatment of type 2 diabetes mellitus. *Archives of Medical Science, 9*(5), 936–943.

Guo, H., Xia, M., Zou, T., Ling, W., Zhong, R., & Zhang, W. (2012). Cyanidin 3-glucoside attenuates obesity- associated insulin resistance and hepatic steatosis in high-fat diet-fed and *db/db* mice via the transcription factor FoxO1. *The Journal of Nutritional Biochemistry, 23*(4), 349–360.

Hameed, A., Hafizur, R. M., Hussain, N., Raza, S. A., Rehman, M., Ashraf, S., et al. (2018). Eriodictyol stimulates insulin secretion through cAMP/PKA signaling pathway in mice islets. *European Journal of Pharmacology, 820*, 245–255.

Han, M. K. (2003). Epigallocatechin gallate, a constituent of green tea suppresses cytokine-induced pancreatic beta-cell damage. *Experimental & Molecular Medicine, 35*(2), 136–139.

Harborne, J. B., & Mabry, T. J. (1982). *The flavonoids: Advances in research* (1st ed.). New York: Chapman & Hall.

Harini, R., Ezhumalai, M., & Pugalendi, K. V. (2012). Anti-hyperglycemic effect of biochanin A, a soy isoflavone, on streptozotocin-diabetic rats. *European Journal of Pharmacology, 676*(1–3), 89–94.

Havsteen, B. H. (2002). The biochemistry and medical significance of the flavonoids. *Pharmacology & Therapeutics, 96*(2–3), 67–202.

Hossain, M. K., Dayem, A. A., Han, J., Yin, Y., Kim, K., Saha, S. K., et al. (2016). Molecular mechanisms of anti-obesity and antidiabetic properties of flavonoids. *International Journal of Molecular Sciences, 17*(4), 569.

Hsia, D. S., Grove, O., & Cefalu, W. T. (2017). An update on sodium-glucose co-transporter-2 inhibitors for the treatment of diabetes mellitus. *Current Opinion in Endocrinology Diabetes and Obesity, 24*(1), 73–79.

Hsu, C. C., Lin, M. H., Cheng, J. T., & Wu, M. C. (2017a). Diosmin, a citrus nutrient, activates imidazoline receptors to alleviate blood glucose and lipids in type-1- like diabetic rats. *Nutrients, 9*(7), 684.

Hsu, C. C., Lin, M. H., Cheng, J. T., & Wu, M. C. (2017b). Antihyperglycaemic action of diosmin, a citrus flavonoid, is induced through endogenous β-endorphin in type- 1 like diabetic rats. *Clinical and Experimental Pharmacology and Physiology, 44*(5), 549–555.

Hsu, C. L., Wu, C. H., Huang, S. L., & Yen, G. C. (2009). Phenolic compounds rutin and O-coumaric acid ameliorate obesity induced by high-fat diet in rats. *Journal of Agricultural and Food Chemistry, 57*(2), 425–431.

Hwang, Y. P., Choi, J. H., Han, E. H., Kim, H. G., Wee, J. H., Jung, K. O., et al. (2011). Purple sweet potato anthocyanins attenuate hepatic lipid accumulation through activating adenosine monophosphate- activated protein kinase in human HepG2 cells and obese mice. *Nutrition Research, 31*(12), 896–906.

Jayaprakasam, B., Vareed, S. K., Olson, L. K., & Nair, M. G. (2005). Insulin secretion by bioactive anthocyanins and anthocyanidins present in fruits. *Journal of Agricultural and Food Chemistry, 53*(1), 28–31.

Johnson, A. M., & Olefsky, J. M. (2013). The origins and drivers of insulin resistance. *Cell, 152*(4), 673–684.

Ju, J. H., Yoon, H. S., Park, H. J., Kim, M. Y., Shin, H. K., Park, K. Y., et al. (2011). Anti-obesity and anti-oxidative effects of purple sweet potato extract in 3T3-L1 adipocytes *in vitro*. *Journal of Medicinal Food, 14*(10), 1097–1106.

Jung, E. J., Lee, M. K., Park, Y. B., Kang, M. A., & Choi, M. S. (2006). Effects of citrus flavonoids on lipid metabolism and glucose-regulating enzyme mRNA levels in type 2 diabetic mice. *The International Journal of Biochemistry & Cell Biology, 38*(7), 1134–1145.

Jung, H. W., Kang, A. N., Kang, S. Y., Park, Y. K., & Song, M. Y. (2017). The root extract of *Pueraria lobata* and its main compound, puerarin, prevent obesity by increasing the energy metabolism in skeletal muscle. *Nutrients, 9*(1), 33.

Kaku, K. (2010). Pathophysiology of type 2 diabetes and its treatment policy. *Japan Medical Association- Journal, 53*(1), 41–46.

Kamalakkannan, N., & Prince, P. S. (2006). Antihyperglycaemic and antioxidant effect of rutin, a polyphenolic flavonoid, in streptozotocin- induced diabetic wistar rats. *Basic and Clinical Pharmacology and Toxicology, 98*(1), 97–103.

Kandasamy, N., & Ashokkumar, N. (2014). Protective effect of bioflavonoid myricetin enhances carbohydrate metabolic enzymes and insulin signalling molecules in streptozotocin- cadmium- induced diabetic nephrotoxic rats. *Toxicology and Applied Pharmacology, 279*(2), 173–185.

Kappel, V. D., Cazarolli, L. H., Pereira, D. F., Postal, B. G., Zamoner, A., Reginatto, F. H., et al. (2013). Involvement of GLUT-4 in the stimulatory effect of rutin on glucose uptake in rat soleus muscle. *Journal of Pharmacy and Pharmacology, 65*(8), 1179–1186.

Kawamura, Y., Tanaka, Y., Kawamori, R., & Maeda, S. (2006). Overexpression of kruppel-like factor 7 regulates adipocytokine gene expressions in human adipocytes and inhibits glucose- induced insulin secretion in pancreatic β-cell line. *Molecular Endocrinology, 20*(4), 844–856.

Kay, C. D. (2006). Aspects of anthocyanin absorption, metabolism and pharmacokinetics in humans. *Nutrition Research Reviews, 19*(1), 137–146.

Kim, E. K., Kwon, K. B., Song, M. Y., Seo, S. W., Park, S. J., Ka, S. O., et al. (2007). Genistein protects pancreatic beta-cells against cytokine- mediated toxicity. *Molecular and Cellular Endocrinology, 278*(1–2), 18–28.

Kim, M. S., Hur, H. J., Kwon, D. Y., & Hwang, J. T. (2012). Tangeretin stimulates glucose uptake via regulation of AMPK signalling pathways in C2C12 myotubes and improves glucose tolerance in high-fat diet- induced obese mice. *Molecular and Cellular Endocrinology, 358*(1), 127–134.

Kobayashi, Y., Suzuki, M., Satsu, H., Arai, S., Hara, Y., & Suzuki, K. (2000). Green tea polyphenols inhibit the sodium-dependent glucose transporter of intestinal epithelial cells by a competitive mechanism. *Journal of Agricultural and Food Chemistry, 48*, 5618–5623.

Kobori, M., Masumoto, S., Akimoto, Y., & Oike, H. (2011). Chronic dietary intake of quercetin alleviates hepatic fat accumulation associated with consumption of a Western- style diet in C57/BL6J mice. *Molecular Nutrition & Food Research, 55*(4), 530–540.

Kobori, M., Masumoto, S., Akimoto, Y., & Takahashi, Y. (2009). Dietary quercetin alleviates diabetic symptoms and reduces streptozotocin-induced disturbance of hepatic gene expression in mice. *Molecular Nutrition & Food Research, 53*, 859–868.

Koyama, Y., Abe, K., Sano, Y., Ishizaki, Y., Njelekela, M., Shoji, Y., et al. (2004). Effects of green tea on gene expressions of hepatic gluconeogenic enzymes *in vivo*. *Planta Medica, 70*(11), 1100–1102.

Kumar, S., & Pandey, A. K. (2013). Chemistry and biological activities of flavonoids: An overview. *Science World Journal*, 162750.

Kwak, S., Ku, S. K., & Bae, J. S. (2014). Fisetin inhibits high glucose-induced vascular inflammation in vitro and in vivo. *Inflammation Research, 63*(9), 779–787.

Kwon, E. Y., & Choi, M. S. (2019). Dietary eriodictyol alleviates adiposity, hepatic steatosis, insulin resistance, and inflammation in diet- induced obese mice. *International Journal of Molecular Sciences, 20*(5), 1227.

Lanthier, N., Molendi-Coste, O., Horsmans, Y., van Rooijen, N., Cani, P. D., & Leelercq, I. A. (2010). Kupffer cell activation is a causal factor for hepatic insulin resistance. *American Journal of Physiology - Gastrointestinal and Liver Physiology, 298*(1), G107–G116.

Lee, H., Li, H., Kweon, M., Choi, Y., Kim, M. J., & Ryu, J. H. (2018). Isobavachalcone from *Angelica keiskei* inhibits adipogenesis and prevents lipid accumulation. *International Journal of Molecular Sciences, 19*(6), 1693.

Lee, M. S., Kim, C. T., Kim, I. H., & Kim, Y. (2009). Inhibitory effects of green tea catechin on the lipid accumulation in 3T3-L1 adipocytes. *Phytotherapy Research, 23*(8), 1088–1091.

Lee, M. S., Kim, C. T., & Kim, Y. (2009). Green tea (−)-epigallocatechin-3-gallate reduces body weight with regulation of multiple genes expression in adipose tissue of diet-induced obese mice. *Annals of Nutrition & Metabolism, 54*(2), 151–157.

Lee, Y. S., Lee, S., Lee, H. S., Kim, B. K., Ohuchi, K., & Shin, K. H. (2005). Inhibitory effects of isorhamnetin-3-*O*-*β*-D-glucoside from *Salicornia herbacea* on rat lens aldose reductase and sorbitol accumulation in streptozotocin- induced diabetic rat tissues. *Biological & Pharmaceutical Bulletin, 28*(5), 916–918.

Li, X., Cai, W., Lee, K., Liu, B., Deng, Y., Chen, Y., et al. (2017). Puerarin attenuates diabetic kidney injury through the suppression of NOX4 expression in podocytes. *Scientific Reports,* 7(1), 14603.

Li, Y., Goto, T., Ikutani, R., Lin, S., Takahashi, N., Takahashi, H., et al. (2016). Xanthoangelol and 4-hydroxyderricin suppress obesity-induced inflammatory responses. *Obesity, 24*(11), 2351–2360.

Li, Y., Goto, T., Yamakuni, K., Takahashi, H., Takahashi, N., Jheng, H. F., et al. (2016). 4-Hydroxyderricin, as a PPARγ agonist, promotes adipogenesis, adiponectin secretion and glucose uptake in 3T3-L1 cells. *Lipids, 51*(7), 787–795.

Li, Y., Zhao, S., Zhang, W., Zhao, P., He, B., Wu, N., et al. (2011). Epigallocatechin-3-O-gallate (EGCG) attenuates FFAs- induced peripheral insulin resistance through AMPK pathway and insulin signalling pathway in vivo. *Diabetes Research and Clinical Practice, 93*(2), 205–214.

Li, Z., Shangguan, Z., Liu, Y., Wang, J., Li, X., Yang, S., et al. (2014). Puerarin protects pancreatic β-cell survival via PI3K/Akt signaling pathway. *Journal of Molecular Endocrinology, 53*, 71–79.

Lin, C. L., & Lin, J. K. (2008). Epigallocatechin gallate (EGCG) attenuates high glucose-induced insulin signaling blockade in human hepG2 hepatoma cells. *Molecular Nutrition & Food Research, 52*(8), 930–939.

Liu, D., Zhen, W., Yang, Z., Carter, J. D., Si, H., & Reynolds, K. A. (2006). Genistein acutely stimulates insulin secretion in pancreatic β-cells through a cAMP- dependent protein kinase pathway. *Diabetes, 55*(4), 1043–1050.

Liu, I. M., Tzeng, T. F., Liou, S. S., & Lan, T. W. (2007a). Myricetin, a naturally occurring flavonol, ameliorates insulin resistance induced by a high fructose diet in rats. *Life Sciences, 81*(21–22), 1479–1488.

Liu, I. M., Tzeng, T. F., Liou, S. S., & Lan, T. W. (2007b). Improvement of insulin sensitivity in obese Zucker rats by myricetin extracted from *Abelmoschus moschatus*. *Planta Medica, 73*(10), 1054–1060.

Liu, W., Tian, R., Hu, W., Jia, Y., Jiang, H., Zhang, J., et al. (2012). Preparation and evaluation of self-microemulsifying drug delivery system of baicalein. *Fitoterapia, 83*(8), 1532–1539.

Liu, W. Y., Liou, S. S., Hong, T. Y., & Liu, I. M. (2017). Protective effects of hesperidin (citrus flavanone) on high glucose- induced oxidative stress and apoptosis in a cellular model for diabetic retinopathy. *Nutrients, 9*(12), 1312.

Liu, Y., Fu, X., Lan, N., Li, S., Zhang, J., Wang, S., et al. (2014a). Luteolin protects against high-fat diet- induced cognitive deficits in obesity mice. *Behavioural Brain Research, 267*, 178–188.

Liu, Y., Li, D., Zhang, Y., Sun, R., & Xia, M. (2014b). Anthocyanin increases adiponectin secretion and protects against diabetes- related endothelial dysfunction. *American Journal of Physiology - Endocrinology And Metabolism, 306*(8), E975–E988.

Liu, Z., & Habener, J. F. (2008). Glucagon-like peptide-1 activation of TCF7L2-dependent Wnt signaling enhances pancreatic beta cell proliferation. *The Journal of Biological Chemistry, 283*(13), 8723–8736.

Luo, C., Yang, H., Tang, C., Yao, G., Kong, L., He, H., et al. (2015). Kaempferol alleviates insulin resistance via hepatic IKK/NF-κB signal in type 2 diabetic rats. *International Immunopharmacology, 28*(1), 744–750.

Maher, P., Dargusch, R., Ehren, J. L., Okada, S., Sharma, K., & Schubert, D. (2011). Fisetin lowers methylglyoxal dependent protein glycation and limits the complications of diabetes. *PLoS One, 6*(6), e21226.

Malik, S., Suchal, K., Khan, S. I., Bhatia, J., Kishore, K., Dinda, A. K., et al. (2017). Apigenin ameliorates streptozotocin-induced diabetic nephropathy in rats via MAPK- NF-κB-TNF-α and TGF-β1- MAPK- fibronectin pathways. *American Journal of Physiology - Renal Physiology, 313*(2), F414–F422.

Manach, C., & Donovan, J. L. (2004). Pharmacokinetics and metabolism of dietary flavonoids in humans. *Free Radical Research, 38*(8), 771–785.

Meezan, E., Meezan, E. M., Jones, K., Moore, R., Barnes, S., & Prasain, J. K. (2005). Contrasting effects of puerarin and daidzin on glucose homeostasis in mice. *Journal of Agricultural and Food Chemistry, 53*(22), 8760–8767.

Messina, M. J. (1999). Legumes and soybeans: Overview of their nutritional profiles and health effects. *The American Journal of Clinical Nutrition, 70*(3 Suppl. 1), 439S–450S.

Mezei, O., Banz, W. J., Steger, R. W., Peluso, M. R., Winters, T. A., & Shay, N. (2003). Soy isoflavones exert anti-diabetic and hypolipidemic effects through the PPAR pathways in obese Zucker rats and murine RAW 264.7 cells. *The Journal of Nutrition, 133*(5), 1238–1243.

Milanski, M., Arruda, A. P., Coope, A., Ignacio-Souza, L. M., Nunez, C. E., Roman, E. A., et al. (2012). Inhibition of hypothalamic inflammation reverses diet- induced insulin resistance in the liver. *Diabetes, 61*(6), 1455–1462.

Moller, D. E., & Kaufman, K. D. (2005). Metabolic syndrome: A clinical and molecular perspective. *Annual Review of Medicine, 56*, 45–62.

Mullen, W., Edwards, C. A., & Crozier, A. (2006). Absorption, excretion and metabolite profiling of methyl-, glucuronyl-, glucosyl- and sulpho-conjugates of quercetin in human plasma and urine after injection of onions. *British Journal of Nutrition, 96*(1), 107–116.

Murea, M., Ma, L., & Freedman, B. I. (2012). Genetic and environmental factors associated with type 2 diabetes and diabetic vascular complications. *The Review of Diabetic Studies, 9*(1), 6–22.

Nejentsev, S., Reijonen, H., Adojaan, B., Kovalchuk, L., Sochnevs, A., Schwartz, E. I., et al. (1997). The effect of HLA-B allele on the IDDM risk defined by DRB1*04 subtypes and DQB1*0302. *Diabetes, 46*(11), 1888–1892.

Ohta, M., Fujinami, A., Kobayashi, N., Amano, A., Ishigami, A., Tokuda, H., et al. (2015). Two chalcones, 4-hydroxyderricin and xanthoangelol, stimulate GLUT4-dependent glucose uptake through the LKB1/AMP-activated protein kinase signaling pathway in 3T3-L1 adipocytes. *Nutrition Research, 35*(7), 618–625.

Ola, M. S., Ahmed, M. M., Ahmad, R., Abuohashish, H. M., Al-Rejaie, S. S., & Alhomida, A. S. (2015). Neuroprotective effects of rutin in streptozotocin- induced diabetic rat retina. *Journal of Molecular Neuroscience, 56*(2), 440–448.

Onda, K., Horike, N., Suzuki, T. I., & Hirano, T. (2013). Polymethoxyflavonoids tangeretin and nobiletin increase glucose uptake in murine adipocytes. *Phytotherapy Research, 27*(2), 312–316.

Osborn, O., & Olefsky, J. M. (2012). The cellular and signaling networks linking the immune system and metabolism in disease. *Nature Medicine, 18*(3), 363–374.

Ottaviani, J. I., Borges, G., Momma, T., Spencer, J. P. E., Keen, C. L., Crozier, A., et al. (2016). The metabolome of [2-^{14}C]-(−)-epicatechin in humans: Implications for the assessment of efficacy safety, and mechanisms of action of polyphenolic bioactives. *Science Reports, 16*, 29034.

Oza, M. J., & Kulkarni, Y. A. (2018a). Biochanin A improves insulin sensitivity and controls hyperglycemia in type 2 diabetes. *Biomedicine & Pharmacotherapy, 107*, 1119–1127.

Oza, M. J., & Kulkarni, Y. A. (2018b). Formononetin treatment in type 2 diabetic rats reduces insulin resistance and hyperglycemia. *Frontiers in Pharmacology, 9*, 739.

Panda, S., & Kar, A. (2007). Apigenin (4',5,7-trihydroxyflavone) regulates hyperglycemia, thyroid dysfunction and lipid peroxidation in alloxan-induced diabetic mice. *Journal of Pharmacy and Pharmacology, 59*(11), 1543–1548.

Paoli, P., Cirri, P., Caselli, A., Ranaldi, F., Bruschi, G., Santi, A., et al. (2013). The insulin-mimetic effect of morin: A promising molecule in diabetes treatment. *Biochimica et Biophysica Acta (BBA), 1830*(4), 3102–3111.

Park, S. A., Choi, M. S., Cho, S. Y., Seo, J. S., Jung, U. J., Kim, M. J., et al. (2006). Genistein and daidzein modulate hepatic glucose and lipid regulating enzyme activities in C57BL/KsJ- *db/db* mice. *Life Sciences, 79*(12), 1207–1213.

Park, H. S., Hur, H. J., Kim, S. H., Park, S. J., Hong, M. J., Sung, M. J., et al. (2016). Biochanin A improves hepatic steatosis and insulin resistance by regulating the hepatic lipid and glucose metabolic pathways in diet-induced obese mice. *Molecular Nutrition & Food Research, 60*(9), 1944–1955.

Paschou, S. A., Papadopoulou-Marketou, N., Chrousos, G. P., & Kanaka-Gantenbein, C. (2018). On type 1 diabetes mellitus pathogenesis. *Endocrine Connections,* 7(1), R38–R46.

Peterson, J. J., Dwyer, J. T., Beecher, G. R., Bhagwat, S. A., Gebhardt, S. E., Haytowitz, D. B., et al. (2006). Flavanones in oranges, tangerines (mandarins), tangors, and tangelos: A compilation and review of the data from the analytical literature. *Journal of Food Composition and Analysis, 19*, S66–S73.

Poretsky, L. (2010). *Principles of diabetes mellitus* (2nd ed.). New York: Springer.

Porras, D., Nistal, E., Martinez Florez, S., Pisonero-Vaquero, S., Olcoz, J. L., Jover, R., et al. (2017). Protective effect of quercetin on high-fat diet- induced non- alcoholic fatty liver disease in mice is mediated by modulating intestinal microbiota imbalance and related gut-liver axis activation. *Free Radical Biology and Medicine, 102*, 188–202.

Prasath, G. S., Pillai, S. I., & Subramanian, S. P. (2014). Fisetin improves glucose homeostasis through the inhibition of gluconeogenic enzymes in hepatic tissues of streptozotocin-induced diabetic rats. *European Journal of Pharmacology, 740*(1), 248–254.

Prasath, G. S., & Subramanian, S. P. (2011). Modulatory effects of fisetin, a bioflavonoid, on hyperglycemia by attenuating the key enzymes of carbohydrate metabolism in hepatic and renal tissues in streptozotocin- induced diabetic rats. *European Journal of Pharmacology, 668*(3), 492–496.

Prasath, G. S., Sundaram, C. S., & Subramanian, S. P. (2013). Fisetin averts oxidative stress in pancreatic tissues of streptozotocin- induced diabetic rats. *Endocrine, 44*(2), 359–368.

Pu, P., Gao, D. M., Mohamed, S., Chen, J., Zhang, J., Zhou, X. Y., et al. (2012a). Naringin ameliorates metabolic syndrome by activating AMP-activated protein kinase in mice fed a high-fat diet. *Archives of Biochemistry and Biophysics, 518*(1), 61–70.

Pu, P., Wang, X. A., Salim, M., Zhu, L. H., Wang, L., Chen, K. J., et al. (2012b). Baicalein, a natural product, selectively activating AMPKα_2 and ameliorates metabolic disorder in diet- induced mice. *Molecular and Cellular Endocrinology, 362*(1–2), 128–138.

Puri, L., & Srinivasan, S. (2010). Anti-hyperglycemic effect of diosmin on hepatic key enzymes of carbohydrate metabolism in streptozotocin- nicotinamide- induced diabetic rats. *Biomedicine & Pharmacotherapy, 64*(7), 477–481.

Qui, G., Tian, W., Huan, M., Chen, J., & Fu, H. (2017). Formononetin exhibits anti-hyperglycemic activity in alloxan- induced type-1 diabetic mice. *Experimental Biology and Medicine (Maywood), 242*(2), 223–230.

Rabinovitch, A., & Suarez- Pinzon, W. L. (1998). Cytokines and their roles in pancreatic islet beta-cells destruction and insulin- dependent diabetes mellitus. *Biochemical Pharmacology, 55*(8), 1139–1149.

Rebello, C. J., Greenway, F. L., Lau, F. H., Lin, Y., Stephens, J. M., Johnson, W. D., et al. (2019). Naringenin promotes thermogenic gene expression in human white adipose tissue. *Obesity(Silver Spring), 27*(1), 103–111.

Redinger, R. N. (2007). The pathophysiology of obesity and its clinical manifestations. *Gastroenterology and Hepatology, 3*(11), 856–863.

Rivera, L., Moron, R., Sanchez, M., Zarzuels, A., & Galisteo, M. (2008). Quercetin ameliorates metabolic syndrome and improves the inflammatory status in obese Zucker rats. *Obesity(Silver Spring), 16*(9), 2081–2087.

Rodriguez-Rodriguez, C., Torres, N., Gutierrez- Uribe, J. A., Noriega, L. G., Torre-Villalvazo, I., Leal- Diaz, A. M., et al. (2015). The effect of isorhamnetin glycosides extracted from Opuntia ficus-indica in a mouse model of diet-induced obesity. *Food & Function, 6*, 805–815.

Rohrborn, D., Wronkowitz, N., & Eckel, J. (2015). DPP4 in diabetes. *Frontiers in Immunology, 6*, 386.

Rotter, V., Nagaev, I., & Smith, U. (2003). Interleukin-6 (IL-6) induces insulin resistance in 3T3-L1 adipocytes and is, like IL-8 and tumor necrosis factor-alpha, overexpressed in human fat cells from insulin- resistant subjects. *The Journal of Biological Chemistry, 278*(46), 45777–45784.

Roy, M., Sen, S., & Chakraborti, A. S. (2008). Action of pelargonidin on hyperglycemia and oxidative damage in diabetic rats: Implication for glycation-induced haemoglobin modification. *Life Sciences, 82*(21–22), 1102–1110.

Sammader, A., Abraham, S. K., & Khuda-Bukhsh, A. R. (2016). Nanopharmaceutical approach using pelargonidin towards enhancement of efficacy for prevention of alloxan-induced DNA damage in L6 cells via activation of PARP and p53. *Environmental Toxicology and Pharmacology, 43*, 27–37.

Sammader, A., Tarafdar, D., Abraham, S. K., Ghosh, K., & Khuda-Bukhsh, A. R. (2017). Nano-pelargonidin protects hyperglycaemic-induced L6 cells against mitochondrial dysfunction. *Planta Medica, 83*(5), 468–475.

Sasaki, R., Nishimura, N., Hoshino, H., Isa, Y., Kadowaki, M., Ichi, T., et al. (2007). Cyanidin 3-glucoside ameliorates hyperglycemia and insulin sensitivity due to down- regulation of retinol binding protein 4 expression in diabetic mice. *Biochemical Pharmacology, 74*(11), 1619–1627.

Seymour, E. M., Tanone, I. I., Urcuyo-Llanes, D. E., Lewis, S. K., Kirakosyan, A., Kondoleon, M. G., et al. (2011). Blueberry intake alters skeletal muscle and adipose tissue peroxisome proliferator- activated receptor activity and reduces insulin resistance in obese rats. *Journal of Medicinal Food, 14*(12), 1511–1518.

Sharma, A. K., Bharti, S., Ojha, S., Bhatia, J., Kumar, N., Ray, R., et al. (2011). Up-regulation of PPARγ, heat shock protein-27 and -72 by naringin attenuates insulin resistance, β-cell dysfunction, hepatic steatosis and kidney damage in a rat model of type 2 diabetes. *British Journal of Nutrition, 106*(11), 1713–1723.

Singh, S., Loke, Y. K., & Furberg, C. D. (2007). Thiazolidinediones and heart failure: A teleo-analysis. *Diabetes Care, 30*(8), 2148–2153.

Song, Y., Park, H. J., Kang, S. N., Jang, S. H., Lee, S. J., Ko, Y. G., et al. (2013). Blueberry peel extracts inhibit adipogenesis in 3T3-L1 cells and reduce high-fat diet- induced obesity. *PLoS One, 8*(7), e69925.

Sorenson, R. L., Brelje, T. C., & Roth, C. (1994). Effect of tyrosine kinase inhibitors on islets of langerhans: Evidence for tyrosine kinases in the regulation of insulin secretion. *Endocrinology, 134*(4), 1975–1978.

Spagnuolo, C., Moccia, S., & Russo, G. L. (2018). Anti-inflammatory effects of flavonoids in neurodegenerative disorders. *European Journal of Medicinal Chemistry, 153*, 105–115.

Stanley Mainzen Prince, P., & Kamalakkannan, N. (2006). Rutin improves glucose homeostasis in streptozotocin diabetic tissues by altering glycolytic and gluconeogenic enzymes. *Journal of Biochemical and Molecular Toxicology, 20*(2), 96–102.

Stonehouse, A. H., Darsow, T., & Maggs, D. G. (2012). Incretin-based therapies. *Journal of Diabetes, 4*(1), 55–67.

Sun, W., Sun, J., Zhang, B., Xing, Y., Yu, X., Li, X., et al. (2017). Baicalein improves insulin resistance via regulating SOCS-3 and enhances the effect of acarbose on diabetes prevention. *Journal of Functional Foods, 37*, 339–353.

Tabak, A. G., Herder, C., Rathmann, W., Brunner, E. J., & Kivimaki, M. (2012). Prediabetes: A high risk state for diabetes development. *Lancet, 379*(9833), 2279–2290.

Takikawa, M., Inoue, S., Horio, F., & Tsuda, T. (2010). Dietary anthocyanin-rich bilberry extract ameliorates hyperglycemia and insulin sensitivity via activation of AMP-activated protein kinase in diabetic mice. *The Journal of Nutrition, 140*(3), 527–533.

Tian, D., Liu, J., Liu, N., Wang, R., Ai, Y., Jin, L., et al. (2016). Daidzin decreases blood glucose and lipid in streptozotocin-induced diabetic mice. *Tropical Journal of Pharmaceutical Research, 15*(11), 2435–2443.

Tong, F., Liu, S., Yan, B., Li, X., Ruan, S., & Yang, S. (2017). Quercetin nanoparticle complex attenuated diabetic nephropathy via regulating the expression level of ICAM-1 on endothelium. *International Journal of Nanomedicine, 12*, 7799–7813.

Tontonoz, P., & Spiegelman, B. M. (2008). Fat and beyond: The diverse biology of PPAR gamma. *Annual Review of Biochemistry, 77*, 289–312.

Tsuda, T., Horio, F., Uchida, K., Aoki, H., & Osawa, T. (2003). Dietary cyanidin 3-*O*-*β*-D-glucoside-rich purple corn color prevents obesity and ameliorates hyperglycemia in mice. *The Journal of Nutrition, 133*(7), 2125–2130.

Tsuda, T., Ueno, Y., Aoki, H., Koda, T., Horio, F., Takahashi, N., et al. (2004). Anthocyanin enhances adipocytokine secretion and adipocyte-specific gene expression in isolated rat adipocytes. *Biochemical and Biophysical Research Communications, 316*(1), 149–157.

Tzeng, T. F., Liou, S. S., & Liu, I. M. (2011). Myricetin ameliorates defective post-receptor insulin signalling via β-endorphin signalling in the skeletal muscle of fructose-fed rats. *Evidence-based Complementary and Alternative Medicine*, 150752, 2011.

Ueda, M., Furuyashiki, T., Yamada, K., Aoki, Y., Sakane, I., Fukuda, I., et al. (2010). Tea catechins modulate the glucose transport system in 3T3-L1 adipocytes. *Food & Function, 1*(2), 167–173.

Ueda, M., Nishiumi, S., Nagayasu, H., Fukuda, I., & Yoshida, K. I. (2008). Epigallocatechin gallate promotes GLUT- 4 translocation in skeletal muscle. *Biochemical and Biophysical Research Communications, 377*(1), 286–290.

Vanitha, P., Uma, C., Suganya, N., Bhakkiyalakshmi, E., Suriyanarayanan, S., Gunasekaran, P., et al. (2014). Modulatory effects of morin on hyperglycemia by attenuating the hepatic key enzymes of carbohydrate metabolism and β-cell function in streptozotocin- induced diabetic rats. *Environmental Toxicology and Pharmacology, 37*(1), 326–335.

Vinayagam, R., & Xu, B. (2015). Anti-diabetic properties of dietary flavonoids: A cellular mechanism review. *Nutrition & Metabolism (London), 12*, 60.

Viollet, B., Guigas, B., Garcia, N. S., Leclerc, J., Foretz, M., & Andreelli, F. (2012). Cellular and molecular mechanisms of metformin: An overview. *Clinical Science (London), 122*(6), 253–270.

Wang, Y. B., Ge, Z. M., Kang, W. Q., Lian, Z. X., Yao, J., & Zhou, C. Y. (2015). Rutin alleviates diabetic cardiomyopathy in a rat model of type 2 diabetes. *Experimental and Therapeutic Medicine, 9*(2), 451–455.

Wein, S., Behm, N., Petersen, R. K., Kristiansen, K., & Wolfram, S. (2010). Quercetin enhances adiponectin secretion by a PPAR-γ independent mechanism. *European Journal of Pharmaceutical Sciences, 41*(1), 16–22.

Williamson, G., Kay, C. D., & Crozier, A. (2018). The bioavailability, transport, and bioactivity of dietary flavonoids: A review from a historical perspective. *Comprehensive Reviews in Food Science and Food Safety, 17*(5), 1054–1112.

Wolfram, S., Raederstorff, D., Preller, M., Wang, Y., Teixeira, S. R., Riegger, C., et al. (2006). Epigallocatechin gallate supplementation alleviates diabetes in rodents. *The Journal of Nutrition, 136*(10), 2512–2518.

World Health Organization. (2018). *Fact sheets on obesity and overweight.* Available from https://www.who.int/news-room/factsheets/detail/obesityandoverweight.

Yan, F., Dai, G., & Zheng, X. (2016). Mulberry anthocyanin extract ameliorates insulin resistance by regulating PI3K/AKT pathway in HepG2 cells and *db/db* mice. *The Journal of Nutritional Biochemistry, 36*, 68–80.

Yang, C. S., Lee, M. J., & Chen, L. (1999). Human salivary tea catechin levels and catechin esterase activities: Implication in human cancer prevention studies. *Cancer Epidemiology Biomarkers & Prevention, 8*(1), 83–89.

Yang, L., Yao, D., Yang, H., Wei, Y., Peng, Y., Ding, Y., et al. (2016). Puerarin protects pancreatic β-cells in obese diabetic mice via activation of GLP-1R signaling. *Molecular Endocrinology, 30*(3), 361–371.

Yoshida, H., Watanabe, W., Oomagari, H., Tsuruta, E., Shida, M., & Kurokawa, M. (2013). Citrus flavonoid naringenin inhibits TLR2 expression in adipocytes. *The Journal of Nutritional Biochemistry, 24*(7), 1276–1284.

Zanatta, L., Rosso, A., Folador, P., Fiqueiredo, M. S. R. B., Pizzolatti, M. G., Leite, L. D., et al. (2008). Insulinomimetic effect of kaempferol 3-neohesperidoside on the rat soleus muscle. *Journal of Natural Products, 71*(4), 532–535.

Zang, M., Xu, S., Maitland-Toolan, K. A., Zuccollo, A., Hou, X., Jiang, B., et al. (2006). Polyphenols stimulate AMP-activated protein kinase, lower lipids, and inhibit accelerated atherosclerosis in diabetic LDL receptor-deficient mice. *Diabetes, 55*(8), 2180–2191.

Zang, Y., Igarashi, K., & Li, Y. (2016). Antidiabetic effects of luteolin and luteolin-7-O-glucoside on KK-A^y mice. *Bioscience Biotechnology and Biochemistry, 80*, 1580–1588.

Zang, Y., Zhang, L., Igarashi, K., & Yu, C. (2015). The anti-obesity and antidiabetic effects of kaempferol glycosides from unripe soybean leaves in high-fat-diet mice. *Food & Function, 6*(3), 834–841.

Zhang, B., Sun, W., Yu, N., Sun, J., Yu, X., Li, X., et al. (2018). Anti-diabetic effect of baicalein is associated with the modulation of gut microbiota in streptozotocin and high-fat-diet induced diabetic rats. *Journal of Functional Foods, 46*, 256–267.
Zhang, T., Yamamoto, N., & Ashida, H. (2014). Chalcones suppress fatty acid-induced lipid accumulation through a LKB1/AMPK signaling pathway in HepG2 cells. *Food & Function, 5*(6), 1134–1141.
Zhang, W. Y., Lee, J. J., Kim, Y., Kim, I. S., Han, J. H., Lee, S. G., et al. (2012). Effects of eriodictyol on glucose uptake and insulin resistance *in vitro*. *Journal of Agricultural and Food Chemistry, 60*(31), 7652–7658.
Zhang, Y., & Liu, D. (2011). Flavonol kaempferol improves chronic hyperglycemia-impaired pancreatic beta-cell viability and insulin secretory function. *European Journal of Pharmacology, 670*(1), 325–332.
Zhang, Y., Niu, F., Sun, J., Xu, F., & Yue, R. (2015). Purple sweet potato (*Ipomoea batatas* L.) color alleviates high-fat-diet-induced obesity in SD rat by mediating leptin's effect and attenuating oxidative stress. *Food Science and Biotechnology, 24*(4), 1523–1532.
Zhang, Y., Zhen, W., Maechler, P., & Liu, D. (2013). Small molecule kaempferol modulates PDX-1 protein expression and subsequently promotes pancreatic β-cell survival and function via CREB. *The Journal of Nutritional Biochemistry, 24*(4), 638–646.
Zhang, Z., Ding, Y., Dai, X., Wang, J., & Li, Y. (2011). Epigallocatechin-3-gallate protects pro-inflammatory cytokine-induced injuries in insulin-producing cells through the mitochondrial pathway. *European Journal of Pharmacology, 670*(1), 311–316.
Zhang, Z. F., Li, Q., Liang, J., Dai, X. Q., Ding, Y., Wang, J. B., et al. (2010). Epigallocatechin-3-O-gallate (EGCG) protects the insulin sensitivity in rat L6 muscle cells exposed to dexamethasone condition. *Phytomedicine, 17*, 14–18.
Zhang, Z. F., Lu, J., Zheng, Y. L., Wu, D. M., Hu, B., Shan, Q., et al. (2013). Purple sweet potato color attenuates hepatic insulin resistance via blocking oxidative stress and endoplasmic reticulum stress in high-fat-diet- treated mice. *The Journal of Nutritional Biochemistry, 24*(6), 1008–1018.
Zygmunt, K., Faubert, B., MacNeil, J., & Tsiani, E. (2010). Naringenin, a citrus flavonoid, increases muscle cell glucose uptake via AMPK. *Biochemical and Biophysical Research Communications, 398*(2), 178–183.

CHAPTER SIX

The impact of microRNAs on alterations of gene regulatory networks in allergic diseases

Indoumady Baskara-Yhuellou and Jörg Tost*
Laboratory for Epigenetics & Environment, Centre National de Recherche en Génomique Humaine, CEA-Institut de Biologie François Jacob, Evry, France
*Corresponding author: E-mail: tost@cng.fr

Contents

Advances in Protein Chemistry and Structural Biology, Volume 120
ISSN 1876-1623
https://doi.org/10.1016/bs.apcsb.2019.11.006

Abstract

Allergic diseases including asthma are worldwide on the rise and contribute significantly to health expenditures. Allergic diseases are prototypic diseases with a strong gene by environment interaction component and epigenetic mechanisms might mediate the effects of the environment on the disease phenotype. MicroRNAs, small non-coding RNAs (miRNAs), regulate gene expression post-transcriptionally. Functional single-stranded miRNAs are generated in multiple steps of enzymatic processing from their precursors and mature miRNAs are included into the RNA-induced silencing complex (RISC). They imperfectly base-pair with the 3′UTR region of targeted genes leading to translational repression or mRNA decay. The cellular context and microenvironment as well the isoform of the mRNA control the dynamics and complexity of the regulatory circuits induced by miRNAs that regulate cell fate decisions and function. MiR-21, miR-146a/b and miR-155 are among the best understood miRNAs of the immune system and implicated in different diseases including allergic diseases. MiRNAs are implicated in the induction of the allergy reinforcing the Th2 phenotype (miR-19a, miR-24, miR-27), while other miRNAs promote regulatory T cells associated with allergen tolerance or unresponsiveness. In the current chapter we describe in detail the biogenesis and regulatory function of miRNAs and summarize current knowledge on miRNAs in allergic diseases and allergy relevant cell fate decisions focusing mainly on immune cells. Furthermore, we evoke the principles of regulatory loops and feedback mechanisms involving miRNAs on examples with relevance for allergic diseases. Finally, we show the potential of miRNAs and exosomes containing miRNAs present in several biological fluids that can be exploited with non-invasive procedures for diagnostic and potentially therapeutic purposes.

List of abbreviations

ACD	Allergic contact dermatitis
AD	Atopic dermatitis
ADAR	Adenosine deaminase
AEC	Alveolar epithelial cells
AGO2	Argonaute protein
AID	Activation-induced deaminase
antagomiR	Synthetic inhibitor of a specific microRNA
APC	Antigen presenting cells
AR	Allergic rhinitis
BALF	Bronchoalveolar lavage fluid
BEC	Bronchial epithelial cells
BIM	Bcl-2-like protein 11
BMDCs	Bone marrow-derived dendritic cells
BSM	Bronchial smooth muscle
BSMC	Bronchial smooth muscle cells
CARD10	Caspase recruitment domain-containing protein 10
CCR4-NOT	CCR4 (carbon catabolite repressor 4)—NOT (Negative on TATA) multi-subunit complex
CREB1	CAMP responsive element binding protein 1
CTLA-4	Cytotoxic T-lymphocyte-associated protein 4

DC	Dendritic cells
DGCR8	DiGeorge syndrome critical region gene 8
DNP	Dinitrophenyl
EAE	Experimental autoimmune encephalomyelitis
EBC	Exhaled breath condensate
EC	Endothelial cells
ENTPD	Ectonucleoside triphosphate diphosphohydrolase
EoE	Eosinophilic esophagitis
EPIT	Epicutaneous immunotherapy
ETS1	Erythroblastosis virus E26 homolog-1
FEV1	Forced expiratory volume in one second (pulmonary function)
FOXO1	Forkhead box protein O1
FOXP3	Forkhead box protein P3 (transcription factor)
FXR1	Fragile X mental retardation syndrome-related protein 1
GATA-3	GATA Binding Protein 3
GW182	Glycine-tryptophan 182 kDa protein
HDAC3	Histone deacetylase c3
HDL	High-density lipoprotein
HDM	House dust mite
HITS-CLIP	High-throughput sequencing of RNA isolated by crosslinking immunoprecipitation
HSC	Hematopoietic stem cells
ICS	Inhaled corticosteroids
IFNγR	Interferon gamma receptor
IgE	Immunoglobulin E
IL	Interleukin
ILC2	Innate lymphoid cells type 2
iTreg	In vitro induced Treg
JAK	Janus kinase
LDT	Laboratory developed tests
lncRNAs	Long non-coding RNAs
miRISC	MiRNA:RISC complex
miRNAs	MicroRNAs
MRE	MiRNA response element
MVBs	Multivesicular bodies
NALF	Nasal lavage fluid
NFκB	Nuclear factor-kappa B
NGS	Next generation sequencing
nTreg	Natural T regulatory cell
OIT	Oral immunotherapy
OVA	Ovalbumin
PABP	PolyA binding proteins
PADC	Poly(A)-deadenylase complex
PAN	Poly(A)-nuclease
PAR-CLIP	Photoactivatable RNA ribonucleoside-enhanced crosslinking and immunoprecipitation
PIAS3	Protein inhibitor of activated STAT3
piRNA	PIWI-interacting RNAs
PTEN	Phosphatase and tensin homolog
pTreg	Peripheral antigen-induced Treg
RA	Retinoic acid
RBL-2H3	Rat basophilic leukemia cells
RBP	RNA binding protein

rER	Rough endoplasmic reticulum
RISC	RNA-induced silencing complex
RORγT	Retinoic acid receptor-related orphan receptor-γt
rRNA	Ribosomal RNAs
RUNX1	Runt-related transcription factor 1
SCIT	Subcutaneous immunotherapy
SG	Stress granules
SLIT	Sublingual immunotherapy
SOCS1	Suppressor of cytokine signaling 1
SPRED2	Sprouty-related, EVH1 domain containing 2
STAT	Signal transducer and activator of transcription
T-BET	T-Box transcription factor 21
TBX21	T-Box transcription factor 21
TF	Transcription factor
TFH	T follicular helper cells
TGFβ	Tumor growth factor β
Th	T helper cells
TLR	Toll like receptor
TNFAIP3	Tumor necrosis factor alpha-induced protein 3
TNFαR	Tumor necrosis factor alpha receptor
TNRC6	Trinucleotide repeat-containing 6 protein
TRAF6	TNF receptor associated factor 6
Treg	Regulatory T cells
tRNAs	Transfer RNAs
TSLP	Thymic stromal lymphopoietin
TUT4	Terminal uridylyltransferase 4
VEGF	Vascular endothelial growth factor
XPO5	Exportin-5-Ran-GTP complex
XRN1	Exoribonuclease

1. Introduction to allergic diseases

Over the last decades, the prevalence and incidence of allergic diseases have rapidly increased around the world especially in developed countries and particularly among children. A number of environmental changes such as increased hygiene (hygiene hypothesis), low vitamin D levels, western diet (infant formula) and a rising prevalence of eczema have been suggested to contribute to this rise especially in a time window characterized by the infant's immaturity of the immune system. The World Allergy Organization (WAO) estimates 10%–40% of the world's population affected by allergies, with 20% of the populations concerned in developed countries. Around 150 million Europeans suffer from chronic allergic diseases and it is estimated that by 2025 half of the entire EU population will be affected (EAACI, 2016; Pawankar, 2014). Annual cost of allergies has risen to $18 billions in the US and 29 billion Euros in Europe (Yu, Freeland, & Nadeau, 2016) and has become a major public health concern.

Allergy is a hypersensitivity reaction inducing symptoms initiated by exposure to innocuous environmental peptides or proteins, so-called allergens, at a dose tolerated by a healthy person (Galli & Tsai, 2012; Johansson et al., 2004; Pulendran & Artis, 2012). At the molecular level, it involves among others Th2 lymphocyte cells and IgE production to allergens, regulatory T cells (Treg), IgG4 and IgA (Stone, Prussin, & Metcalfe, 2010). The severity of symptoms varies from mild reactions, such as eczema, to life threatening reactions such as anaphylactic shock involving multiple organ systems (Sampson et al., 2006). In atopic individuals, i.e. individuals with a predisposition to develop hypersensitivity reactions, and/or individuals with a familial genetic predisposition for enhanced IgE-type immune response, the spectrum of allergic diseases is likely to evolve including the sensitization to multiple allergens, a trajectory that is called the atopic march. The atopic march generally starts early in life with skin allergy (eczema, atopic dermatitis) and followed by food allergy, and later on respiratory allergies (allergic rhinitis and asthma) (Alduraywish et al., 2016; Thomsen, 2015). The prevalence of the different types of allergies is country and lifestyle-dependent, which suggests that environmental factors (microbial exposure, hygiene hypothesis, the infant's microbiota, breastfeeding) and dietary exposure (Western diet), pollution, smoking, and low vitamin D levels are implicated in addition to a certain genetic predisposition (Shroba, Rath, & Barnes, 2018). How the environmental changes can interfere with the induction of tolerance to allergens is not well understood. Common treatments for allergies are allergen avoidance for food allergy, inhaled corticosteroids, long acting beta agonists for respiratory allergies and neutralization of acute systemic reactions with adrenaline.

In asthma, there are different disease subtypes also called endotypes based on the occurrence of various symptoms and molecular and cellular changes. Asthma can in many patients be resolved without intervention or by standard treatments (corticosteroid, bronchodilatators, beta-agonist) while other patients require more aggressive treatments. Currently, inhaled corticosteroid (ICS) therapy remain the first line treatment especially for the eosinophilic subtype, which is present in most severe patients (Caminati, Polk, & Rosenwasser, 2019). Biologics such as monoclonal antibodies targeting IgE, IL-5, IL-4 and IL-13 have become recently available and shown to be effective in subsets of allergic patients, but their high cost limits their routine use (Saco, Pepper, & Casale, 2018; van de Veen & Akdis, 2019). Different forms of immunotherapy have been developed and differ from the antibody-based treatments in that they aim at inducing a sustained state of unresponsiveness (or ideally tolerance) to allergen exposure even after the end of the treatment. Depending on the mode of delivery these are: subcutaneous immunotherapy to pollen (SCIT), sublingual immunotherapy (SLIT) to pollen, hazelnut and peanut, oral immunotherapy to egg and peanut (OIT, pending FDA approval for peanut) or epicutaneous immunotherapy to peanut or to the cow's milk protein casein (EPIT, for food allergy in clinical trials) (Burks et al., 2012; Du Toit et al., 2015; Dupont et al., 2010; Enrique et al., 2005; Fleischer et al., 2013; Oktemer et al., 2016). The principle is to expose the allergic person to gradually increasing amounts of the allergens. The exposition generally triggers a desensitization state (un-sustained unresponsiveness) characterized by a decrease in the extent of the allergic reaction. This improvement is associated with an increased Treg induction from naïve $CD4^+$ T cells and increased IgG4 levels, while the Th2 response and mast cell degranulation decrease. However, response is often only transitory and hypersensitivity is regained at the end of the therapy. Data from preclinical animal models as well as data from some small scale studies on peanut allergy suggest that epigenetic mechanism such as hypomethylation of a conserved non-coding Treg specific region in the *FOXP3* gene, and hypermethylation of the *Gata3* promoter might contribute to the induced state of unresponsiveness and may be considered as a marker for predicting the outcome and success of allergen immunotherapy (Mondoulet et al., 2019; Syed et al., 2014).

DNA methylation, histone modifications, long (lncRNAs) and small noncoding RNAs are part of the epigenetic machinery of a cell. There are several major classes of small RNAs including microRNAs (miRNA), transfer RNAs (tRNAs), ribosomal RNAs (rRNA), small nuclear RNAs and PIWI-interacting RNAs (piRNA). These small RNAs differ by their length, their genesis and their mechanism of action and for PIWI-interacting and miRNAs by their target (genetic elements such as transposons vs. mRNA transcripts) (Ghildiyal & Zamore, 2009; Ishizu, Siomi, & Siomi, 2012; Kim, Han, & Siomi, 2009). In this chapter, we will focus on miRNAs, which is the most studied class of small RNAs in allergic diseases. Their expression pattern is cell-type specific and modified by environmental factors such as life-style (diet, activity or sedentary), aging and influenced by the microbiome. miRNAs are essential for regulating gene expression patterns during numerous physiological mechanisms (cell cycle and proliferation, cell differentiation, homeostasis ...) and their expression status is transmissible during cell division. Their expression pattern is modified in many complex diseases such as cancer, but also in autoimmune and inflammatory diseases including allergies (Garo & Murugaiyan, 2016; Lu & Rothenberg, 2013; Sarver, Sarver, Yuan, & Subramanian, 2018; Tahamtan, Teymoori-Rad, Nakstad, & Salimi, 2018). Currently, their characterization in allergic diseases is not well described and requires further investigation compared to other fields of biomedical research such as cancer. In this chapter, we review in detail the miRNA biogenesis and mechanism of action and explain their implications in the regulation of the immune system and in the physiopathology of allergic diseases including mostly respiratory allergies, but also provide some data on food and cutaneous allergies. Finally, we will discuss the potential of miRNAs as biomarkers for diagnostics and for predicting response to treatment as well as targets for therapeutic development.

2. Description of miRNAs

In 1993, the first microRNA, lin-4, was discovered by two American groups led by Ambros and Ruvkun, which post-transcriptionally regulated the expression of *lin-14* which codes for a protein implicated in the timing of the development of *Caenorhabditis elegans* larvae (Lee, Feinbaum, & Ambros, 1993; Wightman, Ha, & Ruvkun, 1993). This discovery initiated a new era of molecular biology research especially for the post-transcriptional regulation of gene expression. This first example was followed by the identification of numerous miRNAs in various animal model systems as well as in human. Some miRNAs are phylogenetically preserved across species, suggesting their importance during evolution and development (de Rie et al., 2017; Li et al., 2010). MiRNAs are short single-stranded, non-coding RNA formed of 21–23 nucleotides and are usually negative post-transcriptional regulators of gene expression downregulating or repressing the protein level of targeted transcripts. They target in collaboration with others proteins a specific sequence, called the miRNA response element (MRE) of a coding mRNA. The human genome contains currently approximatively 1917 annotated hairpin precursors giving rise to 2654 mature sequences. MiRNAs control around one-third of all human genes. Each miRNA can regulate hundreds of genes and each gene can be regulated by several miRNAs (Griffiths-Jones, Grocock, van Dongen, Bateman, & Enright, 2006; Ana Kozomara, Birgaoanu, & Griffiths-Jones, 2018; Kozomara & Griffiths-Jones, 2014). The work by the Functional Annotation of the Mammalian Genome Consortium showed that 50%

of miRNAs are expressed in a cell-type specific manner while the other half is ubiquitously expressed (de Rie et al., 2017). This finding underlines the importance of miRNAs in the complex regulatory network of gene expression.

Animal models deficient for key components of miRNA genesis (*e.g.* mouse *Dgcr8, Dicer1*) present embryonic lethality highlighting their essential role in mammalian development (Bernstein et al., 2003; Wang, Medvid, Melton, Jaenisch, & Blelloch, 2007). The following paragraphs review the biogenesis of miRNA from the transcription to the maturation by serial cleavage processes and their mechanism of action. Similar to the expression of conventional genes and associated transcripts, expression and function of miRNAs is in normal cells tightly regulated by epigenetic modifications (Krol, Loedige, & Filipowicz, 2010).

2.1 Biogenesis

There are two major pathways for the generation of miRNAs (Fig. 1). In the classical pathway, miRNAs are mostly transcribed in the nucleus by RNA polymerase II (RNA pol II) as part of the introns of a protein-coding or long non-coding RNA gene, or as part of longer transcripts where several miRNAs are regrouped in one transcript such as the miR-17∼92 cluster. The miRNAs can also be expressed as independent transcripts with their own promoter or rarely as part of the exons of coding transcripts sharing the same promoter with the gene (Ballarino et al., 2009; Monteys et al., 2010; Ozsolak et al., 2008; Winter, Jung, Keller, Gregory, & Diederichs, 2009). The resulting transcripts are named pri-miRNAs and fold into a characteristic small hairpin-like precursor. Subsequently, the first processing complex composed of the RNA binding protein DiGeorge syndrome critical region gene 8 (DGCR8) and the ribonuclease type III RNAse Drosha initiate the first step in the maturation process. DGCR8 recognizes particular motifs including N-6 methyladenylated GGAC motifs and cleaves the long pri-miRNAs at the base of the hairpin to give pre-miRNAs of approximately 70 nucleotides (Alarcon, Lee, Goodarzi, Halberg, & Tavazoie, 2015; Denli, Tops, Plasterk, Ketting, & Hannon, 2004). The Exportin-5-Ran-GTP complex (XPO5) transfers the pre-miRNAs from the nucleus to the cytoplasm, where they are further processed. The loop of the hairpin structure is removed by the enzyme Dicer to yield a mature, short double-stranded RNA (Okada et al., 2009; Zhang, Kolb, Jaskiewicz, Westhof, & Filipowicz, 2004). The non-classical pathway includes the same components as the classical biogenesis pathway *i.e.* DGCR8, Exportin-5, Dicer and

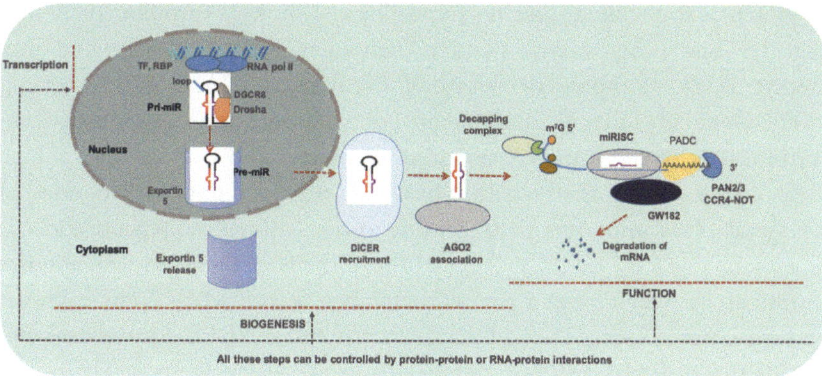

Fig. 1 ***Classical pathway of miRNA biogenesis.*** MiRNAs are transcribed by RNA polymerase II (RNA Pol II) with the help of transcription factors (TF) and RNA Binding Proteins (RBP) to give rise to a small hairpin-like precursor (pri-miR). Subsequently, DiGeorge syndrome critical region gene 8 (DGCR8) and the ribonuclease type III RNAse Drosha cleave the precursor to obtain a pre-miR. Exportin 5 transports the pre-miRNA into the cytoplasm, where it is further matured into a double stranded-RNA by Dicer. The RNA-induced silencing complex (RISC) forms a miRISC complex with one of the strands of the RNA duplex that contains AGO2 and the glycine-tryptophan 182 kDa protein (GW182). For translational inhibition and/or deadenylation and decapping, GW182 recruits the poly(A)-deadenylase complex (PADC) constituted of PAN2/3 and CCR4-NOT. These proteins induce the release of polyA binding proteins from the tail leading to deadenylation and potentially subsequent RNA degradation after removal of the m^7G cap by the action of the decapping complex. All these steps can be inhibited or modulated by various proteins through protein-protein or RNA-protein interactions.

Argonaute 2 protein (AGO2). The latter is implicated in the mechanism of action of miRNAs and will be explained below. The main difference between the pathways resides in the fact that the miRNA genesis can be DGCR8 or Dicer independent and relies instead on the RNaseH-like endonuclease activity of AGO2 (Cheloufi, Dos Santos, Chong, & Hannon, 2010; Cifuentes et al., 2010; Ruby, Jan, & Bartel, 2007; Yang et al., 2010).

2.2 Mechanism of action

The miRNA:mRNA duplex can lead to translational repression, which can be followed by mRNA deadenylation, decapping and degradation (Badis, Saveanu, Fromont-Racine, & Jacquier, 2004). One strand of the RNA duplex, the "guide strand", is incorporated into the RNA-induced silencing complex (RISC) to form the miRNA-induced silencing complex (miRISC) that contains AGO2 and the trinucleotide repeat-containing 6 protein

(TNRC6), also called glycine-tryptophan 182 kDa protein (GW182) (Fig. 1). The other strand, named "passenger strand", is recruited into another RISC complex or degraded (Krol et al., 2010; Winter et al., 2009). Generally, the retained strand contains one or several mismatches in the center of the miRNA response element (MRE) on mRNA which favors partial pairing with the miRNA seed sequence (6–8 nucleotides) at the 5′end. This imperfect pairing leads to translational repression and can be accompanied by mRNA degradation. On the other hand, the absence of mismatches resulting in a perfect match between the MRE and miRNA seed sequence activates the endonuclease activity of AGO2 for the degradation of the miRNA::mRNA duplex (Jo et al., 2015; Jonas & Izaurralde, 2015). The MRE base pairing with the miRNA seed region is generally located in the 3′UTR of a mRNA transcript, where the binding is most efficient due to the absence of translation (Gu, Jin, Zhang, Sarnow, & Kay, 2009). In mammals, repression of translation is the predominant effect compared to mRNA cleavage (Djuranovic, Nahvi, & Green, 2012; Guo, Ingolia, Weissman, & Bartel, 2010). mRNA degradation occurs through the catalytic activity of AGO2 when it is not contained within the miRISC complex, a process that is predominantly found in plants for defense against viruses (Molnar, Melnyk, & Baulcombe, 2011) and has also been shown in experiments with synthetic small interfering RNAs (siRNAs). For translational inhibition and/or deadenylation and decapping, GW182 plays a crucial role in recruiting the poly(A)-deadenylase complex (PADC) constituted of PAN2/3 and CCR4-NOT. PADCs induces the release of polyA binding proteins (PABP) from the tail of the mRNA leading to mRNA recircularization and facilitates translational repression and deadenylation. Finally, the mRNA degradation by the exoribonuclease XRN1 occurs after the removal of the m^7G cap of the target mRNA by the decapping complex (Behm-Ansmant et al., 2006; Braun et al., 2012; Christie, Boland, Huntzinger, Weichenrieder, & Izaurralde, 2013; Hafner, Ascano, & Tuschl, 2011; Jonas & Izaurralde, 2015; Zekri, Kuzuoglu-Ozturk, & Izaurralde, 2013).

While generally associated with repression, miRNAs can also contribute to translational activation in order to regulate/restore cellular homeostasis as *e.g.* has been shown in experiments on human cell lines (THP-1, HEK293) under starving conditions (Vasudevan & Steitz, 2007). In this context, miRNAs interact with their targets in a RISC complex composed of AGO2 and Fragile X mental retardation syndrome-related protein 1 (FXR1) instead of GW182 (Truesdell et al., 2012). The existence of alternative pathways for miRNAs maturation and action indicates that miRNAs represent a dynamic regulatory system aimed at resolving pathological or stress situations and buffer gene expression to facilitate physiological processes. Others mechanisms regulate miRNAs at the level of their transcription, their biogenesis (processing steps) and their suppressive role on transcripts.

2.3 Regulation of miRNA expression and function

The regulatory mechanisms of miRNA expression and function constitute an important part in determining cell fate decision during development, physiological and pathological states. Only a few miRNAs are expressed during all stages of cellular differentiation and maturation. For example, expression of mature and processed let-7 is blocked in embryonic cells, but there is a major increase in expression during differentiation in all cells (Newman, Thomson, & Hammond, 2008; Rybak et al., 2008; Thomson et al., 2006; Viswanathan, Daley, & Gregory, 2008). The regulation of miRNA gene expression is controlled at different levels: transcription, biogenesis *i.e.* processing by Drosha, Dicer and at the functional level through the formation of proper RISC complexes. Regulation is dependent on cell type and developmental state and can be mediated through protein-protein or protein-RNA interactions with RNA binding proteins (RBP).

Similar to protein-coding genes, miRNA promoters contain CpG islands, TATA box sequences and their transcription is performed by RNA polymerase II. Their expression is controlled by factors associated with the polymerase, transcription factors and by epigenetic regulators including chromatin modifiers and DNA methyltransferases/demethylases (Cai, Hagedorn, & Cullen, 2004; Lee et al., 2004). A prominent example for a transcription factor involved in miRNA expression is MYC, which promotes tumorigenesis by activating the miRNA cluster 17~92 as well as miR-9 involved in different cancers (*e.g.* lymphoma and neuroblastoma) and inhibiting tumor suppressor genes such as p53 important for cell cycle arrest and apoptosis (Han, Witmer, Casey, Valle, & Sukumar, 2007; He et al., 2007; Ma et al., 2010). While being regulated by epigenetic modifications, miRs also regulate proteins implicated in the epigenetic process (DNA and histone methylation, acetylation). The interference of miRNAs with the process of translation makes them important modifiers of the epigenome (Feng, Zhang, Wu, & Hu, 2011; Iorio, Piovan, & Croce, 2010; Sato, Tsuchiya, Meltzer, & Shimizu, 2011).

Furthermore, based on their ability of base-pairing with mRNA they can control their own transcription by modulating factors involved in the transcription by single or double regulatory feedback loops. For example, the transcription factor PITX3 and miR-133b are involved in an unilateral negative feedback loop during dopaminergic neuron differentiation (Kim et al., 2007). In allergies, miR-384 is involved in a negative feedback loop with histone deacetylase-3 especially to regulate the inflammatory reaction (Eom et al., 2014b).

More detailed information on regulatory loops of miRNAs implicated in the regulation of immune cell populations relevant for allergy will be given in a later paragraph.

Most mRNA have multiple MREs and are recognized by a number of miRNA::RISC complexes, which is beneficial for increasing miRNA activity. This phenomenon is called miRNA synergism and might be exploited as a therapeutic strategy (Yao et al., 2014). A low expression level or activity of a specific miRNA can be overcome through the synergistic action of multiple miRISC complexes on an individual mRNA target and as consequence increase their retention times and the effect of inhibition (Denzler et al., 2016; Elkayam et al., 2017). This property has been well described during physiological process such as neurogenesis, pregnancy, and embryonic stem cell pluripotency (Pons-Espinal et al., 2017; Sahu & Mallick, 2018) and reveals the need for the upregulation of a set of miRNAs for therapeutic action.

The modulation of several proteins or RBPs involved in the processing of pri-miRNAs to mature transcripts can influence the regulation of miR genesis and function. For example, p53 in collaboration with Drosha induces increased processing of pri-miRNAs to pre-miRNAs during DNA damage (Suzuki et al., 2009). The activity of miRNAs can also be modulated through the regulation of the expression of the miRISC complex elements GW182 and AGO2. Post-transcriptional modifications of the miRNA sequence can also block its synthesis and maturation, like the uridylation by the terminal uridylyltransferase 4 (TUT4) activated by LIN28 on let-7 precursors.

3. Technologies for the analysis of miRNAs: benefits and drawbacks

Progress in understanding miRNA function during normal development and altered expression patterns in various diseases have been made possible by advances in the technologies for the mapping, and characterization of miRNAs as well as increasingly sophisticated functional tools to understand their regulation. The study of miRNAs, includes their identification followed by assessing their function via gain and loss of function assays. Understanding their mechanism of regulation through the identification of their targets is also an essential part. For the establishment of a miRNA expression profile, in many cases a comparative setting with at least two different phenotypes (patients/controls) in commonly used, but of course not mandatory. For the comprehensive profiling of the miR-nome at high-throughput, Next Generation Sequencing (NGS) using a large variety of library preparation kits is performed on several sequencing platforms (Creighton, Reid, & Gunaratne, 2009; Heinicke et al., 2019; Pritchard, Cheng, & Tewari, 2012; Wright et al., 2019). qPCR and the more recently devised Nanostring technology, which is based on a single molecule hybridization, are well suited for analyzing pre-defined miRNA signatures and allow the potential rapid assessment of miRNA expression levels required for clinical routine use (Leshkowitz, Horn-Saban, Parmet, & Feldmesser, 2013; Mestdagh et al., 2014; Seumois et al., 2012).

A spectrum of methods also exists to analyze the function of miRs. These include miRNA-deficient embryonic stem cells and mice models (Park et al., 2012; Prosser, Koike-Yusa, Cooper, Law, & Bradley, 2011), the production of transgenic mice (overexpression of specific miRs), the use of transfectable synthetic antagonists (antagomiRs) or agonists (miR mimics) or viral vectors (Brown & Naldini, 2009; Elmen et al., 2008; Krutzfeldt et al., 2005). In the case of mice deficient for a specific miRNA, the redundancy of miRNA action on the targeted mRNA requires the complementary reciprocal experiment *i.e.* the use of transgenic mice overexpressing the same miRNA to validate the data.

The molecular identification of miRNA targets improves the understanding of the mechanism of action of a specific miRNA which is important, but technically challenging. Several technologies have been developed to address this question including high-throughput sequencing of RNA isolated by crosslinking immunoprecipitation (HITS-CLIP), photoactivatable RNA ribonucleoside-enhanced crosslinking and immunoprecipitation (PAR-CLIP) (Chi, Zang, Mele, & Darnell, 2009; Hafner et al., 2010). It is also possible to compare genome-wide transcriptional profiles of cells lacking and/or overexpressing a specific miRNA and controls (Baek et al., 2008; Thomas, Lieberman, & Lal, 2010). This approach, however, also yields a large number of indirect changes. Finally, and most often used, are bioinformatic target prediction tools based on sequence alignment. Several databases for miRNA annotation and their bioinformatically predicted or experimentally validated targets are publicly available including TargetScan, miRanda and PicTar (Friedman, Farh, Burge, & Bartel, 2009; Hsu et al., 2011; Kozomara & Griffiths-Jones, 2011). Tools such as miRWALK include several of these databases allowing to increase stringency by selecting targets concurrently identified by several algorithms (Sticht, De La Torre, Parveen, & Gretz, 2018).

These different approaches come with their benefits and inconveniences. All target predictions algorithms are based on the short seed region of 6–8 nucleotides pairing with a given miRNA response element (MRE, see above) and the imperfect Watson-Crick complementarity between MRE::miRNA increases the number of false positives (Kim et al., 2009). Almost all algorithms limit the analysis to the 3′UTR sequence and more specifically to conserved sequences between species with the inclusion of validated miRNA target databases in machine learning algorithms (Ekimler & Sahin, 2014) in order to limit the target numbers. Better mathematical and experimental models have to be developed to understand how this complex regulatory network functions and can have consequences at the organism level. The existence of several isoforms for the same gene and the multi-level mechanism involved in the targeting process renders target prediction a complex task (Rajewsky, 2006). Therefore, new molecular approaches have been devised based on the immunoprecipitation of miRISC complexes with an AGO2 antibody followed by the sequencing of the RNA targets by NGS profiling or microarrays (HITS-CLIP, PAR-CLIP), but these are also technically challenging (Lal et al., 2011; Orom, Nielsen, & Lund, 2008). The presence of off-targets is very frequent and requires the validation of each target by functional assays and rescue experiments. As for the mRNA-based sequencing strategies, the absence of distinction between direct targets and indirect effects requires further experiments with reporter gene vector assays and genetic rescue experiments. New methods such as single molecule imaging will be useful for tracking miRNA action in the living cell in the future for a better understanding of its dynamic regulation (Wilbertz et al., 2019).

In summary, all available data point to the fact that the transcription, the sophisticated system of biogenesis, and the function of miRNAs are under the control of a very dynamic system. The regulation of gene expression is dependent on the base pairing between the miRNA seed sequence and mRNA, which is influenced by many factors including RBPs, epigenetic regulation of proteins involved in miRNA synthesis as well as the degree of complementarity between MREs and miRNAs. This binding is also dependent on the cellular context, the subcellular localization or presence in extracellular fluids or exosomes, on the levels of miRNAs and the presence of their target, as well as alternative splicing and polyadenylation of the 3′UTR which alter identity and/or concentration of available MREs (Blazie et al., 2017; Bottini et al., 2017; Nam et al., 2014). MiRNAs increase the complexity of the regulation of protein expression. Network motifs such as negative or positive feedback or feedforward loops are induced during physiological and pathological contexts (Lai, Wolkenhauer, & Vera, 2016). In the following section, we will describe the miRNA expression profiles during immune cell development and in allergic diseases with a description of the most common regulatory loops.

4. Physiological and pathological roles of miRNAs

While their own generation is regulated by a number of proteins and mechanisms, miRNAs fine-tune gene expression by a complex regulatory network influenced by various factors including time, space, concentration and cell type that can have important effects on cell fate decisions and function.

In order to respond properly to a homeostatic condition or pathological state, the immune system has to adapt by producing appropriate cell subtypes (myeloid or lymphoid) with specific activities. The development and function of differentiated immune cells is therefore tightly controlled by several factors: extracellular sensors such as receptors, epigenetic modifications (DNA methylation, histone acetylation and methylation, microRNAs …), transcriptomic processes, factors implicated in splicing, transcription and post-transcriptional modification and long non-coding RNAs (lncRNA) (Cabezas-Wallscheid et al., 2014; Orkin & Zon, 2008). A balance exists between self-renewal and differentiation processes (Wilson et al., 2008; Wilson, Laurenti, & Trumpp, 2009). The importance of miRNAs in the development, maturation and function of immune cells has been demonstrated in loss of function experiments for different genes coding for elements involved in miRNA biogenesis (*Dicer, Drosha, DGCR8* …). An aberrant regulation of miRNAs expression and function has been shown to lead to inappropriate immune responses and the development of chronic diseases (Pauley, Cha, & Chan, 2009; Yu Li & Kowdley, 2012).

The complex interaction between particular well-known miRNAs and immune cell development and function is detailed in the following paragraphs with a special attention to effector T helper $CD4^+$ cells (Th) cell subsets involved in allergic diseases (Baumjohann & Ansel, 2013). Moreover, the review of the currently available literature identifies some recurrent regulation patterns by which miRNAs can operate to induce cellular context–dependent differentiation (Nam et al., 2014; Podshivalova & Salomon, 2013) and function of immune cells under physiological and disease conditions.

4.1 MiRNAs related to the immune system

The best-described miRNAs with prominent roles in the immune system are miR-21, miR-146a/b and miR-155. As mentioned before, the same miRNA can have different and even opposing effects and target different genes depending on the cellular context. In fact, miR-21 and miR-146a/b have been found to work as activators of aberrant proliferation in the context of cancer, but also to act as negative regulators of inflammation in autoimmune and chronic inflammatory diseases, while miR-155 is generally considered pro-inflammatory (Alivernini et al., 2017; Foster et al., 2013; Testa, Pelosi, Castelli, & Labbaye, 2017).

MiR-146a-deficient mice develop autoimmune diseases due to overexpression of the Signal transducer and activator of transcription 1 (STAT-1) (Boldin et al., 2011; Lu et al., 2010; Zhao et al., 2011). On the other hand, overexpression of miR-146a decreased NFκb-induced inflammation and prevented the development of the disease in a mouse model for atherosclerosis (Li, Ching, Luk, & Raffai, 2015). However, as described in more details below, the miR-146 family can also have a pro-inflammatory role. Overexpression of NFκB was also observed in a miR-21-deficient mouse model of periodontitis (Zhou et al., 2018). Mouse models with miR-155 knock-out (KO) mimicking autoimmune disease and *in vitro* models showed a defective production of Th1 and Th17 cells, downregulation of CCL2 and TNFα as well as reduced numbers of proinflammatory macrophage demonstrating the pro-inflammatory role of miR-155 in this context (Nazari-Jahantigh et al., 2012; Yao et al., 2012; Zhang et al., 2014).

These four miRNAs are induced by proinflammatory signals (IL-6, TNFα, IL-1β, LPS, IFNγ) that are recognized by specific receptors (TLR2/4, TNFαR, IFNγR ...), which in turn induce the NFκB and Janus Kinase (JAK)/STAT signaling pathways. NFκB signaling promotes the expression of miR-146a while JAK/STAT signaling promotes the expression of miR-146b, whereas both pathways lead to increased expression of miR-21 and miR-155. Some validated targets of miR-146a and miR-146b include the 3′UTRs of the *TNF receptor associated factor 6 (TRAF6),*

IL-1 receptor-associated kinase 1 (*IRAK1*) and *Caspase recruitment domain-containing protein 10* (*CARD10*), which are part of the upstream NFκB signaling. On the other hand, the miR-21 regulates upstream intermediates of both JAK/STAT and NFκB signaling including *IL-12p35* and *Programmed Cell Death 4* (*PDCD4*). By targeting *IL-12p35*, miR-21 acts as a positive regulator for the induction of the Th2 population and as negative regulator for Th1 cells. In an ovalbumin (OVA)-inducible allergy model miR-21 deficient mice show increased levels of the key functional cytokine for Th1 cells, Ifnγ, and a reduction of lung eosinophilia and Il-4 production in the bronchoalveolar lavage fluid (BALF) (Lu et al., 2011). In a mouse model of inflammation, miR-155 targets the *Suppressor of cytokine signaling 1* (*Socs1*), a negative regulator of the inflammatory NF-κB and JAK-STAT signaling pathways (Nazari-Jahantigh et al., 2012; Pathak et al., 2015). In LPS-activated human monocytes, miR-155, miR-146a and miR-146b are induced resulting in lower *PDCD4* expression due to increased expression of miR-21 expression (Curtale et al., 2013). The relationship is further supported by a KO mouse model for *Pdcd4*, which shows protection from LPS-induced cell death. Furthermore, macrophages in which miR-21 has been deleted present increased PDCD4 expression (Sheedy et al., 2010).

4.2 MiRNAs in the differentiation of effector Th cell subsets

Despite their small proportion among lymphocytes, Th subsets have been clearly identified as key players in several diseases including inflammatory autoimmune diseases, allergies and contribute to cancer exacerbation. Th subtypes are induced in the periphery following activation of naive $CD4^+$ T cells by antigen presenting cells (APCs). The definition and functional abilities of different Th cells depend on the nature of the antigen and the cytokine microenvironment generated by the APCs (O'Shea & Paul, 2010; Zhu, Yamane, & Paul, 2010). The expression of cytokines is not sufficient to distinguish Th subsets as some of them are essential for more than one subset of T helper cells. The expression of a specific pattern of cytokines and transcription factors is required for lineage decisions. Th1 cells regulate immune responses against intracellular pathogens, and viruses (Brown et al., 2015; North & Jung, 2004) and dysfunctional induction of Th1 cells is involved in the pathogenesis of chronic inflammatory and autoimmune diseases (Dardalhon, Korn, Kuchroo, & Anderson, 2008; Leung et al., 2010). Their induction requires the presence of IL-12, which will induce the expression of the transcription factor T-BET (TBX21) via the Signal transducer and activator of transcription 4 (STAT-4). Th1 cells can also induced by STAT-1 trough the IFNγ effector cytokine itself by an autocrine loop (Afkarian et al., 2002). Th2 cells confer protection against extracellular pathogens such as parasites (helminths) and aberrant Th2 response is involved in allergic reactions (Locksley, 2010; Taylor, van der Werf, & Maizels, 2012). Their polarization requires the production of IL-4 in the microenvironment and their function is characterized by the expression of the transcription factor GATA3, induced by STAT-6, and the pro-Th2 cytokines: IL-4, IL-5, and IL-13, which will recruit myeloid and non-hematopoietic cells to the site of inflammation (Ansel, Djuretic, Tanasa, & Rao, 2006). Th1 and Th2 cells are inversely coregulated. When cell fate decision balance the expression toward a Th1 profile, its transcription factor T-bet will suppress Th2-related gene expression and vice versa for Th2 (Ouyang et al., 1998).

Regulatory T cells (Tregs) are responsible for host self-tolerance maintenance. There are two types of Treg cells: natural Treg (nTreg) cells, which develop directly in the thymus and peripheral antigen-induced Treg (pTreg) cells, which differentiate from naïve $CD4^+$ T cells and complete nTreg activity in the periphery. The main difference between the two populations of Tregs is that pTregs are going to acquire a regulatory phenotype in response to pathogens/inflammation (Josefowicz, Lu, & Rudensky, 2012; Josefowicz & Rudensky, 2009; Sakaguchi, Yamaguchi, Nomura, & Ono, 2008). They control inappropriate effector T cell functions such as an aberrant inflammatory response and prevent autoimmune and chronic inflammatory diseases by inhibiting inflammatory cytokines and effector Th cells functions to enable specific immunosuppressive activities including the release of anti-inflammatory cytokines or the competitive consumption of IL-2. Abnormal activities and levels of Tregs are found in autoimmune disease (deficiency of Tregs) or cancer, where increased levels of Tregs impede an appropriate immune response against the tumor cells. Th17 cells as the fourth main population, are important for protection against intracellular and extracellular bacteria and fungi (Li, Wei, et al., 2018; Ma et al., 2008; Patel & Kuchroo, 2015; Ye et al., 2001). They are involved in the pathophysiology of several chronic inflammatory and autoimmune diseases and also in some endotypes of asthma (Chakir et al., 2003; Cua et al., 2003; Tesmer, Lundy, Sarkar, & Fox, 2008).

In vitro induced Tregs (iTregs) require TGFβ and IL-2 for their differentiation, while IL-6, TGFβ, IL-21, IL-23 and IL-1β are required for the differentiation of Th17 cells (Bettelli et al., 2006; Chen et al., 2003; Chung et al., 2009; Zhou et al., 2007). iTregs are identified by the constitutive expression of CD25 (one of the subunits of the IL-2 receptor), the transcription factor FOXP3, which is STAT-5 dependent and induced by the effector cytokines IL-10, TGFβ and IL-2. Th17 are characterized by the transcription factor Retinoic acid receptor-related orphan receptor-γt (RORγT), STAT-3-dependent and generate *in vitro* IL-17A/F and IL-22. Additional Th subsets have been discovered including T follicular helper (TFH) cells, which are activated by IL-21 and IL-6 and show STAT3 mediated signaling partly resembling Th17 cells. Th9 and Th22 cells secrete IL-9 and IL-22 respectively, and share some similarities with Th2 and Th17 cells (Akdis, Palomares, van de Veen, van Splunter, & Akdis, 2012; Duhen, Geiger, Jarrossay, Lanzavecchia, & Sallusto, 2009). Th cells are very heterogeneous and even after they are committed to a specific Th phenotype, they are prone to plasticity in response to various stimuli in the microenvironment (cytokines). Memory iTreg are able to convert into cells with a Th1, Th2 or a Th17-like phenotype in the presence of the associated skewing cytokines, but also the differentiated memory Th can have this property of functional plasticity (Hegazy et al., 2010; Komatsu et al., 2014; Lee et al., 2009; Levine et al., 2017; Panzer et al., 2012; Ueno, Ghosh, Hung, Li, & Jijon, 2015; Yu, Sharma, Edwards, Feigenbaum, & Zhu, 2015). Conversion of Tregs into a Th2-like phenotype impairs oral tolerance and promotes food allergies in a mouse model (Noval Rivas et al., 2015). On the other hand, memory Th2 converted into Tregs show decreased Gata-3 and Irf4 expression and produce little Th2-associated cytokines (Il-4, Il-5, Il-13). Their increased Foxp3 expression correlated with their suppressive activity and transfer of the converted Tregs into a mouse model of asthma improved airway hyperactivity and eosinophilic inflammation (Kim et al., 2010).

The importance of miRNAs for the regulation of Th lineage decision and the maintenance of the Th phenotype has been demonstrated by murine miRNA profiling (181 miRNAs) during *in vitro* Th1 and Th2- skewed culture conditions where several miRNAs were found downregulated (Henriksson et al., 2017; Monticelli et al., 2005). Moreover, the importance of miRNA for the induction and function of the different Th cell populations is emphasized by the fact that $CD4^+$ T cell or Th lineage specific deletion of either *Dicer, Drosha, Dgcr8* or *Ago2* in mouse leads to a defective proliferation and survival of activated $CD4^+$ T cells, a decrease in the number of nTregs and abnormalities in the differentiation of Th cell populations.

Naive Th cells from mice deficient for all miRNAs have decreased levels of the hallmark effector cytokine Tgfβ and the transcription factor forkhead box P3 (Foxp3) which correlates with the suppressive activity of Tregs over conventional $CD4^+$ T cell ($CD4^+CD25^-$). However, under Th1 or Th2 promoting conditions an increase of the cytokine signature characteristic for these subsets was observed (Chong, Rasmussen, Rudensky, & Littman, 2008; Muljo et al., 2005; Tian et al., 2012). These data indicate that miRNA generation is essential for the proliferation of activated $CD4^+$ T cells and induction of nTregs *in vivo* and for iTreg differentiation, but probably not for the other effector Th subsets. However, under non-polarizing conditions a shift of the balance between Th1 and Th2 phenotype compared to wild type mice is observed, which suggests the importance of adequate miRNA expression for the maintenance of naive T cells in order to restrain Th lineage polarization except for Tregs (Bronevetsky et al., 2013; Chong et al., 2008; Muljo et al., 2005). A barrier must be overcome to trigger a particular differentiation toward a specific phenotype that is dependent on the cellular context. Moreover, miRNA assisted T cell polarization is relatively short-lived as the half-life of AGO2 is limited following naïve $CD4^+$ T cell activation (~2 h) due to the ubiquitylation of AGO2 contributing to the renewal of the miRNA repertoire and restrain it to baseline levels (Bronevetsky et al., 2013). Several specific miRNAs are important for the induction of different Th subsets and for the maintenance of the naive state.

MiR-125b is upregulated in naïve $CD4^+$ T cells compared to memory T cells and regulates a number of genes implicated in Th differentiation including *IFNγ* and *IL-10RA* (Rossi et al., 2011). This miRNA might therefore be critical for the regulation network underlying Th subset polarization.

Besides the four miRNAs already mentioned above (miR-21, miR-146a/b and miR-155) involved in the immune system function, the miR-17~92 cluster is an inducer of the Th1 population because its deletion reduced the number of $CD4^+$ T cells producing IFNγ while the opposite effect was observed when the cluster was overexpressed under Th1-polarizing culture conditions (Jiang et al., 2011; Xiao et al., 2008).

In addition to miR-21, whose role as inducer of the Th2 lineage has been described above, miR-24, miR-27, miR-128 and miR-19 have been identified as positive or negative regulators for the expression of Th2 cell-associated cytokines and transcription factors (IL-5, IL-4, GATA3) trough *in vitro* or *ex vivo* gain or loss function assays in mouse models or human samples (Cho et al., 2017; Martinez-Nunez et al., 2014; Pua et al., 2016; Singh et al., 2017). In multiple sclerosis, miR-27 and miR-128 were upregulated in $CD4^+$ T cells from peripheral blood compared to healthy individuals and associated with increased Th1 response and decreased number of Th2 cells (Guerau-de-Arellano et al., 2011). The overexpression of these miRNAs in activated $CD4^+$ T cells decreased characteristics of the Th2 phenotype (IL-4, IL-5 and GATA3) and increased IFNγ production in mouse models and human samples. These effects were reversed when expression of these miRNAs was inhibited in $CD4^+$ T cells from patients yielding characteristics in line with an increased Th2 phenotype. In a mouse model of asthma, the inhibition of miR-24 and miR-27 expression enhanced allergic inflammation and Th2 response (Pua et al., 2016). Furthermore, the differentiation of naïve $CD4^+CD45RA^+$ T cells, from mice deficient for these two miRNAs under Th2 promoting conditions, emphasized the Th2 phenotype and provided further evidence for their specific role in regulating Th2 differentiation. On the other hand, miR-19 gain and loss of function assays for miR-19 in both human and mouse naïve $CD4^+$ T cells under Th2 polarizing culture conditions, showed an enhanced Th2 cell-specific cytokine signature (IL-4, IL-5 IL-13) and -associated inflammation in a mouse model of asthma deficient for miR-19 (Simpson et al., 2014; Singh et al., 2017).

In addition to miR-155, which favors the Th17 lineage through NF-κB signaling, miR-326, targeting *ETS1* a negative regulator of this subpopulation, as well as miR-301a, blocking the STAT-3 inhibitor *PIAS3*, have been shown in *in vitro* and *in vivo* models (knock-out mice for the respective miRs) and patients with inflammatory diseases (multiple sclerosis, encephalomyelitis) to lead to an increased IL-17 production (Mycko et al., 2012; O'Connell, Taganov, Boldin, Cheng, & Baltimore, 2007). In contrast, miR-10a limits Th17 polarization by targeting negative regulators of T-Bet (Takahashi et al., 2012).

MiRNAs play also an important role in the induction and immunosuppressive role of Tregs. The conditional deletion of *Dicer* in the $CD4^+$ T cell lineage or $FOXP3^+$ Treg cells in mice leads to reduced number of pTregs and nTregs and inflammatory disorders such as colitis and inflammation in the colon, lung and liver as well as autoimmune inflammatory disorders (Chong et al., 2008; Cobb et al., 2006; Liston, Lu, O'Carroll, Tarakhovsky, & Rudensky, 2008; Tian et al., 2012; Zhou et al., 2008). In addition, their implication in iTreg differentiation has been demonstrated *in vitro* experiments where $CD4^+$ T cells deficient for *Dicer* and *DROSHA* were cultured in the presence of TGFβ and Retinoic acid (RA) resulting in reduced numbers of FOXP3$^+$ cells (Chong et al., 2008; Cobb et al., 2006). Moreover, screening for miRNAs overexpressed in naïve T cells favoring Treg differentiation revealed a network of miRNAs that drive iTreg differentiation: miR-10b, miR-99a, miR-130a, miR-146b, miR-150 and miR-320 (Warth et al., 2015). Inhibition of miR-150 expression by an antagomiR demonstrated its cooperative action with miR-99a to induce iTreg via the repression of the mammalian target of rapamycin (mTor), a known inhibitor of iTreg differentiation. Others miRNAs reduce iTreg differentiation such as miR-31 acting directly on *FOXP3* expression (Rouas et al., 2009; Zhang et al., 2015). The miR 17~92 cluster restricts Treg cell development and function by targeting the Treg polarizing cytokine TGFβ receptor *in vitro* and in experimental models of autoimmune encephalomyelitis (EAE) by regulating the immunosuppressive cytokine *IL-10* (Jeker & Bluestone, 2013; Jiang et al., 2011). Similarly, miR-15a/16 target *CTLA-4* expression and act directly on its suppressive function, but also on *FOXP3* expression (Liu et al., 2014). Among the well-known immunomiRs, miR-155 surprisingly facilitates Treg homeostasis by targeting *SOSC1*, a negative regulator of IL-2 and STAT-5 signaling, which is crucial for Treg induction as demonstrated by deletion of miR-155 in murine Treg cells. Furthermore, it is a direct target of FOXP3.

As systemic miR-155 KO mouse model as well as Treg mouse model showed reduced Il-2 production correlating with a decreased number of nTreg and iTreg cells even under Treg polarizing culture conditions, but their suppressive activity was maintained (Kohlhaas et al., 2009; Lu, Thai, et al., 2009). MiR-146a induced Tregs by targeting STAT-1 and its deletion in Tregs led to an increase in Treg cells numbers, but Tregs showed a less suppressive function in a mouse model similar to the effects observed in Foxp3$^+$ cells missing elements of the miRISC complex (Liston et al., 2008; Lu et al., 2010).

The induction of differentiated Th subsets (iTreg, Th2, Th17) is a key component in the pathogenesis of allergic inflammation. The three most upregulated miRNAs in allergic inflammation and linked to Th cell polarization are miR-21, miR-146a and miR-155. As described before, miR-21 favors Th2 polarization and miR-155, mostly implicated in inflammatory process induces Th17 and Th1 phenotypes (Lu et al., 2011; Lu, Munitz, & Rothenberg, 2009; Sawant, Wu, Kaplan, & Dent, 2013). Besides these three miRNAs, other miRNAs that have previously been described as limiting Th2 cells and those involved in the induction of iTreg could potentially be exploited in antagomiR or miR mimic-based approaches for treatment in allergic inflammatory diseases, including asthma.

5. MiRNAs in allergic diseases

Allergies are a very heterogeneous group of diseases where the symptoms are largely dependent on the entry route of the respective allergen. There are two distinct phases, the asymptomatic sensitization phase that leads to the induction and proliferation of the T helper Type 2 (Th2) cells from naive $CD4^+$ $CD45RA^+$ T lymphocytes, which are activated following the recognition of the allergen presented by APCs. The cytokine microenvironment induced by APCs, with notably high levels of IL-4, orientates naive $CD4^+$ T cells differentiation toward Th2-cytokine secreting cells including the interleukins IL-4, IL-5, and IL-13 and promotes the production of allergen-specific immunoglobulin (Ig)Es and the recruitment of inflammatory cells (mast cells, basophils, end stage cells: eosinophils, neutrophils, Th17) (Broide, 2001; Stone et al., 2010). Atopy denotes an individual and/or familial predisposition to develop allergies with enhanced IgE-mediated responses to common allergens (Oettgen & Geha, 1999). Recently, it has been demonstrated that type 2 innate lymphoid cells (ILC2) have similar functions as Th2 cells albeit with some differences. They are activated by epithelial cell—derived cytokines IL-33, IL-25, and thymic stromal lymphopoietin (TSLP), do not need antigen presentation and produce Th2 cytokines including IL-9 (Klose & Artis, 2016). ILC2s are important in the early phase of an allergic reaction and collaborate in the induction of allergic inflammation and eosinophilia at least in asthma by favoring Th2 priming via IL-4 production and activate the migration of DC into the draining lymph node (Barlow et al., 2012; Walker & McKenzie, 2013). The second phase is the effector phase, which depends on further recognition of the allergen by IgEs binding on target cells, such as mast cells, resulting in the release of mediators, such as histamine and leukotrienes, which are responsible for allergic symptoms (Fig. 2).

In the next paragraphs, we will review the dysregulation of miRNA expression in allergic diseases. The implication of miRNAs in the aberrant development and function of effector Th cells has been demonstrated in a number of studies in the context of allergic diseases (asthma, rhinitis, atopic dermatitis and food allergy). miR transported by exosomes are also emerging to have an important role during allergic reactions. Other epigenetic

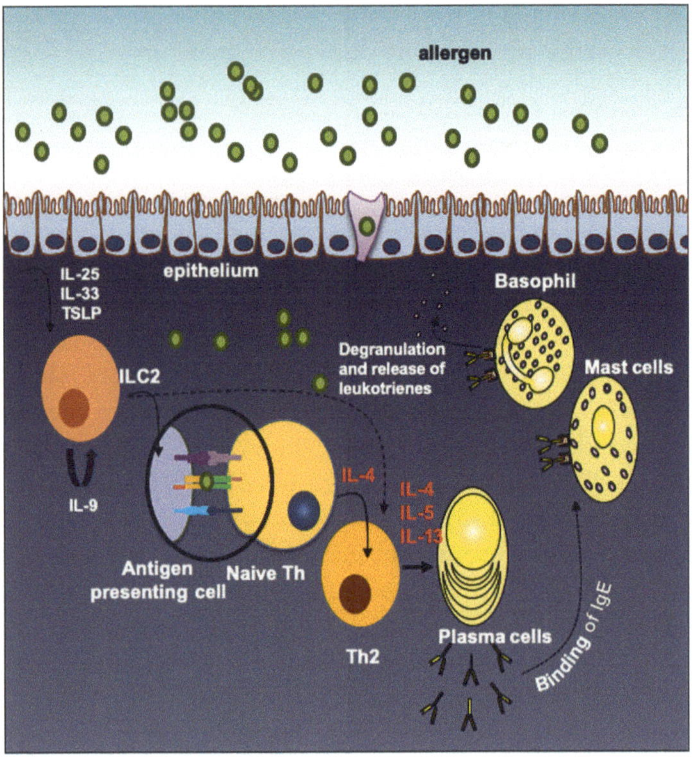

Fig. 2 Pathophysiology of an allergic reponse. Recognition and induction of antigen presenting cells by allergens followed by Th2 differentiation from naive Th cells in a microenvironment containing IL-4. ILC2 cells collaborate with Th2 T cells and DCs to favor allergic inflammation. IgE antibodies are produced by plasma cells following the first encounter with the allergen. The subsequent encounter with the allergen triggers IgE binding on mastocytes and basophils and initiates the release of histamine and leukotrienes, which are responsible for the different symptoms depending on the entry route of the allergen.

modifications such as DNA methylation have also been implicated in the regulation of the allergic inflammatory response and recently extensively reviewed (Potaczek et al., 2017). For example, increased demethylation of a conserved regulatory region of *Foxp3* and gain of methylation at the promoter of *Gata3* specifically in Th2 cells have been found to be induced by epicutaneous immunotherapy and associated with the unresponsiveness to peanut protein sensitization (Mondoulet et al., 2019).

In healthy individuals, pTregs, the negative regulator of other Th subsets, dominate Th2 response in order to prevent the development of allergic diseases. Therefore the induction of this immunosuppressive cell subtype may be useful. Several miRNAs have been described as inducing this subpopulation as miR-100, miR-99a, and miR-10b, which favor pTregs at the expense of Th17 differentiation (Warth et al., 2015). MiR-21, an inducer of Th2 response, also promotes regulatory B cells that - in turn - induce Treg cells (Amu et al., 2010). Overexpression of miR-23b in bone marrow-derived dendritic cells (BMDCs), pulsed with OVA, increased their IL-10 production and induced $CD4^+$ $Foxp3^+$ cells through Notch1 and NF-κB signaling (Zheng et al., 2012). The literature on the function and targets of miRNAs function is often contradictory as effects might differ depending on the cellular-context as described earlier.

5.1 Asthma

Asthma is a heterogeneous multifactorial group of diseases (Agache, Akdis, Jutel, & Virchow, 2012; Lotvall et al., 2011) characterized by inflammatory chronic disorders of the airways that results in airflow obstruction and bronchial hyperresponsiveness. Type 2 cytokines induce the release of cytokines and chemokines that favor the recruitment of immune cells into the lungs such as granulocytes, neutrophils and eosinophils that are responsible for the pro-inflammatory effects and the development of allergic asthma. Given the diversity and complexity of the disease, a classification of patients into subgroups is required. For this purpose, endotypes have been developed to differentiate patients by their clinical symptoms and physiological features. There are also other important endotypes of asthma, which are not related to allergic inflammation (Romanet-Manent et al., 2002), and which will be in this review not further addressed. The well-known four miRNAs miR-155, miR-146a/b and miR-21 are important in allergic diseases given their role in the regulation of the immune system. As available literature is very exhaustive, we only review the most pertinent miRNAs in this chapter. We will focus on these 4 miRNAs implicated in the physiopathology of inflammatory diseases as well as those in the regulation of allergic diseases with a special focus on those involved in type 2 response and respiratory allergic diseases.

5.1.1 MiR-146a/b, miR-21 and miR-155

As mentioned before, cell type and pathological context are important parameters for the regulation and function of miRNAs. Indeed, the same deletion of miR-155 in wild type (wt) mice induces lungs inflammation similar to what is observed in asthma, with a predominant Th2 response (Il-4,Il-5,Il-13), whereas in a mouse model of allergic airway inflammation all the asthma-like effects are reduced upon miR-155 deletion (Malmhall et al., 2014; Okoye et al., 2014; Rodriguez et al., 2007). These results highlight the importance of miR-155 in the prevention of airway inflammatory diseases, but also its requirement for disease maintenance in the context of a pre-existing allergic disease. Furthermore, miR-146a and miR-146b are reduced in $CD4^+$ and $CD8^+$ T-cells of severe asthmatic patients, who are under treatment with glucocorticoids, but not in patients with mild or moderate asthma (Tsitsiou et al., 2012). After an allergic challenge in experimental model of asthma, miR-146a/b, miR-150 and miR-181a were upregulated in splenic $CD4^+$ T-cells (Feng, Shi, Qiu, & Peng, 2012). The observed upregulation of miR-146a raises questions on its supposed anti-inflammatory effect as it seems at least in this context to be implicated in the pathogenesis of the disease. Indeed, miR-146a favors IgE class switch in B cells by upregulating 14-3-3σ expression (Li et al., 2017). Another experimental asthma mouse model showed that increased expression of miR-21, miR-155 and miR-18a correlated with Th2 cytokine levels (Xiao, Noel, Perveen, & Penn, 2016). In human, miR-21 and miR-146a have been found increased in asthmatic patients and specifically in the eosinophilic endotype of asthma (Hammad Mahmoud Hammad et al., 2018; Sawant et al., 2015). However, in severe asthma patients miR-146a is reduced in circulating $CD4^+$ T cells (Tsitsiou et al., 2012). Deregulated expression of several miRNAs in cells from asthma-related tissues (alveolar epithelial cells (AEC), endothelial cells (EC), fibroblasts and bronchial smooth muscle cells (BSMC)) has important impact on the inflammation orchestrated by Th2 cells and thus on the pathogenesis of the disease (Jardim, Dailey, Silbajoris, & Diaz-Sanchez, 2012; Williams et al., 2009). Many deregulated miRNAs target gene products related to the inflammation including *IL8, COX2, IL6*, which we will detail in the following paragraph.

5.1.2 Additional miRNAs in asthma

Experimental mouse models for respiratory allergies (Ovalbumin (OVA), House Dust Mite (HDM) have demonstrated that VEGF plays an important role in Th2 inflammation in the lungs by inhibiting miR-1 expression in the lung endothelium. The intranasal administration of miR-1 was shown to inhibit VEGF-induced Th2 inflammation and notably IL-13 production in the different mouse models. Furthermore, the use of synthetic inhibitors for miR-1 restored the effects of VEGF on lung inflammation (Takyar et al., 2013).

In a mouse model of allergic bronchial asthma, the GTPase Rhoa was upregulated and miR-133a decreased, which was accompanied by hyperresponsiveness of bronchial smooth muscle cells. The causal relationship was confirmed using the antagomiR-mediated inhibition of miR-133a in BSMC cells (Chiba & Misawa, 2010; Chiba, Tanabe, Goto, Sakai, & Misawa, 2009).

Several miRNAs have been identified to play a role in the process of allergic inflammation in different mouse models of asthma (HDM, OVA): miR-221, miR-106a, miR-145 and miR-126. Their inhibition using synthetic antagomirs had anti-inflammatory effects and reduced neutrophilia in the lungs (Collison, Herbert, et al., 2011; Collison, Mattes, Plank, & Foster, 2011; Lacedonia, Palladino, Foschino-Barbaro, Scioscia, & Carpagnano, 2017; Mattes, Collison, Plank, Phipps, & Foster, 2009; Qin et al., 2012; Sharma et al., 2012). However, no change was observed in a model of chronic HDM exposure. MiR-126 was only expressed during the first two weeks following the allergen challenge, which might explain the absence of an effect of the miR-126 antagomiR on the response in the regulation of chronic asthma (Collison, Herbert, et al., 2011; Mattes et al., 2009).

The first discovered miRNA, let-7, directly targets the 3′UTR of *IL13* (Collison, Mattes, et al., 2011; Polikepahad et al., 2010). The intranasal administration of let-7 in HDM-exposed mice reduced the allergic inflammation (Kumar et al., 2011). However, in another study using OVA-exposed mice inhibition of let-7 resolved the Th2-induced inflammation (Il-4, Il-5, Il-13) demonstrating an inflammatory effect of let-7 (Polikepahad et al., 2010) (Table 1).

The control of the inflammatory process and Th2 response through anti-inflammatory cytokines as IL-10 is crucial. MiR-106a directly targets *IL10* and was found to be overexpressed in an OVA-induced allergic asthma model (Sharma et al., 2009, 2012). Its abrogation reduced the chronic inflammatory process and had thus an effect on the severity of the disease. Two interesting miRNAs, miR-24 and mIR-27 encoded in two clusters mirC11 and mirC22 have been identified, which are directly implicated in the Th2 effector function by targeting genes implicated in Th2 cell polarization. Indeed, their inhibition in a mouse model of asthma resulted in increased inflammation accompanied by increased Il-4 and Il-13 production and Th2-associated immunopathology during airway allergic reaction including increased lung pathology, increased IgE and more severe eosinophilia. Their target genes include genes acting upstream Il-4, among which

Table 1 MiRNAs involved in allergic diseases.

miRNA	Disease	Main effects	Targets	Reference
miR-146	Asthma	Overexpressed in splenic $CD4^+$ T lymphocytes	*IRAK1, TRAF6, CARD10, indirectly inhibits COX-2 and IL-1β*	(Feng et al., 2012)
		Reduced in $CD8^+$ and $CD4^+$ T cells in patients treated with inhaled corticosteroid		(Tsitsiou et al., 2012)
		Reduced cytokine induced apoptosis in human bronchial epithelial cells		(Liu et al., 2009; Comer et al., 2014)
		Increased in human asthmatic SMCs		
	AD	Increased expression in the skin and keratinocytes; reduces chronic inflammation	*IRAK1, CARD10, CCL5*	(Rebane et al., 2014)
miR-145	Asthma	(intranasal administration of antagomiRs) Silencing of miR-145 reduced allergic inflammation in an HDM induced mouse model of asthma. Effects similar to steroid treatment	?	(Collison, Mattes, et al., 2011)
miR-155	Asthma	Th2 induction in an OVA-induced mouse model of asthma	*PU.1*	(Malmhall et al., 2014)
		MiR-155 deficient mice develop asthma with increased Th2 cytokines	*c-Maf*	(Rodriguez et al., 2007)
		Increased in BSMCs of asthma patients		(Comer et al., 2015)
		MiR-155 deficient mice in an OVA and HDM asthma model reduce Th2 priming by DCs	*ectonucleoside triphosphate diphosphohydrolases (ENTPD)*	(Zech et al., 2015)

	AD	Its overexpression in naïve Th cells increases proliferation of activated $CD4^+$ T cells and chronic skin inflammation	*CTLA-4*	(Sonkoly et al., 2010)
miR-21	Asthma	Increased in an OVA mouse model for asthma, induced IL-13 in a transgenic mouse model with allergic inflammation	*IL-12p-35*	(Lu, Muntiz, et al., 2009)
		Increased NF-kB activation, and Th1 cytokines and reduced eosinophilia in an OVA-induced mouse model of asthma in miR-21- deficient mice		(Lu et al., 2011)
		AntagomiR-21 reversed the effect by restoring the sensitivity to steroids and suppressed airway hyperresponsiveness in mouse model of asthma insensitive to steroids		(Kim et al., 2017)
		Upregulated in the serum of asthmatic children, potential use as biomarker		(Sawant et al., 2015)
	EoE	Upregulated in the serum of EoE children, potential use as biomarker		(Sawant et al., 2015)
	ACD	Upregulated in patients and mouse model		(Vennegaard et al., 2012)
	AR	Correlation between upregulation of IgE and downregulation of miR-21 in cord blood in AR		(Chen et al., 2010)

(Continued)

Table 1 MiRNAs involved in allergic diseases.—cont'd

miRNA	Disease	Main effects	Targets	Reference
miR-124	Asthma	Induction of type 2 macrophages		(Veremeyko, Siddiqui, Sotnikov, Yung, & Ponomarev, 2013)
miR-106a	Asthma	Regulates inflammation in a mouse model of asthma	*IL-10*	(Sharma et al., 2012)
miR-375	EoE Asthma	Positive regulator of TSLP and thus ILC2s and Th2 cells supported by loss-of- function assay in human *in vitro* model with human bronchial epithelial cells		(Bleck et al., 2013)
miR-125b	CRS with nasal polyps	Upregulation of miR-125 increased production of IFNα/β in human and mouse nasal and bronchial epithelial cells	*4E-BP1*	(Zhang et al., 2012)
miR-150	ACD	Induces tolerance in a mouse model for induced ACD		(Bryniarski et al., 2013)
miR-126	Asthma	Regulates allergic inflammation and eosinophilia in a mouse model of asthma treated with antagomiR-126		(Collison, Herbert et al., 2011)
		Reduces Th2 response and airway hyper responsiveness in mice treated with antagomir against miR-126	*SPI1*	(Mattes et al., 2009)

let-7	Asthma	Inhibits inflammation, AHR in an OVA-induced mouse model of asthma treated with miR mimicslet-7	*IL-13*	(Kumar et al., 2011)
		Inhibits allergic inflammation and IL-13 in mice intravenous administrated antagomiRs for let-7		(Polikepahad et al., 2010)
miR-221	Asthma	Increases expression in asthmatic patients and reduction of allergic inflammation in mice model of asthma where miR-221 is blocked	*Spred-2*	(Qin et al., 2012)
miR-133a	Asthma	Upregulation of *RhoA* in bronchial smooth muscle cells	*RhoA*	(Chiba et al., 2009; Chiba & Misawa, 2010)
miR-24, miR-27	Mouse model of asthma	Limits IL-4 production and Th2 cells; reduces inflammation	*IL-4* Genes upstream IL-4 *Gata3* *Ikzf1*	(Pua et al., 2016)
miR-19	Mouse model of asthma	Promotes Th2 cytokines, Increases allergic inflammation	Signal inhibitors *Pten, Socs1, and Tnfaip3*	(Simpson et al., 2014; Singh et al., 2017).

AD, Atopic dermatitis; *ACD*, allergic contact dermatitis; *CRS*, chronic rhinosinusitis; *AR*, allergic rhinitis; *EoE*, eosinophilic esophagitis.

are two major transcription factors implicated in Th2 differentiation, *Gata3* and *Ikzf* (Cho et al., 2016; Pua et al., 2016). Interestingly, the overexpression of miR-24 in murine T cell reduced differentiation of naïve $CD4^+$ T cells toward the Th2 subtype and promoted other Th cell lineages Th1, Th7 and iTreg (Cho et al., 2016). Further *in vitro* and *in vivo* experiments modulating miRNA expression in both human and mice showed that miR-19 had a positive role in the induction of Th2 by directly targeting inhibitors as *Sosc1*, *Phosphatase and tensin homolog* (*Pten*) and *Tumor necrosis factor alpha-induced protein 3* (*Tnfaip3*) (Simpson et al., 2014; Singh et al., 2017). In miR-19 KO mice, a strong reduction of lung inflammation was observed correlating with a reduction in the level of hallmark cytokines of Th2 and innate lymphoid cells type 2 (ILC2) (Table 1). A miRNA profiling study identified that miR-1248 directly targets IL-5 (Panganiban et al., 2012).

MiRNA signatures implicated in the physiopathology of asthma have been identified in different animal and human studies. Signatures for the classification of allergic diseases are described in more detail in the section on the potential use of miRNAs for diagnostics in allergic diseases. Inhibition of the expression of a set of miRNAs (miR-18a, miR-27a, miR-128a and miR-155) increased inflammatory cytokines (IL-6 and IL-8) in bronchial epithelial cells, which was directly related to the exacerbation of asthma symptoms (Martinez-Nunez et al., 2014). The identification of a miRNA signature in sputum supernatant (miR-223-3p, miR-142-3p, and miR-629-3p) allowed to differentiate severe asthma patients from healthy subjects. The transfection of mimics-mediated activation of miR-629-3p in human bronchial epithelial cells (BEC) increased the production of the proinflammatory cytokine IL-8 implicated in the recruitment and activation of neutrophils and whose level correlate directly with the neutrophilia observed in the sputum of patients with severe asthma (Maes et al., 2016). *In vitro* experiments, using serum from asthmatic children, identified miR-296, miR-16-5p and miR-30d-5p as potential markers for asthma severity because of their respective role in airway hyperresponsiveness (AHR) or on airway smooth muscle maturation, which was demonstrated by *in vitro* assays overexpressing the respective miRNAs (Davis et al., 2017).

In HDM-exposed mice, an increase of the number of exosomes containing miRNAs that target the Th2-signature cytokines Il-5r and Il-13 (miR-346, miR-1827 and miR-574-5p) has been shown in bronchoalveolar lavage fluid (BALF), and their functional relevance was demonstrated after treatment with an inhibitor of exosome release (Gon et al., 2017; Kulshreshtha, Ahmad, Agrawal, & Ghosh, 2013). In asthma, there is a group of patients who are corticosteroid-treatment resistant, which has been correlated with an increased level of miR-21 expression (Elbehidy et al., 2016). These different miRNA signature might in the future be exploited as biomarkers in allergic diseases.

5.2 MiRNAs in other allergies

As described above, miR-21 is an inducer of the Th2 cell population. In line with this function, an upregulation of miR-21 has been observed in different types of allergies including patients with eosinophilic esophagitis (EoE) or allergic contact dermatitis (ACD) and the respective mouse models (Lu, Sherrill, et al., 2012; Vennegaard et al., 2012). Of note is a study that could establish a correlation between low expression of miR-21 and an increase in antenatal cord blood (CB) IgE levels, which was replicated with miR-21 expression data from children with allergic rhinitis (AR). This was the first time that downregulation of miR-21 could be associated with the antenatal production of IgE and the development of AR (Chen et al., 2010). On the other hand, the expression of miR-146a and miR-155 was found increased in human keratinocytes and immune cells from skin lesion from patients with atopic dermatitis (AD). A functional assay deleting miR-146a in a mouse model of AD and in keratinocytes of AD patients promoted chronic inflammation (IFNγ, Il-1β ...), and the overexpression of miR-146a in keratinocytes of the patients reduced IFNγ-activated cells. On the other hand, miR-155 overexpression in Th cells downregulated CTLA-4, an inhibitor of T cell activation accompanied with increased proliferation of T cells (Rebane et al., 2014; Sonkoly et al., 2010). In these studies, miR-146a had the regular anti-inflammatory role while miR-155 sustained inflammation.

In eosinophilic esophagitis (EoE) patients, IL-13 an important cytokine for Th2 effector function, was found to downregulate miR-375 expression, which was negatively correlated with allergic inflammation in the esophagus. The stimulation of esophageal epithelial cells, which are important for airway hyper-reactivity, with IL-13 led to downregulation of the miRNA (Lu, Lim, et al., 2012). miR-375 targets Thymic Stromal Lymphopoietin (TSLP), an important cytokine that favors Th2 induction, and provides evidence for a Th2 amplification loop (Biton et al., 2011; Bleck et al., 2013). In severe eosinophilic chronic rhinosinusitis (CRS) patients, miR-125b was upregulated and inhibited eukaryotic translation initiation factor 4E (EIF4E)-binding protein 1 (4E-BP1) leading to an increase of the inflammatory cytokines IFNα/β in airway epithelial cells, which was confirmed by functional gain- and loss-of-function experiments for miR-125b in human and mouse *in vitro* models (Zhang et al., 2012). In a mouse model of contact sensitivity, a proxy for allergic AD, mice were immunized to haptens conjugated with self-antigens. $CD8^+$ T cells induced tolerance by producing antigen-specific antibody-coated nanovesicles containing miR-150. Deletion of miR-150 led to the absence of the suppressive nanovesicles and the loss of tolerance. This study highlights the importance of miRNA function in regulating other Th cells and allergic inflammation to induce tolerance at distance due to exosome-like vesicles (Bryniarski et al., 2013).

In a mouse model of AR as well as in human patients, let-7 was downregulated in nasal epithelial cells that showed increased levels of inflammatory factors such as IgE, histamines, and TNFα compared to controls. The inflammation was alleviated when let-7 was overexpressed in human nasal epithelial cells, by targeting *SOCS4*, which promoted an enhanced inflammatory response in human epithelial cells (Li, Zhang, et al., 2018). Another study using the same AR mouse model showed that miR-133b was significantly decreased in nasal mucosa compared to control mice correlating with increased IgE production and symptom score (sneezing, rubbing). Intranasal administration of a miR-133b mimic inhibited allergen-specific IgE, reduced infiltration of inflammatory cells into the nasal mucosa and led to decreased Th2 cytokines production in the mucosa but also circulating immune cells probably through targeting *Nlpr3* (Xiao, Jiang, Hu, & Li, 2017).

The literature about miRNAs role in food allergy is limited, a mouse model of food allergen-related intestinal inflammation showed that the cluster miR-17~92 and especially miR-19a was upregulated in B cells from intestinal mucosa compared to control mice, and its deletion prevented IL-4 to negatively regulate IL-10 production by B cells (Liu et al., 2016). Indeed, IL-10-producing B cells have an immunosuppressive role on inflammatory cytokines (IL-4, IFNγ, TNFα) produced by Th cells and macrophages. Furthermore, PBMCs from children aged 4—18 months with cow's milk allergy (CMA) presented with a low miR-193a-5p expression compared to healthy children, which correlated inversely with the upregulation of IL-4. Indeed, inhibition of miR-193a-5p in $CD4^+$ T cells led to the upregulation of IL-4 mRNA and its protein. Furthermore, in children, where CMA resolved, an upregulation of miR-193-5p was identified. Thus, miR-193-5p appears as a negative post-transcriptional regulator of IL-4 expression and might be implicated in the regulation of IgE-mediated CMA response (D'Argenio et al., 2018).

6. General network patterns of miRNA function

MiRNAs use various forms of feedback loops to fine tune gene expression programs in order to regulate protein expression levels and thus their function. A positive feedback loop involves three components: a transcription factor (A) which induces the expression of a miRNA (B), which in turn inhibits a gene (C) that represses A and thereby favors expression of A. This type of regulation may promote the commitment to a particular cell fate by enhancing a particular gene. A negative feedback loop involves a gene A and a miRNA B. The induction of B inhibits A, which has a major role in the expression of the miRNA. In this kind of regulation, levels of target gene are maintained in a defined window of expression to ensure physiological processes, to limit transcriptional bursts and aberrant protein expression or cell fate commitment and to contribute to the complex regulation of gene networks (Blevins et al., 2015; Mukherji et al., 2011; Na & Kim, 2013; Schmiedel et al., 2015; Strovas, Rosenberg, Kuypers, Muscat, & Seelig, 2014). Indeed, below a certain level of the target the repression by the miRNA is maintained, but above the threshold repression decreases. This threshold can be assimilated to a checkpoint for cell fate decision (Mukherji et al., 2011).

Finally, the coherent and incoherent feed-forward loops are formed by three components A, B and C. The coherent one is composed of 3 positive regulators whereas in the incoherent loop B is a negative regulator. In the coherent A can regulate directly C or indirectly through B. This structure delays the induction or repression of the target gene and can thus lead to abnormal cell fate or function. In the incoherent system, the negative regulator B delays the effect of expression of C (Tsang, Zhu, & van Oudenaarden, 2007).

These mechanisms appear simple, but the network of miRNA action on targets has to be considered. Indeed, one miRNA can affect many targets, but also one target can be regulated by several miRNAs and thus bring an important complexity to the system. We will describe these different patterns of regulation during hematopoiesis and the induction of innate or adaptive cells, with a special attention on Th cell subsets in a pathological context particularly during allergic diseases (Table 2).

Table 2 Network motifs in immune cells.

Cell type	Elements implicated in the regulation miR/targets	Effects
HSC	phosphoinositide 3-kinase (PI3K)— Protein kinase B (AKT)—glycogen synthase kinase 3β (GSK3β) pathway-miR126	MiR-126 regulates hematopoiesis by regulating different components of PI3K-AkT-GSK3β signaling and sets a threshold to complete HSC activation (Lechman et al., 2012)
	LIN28B-let-7-HMGBA2	Coherent feed-forward loop important for self-renewal of fetal HSCs and supressed in adult hematopoiesis (Copley et al., 2013; Yuan, Nguyen, Liu, Kanellopoulou, & Muljo, 2012)
	miR-146a -*TRAF6-IRAK1-NFkB* miR-155-AP1-AKT1	MiR-146a is induced by NF-κB and acts trough a negative feedback loop on *TRAF6, IRAK1*
		MiR-155 acts in an incoherent feed-forward loop where it can be both activated by NFκB and AP1 or inhibited by IRAK1 in order to control the duration and intensity of NFκB signaling and thus regulate production of inflammatory cytokines and prevent aberrant inflammatory response (Zhao et al., 2013, 2014)
Macrophages	TLR4- miR-146a-*TRAF6-IRAK1-NFkB* miR-155-AP-1-AKT1	As for HSCs miR-146a participates in a negative feedback loop and miR-155 in an incoherent feed-forward loop to regulate inflammation (O'Connell et al., 2007; Testa et al., 2017)

Granulocytes	C/EBP (inducer) − miR-223 - E2F1 (repressor)-miR-223 miR-223 −NFI-A	Coherent feed-forward loop between E2F1 inhibition, which downregulates miR-223 and NFI-A and lead to mutual repression miR-223-NFI-A These proteins are implicated in the regulation of miR-223 expression, which in turn regulates MEF2C expression important for granulocyte differentiation (Fazi et al., 2005; Johnnidis et al., 2008; Vian et al., 2014).
B cells	miR-21- *BLIMP1-STAT3*	Incoherent feed-forward loop implicated in plasma cell differentiation to prevent aberrant proliferation.(Barnes, Stephenson, Cocco, Tooze, & Doody, 2012)
	miR-146a-NFκB	Same as for macrophages (Testa et al., 2017; Zheng et al., 2018)
	miR-155-AID-miR-181b	Incoherent feed-forward loop to maintain Activation-induced deaminase (AID) at a level to prevent uncontrolled expression and hematopoietic cancer; AID is also under negative feedback loop via miR-181b (de Yebenes et al., 2008; Teng et al., 2008)
T cells	miR-155 - miR-326 − *ETS*1	ETS1 is a negative regulator of Th17; miR-326 and miR-155 target *ETS1*; Positive feedback loop in Th17 cell differentiation in chronic inflammatory disease (Du et al., 2009; Hu et al., 2013)
	miR-21-SMAD7-TGFβ	Positive feedback loop drives Th17 differentiation by targeting SMAD7 a negative regulator of TGFβ (Murugaiyan et al., 2015)
	miR-301-STAT3-IL-23-*PIAS3*	Positive feedback loop favoring STAT-3 signaling through IL-23 by inhibiting signaling of the negative regulator *PIAS3*; important for Th17 differentiation (Mycko et al., 2012)

(Continued)

Table 2 Network motifs in immune cells.—cont'd

Cell type	Elements implicated in the regulation miR/targets	Effects
	miR-20b –miR-155/miR-21/miR-301/STAT-3-RORγT	Negative feedback loop: miR-155, miR-21, miR-301 induce Th17 differentiation, which can be prevented by miR-20b, which in turn inhibits Th17 differentiation by targeting STAT-3 and RORγT (Zhu et al., 2014)
	MiR-181a-*CD69* MiR-17, miR-20a-*CD69*	Incoherent feed-forward loop: negative regulator of TCR signaling by these 3 miRNAs, which target CD69 to limit T-cell activation (Blevins et al., 2015)
	IL2R –STAT-5- miR-182-*FOXO1* (negative regulator of cell proliferation in resting Th cells)	Positive feedback loop that results in proliferation of activated T-cells by IL2R/STAT5- induced miR182 which targets FOXO1 (Stittrich et al., 2010)
	MiR-146a/miR-155 loop with NFκB	Same as macrophages, importance of negative feedback loop for controlling TCR activating NFkB. T cells lacking miR-146a in mice lead to chronic inflammatory autoimmune responses (Yang et al., 2012) and promotes increases TFh differentiation when miR-146a is absent in TCD4+ cells (Hu et al., 2014)
	Cluster miR-17∼92: miR-19b-Pten-miR-17 TGFBRII/CREBI	Coherent feed-forward loop promotes Th1 response and inhibits iTreg differentiation by miR-19b which targets *Pten* and miR-17 for TGFBRII/CREBI (Jiang et al., 2011)

HSC, Hematopoietic stem cells.

6.1 Patterns of regulation during hematopoiesis

MiRNAs involved in a feedback loops during hematopoietic stem cell (HSC) development and function are listed in Table 2. Genes regulated by miRNAs can be implicated in proliferation, survival, signaling pathways and cytokine production (King & Goodell, 2011; Zhao et al., 2014).

6.1.1 Patterns of regulation in innate immune cell development

The well-known miRNAs miR-146a and miR-155 are implicated in macrophage development and function. They act in opposite ways: miR-146a negatively regulates TLR signaling, whereas miR-155 amplifies inflammatory signals as described above (McCoy et al., 2010; O'Connell, Chaudhuri, Rao, & Baltimore, 2009; Taganov, Boldin, Chang, & Baltimore, 2006). The absence of the negative feedback loop caused by deletion of miR-146a in mice leads to myeloproliferation and tumorigenesis (Zhao et al., 2011). The expression levels of miR-17-5p, miR-20a and miR-106a, members of the miR-17~92 cluster and related miR-106~92 families, decrease during the differentiation of monocytes. This leads to derepression of their target the *runt-related transcription factor 1* (*RUNX1*; also known as *AML1*) (Fontana et al., 2007). Increased expression of RUNX1 promotes monocyte differentiation. In addition, RUNX1 can also bind the miR-17~92 and miR-106a~92 promoter regions and repress their expression, suggesting that miR-17~92 and RUNX1 are involved in a mutual negative feedback loop (Rosa et al., 2007).

The patterns of regulation that modulate miRNAs in other innate cells are summarized in Table 2.

6.1.2 Patterns of regulation in adaptive cell development

6.1.2.1 MiRNAs in B cell development

The deletion of Dicer prevents B cell development and induces a block between the pro- and pre-B stage (Koralov et al., 2008), suggesting a crucial role for miRNAs in B cell differentiation. The miR17~92 cluster favors B cell development through downregulation of the proapoptotic protein BIM (Ventura et al., 2008). In contrast to macrophages, miR-146a and 155 participate in another form of network regulation in B cell development and regulation of antibody production by preventing notably autoantibody synthesis (Table 2).

6.1.2.2 MiRNAs in T cell development

The importance of miRNA regulation in the production of mature adult T cells and $CD4^+$ and $CD8^+$ T cell lineage are demonstrated by the decrease of these cells in Dicer deficient-mice (Cobb et al., 2005; Muljo et al., 2005; Rupp et al., 2014; Wu et al., 2007).

Coordinated regulation of multiple targets by many miRNAs changes the developmental pathway of T cells (see sections above and Table 2). The function and maturation of T cells is mostly regulated by miRNAs targeting genes involved in T cell signaling and proliferation. In T cells, miR-146a induces immunosuppressive function of Tregs by targeting *STAT1* a transcription factor specific for Th1 cells.

In summary, these examples reveal complex, diverse and cell-type specific regulation patterns of miRNAs in the different types of immune cells.

6.2 Regulatory network motifs and miRNAs in allergic diseases

Some miRNAs have been identified to participate in regulatory circuits and been implicated in the pathophysiology of allergic diseases. The suppressor of cytokine signaling 1 (SOCS1) has an important role in the regulation of allergic inflammation. It has contradictory effects on the regulation of allergic inflammation. In a mouse model of asthma and in mice sensitized with Dinotrophenyl (DNP)-specific mouse IgE, *SOCS1* had mainly anti-inflammatory effects when it is targeted by miR-155, and limits the promotion of allergic inflammation and Th2 polarization when targeted by miR-19a (Qayum et al., 2016; Simpson et al., 2014). In mast cells from DNP-IgE sensitized mice, a negative feedback loop has been identified involving IL-10, which induces miR-155 expression that represses *SOCS1* promoting IgE mediated anaphylaxis; these effects are reversed when miR-155 is deleted (Qayum et al., 2016). But in mouse mast cells and cellular models activated by DNP-IgE, SOCS1 had a positive role in allergic inflammation by inhibiting the secretion of TGFβ. MiR-122 and miR-384 were identified as negative regulators of SOCS1 and involved in a negative feedback in the regulation of allergic inflammation (Noh et al., 2017). In a mouse model of allergic asthma, miR-396 participated in a negative feedback loop to control the production of Il-5/Il-13 T cells via Stat-6 and Rorγt and showed in functional experiments to downregulate Th2 and Th17 response (Shen, Yu, & Wang, 2018). Other examples of miRNAs that participate in the regulation of allergic inflammation include miR-181a/miR-218 that form a negative feedback loop with transglutaminase II (TGMII) (Eom et al., 2014a) and miR-26 with COX2 (Eom et al., 2014a; Kwon et al., 2015).

All these data demonstrate patterns of regulation by which miRNAs favor pathological states through inappropriate immune cell fate induction and function and suggest promising therapeutic targets. However, this research is still at an early stage and it will need further characterization and modeling of the gene regulatory networks that lead to abnormal immune responses; identification of key nodes can be used as targets for the development of anti-allergic drugs using miRNA analogs or inhibitors.

7. MiRNAs as tool for diagnostics and treatment

The dysregulation of miRNA expression has been causally linked to several human diseases such as chronic inflammatory diseases and allergies including asthma. Their important role for pathogenesis is supported through functional experiments in animal models or *in vitro* as described in detail above. miRNAs are more stable than mRNAs due to their resistance to RNAse activity, which is directly related to their encapsulation in vesicles such as exosomes with or without high density lipoprotein (HDL) or their association with AGO2 for vesicle free miRNAs (Arroyo et al., 2011; Turchinovich & Burwinkel, 2012; Vickers, Palmisano, Shoucri, Shamburek, & Remaley, 2011). This interesting feature make them good candidates for therapeutic development. In the field of cancer, the use of miRNA signatures is already performed by companies as Laboratory Developed Tests (LDT) for diagnosis or prediction of the response to therapy or to a specific drug (Dai et al., 2011; Du et al., 2013; Hayes, Peruzzi, & Lawler, 2014; Li, Lu, Chen, Lu, & Xie, 2013; Vecchione et al., 2013). All these different characteristics could in principle also be exploited for therapeutic purposes and diagnosis in the field of allergy, where they play an important part in the pathophysiology. In this section, we are going to present the therapeutic potential of miRNAs with a special focus on allergy.

7.1 Diagnostic value of circulating miRNAs

MiRNAs are formed in cells and are secreted to modulate neighboring cells or circulate in body fluids (serum, plasma, seminal fluid, synovial, saliva ...) to act at long distance like hormones. They thereby have an important role in the intercellular communication. Extracellular miRNAs can affect target cells and mediate post-transcriptional repression of gene expression (Hergenreider et al., 2012; Mittelbrunn et al., 2011; Pegtel et al., 2010). Moreover, in contrast to mRNAs they are very resistant to drastic conditions such as prolonged exposure to elevated temperature, thawing/freezing cycle and variable pH (Chen et al., 2008; Mitchell et al., 2008). In the case of pulmonary chronic inflammatory diseases, saliva can be easily collected and provide an overview of the cytokine and protein content of the airway microenvironment. For example, the expression level of miRNAs miR-223-3p and miR-142-3p in sputum from severe asthmatic patients is positively correlated with neutrophil numbers and levels of the pro-inflammatory cytokine IL1β (Maes et al., 2016). MiRNAs concentrated in exosomes are present in urine and saliva that can be collected without invasive procedures, which makes them an attractive solution for diagnosis (Azmi, Bao, & Sarkar, 2013; Chen, Hogan, & Ward, 2013; Gallo, Tandon, Alevizos, & Illei, 2012). They are also easily detected in bronchoalveolar lavage fluid (BALF), nasal lavage fluid (NALF), exhaled breath condensate (EBC). However, this type of sampling requires to be performed by well-experienced medical staff in contrast to sputum, serum and plasma. Nonetheless, BALF samples allow the distinction between controls and asthmatic patients (Elbehidy et al., 2016; Levanen et al., 2013; Sinha et al., 2013). MiRNA expression profiles of exosomes from extracellular samples differ from the originating cells and indicate that they are induced and modified by a specific context (Koppers-Lalic et al., 2014). Serum and plasma, which only require a blood sampling, are the most frequently used biofluids and have been shown in numerous studies to contain allergy relevant miRNAs. Increased levels of miR-21 have been found to be inversely correlated with serum IL-12p35 levels and pulmonary function (FEV1), which allows to differentiate between healthy and asthmatic children and predict the response to corticosteroid treatment (Elbehidy et al., 2016). The same differences have been shown for miR-185-5p in peripheral eosinophils (Rodrigo-Munoz et al., 2019). Another study showed that expression of miR-21 is negatively correlated with IgE levels in serum sample from asthmatic patients (Sawant et al., 2015). In others type of allergies such as atopic dermatitis, the expression level of miR-483-5p and miR-203 in serum allowed distinction of healthy from severe allergic children (Lv et al., 2014). Profiling ten miRNAs, selected based on either previous associations with asthma or targeting asthma relevant cytokines, in serum from asthma patients and control individuals defined a signature consisting of miR-1248, miR-26a, let-7a, and let-7d (Panganiban et al., 2012). Of note, the study also identified that miR-1248 directly targets IL-5. Follow-up work showed that miR-125b, miR-16, miR-299-5p, miR-126, miR-206, and miR-133b allowed to determine whether a subject is healthy, allergic (allergic rhinitis) or asthmatic predicting the asthma status correctly for 73 out of 79 patients (92.4%; AUC = 0.97) (Panganiban et al., 2012, 2016). The identification of a miRNA signature in sputum supernatant (miR-223-3p, miR-142-3p, and miR-629-3p) allowed to differentiate severe asthma patients from healthy subjects with miR-629-3p levels correlating directly with the neutrophilia observed in the sputum of patients with severe asthma (Maes et al., 2016). A translational approach, where miRNAs found deregulated in several asthma mouse models were analyzed in human cohorts identified a potential biomarker signature in plasma: miR-21-5p, miR-15a-5p, miR-27a-3p, miR-29c-3p, miR-223-3p, miR-425-5p, and miR-342-3p (Milger et al., 2017). Furthermore, miR-296-5p, miR-16-5p and miR-30d-5p were identified as potential markers for asthma severity because of their respective role in airway hyperresponsiveness (AHR) or on airway smooth muscle maturation (Davis et al., 2017). All these results support the use of circulating miRNAs as a biomarker obtained through non-invasive methods for allergic diseases.

While traditionally regression approaches have been used for the correlation between a specific phenotype and miRNA levels to define a set of miRNAs as biomarkers (Boren et al., 2009), advances in machine-learning algorithms such as multilevel neural networks are increasingly used because of their simplicity to incorporate larger data sets (Keller et al., 2009; Lowery et al., 2009; Yan, Li, Xiong, Xu, & Zheng, 2012). Although, currently available data sets remain limited and therefore still prevent the full potential use of these approaches, first examples defining a signature of six miRNAs for the diagnosis and status of asthmatic patients illustrate their future used in the field of allergy (Panganiban et al., 2016).

As mentioned above, the encapsulation of miRNAs in exosomes increases their resistance and bestows an advantage for clinical use (Almeida, Reis, & Calin, 2011). They are secreted to be incorporated in neighboring cells or act post-transcriptionally at distance after being transported through the bloodstream. Moreover, their endocrine properties are useful in a therapeutic delivery strategy and increased levels of exosomal or free miRNAs have been associated with several diseases (Anel, Gallego-Lleyda, de Miguel, Naval, & Martinez-Lostao, 2019; Fujita, Kadota, Araya, Ochiya, & Kuwano, 2018; Mori, Ludwig, Garcia-Martin, Brandao, & Kahn, 2019; Xu et al., 2018).

7.2 Extracellular vesicles for miRNA-based therapeutics

Extracellular nanovesicles or exosomes, cup-shaped vesicles with a size of around 10–150 nm, result from a series of endocytosis steps. First, small vesicles form in the multivesicular bodies (MVbs) after fusion with the inner membranes of endosomes. These vesicles then fuse with the plasma membrane to yield exosomes, which are released in the extracellular interspace. They contain various types of molecules, proteins, lipids, bioactive agents (prostaglandins, leukotrienes) and RNAs including miRNAs (Keerthikumar et al., 2016; Record, Carayon, Poirot, & Silvente-Poirot, 2014; Subra et al., 2010). The content of the exosomes depends on the cellular state and pathological context as well as the secretion signal. Secretion signals include among others Ca2+ stimulus or IL-4-dependent exocytosis (Gümürdü et al., 2017; Yang et al., 2011). The release of exosomes that contain miRNAs affects target cell function (activation, presentation, immune suppression) according to the nature of cells. They act on neighboring cells, a feature exploitable for local administration of a therapeutic based on exosomes, but can also modulate distant cells. Exosomes can be taken up (with or without internalization at the surface) by APCs to present antigens and reinforce immune surveillance by $CD8^+$ T cells, but also directly induce T cell activation as they carry MHC molecules on their surface (Greening, Gopal, Xu, Simpson, & Chen, 2015; Robbins, Dorronsoro, & Booker, 2016). Exosomes have been associated with several physiological and pathophysiological contexts including immune regulation, inflammation, and lung diseases (asthma) (Fujita et al., 2014; Hough, Chanda, Duncan, Thannickal, & Deshane, 2017; Masyuk, Masyuk, & Larusso, 2013; Vella, Sharples, Nisbet, Cappai, & Hill, 2008). They are present in a variety of body fluids including serum, plasma, saliva, urine, cerebrospinal fluid, breast milk, colostrum, seminal fluid, ovarian liquid fluid (Admyre et al., 2007; da Silveira, Veeramachaneni, Winger, Carnevale, & Bouma, 2012; Gallo et al., 2012; Weber et al., 2010; Zhou et al., 2012). In the context of allergic diseases those in contact with respiratory mucosal surfaces as sputum, BALF and NALF are of particular interest (Kesimer et al., 2009; Lasser et al., 2016; Torregrosa Paredes et al., 2012).

Their increased levels have been associated with allergic diseases, and exosomes can both dampen or exacerbate allergic reactions, probably dependent on their origin and cargo (Canas et al., 2018; Sastre, Cañas, Rodrigo-Muñoz, & Del Pozo, 2017). MiRNA contained in vesicles permit the classification of patients and thus the disease subtypes (Francisco-Garcia et al., 2016; Suzuki et al., 2016). The miRNA content of EV enabled the diagnosis of asthma from BALF and airway epithelial cells, which share almost the same miRNA profiles (Levanen et al., 2013; Solberg et al., 2012). These findings show the potential of exosomal miRNAs for the prediction of disease state for routine use due to the non-invasive procedure of sampling. Likewise, the comparison between miRNA expression profiles in exosomes or vesicle-free from exhaled breath condensate are very similar and allow the distinction between healthy and asthmatic patients (Sinha et al., 2013). In an asthma mouse model with HDM challenge, an increased level of airway extracellular vesicles (EVs) that transported miRNA specific for Th2 cytokines was detected in BALF, while the same miRNAs were decreased in the lungs compared to the control animals (Gon et al., 2017). These effects are reversed when mice were treated with an inhibitor of exosome secretion demonstrating the selective sorting and increased release of Th2 repressive miRNAs in the exosomes.

The development of exosome-based miRNA therapies would facilitate the delivery into the organism by either a direct systemic administration to reach target immune or non-immune cells because of the recirculation between the site of inflammation and lymphoid organs or by local administration to prevent side effects. The utility of exosomes as a transporting tool for the administration of selected miRNAs has raised considerable interest and has been suggested in the fields of cancer and pulmonary diseases (Fattore et al., 2017; Li et al., 2016; Lu et al., 2016; Mohsenin, 2016; Szymczak, Wieczfinska, & Pawliczak, 2016). The easiest way is to customize pre-existing exosomes and load them with specific miRNAs or only to select those with interesting miRNAs. In the cancer field, this has been tested in clinical trials with exosomes from dendritic cells (Ohno, Drummen, & Kuroda, 2016; Pitt et al., 2016). In the field of allergy, miR-155 mimics or inhibitors were loaded into B cell derived exosomes and successfully delivered into macrophage cell lines or *in vivo* in murine hepatocyte cells with less toxicity compared to conventional transfection methods. The exosomes modified with the miRNA synthetics modulated miRNA expression in the target cells, reduced the effects of LPS-induced TNFα production and decreased *SOSC1* inhibition in an LPS-induced model mouse (Momen-Heravi, Bala, Bukong, & Szabo, 2014). Despite some promising proof-of-concept experiments, some elements need further investigation including a better characterization of the type of stimulus that modifies exosome content and whether a modification of the exosome composition from asthma patients is feasible.

7.3 First advances of miRNAs as therapeutic tools

Current examples of miRNA targeting tools in late preclinical development are only found outside the field of allergic diseases. The first miR targeted therapy in phase 2a is SPC3649 (Miravirsen), which blocks miR-122 in order to limit hepatitis C proliferation (Janssen et al., 2013). Subsequently, RG-101 (Regulus Therapeutics), a GalNAc conjugated antisense inhibitor of miR-122 has been devised and is currently tested in phase II trials for the treatment of hepatitis C. The drug induced a prolonged effect in the reduction of the infection with no adverse effects linked to the dose and no escape mutations (Bonneau, Neveu, Kostantin, Tsongalis, & De Guire, 2019). miRNA mimics are short dsRNAs that are loaded into the RISC complexes or precursors with a stem loop, which need to be processed by Dicer to function as miRNAs. Compared to naturally occurring miRs they carry minor modification on their backbone (*e.g.* phosphothioates) or are encapsulated in nanovesicles to increase their stability *in vivo*. Ongoing trials include miRNA mimics for miR-34a (MRX-34, phase I for cancer), miR-29b (remlarsen, phase II cutaneous fibrosis) and let-7a mimics are in pre-clinical development for cancer treatment (Bader, 2012; Gallant-Behm et al., 2019; Liu et al., 2014; Montgomery et al., 2014; Zhang, Liao, & Tang, 2019). An inhibitor for miR-155 is cuurently evaluated in cancer patients (cobomarsen, phase II cutaneous T-cell lymphoma) (Seto et al., 2018).

Another approach using DNA enzyme against *GATA3*, HgD40/SB010 are tested in phase II clinical trial for atopic asthmatics patients (Homburg et al., 2015; Krug et al., 2015).

Alternative strategies of the modulation of miRNA activity include the development of small molecule inhibitors (SMIRs) to target the process of miRNA biogenesis or maturation, but also specific miRNAs (Monroig Pdel, Chen, Zhang, & Calin, 2015), but their mechanism of action remains so far elusive for most examples. Genome editing approaches (Zinc Finger nuclease (ZFN), transcription activation like effector (TALE), and CRISPR/Cas9 systems) leading to permanent modifications on the genome present novel strategies, but they require more efficient delivery systems and present currently still a number of safety challenges.

Administration of miRNA modulators has been restricted to animal models in the field of allergy and asthma. Changes in hallmarks of allergic reactions have been explored in a mouse model of asthma where inhibition of miR-126 was associated with reduced inflammation and in an OVA-induced asthma model where an miR-221 antagomir reduced eosinophil infiltration in BALF (Collison, Herbert, et al., 2011; Mattes et al., 2009).

8. Perspectives and challenges

The altered expression of individual and combinations of miRNAs that are correlated with the pathophysiology of allergic inflammation might be exploited as future therapeutic strategies. In the following we propose some ideas to translate emerging data in potential therapeutic strategies. In asthma, the accessibility of the airways for miRNA modulators by intranasal inoculation makes them a good candidate in this pathological context.

Modulation of a miRNAs targeting key regulators for Th2 polarization such as *GATA3*, or Treg induction favoring *FOXP3* expression might assist in the induction of tolerance and sustained unresponsiveness and reinforce other epigenetic mechanisms such as DNA methylation changes observed in a mouse model for peanut allergy treated with epicutaneous immunotherapy (Mondoulet et al., 2019). In addition, targeting Th2-related cytokines by administration of mimics for miR-1248 (IL-5), let-7 (IL-13) or inhibition of miRNAs known to increase Th2 response i.e. miR-21 (IL-12), miR-145 (IL-5, IL-13) might represent a valuable strategy (Collison, Mattes, et al., 2011; Kumar et al., 2011; Panganiban et al., 2012). Exosomes transporting specific miRNAs do offer an excellent therapeutic strategy and packaging miRNAs in exosomes increases their stability *in vivo* (Almeida et al., 2011). The personalization and selection of already existing exosomes is currently more practical than designing novel exosomes from scratch. A number of different immune cell populations (DC, naïve $CD4^+$ T cells, Th2, B cells, eosinophils, mast cells, macrophages) participate in the pathophysiology of allergic responses and asthma and can secrete exosomes that contribute to the disease (Canas et al., 2017; Mazzeo et al., 2015; Sastre et al., 2017). Mice were found to be protected from allergic reaction after administration of exosomes from tolerized mice to an allergen (Prado et al., 2008). Based on this, selective sorting of exosomes carrying pro-Th2-related miRNAs or relevant for allergy regulation could be re-administered or the pre-existing exosomes could be engineered to target specific cells. Thus, exosome/miRNA-based vaccines could be developed. In addition, exosomes bearing antigens can activate $CD4^+$ T cells to enable the induction of tolerance to a specific allergen via iTreg induction (Thery et al., 2002). A miRNA-based vaccine such as the ones tested for viral infection might be interesting for prophylactic treatment of allergy, by increasing the expression of specific miRNA involved in tolerance induction in target cells (Waring et al., 2018; Yee & Poh, 2016). Furthermore, another important characteristic that can be explored is the plasticity of T cells because of the combined action of miRNAs which have been identified as targets implicated in the abnormal Th response or others immune cell responses (macrophages). The negative post-transcriptional regulation of gene expression by miRNAs is important for the differentiation and regulation of effector Th cells. Slight changes in the balance between the cell populations have an important impact on cell fate considering the number of discovered Th cell subsets and their potential for functional plasticity. Indeed, immune cells already engaged in a phenotype can be reversed by targeting miRNAs implicated in the plasticity, *e.g.* Th2 or Th17 (chronic phase) toward a regulatory phenotype.

Although miRNAs present attractive therapeutic targets, some drawbacks have to be considered such as their expression pattern, which is cell type and pathological context dependent, the limited number of samples analyzed so far in different animal models and patients with allergic diseases often combined with the absence of replication cohorts, as well as methodological differences in the biological material, the processes and methods. The differences between results from human clinical trials, human *in vitro* experimental models and animal models can affect the miRNA expression profiles. Despite a general high degree of conservation, interspecies differences in miRNA expression and a large, but not complete overlap of common targets has to be considered (Rossi et al., 2011).

Special attention should be paid to the selection of the appropriate cellular targets for administration and immune cells should not be the only focus. For example, miR-145 is specifically overexpressed in smooth muscle cells and airway epithelial cells, and has been found to be an attractive target in a mouse model of HDM-induced asthma as the inhibition of this miRNA led to similar effects as treatment with glucocorticoids treatment reducing airway hyperresponsiveness and recruitment of inflammatory cells including neutrophils and eosinophils (Collison, Mattes, et al., 2011).

Most studies use a pre-treatment mice model where specific genes are inactivated with antagomiRs or mimics prior to the allergen challenge, which results in important effects on Th2 response and allergic symptoms. However, when the inhibitory treatment is given when the inflammation is already established, the effects are often much less pronounced. Similarly, miRNA regulate different phases of the disease. For example, miR-126 seems to be important in the acute phase of asthma response, but not during the chronic phase (Collison, Herbert, et al., 2011; Mattes et al., 2009). The use of antagomiRs must be carefully evaluated for the appropriate application in particular stages of the disease and for a defined duration, and not for sustained use, to mimic best disease pathology. Moreover, the type of model used for the modulation of the miRNA expression level might have an important impact on the observed effect. For example, the administration of an antagomiR against miR-21 had no impact on Th2 cytokines, and eosinophilia in an experimental model of asthma; effects which were observed in miR-21$^{-/-}$ deficient mice. Similar effects induced by the different types of miR-21 inhibition were also observed in a model for cardiac diseases (Collison, Mattes, et al., 2011; Lu et al., 2011; Patrick et al., 2010; Thum et al., 2008). Overexpression of let-7, which was found reduced in asthma models, led to a decrease of IL-13 levels *in vivo* and pathological and clinical symptoms associated with allergic airway disease including infiltration of inflammatory cells and airway hyperresponsiveness in one study (Kumar et al., 2011). However, similar results were observed with let-7 antagomiRs administrated before OVA challenge in another mouse model for asthma (Polikepahad et al., 2010).

The complex interaction network of several miRNA to regulate a single process such as inflammatory response or a specific T cell polarization, will require the combined use of several antagomiR and miRNA mimics to have a considerable effect on the phenotype. miRNA synergism might in the future be exploited in the therapeutic strategy (Yao et al., 2014). It will be important to consider the nature and number of the antagomiRs/mimics, but also the number of isoforms for a given mRNA target in order to prevent the induction of compensatory mechanisms by the isoforms not targeted by the miRNAs and off-target effects.

The combined use of unbiased high-throughput technologies such as next-generation sequencing and proteomics, with bioinformatics (machine-learning algorithms) and functional experiments using miRNA mimics and antagomiR against specific mRNA will contribute identifying key nodes in pathways implicated in the pathophysiology of allergies.

In summary, miRNAs are key players in the process of allergic inflammation. Their function and targets are context, disease stage and time dependent. Many factors must be considered given the complex interplay between miRNAs that lead to lineage-specific orientation to control inflammatory response. Among the different miRNAs implicated in the anti-inflammatory process, the induction of tolerance through iTregs is probably key for the control of allergic response. Given the multifactorial nature of allergic diseases, personalized therapies are unavoidable. A better comprehension of the changes to the gene regulatory networks induced by alterations in miRNA expression patterns, but also by other epigenetic modifications such as DNA methylation and chromatin accessibility will help to identify therapeutic strategies and targets for an improved patient management.

References

Admyre, C., Johansson, S. M., Qazi, K. R., Filen, J. J., Lahesmaa, R., Norman, M., et al. (2007). Exosomes with immune modulatory features are present in human breast milk. *Journal of Immunology, 179*(3), 1969–1978. http://dx.doi.org/10.4049/jimmunol.179.3.1969.

Afkarian, M., Sedy, J. R., Yang, J., Jacobson, N. G., Cereb, N., Yang, S. Y., et al. (2002). T-bet is a STAT1-induced regulator of IL-12R expression in naive $CD4^+$ T cells. *Nature Immunology, 3*(6), 549–557. http://dx.doi.org/10.1038/ni794.

Agache, I., Akdis, C., Jutel, M., & Virchow, J. C. (2012). Untangling asthma phenotypes and endotypes. *Allergy, 67*(7), 835–846. http://dx.doi.org/10.1111/j.1398-9995.2012.02832.x.

Akdis, M., Palomares, O., van de Veen, W., van Splunter, M., & Akdis, C. A. (2012). TH17 and TH22 cells: A confusion of antimicrobial response with tissue inflammation versus protection. *The Journal of Allergy and Clinical Immunology, 129*(6), 1438–1449. http://dx.doi.org/10.1016/j.jaci.2012.05.003. quiz1450-1431.

Alarcon, C. R., Lee, H., Goodarzi, H., Halberg, N., & Tavazoie, S. F. (2015). N6-methyladenosine marks primary microRNAs for processing. *Nature, 519*(7544), 482–485. http://dx.doi.org/10.1038/nature14281.

Alduraywish, S. A., Lodge, C. J., Campbell, B., Allen, K. J., Erbas, B., Lowe, A. J., et al. (2016). The march from early life food sensitization to allergic disease: A systematic review and meta-analyses of birth cohort studies. *Allergy, 71*(1), 77–89. http://dx.doi.org/10.1111/all.12784.

Alivernini, S., Gremese, E., McSharry, C., Tolusso, B., Ferraccioli, G., McInnes, I. B., et al. (2017). MicroRNA-155-at the critical Interface of innate and adaptive immunity in arthritis. *Frontiers in Immunology, 8*, 1932. http://dx.doi.org/10.3389/fimmu.2017.01932.

Almeida, M. I., Reis, R. M., & Calin, G. A. (2011). MicroRNA history: Discovery, recent applications, and next frontiers. *Mutation Research, 717*(1–2), 1–8. http://dx.doi.org/10.1016/j.mrfmmm.2011.03.009.

Almqvist, N., Lonnqvist, A., Hultkrantz, S., Rask, C., & Telemo, E. (2008). Serum-derived exosomes from antigen-fed mice prevent allergic sensitization in a model of allergic asthma. *Immunology, 125*(1), 21–27. http://dx.doi.org/10.1111/j.1365-2567.2008.02812.x.

Amu, S., Saunders, S. P., Kronenberg, M., Mangan, N. E., Atzberger, A., & Fallon, P. G. (2010). Regulatory B cells prevent and reverse allergic airway inflammation via FoxP3-positive T regulatory cells in a murine model. *The Journal of Allergy and Clinical Immunology, 125*(5). http://dx.doi.org/10.1016/j.jaci.2010.01.018, 1114-1124.e1118.

Anel, A., Gallego-Lleyda, A., de Miguel, D., Naval, J., & Martinez-Lostao, L. (2019). Role of exosomes in the regulation of T-cell mediated immune responses and in autoimmune disease. *Cells, 8*(2). http://dx.doi.org/10.3390/cells8020154.

Ansel, K. M., Djuretic, I., Tanasa, B., & Rao, A. (2006). Regulation of Th2 differentiation and Il4 locus accessibility. *Annual Review of Immunology, 24*, 607–656. http://dx.doi.org/10.1146/annurev.immunol.23.021704.115821.

Arroyo, J. D., Chevillet, J. R., Kroh, E. M., Ruf, I. K., Pritchard, C. C., Gibson, D. F., et al. (2011). Argonaute2 complexes carry a population of circulating microRNAs independent of vesicles in human plasma. *Proceedings of the National Academy of Sciences of the United States of America, 108*(12), 5003–5008. http://dx.doi.org/10.1073/pnas.1019055108.

Azmi, A. S., Bao, B., & Sarkar, F. H. (2013). Exosomes in cancer development, metastasis, and drug resistance: A comprehensive review. *Cancer and Metastasis Reviews, 32*(3–4), 623–642. http://dx.doi.org/10.1007/s10555-013-9441-9.

Bader, A. G. (2012). MiR-34 - a microRNA replacement therapy is headed to the clinic. *Frontiers in Genetics, 3*, 120. http://dx.doi.org/10.3389/fgene.2012.00120.

Badis, G., Saveanu, C., Fromont-Racine, M., & Jacquier, A. (2004). Targeted mRNA degradation by deadenylation-independent decapping. *Molecular Cell, 15*(1), 5–15. http://dx.doi.org/10.1016/j.molcel.2004.06.028.

Baek, D., Villen, J., Shin, C., Camargo, F. D., Gygi, S. P., & Bartel, D. P. (2008). The impact of microRNAs on protein output. *Nature, 455*(7209), 64–71. http://dx.doi.org/10.1038/nature07242.

Ballarino, M., Pagano, F., Girardi, E., Morlando, M., Cacchiarelli, D., Marchioni, M., et al. (2009). Coupled RNA processing and transcription of intergenic primary microRNAs. *Molecular and Cellular Biology, 29*(20), 5632–5638. http://dx.doi.org/10.1128/mcb.00664-09.

Barlow, J. L., Bellosi, A., Hardman, C. S., Drynan, L. F., Wong, S. H., Cruickshank, J. P., et al. (2012). Innate IL-13-producing nuocytes arise during allergic lung inflammation and contribute to airways hyperreactivity. *The Journal of Allergy and Clinical Immunology, 129*(1). http://dx.doi.org/10.1016/j.jaci.2011.09.041, 191-198.e191-194.

Barnes, N. A., Stephenson, S., Cocco, M., Tooze, R. M., & Doody, G. M. (2012). BLIMP-1 and STAT3 counter regulate microRNA-21 during plasma cell differentiation. *Journal of Immunology, 189*(1), 253–260. http://dx.doi.org/10.4049/jimmunol.1101563.

Baumjohann, D., & Ansel, K. M. (2013). MicroRNA-mediated regulation of T helper cell differentiation and plasticity. *Nature Reviews Immunology, 13*(9), 666–678. http://dx.doi.org/10.1038/nri3494.

Behm-Ansmant, I., Rehwinkel, J., Doerks, T., Stark, A., Bork, P., & Izaurralde, E. (2006). mRNA degradation by miRNAs and GW182 requires both CCR4:NOT deadenylase and DCP1:DCP2 decapping complexes. *Genes & Development, 20*(14), 1885–1898. http://dx.doi.org/10.1101/gad.1424106.

Bernstein, E., Kim, S. Y., Carmell, M. A., Murchison, E. P., Alcorn, H., Li, M. Z., et al. (2003). Dicer is essential for mouse development. *Nature Genetics, 35*(3), 215–217. http://dx.doi.org/10.1038/ng1253.

Bettelli, E., Carrier, Y., Gao, W., Korn, T., Strom, T. B., Oukka, M., et al. (2006). Reciprocal developmental pathways for the generation of pathogenic effector TH17 and regulatory T cells. *Nature, 441*(7090), 235–238. http://dx.doi.org/10.1038/nature04753.

Biton, M., Levin, A., Slyper, M., Alkalay, I., Horwitz, E., Mor, H., et al. (2011). Epithelial microRNAs regulate gut mucosal immunity via epithelium-T cell crosstalk. *Nature Immunology, 12*(3), 239–246. http://dx.doi.org/10.1038/ni.1994.

Blazie, S. M., Geissel, H. C., Wilky, H., Joshi, R., Newbern, J., & Mangone, M. (2017). Alternative polyadenylation directs tissue-specific miRNA targeting in *Caenorhabditis elegans* somatic tissues. *Genetics, 206*(2), 757–774. http://dx.doi.org/10.1534/genetics.116.196774.

Bleck, B., Grunig, G., Chiu, A., Liu, M., Gordon, T., Kazeros, A., et al. (2013). MicroRNA-375 regulation of thymic stromal lymphopoietin by diesel exhaust particles and ambient particulate matter in human bronchial epithelial cells. *Journal of Immunology, 190*(7), 3757–3763. http://dx.doi.org/10.4049/jimmunol.1201165.

Blevins, R., Bruno, L., Carroll, T., Elliott, J., Marcais, A., Loh, C., et al. (2015). MicroRNAs regulate cell-to-cell variability of endogenous target gene expression in developing mouse thymocytes. *PLOS Genetics, 11*(2), e1005020. http://dx.doi.org/10.1371/journal.pgen.1005020.

Boldin, M. P., Taganov, K. D., Rao, D. S., Yang, L., Zhao, J. L., Kalwani, M., et al. (2011). MiR-146a is a significant brake on autoimmunity, myeloproliferation, and cancer in mice. *Journal of Experimental Medicine, 208*(6), 1189–1201. http://dx.doi.org/10.1084/jem.20101823.

Bonneau, E., Neveu, B., Kostantin, E., Tsongalis, G. J., & De Guire, V. (2019). How close are miRNAs from clinical practice? A perspective on the diagnostic and therapeutic market. *Ejifcc, 30*(2), 114–127.

Boren, T., Xiong, Y., Hakam, A., Wenham, R., Apte, S., Chan, G., et al. (2009). MicroRNAs and their target messenger RNAs associated with ovarian cancer response to chemotherapy. *Gynecologic Oncology, 113*(2), 249–255. http://dx.doi.org/10.1016/j.ygyno.2009.01.014.

Bottini, S., Hamouda-Tekaya, N., Mategot, R., Zaragosi, L. E., Audebert, S., Pisano, S., et al. (2017). Post-transcriptional gene silencing mediated by microRNAs is controlled by nucleoplasmic Sfpq. *Nature Communications, 8*(1), 1189. http://dx.doi.org/10.1038/s41467-017-01126-x.

Braun, J. E., Truffault, V., Boland, A., Huntzinger, E., Chang, C. T., Haas, G., et al. (2012). A direct interaction between DCP1 and XRN1 couples mRNA decapping to 5' exonucleolytic degradation. *Nature Structural & Molecular Biology, 19*(12), 1324–1331. http://dx.doi.org/10.1038/nsmb.2413.

Broide, D. H. (2001). Molecular and cellular mechanisms of allergic disease. *The Journal of Allergy and Clinical Immunology, 108*(2 Suppl. l). http://dx.doi.org/10.1067/mai.2001.116436. S65-71.

Bronevetsky, Y., Villarino, A. V., Eisley, C. J., Barbeau, R., Barczak, A. J., Heinz, G. A., et al. (2013). T cell activation induces proteasomal degradation of argonaute and rapid remodeling of the microRNA repertoire. *Journal of Experimental Medicine, 210*(2), 417–432. http://dx.doi.org/10.1084/jem.20111717.

Brown, A. F., Murphy, A. G., Lalor, S. J., Leech, J. M., O'Keeffe, K. M., Mac Aogáin, M., et al. (2015). Memory Th1 cells are protective in invasive *Staphylococcus aureus* infection. *PLOS Pathogens, 11*(11), e1005226. http://dx.doi.org/10.1371/journal.ppat.1005226.

Brown, B. D., & Naldini, L. (2009). Exploiting and antagonizing microRNA regulation for therapeutic and experimental applications. *Nature Reviews Genetics, 10*(8), 578–585. http://dx.doi.org/10.1038/nrg2628.

Bryniarski, K., Ptak, W., Jayakumar, A., Pullmann, K., Caplan, M. J., Chairoungdua, A., et al. (2013). Antigen-specific, antibody-coated, exosome-like nanovesicles deliver suppressor T-cell microRNA-150 to effector T cells to inhibit contact sensitivity. *The Journal of Allergy and Clinical Immunology, 132*(1), 170–181. http://dx.doi.org/10.1016/j.jaci.2013.04.048.

Burks, A. W., Jones, S. M., Wood, R. A., Fleischer, D. M., Sicherer, S. H., Lindblad, R. W., et al. (2012). Oral immunotherapy for treatment of egg allergy in children. *New England Journal of Medicine, 367*(3), 233–243. http://dx.doi.org/10.1056/NEJMoa1200435.

Cabezas-Wallscheid, N., Klimmeck, D., Hansson, J., Lipka, D. B., Reyes, A., Wang, Q., et al. (2014). Identification of regulatory networks in HSCs and their immediate progeny via integrated proteome, transcriptome, and DNA methylome analysis. *Cell Stem Cell, 15*(4), 507–522. http://dx.doi.org/10.1016/j.stem.2014.07.005.

Cai, X., Hagedorn, C. H., & Cullen, B. R. (2004). Human microRNAs are processed from capped, polyadenylated transcripts that can also function as mRNAs. *RNA, 10*(12), 1957–1966. http://dx.doi.org/10.1261/rna.7135204.

Caminati, M., Polk, B., & Rosenwasser, L. J. (2019). What have recent advances in therapy taught us about severe asthma disease mechanisms? *Expert Review of Clinical Immunology*. http://dx.doi.org/10.1080/1744666x.2020.1672536.

Canas, J. A., Sastre, B., Mazzeo, C., Fernandez-Nieto, M., Rodrigo-Munoz, J. M., Gonzalez-Guerra, A., et al. (2017). Exosomes from eosinophils autoregulate and promote eosinophil functions. *Journal of Leukocyte Biology, 101*(5), 1191–1199. http://dx.doi.org/10.1189/jlb.3AB0516-233RR.

Canas, J. A., Sastre, B., Rodrigo-Munoz, J. M., Fernandez-Nieto, M., Barranco, P., Quirce, S., et al. (2018). Eosinophil-derived exosomes contribute to asthma remodelling by activating structural lung cells. *Clinical and Experimental Allergy, 48*(9), 1173–1185. http://dx.doi.org/10.1111/cea.13122.

Chakir, J., Shannon, J., Molet, S., Fukakusa, M., Elias, J., Laviolette, M., et al. (2003). Airway remodeling-associated mediators in moderate to severe asthma: Effect of steroids on TGF-beta, IL-11, IL-17, and type I and type III collagen expression. *The Journal of Allergy and Clinical Immunology, 111*(6), 1293–1298. http://dx.doi.org/10.1067/mai.2003.1557.

Cheloufi, S., Dos Santos, C. O., Chong, M. M., & Hannon, G. J. (2010). A dicer-independent miRNA biogenesis pathway that requires ago catalysis. *Nature, 465*(7298), 584–589. http://dx.doi.org/10.1038/nature09092.

Chen, X., Ba, Y., Ma, L., Cai, X., Yin, Y., Wang, K., et al. (2008). Characterization of microRNAs in serum: A novel class of biomarkers for diagnosis of cancer and other diseases. *Cell Research, 18*(10), 997–1006. http://dx.doi.org/10.1038/cr.2008.282.

Chen, C. Y., Hogan, M. C., & Ward, C. J. (2013). Purification of exosome-like vesicles from urine. *Methods in Enzymology, 524*, 225–241. http://dx.doi.org/10.1016/B978-0-12-397945-2.00013-5.

Chen, R. F., Huang, H. C., Ou, C. Y., Hsu, T. Y., Chuang, H., Chang, J. C., et al. (2010). MicroRNA-21 expression in neonatal blood associated with antenatal immunoglobulin E production and development of allergic rhinitis. *Clinical and Experimental Allergy, 40*(10), 1482–1490. http://dx.doi.org/10.1111/j.1365-2222.2010.03592.x.

Chen, W., Jin, W., Hardegen, N., Lei, K. J., Li, L., Marinos, N., et al. (2003). Conversion of peripheral $CD4^{+}CD25^{-}$ naive T cells to $CD4^{+}CD25^{+}$ regulatory T cells by TGF-beta induction of transcription factor Foxp3. *Journal of Experimental Medicine, 198*(12), 1875–1886. http://dx.doi.org/10.1084/jem.20030152.

Chiba, Y., & Misawa, M. (2010). MicroRNAs and their therapeutic potential for human diseases: MiR-133a and bronchial smooth muscle hyperresponsiveness in asthma. *Journal of Pharmacological Sciences, 114*(3), 264–268. http://dx.doi.org/10.1254/jphs.10r10fm.

Chiba, Y., Tanabe, M., Goto, K., Sakai, H., & Misawa, M. (2009). Down-regulation of miR-133a contributes to up-regulation of rhoa in bronchial smooth muscle cells. *American Journal of Respiratory and Critical Care Medicine, 180*(8), 713–719. http://dx.doi.org/10.1164/rccm.200903-0325OC.

Chi, S. W., Zang, J. B., Mele, A., & Darnell, R. B. (2009). Argonaute HITS-CLIP decodes microRNA-mRNA interaction maps. *Nature, 460*(7254), 479–486. http://dx.doi.org/10.1038/nature08170.

Chong, M. M., Rasmussen, J. P., Rudensky, A. Y., & Littman, D. R. (2008). The RNAseIII enzyme drosha is critical in T cells for preventing lethal inflammatory disease. *Journal of Experimental Medicine, 205*(9), 2005–2017. http://dx.doi.org/10.1084/jem.20081219.

Cho, S., Wu, C. J., Nguyen, D. T., Lin, L. L., Chen, M. C., Khan, A. A., et al. (2017). A novel miR-24-TCF1 Axis in modulating effector T cell responses. *Journal of Immunology, 198*(10), 3919–3926. http://dx.doi.org/10.4049/jimmunol.1601404.

Cho, S., Wu, C. J., Yasuda, T., Cruz, L. O., Khan, A. A., Lin, L. L., et al. (2016). miR-23 approximately 27 approximately 24 clusters control effector T cell differentiation and function. *Journal of Experimental Medicine, 213*(2), 235–249. http://dx.doi.org/10.1084/jem.20150990.

Christie, M., Boland, A., Huntzinger, E., Weichenrieder, O., & Izaurralde, E. (2013). Structure of the PAN3 pseudokinase reveals the basis for interactions with the PAN2 deadenylase and the GW182 proteins. *Molecular Cell, 51*(3), 360–373. http://dx.doi.org/10.1016/j.molcel.2013.07.011.

Chung, Y., Chang, S. H., Martinez, G. J., Yang, X. O., Nurieva, R., Kang, H. S., et al. (2009). Critical regulation of early Th17 cell differentiation by interleukin-1 signaling. *Immunity, 30*(4), 576–587. http://dx.doi.org/10.1016/j.immuni.2009.02.007.

Cifuentes, D., Xue, H., Taylor, D. W., Patnode, H., Mishima, Y., Cheloufi, S., et al. (2010). A novel miRNA processing pathway independent of dicer requires argonaute2 catalytic activity. *Science, 328*(5986), 1694–1698. http://dx.doi.org/10.1126/science.1190809.

Cobb, B. S., Hertweck, A., Smith, J., O'Connor, E., Graf, D., Cook, T., et al. (2006). A role for dicer in immune regulation. *Journal of Experimental Medicine, 203*(11), 2519–2527. http://dx.doi.org/10.1084/jem.20061692.

Cobb, B. S., Nesterova, T. B., Thompson, E., Hertweck, A., O'Connor, E., Godwin, J., et al. (2005). T cell lineage choice and differentiation in the absence of the RNase III enzyme dicer. *Journal of Experimental Medicine, 201*(9), 1367–1373. http://dx.doi.org/10.1084/jem.20050572.

Collison, A., Herbert, C., Siegle, J. S., Mattes, J., Foster, P. S., & Kumar, R. K. (2011a). Altered expression of microRNA in the airway wall in chronic asthma: miR-126 as a potential therapeutic target. *BMC Pulmonary Medicine, 11*, 29. http://dx.doi.org/10.1186/1471-2466-11-29.

Collison, A., Mattes, J., Plank, M., & Foster, P. S. (2011b). Inhibition of house dust mite-induced allergic airways disease by antagonism of microRNA-145 is comparable to glucocorticoid treatment. *The Journal of Allergy and Clinical Immunology, 128*(1). http://dx.doi.org/10.1016/j.jaci.2011.04.005, 160-167.e164.

Comer, B. S., Camoretti-Mercado, B., Kogut, P. C., Halayko, A. J., Solway, J., & Gerthoffer, W. T. (2014). MicroRNA-146a and microRNA-146b expression and anti-inflammatory function in human airway smooth muscle. *American Journal of Physiology. Lung Cellular and Molecular Physiology, 307*(9), L727–734. http://dx.doi.org/10.1152/ajplung.00174.2014.

Comer, B. S., Camoretti-Mercado, B., Kogut, P. C., Halayko, A. J., Solway, J., & Gerthoffer, W. T. (2015). Cyclooxygenase-2 and microRNA-155 expression are elevated in asthmatic airway smooth muscle cells. *American Journal of Respiratory Cell and Molecular Biology, 52*(4), 438–447. http://dx.doi.org/10.1165/rcmb.2014-0129OC.

Copley, M. R., Babovic, S., Benz, C., Knapp, D. J., Beer, P. A., Kent, D. G., et al. (2013). The Lin28b-let-7-Hmga2 axis determines the higher self-renewal potential of fetal haematopoietic stem cells. *Nature Cell Biology, 15*(8), 916–925. http://dx.doi.org/10.1038/ncb2783.

Creighton, C. J., Reid, J. G., & Gunaratne, P. H. (2009). Expression profiling of microRNAs by deep sequencing. *Briefings in Bioinformatics, 10*(5), 490–497. http://dx.doi.org/10.1093/bib/bbp019.

Cua, D. J., Sherlock, J., Chen, Y., Murphy, C. A., Joyce, B., Seymour, B., et al. (2003). Interleukin-23 rather than interleukin-12 is the critical cytokine for autoimmune inflammation of the brain. *Nature, 421*(6924), 744–748. http://dx.doi.org/10.1038/nature01355.

Curtale, G., Mirolo, M., Renzi, T. A., Rossato, M., Bazzoni, F., & Locati, M. (2013). Negative regulation of toll-like receptor 4 signaling by IL-10-dependent microRNA-146b. *Proceedings of the National Academy of Sciences of the United States of America, 110*(28), 11499–11504. http://dx.doi.org/10.1073/pnas.1219852110.

D'Argenio, V., Del Monaco, V., Paparo, L., De Palma, F. D. E., Nocerino, R., D'Alessio, F., et al. (2018). Altered miR-193a-5p expression in children with cow's milk allergy. *Allergy, 73*(2), 379–386. http://dx.doi.org/10.1111/all.13299.

Dai, Y., Xie, C. H., Neis, J. P., Fan, C. Y., Vural, E., & Spring, P. M. (2011). MicroRNA expression profiles of head and neck squamous cell carcinoma with docetaxel-induced multidrug resistance. *Head & Neck, 33*(6), 786–791. http://dx.doi.org/10.1002/hed.21540.

Dardalhon, V., Korn, T., Kuchroo, V. K., & Anderson, A. C. (2008). Role of Th1 and Th17 cells in organ-specific autoimmunity. *Journal of Autoimmunity, 31*(3), 252–256. http://dx.doi.org/10.1016/j.jaut.2008.04.017.

Davis, J. S., Sun, M., Kho, A. T., Moore, K. G., Sylvia, J. M., Weiss, S. T., et al. (2017). Circulating microRNAs and association with methacholine PC20 in the childhood asthma management program (CAMP) cohort. *PLoS One, 12*(7), e0180329. http://dx.doi.org/10.1371/journal.pone.0180329.

Denli, A. M., Tops, B. B., Plasterk, R. H., Ketting, R. F., & Hannon, G. J. (2004). Processing of primary microRNAs by the microprocessor complex. *Nature, 432*(7014), 231–235. http://dx.doi.org/10.1038/nature03049.

Denzler, R., McGeary, S. E., Title, A. C., Agarwal, V., Bartel, D. P., & Stoffel, M. (2016). Impact of microRNA levels, target-site complementarity, and cooperativity on competing endogenous RNA-regulated gene expression. *Molecular Cell, 64*(3), 565–579. http://dx.doi.org/10.1016/j.molcel.2016.09.027.

Djuranovic, S., Nahvi, A., & Green, R. (2012). MiRNA-mediated gene silencing by translational repression followed by mRNA deadenylation and decay. *Science, 336*(6078), 237–240. http://dx.doi.org/10.1126/science.1215691.

Du Toit, G., Roberts, G., Sayre, P. H., Bahnson, H. T., Radulovic, S., Santos, A. F., et al. (2015). Randomized trial of peanut consumption in infants at risk for peanut allergy. *The New England Journal of Medicine, 372*(9), 803–813. http://dx.doi.org/10.1056/NEJMoa1414850.

Du, L., Borkowski, R., Zhao, Z., Ma, X., Yu, X., Xie, X. J., et al. (2013). A high-throughput screen identifies miRNA inhibitors regulating lung cancer cell survival and response to paclitaxel. *RNA Biology, 10*(11), 1700–1713. http://dx.doi.org/10.4161/rna.26541.

Duhen, T., Geiger, R., Jarrossay, D., Lanzavecchia, A., & Sallusto, F. (2009). Production of interleukin 22 but not interleukin 17 by a subset of human skin-homing memory T cells. *Nature Immunology, 10*(8), 857–863. http://dx.doi.org/10.1038/ni.1767.

Du, C., Liu, C., Kang, J., Zhao, G., Ye, Z., Huang, S., et al. (2009). MicroRNA miR-326 regulates TH-17 differentiation and is associated with the pathogenesis of multiple sclerosis. *Nature Immunology, 10*(12), 1252–1259. http://dx.doi.org/10.1038/ni.1798.

Dupont, C., Kalach, N., Soulaines, P., Legoue-Morillon, S., Piloquet, H., & Benhamou, P. H. (2010). Cow's milk epicutaneous immunotherapy in children: A pilot trial of safety, acceptability, and impact on allergic reactivity. *The Journal of Allergy and Clinical Immunology, 125*(5), 1165–1167. http://dx.doi.org/10.1016/j.jaci.2010.02.029.

EAACI. (2016). *Tackling the allergy crisis in Europe (pub. No. 41034)*. Retrieved September 26, 2019, from https://www.eaaci.org/images/media/EAACI_Manifesto_brochure_Interactive.pdf.

Ekimler, S., & Sahin, K. (2014). Computational methods for microRNA target prediction. *Genes (Basel), 5*(3), 671–683. http://dx.doi.org/10.3390/genes5030671.

Elbehidy, R. M., Youssef, D. M., El-Shal, A. S., Shalaby, S. M., Sherbiny, H. S., Sherief, L. M., et al. (2016). MicroRNA–21 as a novel biomarker in diagnosis and response to therapy in asthmatic children. *Molecular Immunology, 71*, 107–114. http://dx.doi.org/10.1016/j.molimm.2015.12.015.

Elkayam, E., Faehnle, C. R., Morales, M., Sun, J., Li, H., & Joshua-Tor, L. (2017). Multivalent recruitment of human argonaute by GW182. *Molecular Cell, 67*(4), 646–658. http://dx.doi.org/10.1016/j.molcel.2017.07.007. e643.

Elmen, J., Lindow, M., Schutz, S., Lawrence, M., Petri, A., Obad, S., et al. (2008). LNA-mediated microRNA silencing in non-human primates. *Nature, 452*(7189), 896–899. http://dx.doi.org/10.1038/nature06783.

Enrique, E., Pineda, F., Malek, T., Bartra, J., Basagana, M., Tella, R., et al. (2005). Sublingual immunotherapy for hazelnut food allergy: A randomized, double-blind, placebo-controlled study with a standardized hazelnut extract. *The Journal of Allergy and Clinical Immunology, 116*(5), 1073–1079. http://dx.doi.org/10.1016/j.jaci.2005.08.027.

Eom, S., Kim, Y., Kim, M., Park, D., Lee, H., Lee, Y. S., et al. (2014a). Transglutaminase II/microRNA-218/-181a loop regulates positive feedback relationship between allergic inflammation and tumor metastasis. *Journal of Biological Chemistry, 289*(43), 29483–29505. http://dx.doi.org/10.1074/jbc.M114.603480.

Eom, S., Kim, Y., Park, D., Lee, H., Lee, Y. S., Choe, J., et al. (2014b). Histone deacetylase-3 mediates positive feedback relationship between anaphylaxis and tumor metastasis. *Journal of Biological Chemistry, 289*(17), 12126–12144. http://dx.doi.org/10.1074/jbc.M113.521245.

Fattore, L., Costantini, S., Malpicci, D., Ruggiero, C. F., Ascierto, P. A., Croce, C. M., et al. (2017). MicroRNAs in melanoma development and resistance to target therapy. *Oncotarget, 8*(13), 22262–22278. http://dx.doi.org/10.18632/oncotarget.14763.

Fazi, F., Rosa, A., Fatica, A., Gelmetti, V., De Marchis, M. L., Nervi, C., et al. (2005). A minicircuitry comprised of microRNA-223 and transcription factors NFI-A and C/EBPalpha regulates human granulopoiesis. *Cell, 123*(5), 819–831. http://dx.doi.org/10.1016/j.cell.2005.09.023.

Feng, M.-J., Shi, F., Qiu, C., & Peng, W.-K. (2012). MicroRNA-181a, -146a and -146b in spleen $CD4^{+}$ T lymphocytes play proinflammatory roles in a murine model of asthma. *International Immunopharmacology, 13*(3), 347–353. http://dx.doi.org/10.1016/j.intimp.2012.05.001.

Feng, Z., Zhang, C., Wu, R., & Hu, W. (2011). Tumor suppressor p53 meets microRNAs. *Journal of Molecular Cell Biology, 3*(1), 44–50. http://dx.doi.org/10.1093/jmcb/mjq040.

Fleischer, D. M., Burks, A. W., Vickery, B. P., Scurlock, A. M., Wood, R. A., Jones, S. M., et al. (2013). Sublingual immunotherapy for peanut allergy: A randomized, double-blind, placebo-controlled multicenter trial. *The Journal of Allergy and Clinical Immunology, 131*(1). http://dx.doi.org/10.1016/j.jaci.2012.11.011, 119-127.e111-117.

Fontana, L., Pelosi, E., Greco, P., Racanicchi, S., Testa, U., Liuzzi, F., et al. (2007). MicroRNAs 17-5p-20a-106a control monocytopoiesis through AML1 targeting and M-CSF receptor upregulation. *Nature Cell Biology, 9*(7), 775–787. http://dx.doi.org/10.1038/ncb1613.

Foster, P. S., Plank, M., Collison, A., Tay, H. L., Kaiko, G. E., Li, J., et al. (2013). The emerging role of microRNAs in regulating immune and inflammatory responses in the lung. *Immunological Reviews, 253*(1), 198–215. http://dx.doi.org/10.1111/imr.12058.

Francisco-Garcia, A., Martinez-Nunez, R. T., Rupani, H., Lau, L. C., Howarth, P. H., & Sanchez-Elsner, T. (2016). LSC Abstract – Altered small RNA cargo in severe asthma exosomes. *European Respiratory Journal, 48*(suppl 60), PP101. http://dx.doi.org/10.1183/13993003.congress-2016.PP101.

Friedman, R. C., Farh, K. K., Burge, C. B., & Bartel, D. P. (2009). Most mammalian mRNAs are conserved targets of microRNAs. *Genome Research, 19*(1), 92–105. http://dx.doi.org/10.1101/gr.082701.108.

Fujita, Y., Kadota, T., Araya, J., Ochiya, T., & Kuwano, K. (2018). Extracellular vesicles: New players in lung immunity. *American Journal of Respiratory Cell and Molecular Biology, 58*(5), 560–565. http://dx.doi.org/10.1165/rcmb.2017-0293TR.

Fujita, Y., Yoshioka, Y., Ito, S., Araya, J., Kuwano, K., & Ochiya, T. (2014). Intercellular communication by extracellular vesicles and their microRNAs in asthma. *Clinical Therapeutics, 36*(6), 873–881. http://dx.doi.org/10.1016/j.clinthera.2014.05.006.

Gallant-Behm, C. L., Piper, J., Lynch, J. M., Seto, A. G., Hong, S. J., Mustoe, T. A., … Marshall, W. S. (2019). A MicroRNA-29 Mimic (Remlarsen) Represses Extracellular Matrix Expression and Fibroplasia in the Skin. *Journal of Investigative Dermatology, 139*(5), 1073–1081.

Galli, S. J., & Tsai, M. (2012). IgE and mast cells in allergic disease. *Nature Medicine, 18*(5), 693–704. http://dx.doi.org/10.1038/nm.2755.

Gallo, A., Tandon, M., Alevizos, I., & Illei, G. G. (2012). The majority of microRNAs detectable in serum and saliva is concentrated in exosomes. *PLoS One,* 7(3), e30679. http://dx.doi.org/10.1371/journal.pone.0030679.

Garo, L. P., & Murugaiyan, G. (2016). Contribution of microRNAs to autoimmune diseases. *Cellular and Molecular Life Sciences, 73*(10), 2041–2051. http://dx.doi.org/10.1007/s00018-016-2167-4.

Ghildiyal, M., & Zamore, P. D. (2009). Small silencing RNAs: An expanding universe. *Nature Reviews Genetics, 10*(2), 94–108. http://dx.doi.org/10.1038/nrg2504.

Gon, Y., Maruoka, S., Inoue, T., Kuroda, K., Yamagishi, K., Kozu, Y., et al. (2017). Selective release of miRNAs via extracellular vesicles is associated with house-dust mite allergen-induced airway inflammation. *Clinical and Experimental Allergy, 47*(12), 1586–1598. http://dx.doi.org/10.1111/cea.13016.

Greening, D. W., Gopal, S. K., Xu, R., Simpson, R. J., & Chen, W. (2015). Exosomes and their roles in immune regulation and cancer. *Seminars in Cell & Developmental Biology, 40,* 72–81. http://dx.doi.org/10.1016/j.semcdb.2015.02.009.

Griffiths-Jones, S., Grocock, R. J., van Dongen, S., Bateman, A., & Enright, A. J. (2006). MiRBase: microRNA sequences, targets and gene nomenclature. *Nucleic Acids Research, 34*(Database issue). http://dx.doi.org/10.1093/nar/gkj112. D140-144.

Guerau-de-Arellano, M., Smith, K. M., Godlewski, J., Liu, Y., Winger, R., Lawler, S. E., et al. (2011). Micro-RNA dysregulation in multiple sclerosis favours pro-inflammatory T-cell-mediated autoimmunity. *Brain, 134*(Pt 12), 3578–3589. http://dx.doi.org/10.1093/brain/awr262.

Gu, S., Jin, L., Zhang, F., Sarnow, P., & Kay, M. A. (2009). Biological basis for restriction of microRNA targets to the 3' untranslated region in mammalian mRNAs. *Nature Structural & Molecular Biology, 16*(2), 144–150. http://dx.doi.org/10.1038/nsmb.1552.

Gümürdü, A., Yildiz, R., Eren, E., Karakülah, G., Ünver, T., Genç, Ş., et al. (2017). MicroRNA exocytosis by large dense-core vesicle fusion. *Scientific Reports,* 7, 45661. http://dx.doi.org/10.1038/srep45661.

Guo, H., Ingolia, N. T., Weissman, J. S., & Bartel, D. P. (2010). Mammalian microRNAs predominantly act to decrease target mRNA levels. *Nature, 466*(7308), 835–840. http://dx.doi.org/10.1038/nature09267.

Hafner, M., Ascano, M., Jr., & Tuschl, T. (2011). New insights in the mechanism of microRNA-mediated target repression. *Nature Structural & Molecular Biology, 18*(11), 1181–1182. http://dx.doi.org/10.1038/nsmb.2170.

Hafner, M., Landthaler, M., Burger, L., Khorshid, M., Hausser, J., Berninger, P., et al. (2010). Transcriptome-wide identification of RNA-binding protein and microRNA target sites by PAR-CLIP. *Cell, 141*(1), 129–141. http://dx.doi.org/10.1016/j.cell.2010.03.009.

Hammad Mahmoud Hammad, R., Hamed, D., Eldosoky, M., Ahmad, A., Osman, H. M., Abd Elgalil, H. M., et al. (2018). Plasma microRNA-21, microRNA-146a and IL-13 expression in asthmatic children. *Innate Immunity, 24*(3), 171–179. http://dx.doi.org/10.1177/1753425918763521.

Han, L., Witmer, P. D., Casey, E., Valle, D., & Sukumar, S. (2007). DNA methylation regulates microRNA expression. *Cancer Biology & Therapy, 6*(8), 1284–1288. http://dx.doi.org/10.4161/cbt.6.8.4486.

Hayes, J., Peruzzi, P. P., & Lawler, S. (2014). MicroRNAs in cancer: Biomarkers, functions and therapy. *Trends in Molecular Medicine, 20*(8), 460–469. http://dx.doi.org/10.1016/j.molmed.2014.06.005.

Hegazy, A. N., Peine, M., Helmstetter, C., Panse, I., Frohlich, A., Bergthaler, A., et al. (2010). Interferons direct Th2 cell reprogramming to generate a stable GATA-3($^+$) T-bet($^+$) cell subset with combined Th2 and Th1 cell functions. *Immunity, 32*(1), 116–128. http://dx.doi.org/10.1016/j.immuni.2009.12.004.

He, L., He, X., Lim, L. P., de Stanchina, E., Xuan, Z., Liang, Y., et al. (2007). A microRNA component of the p53 tumour suppressor network. *Nature, 447*(7148), 1130–1134. http://dx.doi.org/10.1038/nature05939.

Heinicke, F., Zhong, X., Zucknick, M., Breidenbach, J., Sundaram, A. Y. M., S, T. F., et al. (2019). Systematic assessment of commercially available low-input miRNA library preparation kits. *RNA Biology*, 1–12. http://dx.doi.org/10.1080/15476286.2019.1667741.

Henriksson, J., Chen, X., Gomes, T., Ullah, U., Meyer, K. B., Miragaia, R., et al. (2017). Genome-wide CRISPR screens in T helper cells reveal pervasive cross-talk between activation and differentiation. *bioRxiv*, 196022. http://dx.doi.org/10.1101/196022.

Hergenreider, E., Heydt, S., Treguer, K., Boettger, T., Horrevoets, A. J., Zeiher, A. M., et al. (2012). Atheroprotective communication between endothelial cells and smooth muscle cells through miRNAs. *Nature Cell Biology, 14*(3), 249–256. http://dx.doi.org/10.1038/ncb2441.

Homburg, U., Renz, H., Timmer, W., Hohlfeld, J. M., Seitz, F., Luer, K., et al. (2015). Safety and tolerability of a novel inhaled GATA3 mRNA targeting DNAzyme in patients with TH2-driven asthma. *The Journal of Allergy and Clinical Immunology, 136*(3), 797–800. http://dx.doi.org/10.1016/j.jaci.2015.02.018.

Hough, K. P., Chanda, D., Duncan, S. R., Thannickal, V. J., & Deshane, J. S. (2017). Exosomes in immunoregulation of chronic lung diseases. *Allergy, 72*(4), 534–544. http://dx.doi.org/10.1111/all.13086.

Hsu, S. D., Lin, F. M., Wu, W. Y., Liang, C., Huang, W. C., Chan, W. L., et al. (2011). miRTarBase: a database curates experimentally validated microRNA-target interactions. *Nucleic Acids Research, 39*(Database issue). http://dx.doi.org/10.1093/nar/gkq1107. D163-169.

Hu, R., Huffaker, T. B., Kagele, D. A., Runtsch, M. C., Bake, E., Chaudhuri, A. A., et al. (2013). MicroRNA-155 confers encephalogenic potential to Th17 cells by promoting effector gene expression. *Journal of Immunology, 190*(12), 5972–5980. http://dx.doi.org/10.4049/jimmunol.1300351.

Hu, R., Kagele, D. A., Huffaker, T. B., Runtsch, M. C., Alexander, M., Liu, J., et al. (2014). miR-155 promotes T follicular helper cell accumulation during chronic, low-grade inflammation. *Immunity, 41*(4), 605–619. http://dx.doi.org/10.1016/j.immuni.2014.09.015.

Iorio, M. V., Piovan, C., & Croce, C. M. (2010). Interplay between microRNAs and the epigenetic machinery: An intricate network. *Biochimica et Biophysica Acta, 1799*(10–12), 694–701. http://dx.doi.org/10.1016/j.bbagrm.2010.05.005.

Ishizu, H., Siomi, H., & Siomi, M. C. (2012). Biology of PIWI-interacting RNAs: New insights into biogenesis and function inside and outside of germlines. *Genes & Development, 26*(21), 2361–2373. http://dx.doi.org/10.1101/gad.203786.112.

Janssen, H. L., Reesink, H. W., Lawitz, E. J., Zeuzem, S., Rodriguez-Torres, M., Patel, K., et al. (2013). Treatment of HCV infection by targeting microRNA. *The New England Journal of Medicine, 368*(18), 1685–1694. http://dx.doi.org/10.1056/NEJMoa1209026.

Jardim, M. J., Dailey, L., Silbajoris, R., & Diaz-Sanchez, D. (2012). Distinct microRNA expression in human airway cells of asthmatic donors identifies a novel asthma-associated gene. *American Journal of Respiratory Cell and Molecular Biology, 47*(4), 536–542. http://dx.doi.org/10.1165/rcmb.2011-0160OC.

Jeker, L. T., & Bluestone, J. A. (2013). MicroRNA regulation of T-cell differentiation and function. *Immunological Reviews, 253*(1), 65–81. http://dx.doi.org/10.1111/imr.12061.

Jiang, S., Li, C., Olive, V., Lykken, E., Feng, F., Sevilla, J., et al. (2011). Molecular dissection of the miR-17-92 cluster's critical dual roles in promoting Th1 responses and preventing inducible treg differentiation. *Blood, 118*(20), 5487–5497. http://dx.doi.org/10.1182/blood-2011-05-355644.

Johansson, S. G., Bieber, T., Dahl, R., Friedmann, P. S., Lanier, B. Q., Lockey, R. F., et al. (2004). Revised nomenclature for allergy for global use: Report of the nomenclature review committee of the world allergy organization, October 2003. *The Journal of Allergy and Clinical Immunology, 113*(5), 832–836. http://dx.doi.org/10.1016/j.jaci.2003.12.591.

Johnnidis, J. B., Harris, M. H., Wheeler, R. T., Stehling-Sun, S., Lam, M. H., Kirak, O., et al. (2008). Regulation of progenitor cell proliferation and granulocyte function by microRNA-223. *Nature, 451*(7182), 1125–1129. http://dx.doi.org/10.1038/nature06607.

Jonas, S., & Izaurralde, E. (2015). Towards a molecular understanding of microRNA-mediated gene silencing. *Nature Reviews Genetics, 16*(7), 421–433. http://dx.doi.org/10.1038/nrg3965.

Josefowicz, S. Z., Lu, L. F., & Rudensky, A. Y. (2012). Regulatory T cells: Mechanisms of differentiation and function. *Annual Review of Immunology, 30*, 531–564. http://dx.doi.org/10.1146/annurev.immunol.25.022106.141623.

Josefowicz, S. Z., & Rudensky, A. (2009). Control of regulatory T cell lineage commitment and maintenance. *Immunity, 30*(5), 616–625. http://dx.doi.org/10.1016/j.immuni.2009.04.009.

Jo, M. H., Shin, S., Jung, S. R., Kim, E., Song, J. J., & Hohng, S. (2015). Human argonaute 2 has diverse reaction pathways on target RNAs. *Molecular Cell, 59*(1), 117–124. http://dx.doi.org/10.1016/j.molcel.2015.04.027.

Keerthikumar, S., Chisanga, D., Ariyaratne, D., Al Saffar, H., Anand, S., Zhao, K., et al. (2016). ExoCarta: A web-based compendium of exosomal cargo. *Journal of Molecular Biology, 428*(4), 688–692. http://dx.doi.org/10.1016/j.jmb.2015.09.019.

Keller, A., Leidinger, P., Lange, J., Borries, A., Schroers, H., Scheffler, M., et al. (2009). Multiple sclerosis: microRNA expression profiles accurately differentiate patients with relapsing-remitting disease from healthy controls. *PLoS One, 4*(10), e7440. http://dx.doi.org/10.1371/journal.pone.0007440.

Kesimer, M., Scull, M., Brighton, B., DeMaria, G., Burns, K., O'Neal, W., et al. (2009). Characterization of exosome-like vesicles released from human tracheobronchial ciliated epithelium: A possible role in innate defense. *FASEB Journal, 23*(6), 1858–1868. http://dx.doi.org/10.1096/fj.08-119131.

Kim, V. N., Han, J., & Siomi, M. C. (2009). Biogenesis of small RNAs in animals. *Nature Reviews Molecular Cell Biology, 10*(2), 126–139. http://dx.doi.org/10.1038/nrm2632.

Kim, R. Y., Horvat, J. C., Pinkerton, J. W., Starkey, M. R., Essilfie, A. T., Mayall, J. R., et al. (2017). MicroRNA-21 drives severe, steroid-insensitive experimental asthma by amplifying phosphoinositide 3-kinase-mediated suppression of histone deacetylase 2. *The Journal of Allergy and Clinical Immunology, 139*(2), 519–532. http://dx.doi.org/10.1016/j.jaci.2016.04.038.

Kim, J., Inoue, K., Ishii, J., Vanti, W. B., Voronov, S. V., Murchison, E., et al. (2007). A microRNA feedback circuit in midbrain dopamine neurons. *Science, 317*(5842), 1220–1224. http://dx.doi.org/10.1126/science.1140481.

Kim, B. S., Kim, I. K., Park, Y. J., Kim, Y. S., Kim, Y. J., Chang, W. S., et al. (2010). Conversion of Th2 memory cells into Foxp3$^+$ regulatory T cells suppressing Th2-mediated allergic asthma. *Proceedings of the National Academy of Sciences of the United States of America, 107*(19), 8742–8747. http://dx.doi.org/10.1073/pnas.0911756107.

King, K. Y., & Goodell, M. A. (2011). Inflammatory modulation of HSCs: Viewing the HSC as a foundation for the immune response. *Nature Reviews Immunology, 11*(10), 685–692. http://dx.doi.org/10.1038/nri3062.

Klose, C. S., & Artis, D. (2016). Innate lymphoid cells as regulators of immunity, inflammation and tissue homeostasis. *Nature Immunology, 17*(7), 765–774. http://dx.doi.org/10.1038/ni.3489.

Kohlhaas, S., Garden, O. A., Scudamore, C., Turner, M., Okkenhaug, K., & Vigorito, E. (2009). Cutting edge: The Foxp3 target miR-155 contributes to the development of regulatory T cells. *Journal of Immunology, 182*(5), 2578–2582. http://dx.doi.org/10.4049/jimmunol.0803162.

Komatsu, N., Okamoto, K., Sawa, S., Nakashima, T., Oh-hora, M., Kodama, T., et al. (2014). Pathogenic conversion of Foxp3$^+$ T cells into TH17 cells in autoimmune arthritis. *Nature Medicine, 20*(1), 62–68. http://dx.doi.org/10.1038/nm.3432.

Koppers-Lalic, D., Hackenberg, M., Bijnsdorp, I. V., van Eijndhoven, M. A. J., Sadek, P., Sie, D., et al. (2014). Nontemplated nucleotide additions distinguish the small RNA composition in cells from exosomes. *Cell Reports, 8*(6), 1649–1658. http://dx.doi.org/10.1016/j.celrep.2014.08.027.

Koralov, S. B., Muljo, S. A., Galler, G. R., Krek, A., Chakraborty, T., Kanellopoulou, C., et al. (2008). Dicer ablation affects antibody diversity and cell survival in the B lymphocyte lineage. *Cell, 132*(5), 860–874. http://dx.doi.org/10.1016/j.cell.2008.02.020.

Kozomara, A., Birgaoanu, M., & Griffiths-Jones, S. (2018). miRBase: from microRNA sequences to function. *Nucleic Acids Research, 47*(D1), D155–D162. http://dx.doi.org/10.1093/nar/gky1141.

Kozomara, A., & Griffiths-Jones, S. (2011). miRBase: integrating microRNA annotation and deep-sequencing data. *Nucleic Acids Research, 39*(Database issue). http://dx.doi.org/10.1093/nar/gkq1027. D152-157.

Kozomara, A., & Griffiths-Jones, S. (2014). miRBase: annotating high confidence microRNAs using deep sequencing data. *Nucleic Acids Research, 42*(Database issue). http://dx.doi.org/10.1093/nar/gkt1181. D68-73.

Krol, J., Loedige, I., & Filipowicz, W. (2010). The widespread regulation of microRNA biogenesis, function and decay. *Nature Reviews Genetics, 11*(9), 597–610. http://dx.doi.org/10.1038/nrg2843.

Krug, N., Hohlfeld, J. M., Kirsten, A. M., Kornmann, O., Beeh, K. M., Kappeler, D., et al. (2015). Allergen-induced asthmatic responses modified by a GATA3-specific DNAzyme. *The New England Journal of Medicine, 372*(21), 1987–1995. http://dx.doi.org/10.1056/NEJMoa1411776.

Krutzfeldt, J., Rajewsky, N., Braich, R., Rajeev, K. G., Tuschl, T., Manoharan, M., et al. (2005). Silencing of microRNAs in vivo with 'antagomirs'. *Nature, 438*(7068), 685–689. http://dx.doi.org/10.1038/nature04303.

Kulshreshtha, A., Ahmad, T., Agrawal, A., & Ghosh, B. (2013). Proinflammatory role of epithelial cell-derived exosomes in allergic airway inflammation. *The Journal of Allergy and Clinical Immunology, 131*(4). http://dx.doi.org/10.1016/j.jaci.2012.12.1565, 1194-1203, 1203.e1191-1114.

Kumar, M., Ahmad, T., Sharma, A., Mabalirajan, U., Kulshreshtha, A., Agrawal, A., et al. (2011). Let-7 microRNA-mediated regulation of IL-13 and allergic airway inflammation. *The Journal of Allergy and Clinical Immunology, 128*(5). http://dx.doi.org/10.1016/j.jaci.2011.04.034, 1077-1085.e1071-1010.

Kwon, Y., Kim, Y., Eom, S., Kim, M., Park, D., Kim, H., et al. (2015). MicroRNA-26a/-26b-COX-2-MIP-2 loop regulates allergic inflammation and allergic inflammation-promoted enhanced tumorigenic and Metastatic potential of cancer cells. *Journal of Biological Chemistry, 290*(22), 14245–14266. http://dx.doi.org/10.1074/jbc.M115.645580.

Lacedonia, D., Palladino, G. P., Foschino-Barbaro, M. P., Scioscia, G., & Carpagnano, G. E. (2017). Expression profiling of miRNA-145 and miRNA-338 in serum and sputum of patients with COPD, asthma, and asthma-COPD overlap syndrome phenotype. *International Journal of Chronic Obstructive Pulmonary Disease, 12*, 1811–1817. http://dx.doi.org/10.2147/copd.s130616.

Lai, X., Wolkenhauer, O., & Vera, J. (2016). Understanding microRNA-mediated gene regulatory networks through mathematical modelling. *Nucleic Acids Research, 44*(13), 6019–6035. http://dx.doi.org/10.1093/nar/gkw550.

Lal, A., Thomas, M. P., Altschuler, G., Navarro, F., O'Day, E., Li, X. L., et al. (2011). Capture of microRNA-bound mRNAs identifies the tumor suppressor miR-34a as a regulator of growth factor signaling. *PLOS Genetics*, 7(11), e1002363. http://dx.doi.org/10.1371/journal.pgen.1002363.

Lasser, C., O'Neil, S. E., Shelke, G. V., Sihlbom, C., Hansson, S. F., Gho, Y. S., et al. (2016). Exosomes in the nose induce immune cell trafficking and harbour an altered protein cargo in chronic airway inflammation. *Journal of Translational Medicine, 14*(1), 181. http://dx.doi.org/10.1186/s12967-016-0927-4.

Lechman, E. R., Gentner, B., van Galen, P., Giustacchini, A., Saini, M., Boccalatte, F. E., et al. (2012). Attenuation of miR-126 activity expands HSC in vivo without exhaustion. *Cell Stem Cell, 11*(6), 799–811. http://dx.doi.org/10.1016/j.stem.2012.09.001.

Lee, R. C., Feinbaum, R. L., & Ambros, V. (1993). The *C. elegans* heterochronic gene lin-4 encodes small RNAs with antisense complementarity to lin-14. *Cell, 75*(5), 843–854. http://dx.doi.org/10.1016/0092-8674(93)90529-y.

Lee, Y., Kim, M., Han, J., Yeom, K. H., Lee, S., Baek, S. H., et al. (2004). MicroRNA genes are transcribed by RNA polymerase II. *The EMBO Journal, 23*(20), 4051–4060. http://dx.doi.org/10.1038/sj.emboj.7600385.

Lee, Y. K., Turner, H., Maynard, C. L., Oliver, J. R., Chen, D., Elson, C. O., et al. (2009). Late developmental plasticity in the T helper 17 lineage. *Immunity, 30*(1), 92–107. http://dx.doi.org/10.1016/j.immuni.2008.11.005.

Leshkowitz, D., Horn-Saban, S., Parmet, Y., & Feldmesser, E. (2013). Differences in microRNA detection levels are technology and sequence dependent. *RNA, 19*(4), 527–538. http://dx.doi.org/10.1261/rna.036475.112.

Leung, S., Liu, X., Fang, L., Chen, X., Guo, T., & Zhang, J. (2010). The cytokine milieu in the interplay of pathogenic Th1/Th17 cells and regulatory T cells in autoimmune disease. *Cellular & Molecular Immunology*, 7(3), 182–189. http://dx.doi.org/10.1038/cmi.2010.22.

Levanen, B., Bhakta, N. R., Torregrosa Paredes, P., Barbeau, R., Hiltbrunner, S., Pollack, J. L., et al. (2013). Altered microRNA profiles in bronchoalveolar lavage fluid exosomes in asthmatic patients. *The Journal of Allergy and Clinical Immunology, 131*(3), 894–903. http://dx.doi.org/10.1016/j.jaci.2012.11.039.

Levine, A. G., Mendoza, A., Hemmers, S., Moltedo, B., Niec, R. E., Schizas, M., et al. (2017). Stability and function of regulatory T cells expressing the transcription factor T-bet. *Nature, 546*(7658), 421–425. http://dx.doi.org/10.1038/nature22360.

Li, S. C., Chan, W. C., Hu, L. Y., Lai, C. H., Hsu, C. N., & Lin, W. C. (2010). Identification of homologous microRNAs in 56 animal genomes. *Genomics, 96*(1), 1–9. http://dx.doi.org/10.1016/j.ygeno.2010.03.009.

Li, Y., Chen, P., Zu, L., Liu, B., Wang, M., & Zhou, Q. (2016). MicroRNA-338-3p suppresses metastasis of lung cancer cells by targeting the EMT regulator Sox4. *American Journal of Cancer Research, 6*(2), 127–140.

Li, K., Ching, D., Luk, F. S., & Raffai, R. L. (2015). Apolipoprotein E enhances microRNA-146a in monocytes and macrophages to suppress nuclear factor-kappaB-driven inflammation and atherosclerosis. *Circulation Research, 117*(1), e1–e11. http://dx.doi.org/10.1161/circresaha.117.305844.

Li, F., Huang, Y., Huang, Y. Y., Kuang, Y. S., Wei, Y. J., Xiang, L., et al. (2017). MicroRNA-146a promotes IgE class switch in B cells via upregulating 14-3-3sigma expression. *Molecular Immunology, 92*, 180–189. http://dx.doi.org/10.1016/j.molimm.2017.10.023.

Li, Y., & Kowdley, K. V. (2012). MicroRNAs in common human diseases. *Genomics, Proteomics & Bioinformatics, 10*(5), 246–253. http://dx.doi.org/10.1016/j.gpb.2012.07.005.

Li, X., Lu, Y., Chen, Y., Lu, W., & Xie, X. (2013). MicroRNA profile of paclitaxel-resistant serous ovarian carcinoma based on formalin-fixed paraffin-embedded samples. *BMC Cancer, 13*, 216. http://dx.doi.org/10.1186/1471-2407-13-216.

Liston, A., Lu, L. F., O'Carroll, D., Tarakhovsky, A., & Rudensky, A. Y. (2008). Dicer-dependent microRNA pathway safeguards regulatory T cell function. *Journal of Experimental Medicine, 205*(9), 1993–2004. http://dx.doi.org/10.1084/jem.20081062.

Liu, X., Nelson, A., Wang, X., Kanaji, N., Kim, M., Sato, T., et al. (2009). MicroRNA-146a modulates human bronchial epithelial cell survival in response to the cytokine-induced apoptosis. *Biochemical and Biophysical Research Communications, 380*(1), 177–182. http://dx.doi.org/10.1016/j.bbrc.2009.01.066.

Liu, X., Robinson, S. N., Setoyama, T., Tung, S. S., D'Abundo, L., Shah, M. Y., et al. (2014). FOXP3 is a direct target of miR15a/16 in umbilical cord blood regulatory T cells. *Bone Marrow Transplantation, 49*(6), 793–799. http://dx.doi.org/10.1038/bmt.2014.57.

Liu, Y. M., Xia, Y., Dai, W., Han, H. Y., Dong, Y. X., Cai, J., … Guan, J. (2014). Cholesterol-conjugated let-7a mimics: antitumor efficacy on hepatocellular carcinoma in vitro and in a preclinical orthotopic xenograft model of systemic therapy. *BMC Cancer, 14*, 889. http://dx.doi.org/10.1186/1471-2407-14-889.

Liu, Z. Q., Yang, G., Geng, X. R., Liu, J. Q., Mo, L. H., Liu, Z. G., et al. (2016). Micro RNA-17-92 cluster mediates interleukin-4-suppressed IL-10 expression in B cells. *American Journal of Translational Research, 8*(5), 2317–2324.

Li, Y., Wei, C., Xu, H., Jia, J., Wei, Z., Guo, R., et al. (2018b). The immunoregulation of Th17 in host against intracellular bacterial infection. *Mediators of Inflammation*, 2018, 6587296. http://dx.doi.org/10.1155/2018/6587296.

Li, L., Zhang, S., Jiang, X., Liu, Y., Liu, K., & Yang, C. (2018a). MicroRNA-let-7e regulates the progression and development of allergic rhinitis by targeting suppressor of cytokine signaling 4 and activating Janus kinase 1/signal transducer and activator of transcription 3 pathway. *Experimental and Therapeutic Medicine, 15*(4), 3523–3529. http://dx.doi.org/10.3892/etm.2018.5827.

Locksley, R. M. (2010). Asthma and allergic inflammation. *Cell, 140*(6), 777–783. http://dx.doi.org/10.1016/j.cell.2010.03.004.

Lotvall, J., Akdis, C. A., Bacharier, L. B., Bjermer, L., Casale, T. B., Custovic, A., et al. (2011). Asthma endotypes: A new approach to classification of disease entities within the asthma syndrome. *The Journal of Allergy and Clinical Immunology, 127*(2), 355–360. http://dx.doi.org/10.1016/j.jaci.2010.11.037.

Lowery, A. J., Miller, N., Devaney, A., McNeill, R. E., Davoren, P. A., Lemetre, C., et al. (2009). MicroRNA signatures predict oestrogen receptor, progesterone receptor and HER2/neu receptor status in breast cancer. *Breast Cancer Research: BCR, 11*(3), R27. http://dx.doi.org/10.1186/bcr2257.

Lu, L. F., Boldin, M. P., Chaudhry, A., Lin, L. L., Taganov, K. D., Hanada, T., et al. (2010). Function of miR-146a in controlling treg cell-mediated regulation of Th1 responses. *Cell, 142*(6), 914–929. http://dx.doi.org/10.1016/j.cell.2010.08.012.

Lu, Y. C., Chang, J. T., Chan, E. C., Chao, Y. K., Yeh, T. S., Chen, J. S., et al. (2016). miR-196, an emerging cancer biomarker for digestive tract cancers. *Journal of Cancer, 7*(6), 650–655. http://dx.doi.org/10.7150/jca.13460.

Lu, T. X., Hartner, J., Lim, E. J., Fabry, V., Mingler, M. K., Cole, E. T., et al. (2011). MicroRNA-21 limits in vivo immune response-mediated activation of the IL-12/IFN-gamma pathway, Th1 polarization, and the severity of delayed-type hypersensitivity. *Journal of Immunology, 187*(6), 3362–3373. http://dx.doi.org/10.4049/jimmunol.1101235.

Lu, T. X., Lim, E. J., Wen, T., Plassard, A. J., Hogan, S. P., Martin, L. J., et al. (2012a). MiR-375 is downregulated in epithelial cells after IL-13 stimulation and regulates an IL-13-induced epithelial transcriptome. *Mucosal Immunology, 5*(4), 388–396. http://dx.doi.org/10.1038/mi.2012.16.

Lu, T. X., Munitz, A., & Rothenberg, M. E. (2009b). MicroRNA-21 is up-regulated in allergic airway inflammation and regulates IL-12p35 expression. *Journal of Immunology, 182*(8), 4994–5002. http://dx.doi.org/10.4049/jimmunol.0803560.

Lu, T. X., & Rothenberg, M. E. (2013). Diagnostic, functional, and therapeutic roles of microRNA in allergic diseases. *The Journal of Allergy and Clinical Immunology, 132*(1), 3–13. http://dx.doi.org/10.1016/j.jaci.2013.04.039. quiz 14.

Lu, T. X., Sherrill, J. D., Wen, T., Plassard, A. J., Besse, J. A., Abonia, J. P., et al. (2012b). MicroRNA signature in patients with eosinophilic esophagitis, reversibility with glucocorticoids, and assessment as disease biomarkers. *The Journal of Allergy and Clinical Immunology, 129*(4). http://dx.doi.org/10.1016/j.jaci.2012.01.060, 1064-1075.e1069.

Lu, L. F., Thai, T. H., Calado, D. P., Chaudhry, A., Kubo, M., Tanaka, K., et al. (2009a). Foxp3-dependent microRNA155 confers competitive fitness to regulatory T cells by targeting SOCS1 protein. *Immunity, 30*(1), 80–91. http://dx.doi.org/10.1016/j.immuni.2008.11.010.

Lv, Y., Qi, R., Xu, J., Di, Z., Zheng, H., Huo, W., et al. (2014). Profiling of serum and urinary microRNAs in children with atopic dermatitis. *PLoS One, 9*(12), e115448. http://dx.doi.org/10.1371/journal.pone.0115448.

Ma, C. S., Chew, G. Y., Simpson, N., Priyadarshi, A., Wong, M., Grimbacher, B., et al. (2008). Deficiency of Th17 cells in hyper IgE syndrome due to mutations in STAT3. *Journal of Experimental Medicine, 205*(7), 1551–1557. http://dx.doi.org/10.1084/jem.20080218.

Maes, T., Cobos, F. A., Schleich, F., Sorbello, V., Henket, M., De Preter, K., et al. (2016). Asthma inflammatory phenotypes show differential microRNA expression in sputum. *The Journal of Allergy and Clinical Immunology, 137*(5), 1433–1446. http://dx.doi.org/10.1016/j.jaci.2016.02.018.

Malmhall, C., Alawieh, S., Lu, Y., Sjostrand, M., Bossios, A., Eldh, M., et al. (2014). MicroRNA-155 is essential for T(H)2-mediated allergen-induced eosinophilic inflammation in the lung. *The Journal of Allergy and Clinical Immunology, 133*(5). http://dx.doi.org/10.1016/j.jaci.2013.11.008, 1429-1438, 1438.e1421-1427.

Martinez-Nunez, R. T., Bondanese, V. P., Louafi, F., Francisco-Garcia, A. S., Rupani, H., Bedke, N., et al. (2014). A microRNA network dysregulated in asthma controls IL-6 production in bronchial epithelial cells. *PLoS One, 9*(10), e111659. http://dx.doi.org/10.1371/journal.pone.0111659.

Masyuk, A. I., Masyuk, T. V., & Larusso, N. F. (2013). Exosomes in the pathogenesis, diagnostics and therapeutics of liver diseases. *Journal of Hepatology, 59*(3), 621–625. http://dx.doi.org/10.1016/j.jhep.2013.03.028.

Mattes, J., Collison, A., Plank, M., Phipps, S., & Foster, P. S. (2009). Antagonism of microRNA-126 suppresses the effector function of TH2 cells and the development of allergic airways disease. *Proceedings of the National Academy of Sciences of the United States of America, 106*(44), 18704–18709. http://dx.doi.org/10.1073/pnas.0905063106.

Mazzeo, C., Canas, J. A., Zafra, M. P., Rojas Marco, A., Fernandez-Nieto, M., Sanz, V., ... Del Pozo, V. (2015). Exosome secretion by eosinophils: A possible role in asthma pathogenesis. *The Journal of Allergy and Clinical Immunology, 135*(6), 1603–1613. http://dx.doi.org/10.1016/j.jaci.2014.11.026.

Ma, L., Young, J., Prabhala, H., Pan, E., Mestdagh, P., Muth, D., et al. (2010). miR-9, a MYC/MYCN-activated microRNA, regulates E-cadherin and cancer metastasis. *Nature Cell Biology, 12*(3), 247–256. http://dx.doi.org/10.1038/ncb2024.

McCoy, C. E., Sheedy, F. J., Qualls, J. E., Doyle, S. L., Quinn, S. R., Murray, P. J., et al. (2010). IL-10 inhibits miR-155 induction by toll-like receptors. *Journal of Biological Chemistry, 285*(27), 20492–20498. http://dx.doi.org/10.1074/jbc.M110.102111.

Mestdagh, P., Hartmann, N., Baeriswyl, L., Andreasen, D., Bernard, N., Chen, C., et al. (2014). Evaluation of quantitative miRNA expression platforms in the microRNA quality control (miRQC) study. *Nature Methods, 11*(8), 809–815. http://dx.doi.org/10.1038/nmeth.3014.

Milger, K., Gotschke, J., Krause, L., Nathan, P., Alessandrini, F., Tufman, A., et al. (2017). Identification of a plasma miRNA biomarker signature for allergic asthma: A translational approach. *Allergy, 72*(12), 1962–1971. http://dx.doi.org/10.1111/all.13205.

Mitchell, P. S., Parkin, R. K., Kroh, E. M., Fritz, B. R., Wyman, S. K., Pogosova-Agadjanyan, E. L., et al. (2008). Circulating microRNAs as stable blood-based markers for cancer detection. *Proceedings of the National Academy of Sciences of the United States of America, 105*(30), 10513–10518. http://dx.doi.org/10.1073/pnas.0804549105.

Mittelbrunn, M., Gutierrez-Vazquez, C., Villarroya-Beltri, C., Gonzalez, S., Sanchez-Cabo, F., Gonzalez, M. A., et al. (2011). Unidirectional transfer of microRNA-loaded exosomes from T cells to antigen-presenting cells. *Nature Communications, 2*, 282. http://dx.doi.org/10.1038/ncomms1285.

Mohsenin, V. (2016). The emerging role of microRNAs in hypoxia-induced pulmonary hypertension. *Sleep Breath, 20*(3), 1059–1067. http://dx.doi.org/10.1007/s11325-016-1351-y.

Molnar, A., Melnyk, C., & Baulcombe, D. C. (2011). Silencing signals in plants: A long journey for small RNAs. *Genome Biology, 12*(1), 215. http://dx.doi.org/10.1186/gb-2010-11-12-219.

Momen-Heravi, F., Bala, S., Bukong, T., & Szabo, G. (2014). Exosome-mediated delivery of functionally active miRNA-155 inhibitor to macrophages. *Nanomedicine, 10*(7), 1517–1527. http://dx.doi.org/10.1016/j.nano.2014.03.014.

Mondoulet, L., Dioszeghy, V., Busato, F., Plaquet, C., Dhelft, V., Bethune, K., et al. (2019). Gata3 hypermethylation and Foxp3 hypomethylation are associated with sustained protection and bystander effect following epicutaneous immunotherapy in peanut-sensitized mice. *Allergy, 74*(1), 152–164. http://dx.doi.org/10.1111/all.13479.

Monroig Pdel, C., Chen, L., Zhang, S., & Calin, G. A. (2015). Small molecule compounds targeting miRNAs for cancer therapy. *Advanced Drug Delivery Reviews, 81*, 104–116. http://dx.doi.org/10.1016/j.addr.2014.09.002.

Monteys, A. M., Spengler, R. M., Wan, J., Tecedor, L., Lennox, K. A., Xing, Y., et al. (2010). Structure and activity of putative intronic miRNA promoters. *RNA, 16*(3), 495–505. http://dx.doi.org/10.1261/rna.1731910.

Montgomery, R. L., Yu, G., Latimer, P. A., Stack, C., Robinson, K., Dalby, C. M., et al. (2014). MicroRNA mimicry blocks pulmonary fibrosis. *EMBO Molecular Medicine, 6*(10), 1347–1356. http://dx.doi.org/10.15252/emmm.201303604.
Monticelli, S., Ansel, K. M., Xiao, C., Socci, N. D., Krichevsky, A. M., Thai, T. H., et al. (2005). MicroRNA profiling of the murine hematopoietic system. *Genome Biology, 6*(8), R71. http://dx.doi.org/10.1186/gb-2005-6-8-r71.
Mori, M. A., Ludwig, R. G., Garcia-Martin, R., Brandao, B. B., & Kahn, C. R. (2019). Extracellular miRNAs: From biomarkers to mediators of physiology and disease. *Cell Metabolism, 30*(4), 656–673. http://dx.doi.org/10.1016/j.cmet.2019.07.011.
Mukherji, S., Ebert, M. S., Zheng, G. X., Tsang, J. S., Sharp, P. A., & van Oudenaarden, A. (2011). MicroRNAs can generate thresholds in target gene expression. *Nature Genetics, 43*(9), 854–859. http://dx.doi.org/10.1038/ng.905.
Muljo, S. A., Ansel, K. M., Kanellopoulou, C., Livingston, D. M., Rao, A., & Rajewsky, K. (2005). Aberrant T cell differentiation in the absence of dicer. *Journal of Experimental Medicine, 202*(2), 261–269. http://dx.doi.org/10.1084/jem.20050678.
Murugaiyan, G., da Cunha, A. P., Ajay, A. K., Joller, N., Garo, L. P., Kumaradevan, S., et al. (2015). MicroRNA-21 promotes Th17 differentiation and mediates experimental autoimmune encephalomyelitis. *Journal of Clinical Investigation, 125*(3), 1069–1080. http://dx.doi.org/10.1172/jci74347.
Mycko, M. P., Cichalewska, M., Machlanska, A., Cwiklinska, H., Mariasiewicz, M., & Selmaj, K. W. (2012). MicroRNA-301a regulation of a T-helper 17 immune response controls autoimmune demyelination. *Proceedings of the National Academy of Sciences of the United States of America, 109*(20), E1248–E1257. http://dx.doi.org/10.1073/pnas.1114325109.
Na, Y. J., & Kim, J. H. (2013). Understanding cooperativity of microRNAs via microRNA association networks. *BMC Genomics, 14*(Suppl. 5), S17. http://dx.doi.org/10.1186/1471-2164-14-s5-s17.
Nam, J. W., Rissland, O. S., Koppstein, D., Abreu-Goodger, C., Jan, C. H., Agarwal, V., et al. (2014). Global analyses of the effect of different cellular contexts on microRNA targeting. *Molecular Cell, 53*(6), 1031–1043. http://dx.doi.org/10.1016/j.molcel.2014.02.013.
Nazari-Jahantigh, M., Wei, Y., Noels, H., Akhtar, S., Zhou, Z., Koenen, R. R., et al. (2012). MicroRNA-155 promotes atherosclerosis by repressing Bcl6 in macrophages. *Journal of Clinical Investigation, 122*(11), 4190–4202. http://dx.doi.org/10.1172/jci61716.
Newman, M. A., Thomson, J. M., & Hammond, S. M. (2008). Lin-28 interaction with the let-7 precursor loop mediates regulated microRNA processing. *RNA, 14*(8), 1539–1549. http://dx.doi.org/10.1261/rna.1155108.
Noh, K., Kim, M., Kim, Y., Kim, H., Kim, H., Byun, J., et al. (2017). miR-122-SOCS1-JAK2 axis regulates allergic inflammation and allergic inflammation-promoted cellular interactions. *Oncotarget, 8*(38), 63155–63176. http://dx.doi.org/10.18632/oncotarget.19149.
North, R. J., & Jung, Y. J. (2004). Immunity to tuberculosis. *Annual Review of Immunology, 22*, 599–623. http://dx.doi.org/10.1146/annurev.immunol.22.012703.104635.
Noval Rivas, M., Burton, O. T., Wise, P., Charbonnier, L. M., Georgiev, P., Oettgen, H. C., et al. (2015). Regulatory T cell reprogramming toward a Th2-cell-like lineage impairs oral tolerance and promotes food allergy. *Immunity, 42*(3), 512–523. http://dx.doi.org/10.1016/j.immuni.2015.02.004.
O'Connell, R. M., Chaudhuri, A. A., Rao, D. S., & Baltimore, D. (2009). Inositol phosphatase SHIP1 is a primary target of miR-155. *Proceedings of the National Academy of Sciences of the United States of America, 106*(17), 7113–7118. http://dx.doi.org/10.1073/pnas.0902636106.

O'Connell, R. M., Taganov, K. D., Boldin, M. P., Cheng, G., & Baltimore, D. (2007). MicroRNA-155 is induced during the macrophage inflammatory response. *Proceedings of the National Academy of Sciences of the United States of America, 104*(5), 1604–1609. http://dx.doi.org/10.1073/pnas.0610731104.

O'Shea, J. J., & Paul, W. E. (2010). Mechanisms underlying lineage commitment and plasticity of helper $CD4^+$ T cells. *Science, 327*(5969), 1098–1102. http://dx.doi.org/10.1126/science.1178334.

Oettgen, H. C., & Geha, R. S. (1999). IgE in asthma and atopy: Cellular and molecular connections. *Journal of Clinical Investigation, 104*(7), 829–835. http://dx.doi.org/10.1172/JCI8205.

Ohno, S., Drummen, G. P., & Kuroda, M. (2016). Focus on extracellular vesicles: Development of extracellular vesicle-based therapeutic systems. *International Journal of Molecular, 17*(2), 172. http://dx.doi.org/10.3390/ijms17020172.

Okada, C., Yamashita, E., Lee, S. J., Shibata, S., Katahira, J., Nakagawa, A., et al. (2009). A high-resolution structure of the pre-microRNA nuclear export machinery. *Science, 326*(5957), 1275–1279. http://dx.doi.org/10.1126/science.1178705.

Okoye, I. S., Czieso, S., Ktistaki, E., Roderick, K., Coomes, S. M., Pelly, V. S., et al. (2014). Transcriptomics identified a critical role for Th2 cell-intrinsic miR-155 in mediating allergy and antihelminth immunity. *Proceedings of the National Academy of Sciences of the United States of America, 111*(30), E3081–3090. http://dx.doi.org/10.1073/pnas.1406322111.

Oktemer, T., Altintoprak, N., Muluk, N. B., Senturk, M., Kar, M., Bafaqeeh, S. A., et al. (2016). Clinical efficacy of immunotherapy in allergic rhinitis. *American Journal of Rhinology & Allergy, 30*(5), 4–7. http://dx.doi.org/10.2500/ajra.2016.30.4368.

Orkin, S. H., & Zon, L. I. (2008). Hematopoiesis: An evolving paradigm for stem cell biology. *Cell, 132*(4), 631–644. http://dx.doi.org/10.1016/j.cell.2008.01.025.

Orom, U. A., Nielsen, F. C., & Lund, A. H. (2008). MicroRNA-10a binds the 5'UTR of ribosomal protein mRNAs and enhances their translation. *Molecular Cell, 30*(4), 460–471. http://dx.doi.org/10.1016/j.molcel.2008.05.001.

Ouyang, W., Ranganath, S. H., Weindel, K., Bhattacharya, D., Murphy, T. L., Sha, W. C., et al. (1998). Inhibition of Th1 development mediated by GATA-3 through an IL-4-independent mechanism. *Immunity, 9*(5), 745–755. http://dx.doi.org/10.1016/s1074-7613(00)80671-8.

Ozsolak, F., Poling, L. L., Wang, Z., Liu, H., Liu, X. S., Roeder, R. G., et al. (2008). Chromatin structure analyses identify miRNA promoters. *Genes & Development, 22*(22), 3172–3183. http://dx.doi.org/10.1101/gad.1706508.

Panganiban, R. P., Pinkerton, M. H., Maru, S. Y., Jefferson, S. J., Roff, A. N., & Ishmael, F. T. (2012). Differential microRNA epression in asthma and the role of miR-1248 in regulation of IL-5. *American Journal of Clinical and Experimental Immunology, 1*(2), 154–165.

Panganiban, R. P., Wang, Y., Howrylak, J., Chinchilli, V. M., Craig, T. J., August, A., et al. (2016). Circulating microRNAs as biomarkers in patients with allergic rhinitis and asthma. *The Journal of Allergy and Clinical Immunology, 137*(5), 1423–1432. http://dx.doi.org/10.1016/j.jaci.2016.01.029.

Panzer, M., Sitte, S., Wirth, S., Drexler, I., Sparwasser, T., & Voehringer, D. (2012). Rapid in vivo conversion of effector T cells into Th2 cells during helminth infection. *The Journal of Immunology, 188*(2), 615–623. http://dx.doi.org/10.4049/jimmunol.1101164.

Park, C. Y., Jeker, L. T., Carver-Moore, K., Oh, A., Liu, H. J., Cameron, R., et al. (2012). A resource for the conditional ablation of microRNAs in the mouse. *Cell Reports, 1*(4), 385–391. http://dx.doi.org/10.1016/j.celrep.2012.02.008.

Patel, D. D., & Kuchroo, V. K. (2015). Th17 cell pathway in human immunity: Lessons from genetics and therapeutic interventions. *Immunity, 43*(6), 1040–1051. http://dx.doi.org/10.1016/j.immuni.2015.12.003.

Pathak, S., Grillo, A. R., Scarpa, M., Brun, P., D'Incà, R., Nai, L., et al. (2015). MiR-155 modulates the inflammatory phenotype of intestinal myofibroblasts by targeting SOCS1 in ulcerative colitis. *Experimental & Molecular Medicine, 47*(5). http://dx.doi.org/10.1038/emm.2015.21. e164-e164.

Patrick, D. M., Montgomery, R. L., Qi, X., Obad, S., Kauppinen, S., Hill, J. A., et al. (2010). Stress-dependent cardiac remodeling occurs in the absence of microRNA-21 in mice. *Journal of Clinical Investigation, 120*(11), 3912–3916. http://dx.doi.org/10.1172/jci43604.

Pauley, K. M., Cha, S., & Chan, E. K. (2009). MicroRNA in autoimmunity and autoimmune diseases. *Journal of Autoimmunity, 32*(3–4), 189–194. http://dx.doi.org/10.1016/j.jaut.2009.02.012.

Pawankar, R. (2014). Allergic diseases and asthma: A global public health concern and a call to action. *World Allergy Organization Journal, 7*(1), 1–3. http://dx.doi.org/10.1186/1939-4551-7-12.

Pegtel, D. M., Cosmopoulos, K., Thorley-Lawson, D. A., van Eijndhoven, M. A., Hopmans, E. S., Lindenberg, J. L., et al. (2010). Functional delivery of viral miRNAs via exosomes. *Proceedings of the National Academy of Sciences of the United States of America, 107*(14), 6328–6333. http://dx.doi.org/10.1073/pnas.0914843107.

Pitt, J. M., André, F., Amigorena, S., Soria, J. C., Eggermont, A., Kroemer, G., et al. (2016). Dendritic cell-derived exosomes for cancer therapy. *Journal of Clinical Investigation, 126*(4), 1224–1232. http://dx.doi.org/10.1172/jci81137.

Podshivalova, K., & Salomon, D. R. (2013). MicroRNA regulation of T-lymphocyte immunity: Modulation of molecular networks responsible for T-cell activation, differentiation, and development. *Critical Reviews in Immunology, 33*(5), 435–476.

Polikepahad, S., Knight, J. M., Naghavi, A. O., Oplt, T., Creighton, C. J., Shaw, C., et al. (2010). Proinflammatory role for let-7 microRNAS in experimental asthma. *Journal of Biological Chemistry, 285*(39), 30139–30149. http://dx.doi.org/10.1074/jbc.M110.145698.

Pons-Espinal, M., de Luca, E., Marzi, M. J., Beckervordersandforth, R., Armirotti, A., Nicassio, F., et al. (2017). Synergic functions of miRNAs determine neuronal fate of adult neural stem cells. *Stem Cell Reports, 8*(4), 1046–1061. http://dx.doi.org/10.1016/j.stemcr.2017.02.012.

Potaczek, D. P., Harb, H., Michel, S., Alhamwe, B. A., Renz, H., & Tost, J. (2017). Epigenetics and allergy: From basic mechanisms to clinical applications. *Epigenomics, 9*(4), 539–571. http://dx.doi.org/10.2217/epi-2016-0162.

Prado, N., Marazuela, E. G., Segura, E., Fernandez-Garcia, H., Villalba, M., Thery, C., et al. (2008). Exosomes from bronchoalveolar fluid of tolerized mice prevent allergic reaction. *Journal of Immunology, 181*(2), 1519–1525. http://dx.doi.org/10.4049/jimmunol.181.2.1519.

Pritchard, C. C., Cheng, H. H., & Tewari, M. (2012). MicroRNA profiling: Approaches and considerations. *Nature Reviews Genetics, 13*(5), 358–369. http://dx.doi.org/10.1038/nrg3198.

Prosser, H. M., Koike-Yusa, H., Cooper, J. D., Law, F. C., & Bradley, A. (2011). A resource of vectors and ES cells for targeted deletion of microRNAs in mice. *Nature Biotechnology, 29*(9), 840–845. http://dx.doi.org/10.1038/nbt.1929.

Pua, H. H., Steiner, D. F., Patel, S., Gonzalez, J. R., Ortiz-Carpena, J. F., Kageyama, R., et al. (2016). MicroRNAs 24 and 27 suppress allergic inflammation and target a network of regulators of T helper 2 cell-associated cytokine production. *Immunity, 44*(4), 821–832. http://dx.doi.org/10.1016/j.immuni.2016.01.003.

Pulendran, B., & Artis, D. (2012). New paradigms in type 2 immunity. *Science, 337*(6093), 431–435. http://dx.doi.org/10.1126/science.1221064.

Qayum, A. A., Paranjape, A., Abebayehu, D., Kolawole, E. M., Haque, T. T., McLeod, J. J., et al. (2016). IL-10-Induced miR-155 targets SOCS1 to enhance IgE-mediated mast cell function. *Journal of Immunology, 196*(11), 4457–4467. http://dx.doi.org/10.4049/jimmunol.1502240.

Qin, H. B., Xu, B., Mei, J. J., Li, D., Liu, J. J., Zhao, D. Y., et al. (2012). Inhibition of miRNA-221 suppresses the airway inflammation in asthma. *Inflammation, 35*(4), 1595–1599. http://dx.doi.org/10.1007/s10753-012-9474-1.

Rajewsky, N. (2006). MicroRNA target predictions in animals. *Nature Genetics, 38*(Suppl. l), S8–S13. http://dx.doi.org/10.1038/ng1798.

Rebane, A., Runnel, T., Aab, A., Maslovskaja, J., Ruckert, B., Zimmermann, M., et al. (2014). MicroRNA-146a alleviates chronic skin inflammation in atopic dermatitis through suppression of innate immune responses in keratinocytes. *The Journal of Allergy and Clinical Immunology, 134*(4). http://dx.doi.org/10.1016/j.jaci.2014.05.022, 836-847.e811.

Record, M., Carayon, K., Poirot, M., & Silvente-Poirot, S. (2014). Exosomes as new vesicular lipid transporters involved in cell-cell communication and various pathophysiologies. *Biochimica et Biophysica Acta, 1841*(1), 108–120. http://dx.doi.org/10.1016/j.bbalip.2013.10.004.

de Rie, D., Abugessaisa, I., Alam, T., Arner, E., Arner, P., Ashoor, H., et al. (2017). An integrated expression atlas of miRNAs and their promoters in human and mouse. *Nature Biotechnology, 35*(9), 872–878. http://dx.doi.org/10.1038/nbt.3947.

Robbins, P. D., Dorronsoro, A., & Booker, C. N. (2016). Regulation of chronic inflammatory and immune processes by extracellular vesicles. *Journal of Clinical Investigation, 126*(4), 1173–1180. http://dx.doi.org/10.1172/jci81131.

Rodrigo-Munoz, J. M., Canas, J. A., Sastre, B., Rego, N., Greif, G., Rial, M., … Del Pozo, V. (2019). Asthma diagnosis using integrated analysis of eosinophil microRNAs. *Allergy, 74*(3), 507–517. http://dx.doi.org/10.1111/all.13570.

Rodriguez, A., Vigorito, E., Clare, S., Warren, M. V., Couttet, P., Soond, D. R., et al. (2007). Requirement of bic/microRNA-155 for normal immune function. *Science, 316*(5824), 608–611. http://dx.doi.org/10.1126/science.1139253.

Romanet-Manent, S., Charpin, D., Magnan, A., Lanteaume, A., Vervloet, D., & Group, E. C. (2002). Allergic vs nonallergic asthma: What makes the difference? *Allergy, 57*(7), 607–613. http://dx.doi.org/10.1034/j.1398-9995.2002.23504.x.

Rosa, A., Ballarino, M., Sorrentino, A., Sthandier, O., De Angelis, F. G., Marchioni, M., et al. (2007). The interplay between the master transcription factor PU.1 and miR-424 regulates human monocyte/macrophage differentiation. *Proceedings of the National Academy of Sciences of the United States of America, 104*(50), 19849–19854. http://dx.doi.org/10.1073/pnas.0706963104.

Rossi, R. L., Rossetti, G., Wenandy, L., Curti, S., Ripamonti, A., Bonnal, R. J., et al. (2011). Distinct microRNA signatures in human lymphocyte subsets and enforcement of the naive state in $CD4^+$ T cells by the microRNA miR-125b. *Nature Immunology, 12*(8), 796–803. http://dx.doi.org/10.1038/ni.2057.

Rouas, R., Fayyad-Kazan, H., El Zein, N., Lewalle, P., Rothe, F., Simion, A., et al. (2009). Human natural treg microRNA signature: Role of microRNA-31 and microRNA-21 in FOXP3 expression. *Eur Journal of Immunology, 39*(6), 1608–1618. http://dx.doi.org/10.1002/eji.200838509.

Ruby, J. G., Jan, C. H., & Bartel, D. P. (2007). Intronic microRNA precursors that bypass drosha processing. *Nature, 448*(7149), 83–86. http://dx.doi.org/10.1038/nature05983.

Rupp, L. J., Brady, B. L., Carpenter, A. C., De Obaldia, M. E., Bhandoola, A., Bosselut, R., et al. (2014). The microRNA biogenesis machinery modulates lineage commitment during alphabeta T cell development. *Journal of Immunology, 193*(8), 4032–4042. http://dx.doi.org/10.4049/jimmunol.1401359.

Rybak, A., Fuchs, H., Smirnova, L., Brandt, C., Pohl, E. E., Nitsch, R., et al. (2008). A feedback loop comprising lin-28 and let-7 controls pre-let-7 maturation during neural stem-cell commitment. *Nature Cell Biology, 10*(8), 987–993. http://dx.doi.org/10.1038/ncb1759.

Saco, T. V., Pepper, A., & Casale, T. B. (2018). Uses of biologics in allergic diseases: What to choose and when. *Annals of Allergy, Asthma & Immunology, 120*(4), 357–366. http://dx.doi.org/10.1016/j.anai.2018.02.029.

Sahu, M., & Mallick, B. (2018). Deciphering synergistic regulatory networks of microRNAs in hESCs and fibroblasts. *International Journal if Biological Macromolecules, 113*, 1279–1286. http://dx.doi.org/10.1016/j.ijbiomac.2018.03.061.

Sakaguchi, S., Yamaguchi, T., Nomura, T., & Ono, M. (2008). Regulatory T cells and immune tolerance. *Cell, 133*(5), 775–787. http://dx.doi.org/10.1016/j.cell.2008.05.009.

Sampson, H. A., Munoz-Furlong, A., Campbell, R. L., Adkinson, N. F., Jr., Bock, S. A., Branum, A., et al. (2006). Second symposium on the definition and management of anaphylaxis: Summary report–second national Institute of allergy and infectious disease/food allergy and anaphylaxis network symposium. *The Journal of Allergy and Clinical Immunology, 117*(2), 391–397. http://dx.doi.org/10.1016/j.jaci.2005.12.1303.

Sarver, A. L., Sarver, A. E., Yuan, C., & Subramanian, S. (2018). Omcd: OncomiR cancer database. *BMC Cancer, 18*(1), 1223. http://dx.doi.org/10.1186/s12885-018-5085-z.

Sastre, B., Cañas, J. A., Rodrigo-Muñoz, J. M., & Del Pozo, V. (2017). Novel modulators of asthma and allergy: Exosomes and microRNAs. *Frontiers in Immunology, 8*, 826. http://dx.doi.org/10.3389/fimmu.2017.00826.

Sato, F., Tsuchiya, S., Meltzer, S. J., & Shimizu, K. (2011). MicroRNAs and epigenetics. *The FEBS Journal, 278*(10), 1598–1609. http://dx.doi.org/10.1111/j.1742-4658.2011.08089.x.

Sawant, D. V., Wu, H., Kaplan, M. H., & Dent, A. L. (2013). The Bcl6 target gene microRNA-21 promotes Th2 differentiation by a T cell intrinsic pathway. *Molecular Immunology, 54*(3–4), 435–442. http://dx.doi.org/10.1016/j.molimm.2013.01.006.

Sawant, D. V., Yao, W., Wright, Z., Sawyers, C., Tepper, R. S., Gupta, S. K., et al. (2015). Serum microRNA-21 as a biomarker for allergic inflammatory disease in children. *MicroRNA, 4*(1), 36–40.

Schmiedel, J. M., Klemm, S. L., Zheng, Y., Sahay, A., Bluthgen, N., Marks, D. S., et al. (2015). Gene expression. MicroRNA control of protein expression noise. *Science, 348*(6230), 128–132. http://dx.doi.org/10.1126/science.aaa1738.

Seto, A. G., Beatty, X., Lynch, J. M., Hermreck, M., Tetzlaff, M., Duvic, M., & Jackson, A. L. (2018). Cobomarsen, an oligonucleotide inhibitor of miR-155, co-ordinately regulates multiple survival pathways to reduce cellular proliferation and survival in cutaneous T-cell lymphoma. *British Journal of Haematology, 183*(3), 428–444. http://dx.doi.org/10.1111/bjh.15547.

Seumois, G., Vijayanand, P., Eisley, C. J., Omran, N., Kalinke, L., North, M., et al. (2012). An integrated nano-scale approach to profile miRNAs in limited clinical samples. *American Journal of Clinical and Experimental Immunology, 1*(2), 70–89.

Sharma, A., Kumar, M., Ahmad, T., Mabalirajan, U., Aich, J., Agrawal, A., et al. (2012). Antagonism of mmu-mir-106a attenuates asthma features in allergic murine model. *Journal of Applied Physiology (1985), 113*(3), 459–464. http://dx.doi.org/10.1152/japplphysiol.00001.2012.

Sharma, A., Kumar, M., Aich, J., Hariharan, M., Brahmachari, S. K., Agrawal, A., et al. (2009). Post transcriptional regulation of interleukin-10 expression by hsa-miR-106a. *Proceedings of the National Academy of Sciences of the United States of America, 106*(14), 5761–5766. http://dx.doi.org/10.1073/pnas.0808743106.

Sheedy, F. J., Palsson-McDermott, E., Hennessy, E. J., Martin, C., O'Leary, J. J., Ruan, Q., et al. (2010). Negative regulation of TLR4 via targeting of the proinflammatory tumor suppressor PDCD4 by the microRNA miR-21. *Nature Immunology, 11*(2), 141–147. http://dx.doi.org/10.1038/ni.1828.

Shen, C.-B., Yu, L., & Wang, S. (2018). miR-396 ameliorate allergic inflammation in a mouse model of asthma. *The Journal of Immunology, 200*(1 Suppl. ment), 44.23-44.23.

Shroba, J., Rath, N., & Barnes, C. (2018). Possible role of environmental factors in the development of food allergies. *Clinical Reviews in Allergy & Immunology*. http://dx.doi.org/10.1007/s12016-018-8703-2.

da Silveira, J. C., Veeramachaneni, D. N., Winger, Q. A., Carnevale, E. M., & Bouma, G. J. (2012). Cell-secreted vesicles in equine ovarian follicular fluid contain miRNAs and proteins: A possible new form of cell communication within the ovarian follicle. *Biology of Reproduction, 86*(3), 71. http://dx.doi.org/10.1095/biolreprod.111.093252.

Simpson, L. J., Patel, S., Bhakta, N. R., Choy, D. F., Brightbill, H. D., Ren, X., et al. (2014). A microRNA upregulated in asthma airway T cells promotes TH2 cytokine production. *Nature Immunology, 15*(12), 1162–1170. http://dx.doi.org/10.1038/ni.3026.

Singh, P. B., Pua, H. H., Happ, H. C., Schneider, C., von Moltke, J., Locksley, R. M., et al. (2017). MicroRNA regulation of type 2 innate lymphoid cell homeostasis and function in allergic inflammation. *The Journal of Experimental Medicine, 214*(12), 3627–3643. http://dx.doi.org/10.1084/jem.20170545.

Sinha, A., Yadav, A. K., Chakraborty, S., Kabra, S. K., Lodha, R., Kumar, M., et al. (2013). Exosome-enclosed microRNAs in exhaled breath hold potential for biomarker discovery in patients with pulmonary diseases. *The Journal of Allergy and Clinical Immunology, 132*(1), 219–222. http://dx.doi.org/10.1016/j.jaci.2013.03.035.

Solberg, O. D., Ostrin, E. J., Love, M. I., Peng, J. C., Bhakta, N. R., Hou, L., et al. (2012). Airway epithelial miRNA expression is altered in asthma. *American Journal of Respiratory and Critical Care Medicine, 186*(10), 965–974. http://dx.doi.org/10.1164/rccm.201201-0027OC.

Sonkoly, E., Janson, P., Majuri, M. L., Savinko, T., Fyhrquist, N., Eidsmo, L., et al. (2010). MiR-155 is overexpressed in patients with atopic dermatitis and modulates T-cell proliferative responses by targeting cytotoxic T lymphocyte-associated antigen 4. *The Journal of Allergy and Clinical Immunology, 126*(3). http://dx.doi.org/10.1016/j.jaci.2010.05.045, 581-589.e581-520.

Sticht, C., De La Torre, C., Parveen, A., & Gretz, N. (2018). miRWalk: An online resource for prediction of microRNA binding sites. *PLoS One, 13*(10), e0206239. http://dx.doi.org/10.1371/journal.pone.0206239.

Stittrich, A. B., Haftmann, C., Sgouroudis, E., Kuhl, A. A., Hegazy, A. N., Panse, I., et al. (2010). The microRNA miR-182 is induced by IL-2 and promotes clonal expansion of activated helper T lymphocytes. *Nature Immunology, 11*(11), 1057–1062. http://dx.doi.org/10.1038/ni.1945.

Stone, K. D., Prussin, C., & Metcalfe, D. D. (2010). IgE, mast cells, basophils, and eosinophils. *The Journal of Allergy and Clinical Immunology, 125*(2 Suppl. 2). http://dx.doi.org/10.1016/j.jaci.2009.11.017. S73-80.

Strovas, T. J., Rosenberg, A. B., Kuypers, B. E., Muscat, R. A., & Seelig, G. (2014). MicroRNA-based single-gene circuits buffer protein synthesis rates against perturbations. *ACS Synthetic Biology, 3*(5), 324–331. http://dx.doi.org/10.1021/sb4001867.

Subra, C., Grand, D., Laulagnier, K., Stella, A., Lambeau, G., Paillasse, M., et al. (2010). Exosomes account for vesicle-mediated transcellular transport of activatable phospholipases and prostaglandins. *Journal of Lipid Research, 51*(8), 2105–2120. http://dx.doi.org/10.1194/jlr.M003657.

Suzuki, H. I., Yamagata, K., Sugimoto, K., Iwamoto, T., Kato, S., & Miyazono, K. (2009). Modulation of microRNA processing by p53. *Nature, 460*(7254), 529–533. http://dx.doi.org/10.1038/nature08199.

Suzuki, M., Konno, S., Makita, H., Shimizu, K., Kimura, H., Kimura, H., & Nishimura, M. (2016). Altered circulating exosomal RNA profiles detected by next-generation sequencing in patients with severe asthma. *European Respiratory Journal, 48*, PA3410. http://dx.doi.org/10.1183/13993003.congress-2016.PA3410.

Syed, A., Garcia, M. A., Lyu, S. C., Bucayu, R., Kohli, A., Ishida, S., et al. (2014). Peanut oral immunotherapy results in increased antigen-induced regulatory T-cell function and hypomethylation of forkhead box protein 3 (FOXP3). *The Journal of Allergy and Clinical Immunology, 133*(2), 500–510. http://dx.doi.org/10.1016/j.jaci.2013.12.1037.

Szymczak, I., Wieczfinska, J., & Pawliczak, R. (2016). Molecular background of miRNA role in asthma and COPD: An updated insight. *BioMed Research International*, 2016, 1–10. http://dx.doi.org/10.1155/2016/7802521.

Taganov, K. D., Boldin, M. P., Chang, K. J., & Baltimore, D. (2006). NF-kappaB-dependent induction of microRNA miR-146, an inhibitor targeted to signaling proteins of innate immune responses. *Proceedings of the National Academy of Sciences of the United States of America, 103*(33), 12481–12486. http://dx.doi.org/10.1073/pnas.0605298103.

Tahamtan, A., Teymoori-Rad, M., Nakstad, B., & Salimi, V. (2018). Anti-inflammatory MicroRNAs and their potential for inflammatory diseases treatment. *Frontiers in Immunology, 9*, 1377. http://dx.doi.org/10.3389/fimmu.2018.01377.

Takahashi, H., Kanno, T., Nakayamada, S., Hirahara, K., Sciume, G., Muljo, S. A., et al. (2012). TGF-beta and retinoic acid induce the microRNA miR-10a, which targets Bcl-6 and constrains the plasticity of helper T cells. *Nature Immunology, 13*(6), 587–595. http://dx.doi.org/10.1038/ni.2286.

Takyar, S., Vasavada, H., Zhang, J. G., Ahangari, F., Niu, N., Liu, Q., et al. (2013). VEGF controls lung Th2 inflammation via the miR-1-Mpl (myeloproliferative leukemia virus oncogene)-P-selectin axis. *Journal of Experimental Medicine, 210*(10), 1993–2010. http://dx.doi.org/10.1084/jem.20121200.

Taylor, M. D., van der Werf, N., & Maizels, R. M. (2012). T cells in helminth infection: The regulators and the regulated. *Trends in Immunology, 33*(4), 181–189. http://dx.doi.org/10.1016/j.it.2012.01.001.

Teng, G., Hakimpour, P., Landgraf, P., Rice, A., Tuschl, T., Casellas, R., et al. (2008). MicroRNA-155 is a negative regulator of activation-induced cytidine deaminase. *Immunity, 28*(5), 621–629. http://dx.doi.org/10.1016/j.immuni.2008.03.015.

Tesmer, L. A., Lundy, S. K., Sarkar, S., & Fox, D. A. (2008). Th17 cells in human disease. *Immunological Reviews, 223*, 87–113. http://dx.doi.org/10.1111/j.1600-065X.2008.00628.x.

Testa, U., Pelosi, E., Castelli, G., & Labbaye, C. (2017). miR-146 and miR-155: Two key modulators of immune response and tumor development. *Noncoding RNA, 3*(3). http://dx.doi.org/10.3390/ncrna3030022.

Thery, C., Duban, L., Segura, E., Veron, P., Lantz, O., & Amigorena, S. (2002). Indirect activation of naive $CD4^+$ T cells by dendritic cell-derived exosomes. *Nature Immunology, 3*(12), 1156–1162. http://dx.doi.org/10.1038/ni854.

Thomas, M., Lieberman, J., & Lal, A. (2010). Desperately seeking microRNA targets. *Nature Structural & Molecular Biology, 17*(10), 1169–1174. http://dx.doi.org/10.1038/nsmb.1921.

Thomsen, S. F. (2015). Epidemiology and natural history of atopic diseases. *European Clinical Respiratory Journal, 2*. http://dx.doi.org/10.3402/ecrj.v2.24642.

Thomson, J. M., Newman, M., Parker, J. S., Morin-Kensicki, E. M., Wright, T., & Hammond, S. M. (2006). Extensive post-transcriptional regulation of microRNAs and its implications for cancer. *Genes & Development, 20*(16), 2202–2207. http://dx.doi.org/10.1101/gad.1444406.

Thum, T., Gross, C., Fiedler, J., Fischer, T., Kissler, S., Bussen, M., et al. (2008). MicroRNA-21 contributes to myocardial disease by stimulating MAP kinase signalling in fibroblasts. *Nature, 456*(7224), 980–984. http://dx.doi.org/10.1038/nature07511.

Tian, L., De Hertogh, G., Fedeli, M., Staats, K. A., Schonefeldt, S., Humblet-Baron, S., et al. (2012). Loss of T cell microRNA provides systemic protection against autoimmune pathology in mice. *Journal of Autoimmunity, 38*(1), 39–48. http://dx.doi.org/10.1016/j.jaut.2011.12.004.

Torregrosa Paredes, P., Esser, J., Admyre, C., Nord, M., Rahman, Q. K., Lukic, A., et al. (2012). Bronchoalveolar lavage fluid exosomes contribute to cytokine and leukotriene production in allergic asthma. *Allergy, 67*(7), 911–919. http://dx.doi.org/10.1111/j.1398-9995.2012.02835.x.

Truesdell, S. S., Mortensen, R. D., Seo, M., Schroeder, J. C., Lee, J. H., LeTonqueze, O., et al. (2012). MicroRNA-mediated mRNA translation activation in quiescent cells and oocytes involves recruitment of a nuclear microRNP. *Scientific Reports, 2*, 842. http://dx.doi.org/10.1038/srep00842.

Tsang, J., Zhu, J., & van Oudenaarden, A. (2007). MicroRNA-mediated feedback and feed-forward loops are recurrent network motifs in mammals. *Molecular Cell, 26*(5), 753–767. http://dx.doi.org/10.1016/j.molcel.2007.05.018.

Tsitsiou, E., Williams, A. E., Moschos, S. A., Patel, K., Rossios, C., Jiang, X., et al. (2012). Transcriptome analysis shows activation of circulating $CD8^+$ T cells in patients with severe asthma. *The Journal of Allergy and Clinical Immunology, 129*(1), 95–103. http://dx.doi.org/10.1016/j.jaci.2011.08.011.

Turchinovich, A., & Burwinkel, B. (2012). Distinct AGO1 and AGO2 associated miRNA profiles in human cells and blood plasma. *RNA Biology, 9*(8), 1066–1075. http://dx.doi.org/10.4161/rna.21083.

Ueno, A., Ghosh, A., Hung, D., Li, J., & Jijon, H. (2015). Th17 plasticity and its changes associated with inflammatory bowel disease. *World Journal of Gastroenterology, 21*(43), 12283–12295. http://dx.doi.org/10.3748/wjg.v21.i43.12283.

Vasudevan, S., & Steitz, J. A. (2007). AU-rich-element-mediated upregulation of translation by FXR1 and Argonaute 2. *Cell, 128*(6), 1105–1118. http://dx.doi.org/10.1016/j.cell.2007.01.038.

Vecchione, A., Belletti, B., Lovat, F., Volinia, S., Chiappetta, G., Giglio, S., et al. (2013). A microRNA signature defines chemoresistance in ovarian cancer through modulation of angiogenesis. *Proceedings of the National Academy of Sciences of the United States of America, 110*(24), 9845–9850. http://dx.doi.org/10.1073/pnas.1305472110.

van de Veen, W., & Akdis, M. (2019). The use of biologics for immune modulation in allergic disease. *Journal of Clinical Investigation, 130*, 1452–1462. http://dx.doi.org/10.1172/jci124607.

Vella, L. J., Sharples, R. A., Nisbet, R. M., Cappai, R., & Hill, A. F. (2008). The role of exosomes in the processing of proteins associated with neurodegenerative diseases. *European Biophysics Journal, 37*(3), 323–332. http://dx.doi.org/10.1007/s00249-007-0246-z.

Vennegaard, M. T., Bonefeld, C. M., Hagedorn, P. H., Bangsgaard, N., Lovendorf, M. B., Odum, N., et al. (2012). Allergic contact dermatitis induces upregulation of identical microRNAs in humans and mice. *Contact Dermatitis, 67*(5), 298–305. http://dx.doi.org/10.1111/j.1600-0536.2012.02083.x.

Ventura, A., Young, A. G., Winslow, M. M., Lintault, L., Meissner, A., Erkeland, S. J., et al. (2008). Targeted deletion reveals essential and overlapping functions of the miR-17 through 92 family of miRNA clusters. *Cell, 132*(5), 875–886. http://dx.doi.org/10.1016/j.cell.2008.02.019.

Veremeyko, T., Siddiqui, S., Sotnikov, I., Yung, A., & Ponomarev, E. D. (2013). IL-4/IL-13-dependent and independent expression of miR-124 and its contribution to M2 phenotype of monocytic cells in normal conditions and during allergic inflammation. *PLoS One, 8*(12), e81774. http://dx.doi.org/10.1371/journal.pone.0081774.

Vian, L., Di Carlo, M., Pelosi, E., Fazi, F., Santoro, S., Cerio, A. M., et al. (2014). Transcriptional fine-tuning of microRNA-223 levels directs lineage choice of human hematopoietic progenitors. *Cell Death & Differentiation, 21*(2), 290–301. http://dx.doi.org/10.1038/cdd.2013.145.

Vickers, K. C., Palmisano, B. T., Shoucri, B. M., Shamburek, R. D., & Remaley, A. T. (2011). MicroRNAs are transported in plasma and delivered to recipient cells by high-density lipoproteins. *Nature Cell Biology, 13*(4), 423–433. http://dx.doi.org/10.1038/ncb2210.

Viswanathan, S. R., Daley, G. Q., & Gregory, R. I. (2008). Selective blockade of microRNA processing by Lin28. *Science, 320*(5872), 97–100. http://dx.doi.org/10.1126/science.1154040.

Walker, J. A., & McKenzie, A. N. (2013). Development and function of group 2 innate lymphoid cells. *Current Opinion In Immunology, 25*(2), 148–155. http://dx.doi.org/10.1016/j.coi.2013.02.010.

Wang, Y., Medvid, R., Melton, C., Jaenisch, R., & Blelloch, R. (2007). DGCR8 is essential for microRNA biogenesis and silencing of embryonic stem cell self-renewal. *Nature Genetics, 39*(3), 380–385. http://dx.doi.org/10.1038/ng1969.

Waring, B. M., Sjaastad, L. E., Fiege, J. K., Fay, E. J., Reyes, I., Moriarity, B., et al. (2018). MicroRNA-based attenuation of Influenza virus across Susceptible hosts. *Journal of Virology, 92*(2). http://dx.doi.org/10.1128/jvi.01741-17.

Warth, S. C., Hoefig, K. P., Hiekel, A., Schallenberg, S., Jovanovic, K., Klein, L., et al. (2015). Induced miR-99a expression represses Mtor cooperatively with miR-150 to promote regulatory T-cell differentiation. *The EMBO Journal, 34*(9), 1195–1213. http://dx.doi.org/10.15252/embj.201489589.

Weber, J. A., Baxter, D. H., Zhang, S., Huang, D. Y., Huang, K. H., Lee, M. J., et al. (2010). The microRNA spectrum in 12 body fluids. *Clinical Chemistry, 56*(11), 1733–1741. http://dx.doi.org/10.1373/clinchem.2010.147405.

Wightman, B., Ha, I., & Ruvkun, G. (1993). Posttranscriptional regulation of the heterochronic gene lin-14 by lin-4 mediates temporal pattern formation in *C. elegans*. *Cell, 75*(5), 855–862. http://dx.doi.org/10.1016/0092-8674(93)90530-4.

Wilbertz, J. H., Voigt, F., Horvathova, I., Roth, G., Zhan, Y., & Chao, J. A. (2019). Single-molecule imaging of mRNA localization and regulation during the integrated stress response. *Molecular Cell, 73*(5). http://dx.doi.org/10.1016/j.molcel.2018.12.006, 946-958.e947.

Williams, A. E., Larner-Svensson, H., Perry, M. M., Campbell, G. A., Herrick, S. E., Adcock, I. M., et al. (2009). MicroRNA expression profiling in mild asthmatic human airways and effect of corticosteroid therapy. *PLoS One, 4*(6), e5889. http://dx.doi.org/10.1371/journal.pone.0005889.

Wilson, A., Laurenti, E., Oser, G., van der Wath, R. C., Blanco-Bose, W., Jaworski, M., et al. (2008). Hematopoietic stem cells reversibly switch from dormancy to self-renewal during homeostasis and repair. *Cell, 135*(6), 1118–1129. http://dx.doi.org/10.1016/j.cell.2008.10.048.

Wilson, A., Laurenti, E., & Trumpp, A. (2009). Balancing dormant and self-renewing hematopoietic stem cells. *Current Opinion In Genetics & Development, 19*(5), 461–468. http://dx.doi.org/10.1016/j.gde.2009.08.005.

Winter, J., Jung, S., Keller, S., Gregory, R. I., & Diederichs, S. (2009). Many roads to maturity: microRNA biogenesis pathways and their regulation. *Nature Cell Biology, 11*(3), 228–234. http://dx.doi.org/10.1038/ncb0309-228.

Wright, C., Rajpurohit, A., Burke, E. E., Williams, C., Collado-Torres, L., Kimos, M., et al. (2019). Comprehensive assessment of multiple biases in small RNA sequencing reveals significant differences in the performance of widely used methods. *BMC Genomics, 20*(1), 513. http://dx.doi.org/10.1186/s12864-019-5870-3.

Wu, H., Neilson, J. R., Kumar, P., Manocha, M., Shankar, P., Sharp, P. A., et al. (2007). miRNA profiling of naive, effector and memory CD8 T cells. *PLoS One, 2*(10), e1020. http://dx.doi.org/10.1371/journal.pone.0001020.

Xiao, L., Jiang, L., Hu, Q., & Li, Y. (2017). MicroRNA-133b ameliorates allergic inflammation and symptom in murine model of allergic rhinitis by targeting Nlrp3. *Cellular Physiology and Biochemistry, 42*(3), 901–912. http://dx.doi.org/10.1159/000478645.

Xiao, R., Noel, A., Perveen, Z., & Penn, A. L. (2016). In utero exposure to second-hand smoke activates pro-asthmatic and oncogenic miRNAs in adult asthmatic mice. *Environmental and Molecular Mutagenesis, 57*(3), 190–199. http://dx.doi.org/10.1002/em.21998.

Xiao, C., Srinivasan, L., Calado, D. P., Patterson, H. C., Zhang, B., Wang, J., et al. (2008). Lymphoproliferative disease and autoimmunity in mice with increased miR-17-92 expression in lymphocytes. *Nature Immunology, 9*(4), 405–414. http://dx.doi.org/10.1038/ni1575.

Xu, R., Rai, A., Chen, M., Suwakulsiri, W., Greening, D. W., & Simpson, R. J. (2018). Extracellular vesicles in cancer - implications for future improvements in cancer care. *Nature Reviews Clinical Oncology, 15*(10), 617–638. http://dx.doi.org/10.1038/s41571-018-0036-9.

Yang, L., Boldin, M. P., Yu, Y., Liu, C. S., Ea, C. K., Ramakrishnan, P., et al. (2012). miR-146a controls the resolution of T cell responses in mice. *Journal of Experimental Medicine, 209*(9), 1655–1670. http://dx.doi.org/10.1084/jem.20112218.

Yang, M., Chen, J., Su, F., Yu, B., Su, F., Lin, L., et al. (2011). Microvesicles secreted by macrophages shuttle invasion-potentiating microRNAs into breast cancer cells. *Molecular Cancer, 10*, 117. http://dx.doi.org/10.1186/1476-4598-10-117.

Yang, J. S., Maurin, T., Robine, N., Rasmussen, K. D., Jeffrey, K. L., Chandwani, R., et al. (2010). Conserved vertebrate mir-451 provides a platform for Dicer-independent, Ago2-mediated microRNA biogenesis. *Proceedings of the National Academy of Sciences of the United States of America, 107*(34), 15163–15168. http://dx.doi.org/10.1073/pnas.1006432107.

Yan, Z., Li, J., Xiong, Y., Xu, W., & Zheng, G. (2012). Identification of candidate colon cancer biomarkers by applying a random forest approach on microarray data. *Oncology Reports, 28*(3), 1036–1042. http://dx.doi.org/10.3892/or.2012.1891.

Yao, W., Guo, G., Zhang, Q., Fan, L., Wu, N., & Bo, Y. (2014). The application of multiple miRNA response elements enables oncolytic adenoviruses to possess specificity to glioma cells. *Virology, 458–459*, 69–82. http://dx.doi.org/10.1016/j.virol.2014.04.007.

Yao, R., Ma, Y. L., Liang, W., Li, H. H., Ma, Z. J., Yu, X., et al. (2012). MicroRNA-155 modulates Treg and Th17 cells differentiation and Th17 cell function by targeting SOCS1. *PLoS One, 7*(10), e46082. http://dx.doi.org/10.1371/journal.pone.0046082.

de Yebenes, V. G., Belver, L., Pisano, D. G., Gonzalez, S., Villasante, A., Croce, C., et al. (2008). miR-181b negatively regulates activation-induced cytidine deaminase in B cells. *Journal of Experimental Medicine, 205*(10), 2199–2206. http://dx.doi.org/10.1084/jem.20080579.

Yee, P. T., & Poh, C. L. (2016). Development of novel miRNA-based vaccines and antivirals against enterovirus 71. *Current Pharmaceutical Design, 22*(44), 6694–6700. http://dx.doi.org/10.2174/1381612822666160720165613.

Ye, P., Rodriguez, F. H., Kanaly, S., Stocking, K. L., Schurr, J., Schwarzenberger, P., et al. (2001). Requirement of interleukin 17 receptor signaling for lung CXC chemokine and granulocyte colony-stimulating factor expression, neutrophil recruitment, and host defense. *Journal of Experimental Medicine, 194*(4), 519–527. http://dx.doi.org/10.1084/jem.194.4.519.

Yuan, J., Nguyen, C. K., Liu, X., Kanellopoulou, C., & Muljo, S. A. (2012). Lin28b reprograms adult bone marrow hematopoietic progenitors to mediate fetal-like lymphopoiesis. *Science, 335*(6073), 1195–1200. http://dx.doi.org/10.1126/science.1216557.

Yu, W., Freeland, D. M. H., & Nadeau, K. C. (2016). Food allergy: Immune mechanisms, diagnosis and immunotherapy. *Nature Reviews Immunology, 16*, 751. http://dx.doi.org/10.1038/nri.2016.111.

Yu, F., Sharma, S., Edwards, J., Feigenbaum, L., & Zhu, J. (2015). Dynamic expression of transcription factors T-bet and GATA-3 by regulatory T cells maintains immunotolerance. *Nature Immunology, 16*(2), 197–206. http://dx.doi.org/10.1038/ni.3053.

Zech, A., Ayata, C. K., Pankratz, F., Meyer, A., Baudiss, K., Cicko, S., et al. (2015). MicroRNA-155 modulates P2R signaling and Th2 priming of dendritic cells during allergic airway inflammation in mice. *Allergy, 70*(9), 1121–1129. http://dx.doi.org/10.1111/all.12643.

Zekri, L., Kuzuoglu-Ozturk, D., & Izaurralde, E. (2013). GW182 proteins cause PABP dissociation from silenced miRNA targets in the absence of deadenylation. *The EMBO Journal, 32*(7), 1052–1065. http://dx.doi.org/10.1038/emboj.2013.44.

Zhang, J., Cheng, Y., Cui, W., Li, M., Li, B., & Guo, L. (2014). MicroRNA-155 modulates Th1 and Th17 cell differentiation and is associated with multiple sclerosis and experimental autoimmune encephalomyelitis. *Journal of Neuroimmunology, 266*(1–2), 56–63. http://dx.doi.org/10.1016/j.jneuroim.2013.09.019.

Zhang, L., Ke, F., Liu, Z., Bai, J., Liu, J., Yan, S., et al. (2015). MicroRNA-31 negatively regulates peripherally derived regulatory T-cell generation by repressing retinoic acid-inducible protein 3. *Nature Communications, 6*(1), 7639. http://dx.doi.org/10.1038/ncomms8639.

Zhang, H., Kolb, F. A., Jaskiewicz, L., Westhof, E., & Filipowicz, W. (2004). Single processing center models for human dicer and bacterial RNase III. *Cell, 118*(1), 57–68. http://dx.doi.org/10.1016/j.cell.2004.06.017.

Zhang, L., Liao, Y., & Tang, L. (2019). MicroRNA-34 family: a potential tumor suppressor and therapeutic candidate in cancer. *Journal of Experimental & Clinical Cancer Research, 38*(1), 53. http://dx.doi.org/10.1186/s13046-019-1059-5.

Zhang, X. H., Zhang, Y. N., Li, H. B., Hu, C. Y., Wang, N., Cao, P. P., et al. (2012). Overexpression of miR-125b, a novel regulator of innate immunity, in eosinophilic chronic rhinosinusitis with nasal polyps. *American Journal of Respiratory and Critical Care Medicine, 185*(2), 140–151. http://dx.doi.org/10.1164/rccm.201103-0456OC.

Zhao, J. L., Ma, C., O'Connell, R. M., Mehta, A., DiLoreto, R., Heath, J. R., et al. (2014). Conversion of danger signals into cytokine signals by hematopoietic stem and progenitor cells for regulation of stress-induced hematopoiesis. *Cell Stem Cell, 14*(4), 445–459. http://dx.doi.org/10.1016/j.stem.2014.01.007.

Zhao, J. L., Rao, D. S., Boldin, M. P., Taganov, K. D., O'Connell, R. M., & Baltimore, D. (2011). NF-kappaB dysregulation in microRNA-146a-deficient mice drives the development of myeloid malignancies. *Proceedings of the National Academy of Sciences of the United States of America, 108*(22), 9184–9189. http://dx.doi.org/10.1073/pnas.1105398108.

Zhao, J. L., Rao, D. S., O'Connell, R. M., Garcia-Flores, Y., & Baltimore, D. (2013). MicroRNA-146a acts as a guardian of the quality and longevity of hematopoietic stem cells in mice. *elife, 2*, e00537. http://dx.doi.org/10.7554/eLife.00537.

Zheng, J., Jiang, H. Y., Li, J., Tang, H. C., Zhang, X. M., Wang, X. R., et al. (2012). MicroRNA-23b promotes tolerogenic properties of dendritic cells in vitro through inhibiting Notch1/NF-?B signalling pathways. *Allergy, 67*, 362–370. http://dx.doi.org/10.1111/j.1398-9995.2011.02776.x.

Zheng, B., Xi, Z., Liu, R., Yin, W., Sui, Z., Ren, B., et al. (2018). The function of microRNAs in B-cell development, lymphoma, and their potential in clinical practice. *Frontiers in Immunology, 9*(936). http://dx.doi.org/10.3389/fimmu.2018.00936.

Zhou, L., Ivanov, II, Spolski, R., Min, R., Shenderov, K., Egawa, T., et al. (2007). IL-6 programs T(H)-17 cell differentiation by promoting sequential engagement of the IL-21 and IL-23 pathways. *Nature Immunology, 8*(9), 967–974. http://dx.doi.org/10.1038/ni1488.

Zhou, X., Jeker, L. T., Fife, B. T., Zhu, S., Anderson, M. S., McManus, M. T., et al. (2008). Selective miRNA disruption in T reg cells leads to uncontrolled autoimmunity. *Journal of Experimental Medicine, 205*(9), 1983–1991. http://dx.doi.org/10.1084/jem.20080707.

Zhou, Q., Li, M., Wang, X., Li, Q., Wang, T., Zhu, Q., et al. (2012). Immune-related microRNAs are abundant in breast milk exosomes. *International Journal of Biological Sciences, 8*(1), 118–123. http://dx.doi.org/10.7150/ijbs.8.118.

Zhou, W., Su, L., Duan, X., Chen, X., Hays, A., Upadhyayula, S., et al. (2018). MicroRNA-21 down-regulates inflammation and inhibits periodontitis. *Molecular Immunology, 101*, 608–614. http://dx.doi.org/10.1016/j.molimm.2018.05.008.

Zhu, E., Wang, X., Zheng, B., Wang, Q., Hao, J., Chen, S., et al. (2014). miR-20b suppresses Th17 differentiation and the pathogenesis of experimental autoimmune encephalomyelitis by targeting RORgammat and STAT3. *Journal of Immunology, 192*(12), 5599–5609. http://dx.doi.org/10.4049/jimmunol.1303488.

Zhu, J., Yamane, H., & Paul, W. E. (2010). Differentiation of effector CD4 T cell populations. *Annual Review of Immunology, 28*(1), 445–489. http://dx.doi.org/10.1146/annurev-immunol-030409-101212.

CHAPTER SEVEN

Disease modifying drugs for rheumatological diseases: a brief history of everything

Joanna L. Giles*, Oktawia J. Polak and John Landon
MicroPharm Ltd, Newcastle Emlyn, Carmarthenshire, United Kingdom
*Corresponding author: E-mail: Joanna.Giles@micropharm.co.uk

Contents

Abstract

The rheumatological diseases are a group of chronic, painful, degenerative and debilitating conditions with an increasing prevalence across the globe. The pathogenesis of these disorders is complex, overlapping and not fully understood. As such, it is difficult and time consuming to achieve correct diagnosis and complete remission for an individual patient.

In this review we describe the most common forms of inflammatory arthritis and discuss how the management and treatment options for these rheumatic diseases

Advances in Protein Chemistry and Structural Biology, Volume 120
ISSN 1876-1623
https://doi.org/10.1016/bs.apcsb.2019.11.007

have developed over time. We outline the successes and the limitations of current treatment regimens and discuss the economic burden of the current options.

With advancements in understanding of disease mechanisms, we discuss the importance of the biologics revolution in the context of rheumatological disease and how the development of biosimilars and small molecule inhibitors will impact current treatment options in order to alleviate some of the cost burden of biological therapies.

The ideal treatment strategy for the future would involve personalized and predictive medicine where by treatments can be tailored to an individual patient's needs in order to achieve fast and successful remission with no adverse events.

1. Introduction

With an aging population, the development of effective, safe and cost-efficient treatments for rheumatological disease remains a pinnacle of medical research. The ultimate goal of this research is to achieve effective control of disease, sustained amelioration, preservation of joint structure and overall improvement in patient wellbeing.

Generally, an arthritic condition presents as degeneration, deformity and pain in one or more joints, tendons, ligaments, bones and/or muscles; it can be categorized by how many joints are involved, ranging from monoarthritis (where one joint alone is affected) to polyarthritis (where 6 or more joints are affected). However, often the symptoms are far more complex than this and may develop to systemic involvement and disease of other organs such as the skin, liver, eyes, and heart. Overall, the rheumatic diseases are considered to comprise of over 100 different types of arthritis. They are frequently considered as a disease spectrum — as there is often an overlap in genetics, phenotype and clinical features. The most common forms of arthritis are outlined in Table 1. If left untreated, the rheumatic disorders are debilitating and painful conditions which are detrimental to a patient's quality of life and can lead to a reduced mortality.

With such an overlap of disease characteristics and blurred boundaries of disease categorization, diagnosis of a patient can pose difficulties and take several months or even years. A lengthy diagnosis will lead to uncertainties in treatment selection and extended treatment and recovery times. These factors not only impact the direct cost to the healthcare system but also affect indirect costs such as unemployment. The morale, mental wellbeing and outlook for the patient are also affected. There is a notable link between the chronic pain of arthritis and depression-undoubtably exacerbated by

Table 1 Worldwide prevalence of the most common rheumatic disorders.

Type of arthritis	Symptoms	Estimated world-wide prevalence	Reference
Osteoarthritis	Joint pain and stiffness due to cartilage degradation.	9.6%–18%	Haq, 2003
Rheumatoid Arthritis	Morning stiffness, joint pain, swelling and inflammation in 3 or more joints.	0.3%–1%	Hayter & Cook, 2012
Gout	Swollen joints, intense pain often in the big toe.	1%	Shi et al., 2010
Psoriatic Arthritis	Inflammation of synovial tissue, joint pain and swelling, fatigue, scaly skin rash.	Up to 0.45%	Liu, 2014
Ankylosing Spondylitis	Bamboo spine (stiff backbone), lower back pain, inflammation in entheses, fatigue.	0.18%	Dean et al., 2014
Juvenile Idiopathic Arthritis	Joint pain and inflammation in one or more joints, manifesting before the age of 16.	Up to 0.15%	Weiss & Ilowite, 2007
Systemic Lupus Erythematosus	Joint stiffness, extreme fatigue, skin rash, inflammation of many organs, organ damage.	0.07%	Dall'Era et al., 2019

difficulties in reaching a correct diagnosis (Atkinson, Slater, Patterson, Grant, & Garfin, 1991; McBain, Shipley, Newman, & members of the Appearance Research Collaboration (ARC), 2013; Murphy, Sacks, Brady, Hootman, & Chapman, 2012). Early diagnosis and rapid treatment have been shown to prevent further joint damage, drastically improve pathology and save health-care providers substantial amounts of money (Nisar, 2019). A clearer understanding of both disease and treatment mechanisms therefore remain of high importance.

Rheumatic disorders are classified as polygenic. Inflammation is prominent in many rheumatological conditions; however, the exact mechanisms and triggers of inflammation remain elusive. Genome wide association studies (GWAS) have established genetic links, and research has also shown environment, phycological stress, gender, infection and the microbiome to have some role in disease progression.

The most common form of arthritis is Osteoarthritis (OA), which involves the chronic thinning of cartilage as a result of friction between joints. It is estimated that up to 15% of individuals over 60 suffer with OA (Haq, 2003). OA is not generally considered to be an inflammatory form of arthritis and is instead typically caused by wear and tear of cartilage over time, sometimes triggered by injury and with some genetic linkage. It is therefore classified as degenerative arthritis as opposed to inflammatory, and as such will not be discussed further in this review. Similarly, gout is a painful form of arthritis with a very high prevalence, affecting approximately 1% of the population, the majority being males (Shi, Mucsi, & Ng, 2010). The disease is classified as a metabolic form of arthritis associated with a deposit of uric acid (UA), in the form of monosodium urate (MSU) crystals, within the joints and periarticular tissues which is usually linked with hyperuricemia (Choi, Mount, & Reginato, 2005). When the UA concentration in serum exceeds its physiological saturation limit in body fluids, it undergoes crystallization to MSU crystals, which in turn trigger the innate defense mechanisms activation, inflammation and immune activation (Roddy & Doherty, 2010). Despite this inflammatory element to the disease, it is most commonly considered a metabolic disorder and will not be discussed further here.

The disease burden of inflammatory arthritis is an increasing concern with over 1% of the entire population suffering with Rheumatoid Arthritis (RA) alone (Hayter & Cook, 2012). The world-wide prevalence of the most common forms of arthritis are summarized in Table 1. As reported by the NHS, there are currently over 10 million people suffering from arthritis or related joint conditions to date in the UK alone. Moreover, Hootman, Helmick, Barbour, Theis, and Boring (2016) reported that there were 52.5 million adults in the US diagnosed with arthritis in 2010–12 and it has been predicted that this number will rise to 78.4 million by 2040. These values are considered to be underestimates due to the nature of the questions in the NHIS questionnaires used for diagnosis of disease. In order to improve the data already available, Jafarzadeh and Felson have suggested an alternative approach for estimating the disease prevalence by using expanded survey information on joint symptoms, which resulted in an increase in estimated arthritis prevalence with 91.2 million adults suggested to be affected by arthritis compared with 54.4 million adults previously estimated by the National Health Interview Survey (NHIS) in 2015 in the US (Jafarzadeh & Felson, 2018). As prevalence of rheumatological disease is clearly increasing, and as the diseases are all irreversible, it is unsurprizing that there remains a large focus on research into disease mechanisms and treatment options as will be discussed here.

The prevalence of inflammatory rheumatisms increases with age in both female and male populations, therefore the gradual increase in the population aged 60 years and older in developed countries shown by demographic studies may partially justify the increasing prevalence (2). The high incidence rate of arthritis and the fact that disease is irreversible reflects the significant cost of the disease to public health systems, including drug costs, research expenditure, medical care as well as pensions and benefits for people with disabilities associated with the disease. The cost of arthritis in the US alone is estimated to be over $128 billion per year (4). Today, health systems in many countries are facing issues respecting the disease management, therefore, so much research is still being carried out in order to devise more effective and efficient therapies.

2. Types of arthritis

2.1 Rheumatoid arthritis

RA is typically characterized by chronic bone and synovial inflammation and destruction, presented as tenderness, pain and swelling of the joints, and is usually distinguished from other forms of arthritis through the measurement of autoantibodies against citrullinated peptide or IgG (also known as Rheumatoid factor) (Firestein, 2003). The synovial fluid of an RA patient will be rich in B-cells, T-cells and many inflammatory cytokines (Choy &

Panayi, 2001). At present, the primary means of treatments target the inflammatory characteristics of disease, once presented at clinic. Treatment, therefore, is not usually correctly administered until long after initial disease onset.

Despite inflammation being a clear hallmark of disease, no one specific inflammatory pathway has been identified as the root cause of disease.

GWAS have been illuminating, but complex, when studying the heritability of RA. This has been recently well reviewed by Karami, Aslani, Jamshidi, Garshasbi, and Mahmoudi (2019). There is a strong link to some of the HLA genes (namely HLA-DRB1*01 and HLA-DRB1*04) with other HLA motifs conferring protection. These suggest a vital role for the adaptive arm of the immune system. However, there is strong evidence for involvement of other non-HLA genes with IL-23R and PTPN22 (Protein tyrosine phosphatase non-receptor type 22 (PTPN22) showing the strongest links. Whereby, there is strong overlap of associations with other autoimmune diseases where PTPN22 is also linked to SLE, type 1 diabetes, Grave's disease, Myasthenia Gravis and systemic sclerosis.

The third most strongly associated region for RA susceptibility is the TNF receptor associated factor 1 (TRAF1) region, whereby the TRAF-1-C5 region also shows a strong implication in disease. TRAF1 is a key player in cytokine signaling and this link to RA is unsurprizing given the success of anti-TNFα biologics in RA treatment (which will be discusses later). TNFα plays a prominent role in disease, as it has been linked to enhanced inflammation and bone degradation through increased action of metalloproteinases and collagenase (Strand, Kimberly, & Isaacs, 2007).

To further cloud our understanding, the studies show that ethnicity also affects the results of the GWAS profile. On the whole, the GWAS studies have been undoubtably useful in directing scientific research on mechanisms of disease, but have also shown that rheumatological disease are indeed multifactorial and complex with cross-over of several key inflammatory pathways.

Aside from GWAS, animal models have implicated a heavy involvement of T-cells, B-cells and multiple cytokines. The cytokines identified as biomarkers of disease include TNFα and IL-6 (McInnes, Buckley, & Isaacs, 2016) which are both heavily involved in inflammation and link to multiple inflammatory effects. Difficulties in identifying the key pathways involved in disease may well be due to the high overlap in disease characteristics. But it is undoubtably crucial to identify the early biomarkers in order to achieve rapid and targeted treatment to ultimately minimize disease progression.

2.2 Spondyloarthritis

Spondyloarthritis (SpA) is a subgroup of inter-related inflammatory rheumatic conditions that includes, but is not limited to, Psoriatic Arthritis (PsA), Ankylosing Spondylitis (AS) and Reactive Arthritis (ReA) (Liu, 2014). In each of these conditions, both peripheral and axial joints can be affected, however the disease itself is phenotypically quite different from RA. SpA are considered as forms of oligo-arthritis with a high involvement of joints in the lower limbs and entheses (enthesitis) (Heuft-Dorenbosch, 2003). PsA is typically characterized by arthritic joints as well as enthesitis, dactylitis, spondylitis and psoriasis. The most distinct form is AS, which is commonly referred to as "Bamboo spine" due to the progressive ankylosis of the spine and sacroiliac joints, leading to a fusion of joints in the lower part of the body. AS has an estimated worldwide prevalence of 0.18% (Dean et al., 2014).

GWAS have implicated over 100 loci to be linked to the susceptibility of AS (International Genetics of Ankylosing Spondylitis Consortium (IGAS), 2013) but the best understood genetic association to any rheumatological disease is the link between the HLA-B27 locus and spondyloarthritis first identified in 1973 (Brewerton et al., 1973; Schlosstein, Terasaki, Bluestone, & Pearson, 1973). Both transgenic rats and mice expressing HLA-B27 have been shown to develop a disease mimicking spondyloarthritis (Khare, 1995; Khare, Hansen, Luthra, & David, 1996).

The HLA class I molecules (such as HLA-B27) serve to present antigenic peptide to $CD8^{+}$ T cells in order to invoke an appropriate immune response. There are two main hypotheses for the pathogenic role of HLA-B27 namely the "arthritogenic peptide hypothesis" and the "B27 misfolding hypothesis" (McHugh & Bowness, 2012). The arthritogenic peptide hypothesis describes how auto-reactivity ensues after exposure to certain bacterial proteins, which lead to cross-reactivity with self-protein in the joins, leading to a form of "molecular mimicry" whereby joints are unintentionally destroyed. The "B27 misfolding hypothesis" stems from the molecular makeup of the HLA-B27 molecule itself. HLA-B27 has a free cysteine residue at position 67, which contributes to an increased likelihood of misfolding during processing in the endoplasmic reticulum. This misfolding and subsequent presentation at the cell surface has been shown to lead to augmented binding properties with innate immune ligands (Giles et al., 2012; Wong-Baeza et al., 2013) which will consequently augment the downstream immune response.

As discussed for RA, there is a pivotal role of TNFα in disease progression in SpA such as, where anti-TNF biological treatment has been shown to be highly successful (Maxwell et al., 2015). Biologics targeting IL-17A have also implicated IL-17 to have a key role in AS, where the current NICE approved DMARD treatments are outlined in Table 2.

2.3 Systemic lupus erythematosus

According to the Lupus Foundation of America, over 5 million people (predominantly women) are affected with Systemic Lupus Erythematosus (SLE) worldwide, and disease can range from mild to lethal. SLE is the most heterogeneous of the rheumatic diseases, and is often misdiagnosed owing to the high overlap of symptoms with other autoimmune diseases. SLE is a chronic condition, presenting with a high degree of inflammation and multiple organ damage. As with other forms of arthritis, SLE patients will suffer with joint pain and inflammation (Dall'Era et al., 2019). Often, this pain will be accompanied by a rash and tiredness. In the most severe cases of SLE organs such as the heart, lungs and brain are affected and it is in these cases where the disease becomes life threatening. Pathogenesis is mediated by loss of self-tolerance and subsequent auto-immune attack, provoked by cytokine dysregulation (Liu & Davidson, 2012). Many patients will have circulating autoantibodies present in their blood plasma.

The mechanism of disease in SLE is not well characterized, as there has been shown to be a complex interplay between both the innate and adaptive immune system.

Several immunological genes have been linked to disease, where GWAS studies have implicated over 40 gene loci (Niewold, 2015). As such there are several targets for biological treatments of SLE. However, there is a strong environmental influence and this has been demonstrated in monozygotic twin studies (Ghodke-Puranik & Niewold, 2015).

Establishing the correct treatment protocol is difficult, as pathogenesis is so complex and the disease presents in a heterogenic manner. Alleviation of symptoms will mostly rely on managing inflammation in order to prevent organ destruction.

2.4 Juvenile idiopathic arthritis

Symptoms of Juvenile idiopathic arthritis (JIA) mirror those of other forms of arthritis and include stiffness, damage and inflammation of one or more joint. JIA, however, manifests in children and adolescents before the age of 16 years and lasts no less than six weeks. The estimated prevalence of

Table 2 Overview of the NICE approved DMARDs.

DMARD	Target	Disease	Administration	Side-effects
Infliximab	TNFα	RA, PsA, AS	IV infusion	Higher susceptibility to infection, headaches, spasms, abdominal pain.
Adalimumab	TNFα	RA, PsA, AS, JIA	SC injection	Higher susceptibility to infection, anemia, anxiety, insomnia, vertigo, spasms
Etanercept	TNFα	RA, PsA, AS, JIA	SC injection	Higher susceptibility to infection, cystitis, fever, swelling, sepsis
Cerolizumab pegol	TNFα	RA, PsA, AS	SC injection	Higher susceptibility to infection, abscesses, fever, headaches, nausea, skin reactions*Δ
Golimumab	TNFα	RA, PsA, AS, JIA	SC injection	Higher susceptibility to infection*
Rituximab	CD20 (B-cells)	RA, SLE	IV infusion	Higher susceptibility to infection, reactivation of latent infection
Belimumab	BLyS (B-cells)	SLE	IV infusion	Higher susceptibility to infection, tiredness, muscle pain
Abatacept	CD80 and CD86 (T-cells)	RA, PsA, JIA,	IV infusion	Higher susceptibility to infection, nausea, vomiting
Tocilizumab	IL-6	RA, JIA	IV infusion/SC injection	Higher susceptibility to infection, Gastrointestinal perforation, headaches, itching*
Sarilumab	IL-6	RA	SC injection	Higher susceptibility to infection, cold sores*
Tofacitinib	JAK	RA, PsA	Oral administration	Higher susceptibility to infection, nausea, increased chance of clots (DVT)
Baricitinib	JAK	RA, SLE	Oral administration	Higher susceptibility to infection, headaches, increased chance of clots (DVT)
Secukinumab	IL-17	PsA, AS	SC injection	Higher susceptibility to infection, diarrhea, headaches
Ustekinumab	IL-12 and IL-23	PsA	SC injection	Higher susceptibility to infection

JIA in developed countries varies between 8 and 150 per 100 000 (Weiss & Ilowite, 2007). This chronic, autoimmune, inflammatory disease often leads to short- and long-term disability and, in 5%–8% of patients suffering from systemic JIA, the disease progresses into life-threatening macrophage activation syndrome (Ravelli & Martini, 2007). The cause and mechanism of disease is not completely understood, but as with other forms of arthritis both inflammatory and genetic mechanisms are thought to contribute to pathogenesis.

Patient samples have shown that many types of immune cell (including both B-cells and T-cells) are present in the synovial fluid, suggesting a strong role for the adaptive immune response (Murray et al., 1996). Due to the high abundance of Th1 cells in the synovial fluid of JIA patients, this subset of T-cells has historically thought to be a key player in disease progression (Gattorno et al., 1997). More recently, interest in the Th17 subset has gained momentum due to several studies implicating the cells in both inflammatory and autoimmune disease (Maggi, Cosmi, Simonini, Annunziato, & Cimaz, 2016). The relatively low abundance of Th17 cells measured at the sites of inflammation may be explained by either a reduced ability to expand by clonal expansion or through their plasticity (their ability to differentiate into Th1 cells through the action of TNFα). As with AS, increased levels of IL-17 are thought to contribute to pathogenesis (Nistala et al., 2010).

Several proteases which are mediated by these immune cells have been identified as responsible for the destruction observed in the joints (Marco Gattorno et al., 2002; Peake et al., 2005) where genetic studies have shown that Matrix Metalloproteinases (MMPs) MMP1 and MMP3, are linked to overproduction of MMPs (Abd-Allah et al., 2015).

The Th17 and Th17/Th1 cells are postulated to exert their pathogenic effect through the regulation of MMPs in the synovial tissues. These, and other enzymes are capable of tissue infiltration and deterioration, and contribute greatly to the damage seen in JIA joints.

As shown in Table 2, several biological drugs are in use for the treatment and disease modification of JIA. Both IL-6 inhibitors and TNFα inhibitors will inhibit the production of MMPs and thus ameliorate disease by reducing the damage occurring in the joints.

As already encountered with other forms of arthritis, clinicians must first characterize the inflammatory processes at play in an individual patient in order to predict the best line of treatment. Therefore, it is again of high importance to identify suitable biomarkers in order to achieve fast and accurate diagnosis and subsequent therapy.

3. Treatment protocols for arthritis

The current treatment regime for a patient presenting with inflammatory arthritis often involves early introduction of conventional DMARDs (cDMARDs) (disease modifying anti-rheumatic drugs) such as Methotrexate (MTX), leflunomide or sulfasalazine and then progressing to biological DMARDs (bDMARDs) such as infliximab (Fig. 1). MTX is still considered the gold-standard for treatment of RA (Saag et al., 2008) which has been used safely and inexpensively for over 30 years. Pain management in the form of Non-Steroidal Anti-Inflammatory drugs are also recommended, which are invaluable in the alleviation of inflammation, but do not improve the disease course. Inflamed and painful joints can also be treated directly by injection of glucocorticosteroids to the affected joints. Treatment regimens depend greatly on the individual symptoms and joint or organ damage already presented. Treatment protocols can be difficult to establish due to the difficulties in reaching a correct diagnosis and the unpredictability of how a patient will respond to a particular treatment.

Since the first use of injectable gold in 1930 (Kean, Forestier, Kassam, Buchanan, & Rooney, 1985) synthetic DMARDs have shown to provide prominent therapeutic effects. The term DMARD describes clearly the function of these drugs. The "Disease Modifying" term stems from the

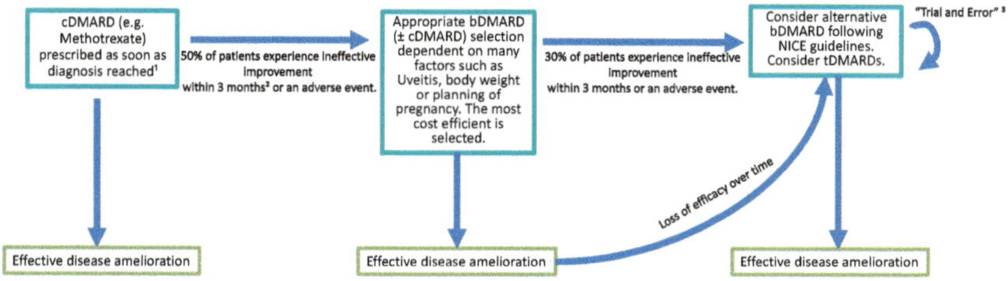

Fig. 1 ***Issues with the Current Treatment Schedule for RA patients in the UK.***[1] The correct diagnosis of rheumatic disease can be timely due to the overlapping nature of presented symptoms. As the disease is irreversible and destructive, a lengthy diagnosis often leads to increased joint damage in the patient. [2] A lengthy trial for an ineffective drug is expensive to the healthcare provider and frustrating for the patient. [3] A lack of understanding of treatment mechanisms leads to limited guidelines on the selection of bDMARD for each patient. This is an inappropriate and time-consuming method where a patient may see no benefit over several months and will delay selection of appropriate treatments. This method is costly and ineffective.

idea that the drug is able to prevent bone erosion, and thus alter the progression of disease. The "Anti-Rheumatic" term relates to the fact that these drugs were first used to treat RA (Gumpel, 1976).

MTX is structurally similar to folic acid, and as such is able to inhibit several enzymes that metabolize folic acid. The inhibition of dihydrofolate reductase is of mechanistic importance in its use as a DMARD, as this enzyme converts folic acid into reduced folate cofactors which are required for DNA and RNA synthesis. MTX, thereby, indirectly inhibits cellular proliferation (Hoffbrand & Tripp, 1972) and this accounts for the efficacy observed it's use as an anti-cancer drug. Prior to its use in RA, the first clinical trials of MTX treated cancer patients in 1953 (Bleyer, 1978). It is important to note that the doses used for anti-cancer therapy are often high-doses in the range of 1000—5000 mg per week, whereas the doses prescribed for arthritic conditions, such as RA are deemed "Low-dose" in the range of 25—30 mg per week (Malaviya, 2016). The mechanism of action of low-dose MTX is thought to center around reduced cell proliferation, increased T-cell apoptosis, altered cytokine production and inhibition of neutrophils (Chan & Cronstein, 2010) (Cronstein, 2005). Due to the ability of MTX to inhibit cell proliferation, it is unsurprizing that a major adverse effect observed in some patients is cellular toxicity, often related to hepatotoxicity and fibrosis which are manifested in the clinic as elevated liver markers (Ruperto et al., 2004), respiratory infections, and gastrointestinal symptoms (Brunner et al., 2005).

Although cDMARDs are well tolerated and have provided good clinical outcomes overall, the adverse responses described remain a cause for concern. Adverse events include (but are not limited to) nausea, vomiting, anemia, fever and depression (Smolen et al., 2018). In addition, a high proportion of patients are observed not to respond at all (these patients are classified as "non-responders") or the response is observed to diminish quickly over time. The reasons that MTX is effective for some patients but not for others, is not clear. The search for genetic biomarkers to aid with predictive treatment responses has been fruitful for azathioprine with variation in the TPMP gene being linked with response outcomes (Ansari et al., 2002). Clear genetic biomarkers to determine response to MTX have not yet been determined with many inconsistent findings. SNPs in genes coding for proteins in MTX metabolism have been the active area of research, but at present there is no one SNP that can be used for predictive treatments (Mikkelsen et al., 2011).

The most effective route of treatment through cDMARDs (in the form of monotherapy or combination therapy) is still heavily debated among clinicians and highly dependent on the patient. Despite relentless ongoing research, the current approach to treatment selection for many of the rheumatological diseases is frequently based on a "trial and error" basis (Fig. 1).

Since a significant proportion of patients were found not to respond to such cDMARDs, this led to the development of a different group of drugs that are known as the biological DMARDs (bDMARDs).

The development of these bDMARDs, also known as biologics, biopharmaceuticals or biotherapeutics, has revolutionized the pharmaceutical industry. bDMARDs are large and complex proteins manufactured by means of natural resources, such as human, animal and microorganism and can be produced by employing biotechnology methods, typically recombinant DNA (rDNA) technology (Lybecker, 2016). This group of molecules are currently used in the diagnosis and treatment of many serious, chronic and life-threatening diseases and medical conditions, including snake envenomation, cancers, rare infectious diseases such as polio and, of course, many rheumatic diseases.

4. The development of immunotherapeutics

The potential of immunoglobins to treat human disease was demonstrated by von Behring and Kitasato in the 19th century. In their study, they immunized rabbits with attenuated bacteria causing tetanus and injected their sera into formerly infected animals (Kitasato, 1890). The experiments proved the therapeutic properties of administered sera and pioneered the research in the field of passive immunization. In a short time, biological therapy against diphtheria, botulism or snake venoms were researched and the approach was tested on humans. As a result, the first successful biological therapy in humans was the blood serum therapy against diphtheria, which relied on the administration of the blood products of the immunized animals. Serum therapy was soon applied to treat many infectious diseases, including meningococcal meningitis, scarlet fever and pneumococcal pneumonia (Reynolds, de Zafra, Kim, & Gelzleichter, 2013).

Although the serum therapy ushered in the era of biologics, administration of heterologous sera resulted in high risk of allergies or life-threatening anaphylactic shock in patients. Hence, with the development of antibiotics, the use of animal serum therapy was limited to toxin-mediated conditions, such as diphtheria, botulism, tetanus and venomous bites. Development of the process for large-scale human plasma fractionation using cold ethanol was the cornerstone of immunoglobulins (Ig) purification for clinical use and led to the replacement of the proactive and therapeutic animal sera by human gamma globulins (Berger, 2002). This form of biological therapy was injected intramuscularly due to the side effects caused by Ig aggregations when administered intravenously. Further advances in immunoglobulin (Ig) preparations, such as the use of pepsin to inhibit their aggregation and Ig isolation at low pH, reduced the risk of side effects and subsequently, allowed their intravenous administration, which extended their application in the management of immunodeficiency syndromes as well as inflammatory and autoimmune conditions.

Therapeutic antibodies may be divided into polyclonal and monoclonal. Polyclonal antibodies (PAbs) can be obtained by animal immunization with the antigen of choice and are products of multiple B-cells, which results in their poly-specificity. Monoclonal antibodies (MAbs) are obtained by the fusion of B-cells with myeloma cells and are products of a single immunoglobulin-producing clone, therefore, can only recognize a single epitope on the molecule of choice. Although, in the last few decades, both PAb and MAb have had a great positive impact on healthcare management the development of therapeutics based on antibody administration is primary focused on MAb.

The method of *in vitro* production of murine MAbs of predefined specificity from hybridomas, cell lines made by fusion of murine myeloma and murine spleen cells from an immunized donor, was first described by Köhler and Milstein (1975). This ground-breaking technology underwent rapid development and as soon as in 1980 the first patient was treated by means of murine MAbs (Stern & Herrmann, 2005). Although the therapy turned out to be safe and well-tolerated, in vivo results were clinically ineffective due to the circulating tumor antigens and the failure in delivering appropriate amount of the antibodies (Nadler, Stashenko & Hardy, 1980). Further studies revealed several drawbacks of murine MAbs, such as their short half-life in humans, association with allergic reactions due to the production of anti-drug antibodies, limited induction of the of complement mediated and antibody-dependant cellular cytotoxicity and limitations of the manufacture process (Ober, Radu, Ghetie, & Ward, 2001). In order to overcome described limitations, MAbs have been improved by a number of modifications and technologic refinements resulting in the development of recombinant chimeric murine-human antibodies, humanized MAbs as well as fully humanized MAbs. Novel methods of MAbs manufacture allowed the efficient production of large quantities of biotherapeutics with reduced cost as well as decreased immunogenicity, therefore, expanding their application. Furthermore, advances in understanding of disease processes and molecular cell mechanisms responsible for disease onset have also contributed to the expansion of MAb-based therapies. In Europe and the United States, there were 30 therapeutic MAbs on the marked by 2010 and the number of new ones in clinical trials growing, with the approval success rate ranging from 23% to 63% (Buss, Henderson, McFarlane, Shenton, & de Haan, 2012; Reichert & Paquette, 2003).

There are many requirements that both biologicals and traditional chemically synthesized drugs must fulfill in order to enter clinical trials and ultimately get approval of an appropriate regulatory agency, resulting in their actual manufacture and marketing. The requirements of every drug can be divided into three major categories, namely: effectiveness, safety and cost-efficiency, all being interconnected with each other (Fig. 2). As well as the imperative requirements of a drug, there are also other factors to consider. These include milder side-effects, the risks of not having a treatment option, the demand for the drug and whether the drug can be produced on a large scale. These will all depend on the drug in question.

Factors influencing the cornerstones of biological development are complex and will be discussed with regard to therapeutic antibodies.

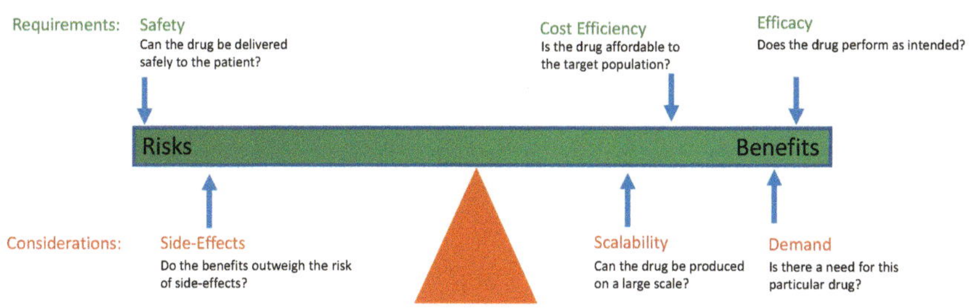

Fig. 2 Balancing the requirements of a novel therapeutic.

The effectiveness of a therapeutic antibody product depends on the strength of the response which it is able to produce in vivo i.e. the efficacy. Efficacy in turn depends on avidity (for PAbs) or affinity (for Mabs) of the antibody to the target antigenic epitopes. To be effective, the antibody must bind to the target in order to achieve a measurable improvement in disease.

All drugs can produce side effects, but the benefit-harm balance must be kept at all times. Drugs treating serious and life-threatening diseases and medical conditions often exert more severe side effects but as long as they bring more benefits than harm, and there are no substitutes on the market the likelihood of their approval is high. For example, when treating a fatal snake envenomation with a biological therapeutic, the concern over side-effects is minimized as the risk of death is high. On the other hand, if a patient presents with a non-life-threatening flare of psoriasis, but reactivation of latent-TB is possible with a particular biological treatment, then that biological treatment option would be avoided.

As will be discussed later, different individuals do not respond in the same way to the same biological treatments. This is not yet predictable, and the reasons for effective versus non-effective responses to treatments are not yet understood. The identification of predictive biomarkers would certainly be desirable in the field of biological therapy, this is not likely to be achieved in the near future.

As with all drugs, the cost of production of a biological drug is a huge factor in their development and approval. Not all biologicals can be easily scaled up for production, and this will also affect the likelihood of being released to the market. The overall demand for the drug will govern whether the drug will likely be developed; the bigger pharmaceutical companies will target more common medical conditions, focusing on the developed world. In these instances, biological therapies are subject to strict manufacturing and testing processes with many production steps. The more rigorous the testing and production, the more expensive the drug. Such drugs are currently far too expensive and inaccessible to the developing world.

5. NICE approved targeted therapies in rheumatic conditions

The use of biological DMARDs (bDMARDs) have revolutionized the treatment of rheumatic diseases, whereby these molecules' functionality is based upon direct targeting of the immune system and pathways involved in pathogenesis. A wealth of clinical trial data and patient reports have provided support for their safety, efficacy and sustainability of these biological therapies. As detailed in Table 2, there are currently several biological treatment options in rheumatological disease, with several immune targets to choose from. There is also flexibility with treatment strategies, where monotherapy or combination therapy with cDMARDs are also options. In the context of rheumatic diseases, the European League Against Rheumatism (EULAR) committee do not distinguish between the recommended bDMARDs with respect to efficacy and all are thought to provide comparable clinical outcome. Here we summarize the targeted therapies which are approved by NICE (see Table 2) and discuss the future of these particular treatments with the advent of biosimilars.

5.1 Anti-TNF drugs

TNFα was first measured in arthritic joints, synovial fluid and patient serum in the late 1980s (Hopkins & Meager, 1988; Saxne, Palladino, Heinegård, Talal, & Wollheim, 1988; Yocum et al., 1989) and has been implicated in the pathogenesis of many rheumatological diseases (Marra, Bansback, Anis, & Shojania, 2011). As such the anti-TNFs were the first group of biologics to be used in a clinical setting (Elliott et al., 1994). These treatments remain the most commonly used for the alleviation of several rheumatic diseases. TNFα is a pleiotropic cytokine released by activated macrophages, T-cells and other immune cells following infection (Ellerin, Rubin, & Weinblatt, 2003). TNFα then recruits other inflammatory mediators to the site of infection and is thus implicated in many immunological pathways and linked to many biological effects including promotion of monocyte differentiation, enhancing B-cell and T-cell proliferation, aberrant activation of osteoclast formation and a general increase of inflammation. The TNFα molecule is a 26 kDa transmembrane protein (mTNF), cleaved by TNF-converting enzyme (TACE) into a soluble TNF (sTNF) monomer of 17 kDa (Kriegler, Perez, DeFay, Albert, & Lu, 1988). sTNF and mTNF may exist as monomers or homotrimers of 51 kDa. Typically, TNFα binds to its receptors TNFR1, which is ubiquitously expressed, and TNFR2, which is only expressed on immune cells. The binding of TNFα to TRAF1 initiates a complex downstream cascade that leads to apoptosis and inflammation (Brenner, Blaser, & Mak, 2015). TNFR2 preferentially binds to mTNFα (Grell, 1995) which leads to an increase in NFκB, consequently further enhancing inflammation.

The mechanism of action of TNFα is highly complex and involves a vast number of cells and pathways [Reviewed (Mitoma, Horiuchi, Tsukamoto, & Ueda, 2018)], so aside from limiting the effects of sTNFα by binding and neutralization, the exact mechanism of action of the anti-TNF biologics is not completely defined.

The discovery and subsequent first uses of the anti-TNFs were a real breakthrough in the treatment of chronic rheumatic diseases and have been used routinely, safely and effectively since the 2000s. The concept of targeting TNFα in inflammatory disease came from the first animal studies conducted in 1985 by Beutler (Beutler, Milsark, & Cerami, 1985) whereby mice injected with monoclonal anti-TNF were protected against the lethal effects of endotoxin lipopolysaccharide. Later measurements of elevated TNFα in sepsis patient serum, and experiments using anti-TNF antibodies to prevent septic shock in baboons (Tracey et al., 1987) strengthened the drive to develop inhibitors of TNFα in order to ameliorate chronic inflammation. Furthermore, a polymorphism in exon 6 of TNFR2 has been linked to SLE (Barton, John, Ollier, Silman, & Worthington, 2001).

Since the year 2000, there have been 5 anti-TNFα biologics which have been approved for use in rheumatological disease (Table 2) [Reviewed (Luchetti, Benfaremo, & Gabrielli, 2018)]. These are Infliximab, Adalimumab, Golimumab, Etanercept and Cerolizumab pegol. All 5 biologicals are administered systemically by injection where Infliximab requires intravenous injection and requires a lengthy hospital visit for each treatment. This systemic administration can be inconvenient to the patient, costly to healthcare facilities and can also lead to systemic undesirable side-effects.

Infliximab was the first TNFα blocker to show an improvement in disease state of rheumatic diseases in clinical trials (Elliott et al., 1994). A pivotal randomized clinical trial in the study of anti-TNF drugs was a study using infliximab (Maini et al., 1999) where 428 patients with active RA were randomized to infliximab or placebo and significantly higher clinical response was observed at 30 weeks with patients taking infliximab compared to MTX alone. Infliximab is a monoclonal, chimeric human IgG1 which is linked to murine Ig variable regions and is administered intravenously. The safety and long-term efficacy of infliximab has also been extensively studied and shown to generally be considered as safe with some adverse events to consider (Braun et al., 2005; Elkayam & Elalouf, 2015) (Table 2).

Adalimumab and Golimumab were developed later and are also monoclonal but fully humanized, thus are potentially less immunogenic than Infliximab. Etanercept is an engineered human dimeric fusion protein, consisting of recombinant TNFα receptor and the Fc portion of IgG1. Like Adalimumab and Golimumab, Etanercept is administered by injection subcutaneously and has been shown to have a higher affinity for TNFα, but has a shorter half-life. The most recently approved anti-TNF biological is Cerolizumab pegol. This molecule is a FAB' fragment which contains both human and murine amino acid sequences, but the hinge is modified by polyethylene glycol, which reduces the immunogenicity and improves the solubility and stability. No anti-TNFα treatment is clinically preferred over another.

Any adverse event, or side effect is a primary concern when considering the safety of biologics. The main safety concern is a higher risk of opportunistic infection due to suppression of the immune system. Upper respiratory tract infections (including pneumonia) are reported as a common side effects of anti-TNF therapy, as well as reactivation of latent tuberculosis (TB). It is therefore recommended that all patients are screened for TB and Hepatitis-B prior to prescription of any anti-TNF therapy. Concerns over malignancies as an adverse event have been considered, where a higher rate of lung cancer and lymphoma have been recorded for RA patients receiving biological therapy compared to cDMARDs (Gridley et al., 1993). On the other hand, rates of breast and colon cancer were reported to be significantly decreased. Infusion reactions are also possible when receiving biological therapy. These manifest as headaches, pruritus, flushing and hypertension. These effects are unpleasant, but are self-limiting and mild. Other reported adverse events include the onset of lupus-like syndrome, demyelinating syndrome and the development of blocking antibodies. No single anti-TNFα treatment is thought to cause more adverse events than any other. Therefore, although anti-TNFs are hailed as the gold standard treatment for many immune mediated diseases, including many rheumatological diseases, it's clear that these treatments are far from perfect.

Despite the significant improvements observed in patients from many clinical trials, and resultant regulatory approval for diseases including RA, PsA, AS and JIA, not all patients achieve improved disease states. A significant percentage of patients do not respond to the first anti-TNF biologic administered. 40% of patients fall into the category of "Non-responders" for PsA and RA. Sometimes, these patients will respond to a second anti-TNF biological, but sometimes they will be "Secondary Non-responders". Furthermore, there is an observed issue of loss efficacy, whereby a biologic which was providing good clinical outcomes reduces in efficacy in that patient over time. These lack of response observations may be due to development of immunogenicity, intolerability or structural damage. In these cases, the use of other biological drugs against other targets prove highly beneficial.

5.2 B-cell depletion

Rituximab and Belimumab are two types of B-cell depleting monoclonal bDMARDs prescribed for RA (Rituximab) and SLE (Rituximab and SLE). Rituximab will be prescribed for patients that do not respond to cDMARDs such as MTX, and will have usually not responded adequately to an anti-TNF therapy (Luchetti et al., 2018). Such therapies are also used for the treatment of B-cell malignancies.

Rituximab is a chimeric mouse/human MAb, administered via IV, that targets CD-20 protein on the surface of B-cells (Weiner, 2010). The bDMARD functions by depleting circulating B-cells through several mechanisms. Rituximab is able to trigger the apoptosis of B-cells directly, trigger cytotoxicity of Rituximab coated cells via the complement pathway and also trigger killing through antibody-depended cellular cytotoxicity (Clark, 2005). Belimumab is a human MAb, again administered by IV infusion, that targets BLYS (B-lymphocyte Stimulator) in order to block binding to of BLYS to its receptor on the B-cell and thus prevents their survival. As a result of the B-cell depletion, such bDMARDs are also able to diminish the activation of T-cells.

Excellent clinical responses with no significant adverse events led to the approval of Rituximab for treatment of RA in 2006 (Palylyk-Colwell & McGahan, 2006), and the approval of Belimumab for SLE in 2011 (Sanz, Yasothan, & Kirkpatrick, 2011). Despite suitable safety profiles, the question of safety is coming back into light for Rituximab. This is due to the facts that optimum dosage is still yet to be established and it is known to be unsuitable for administration to pregnant women. There remain the concerns over risk of de-novo infection and potential reactivation of HBV and HCV infection due to the compromised humoral immune response, although the risk of this should be minimized thorough consultation with the patient before prescription. B-cell depletion also poses a risk of disruption to vaccination programmes whereby the development of new specific antibodies may be impacted if a patient is treated with a B-cell bDMARD within 6 months of vaccination. Any required vaccinations must therefore be planned prior to administration of the bDMARD. Finally, there is evidence of emergence of new immune-related disorders, including psoriasis, when undergoing treatment with Rituximab (Dass, Vital, & Emery, 2007) despite demonstrating a good response to the underlying target disease. As such, treatments using rituximab can sometimes yield unpredictable responses and is by no means the ideal treatment option.

5.3 T-cell depletion

The only T-cell depleting antibody currently approved for use in rheumatic disease is Abatacept, which is a soluble, recombinant, fully-humanized fusion protein directed to CD80/CD86 (Bonelli & Scheinecker, 2018). Naïve T-cells require 2 separate signals in order to become fully activated. The first signal is the binding between the T-cell Receptor and the Major Histocompatibility complex (MHC) II bound to peptide on the antigen presenting cell. The second co-stimulatory signal is the interaction between CD28 on the T cell and the CD86 or CD80 molecule on the antigen presenting cells. Abatacept mediates its function by interfering with the CD28 signaling pathway, preventing propagation of the second activation signal (Sarzi-Puttini, Ceribelli, Marotto, Batticciotto, & Atzeni, 2019). Therefore, during chronic inflammation there is a reduced activation of T-effector cells and, overall, T-cell activation and propagation will diminish. Further to this, there is some evidence that T-regulatory cells and macrophages may also be affected (Munn, Sharma, & Mellor, 2004; Álvarez-Quiroga et al., 2011), although this may not be the primary mechanism of interest in autoimmune amelioration.

The first evidence in mice of the successful blockade of the CD28 ligand and consequent disease amelioration came from the study by Webb, Walmsley, and Feldmann (1996). Here, a collagen-induced arthritis mouse model was effectively treated by blocking the CD86 and CD80 axis. This effect was measured by reduced expression of Interferon γ from the lymph nodes. In humans, this has been translated to successful clinical trials for patients suffering with RA and JIA (Kremer et al., 2006), but with mixed results with PsA (Ursini, Russo, De Giorgio, De Sarro, & D'Angelo, 2018). Furthermore, safety profiles which are similar to that of Infliximab have been obtained for the use of Abatacept (Schiff et al., 2008) and is hence a useful option when considering treatment options in patients.

5.4 IL-6 blockade

Elevated levels of IL-6 have been measured in both the sera and synovial fluid of RA patients and in the sera of JIA patients (Houssiau, Devogelaer, Damme, Deuxchaisnes, & Snick, 1988), symptoms of these elevated levels are unpleasant and include fatigue, anemia and acute phase reactions. It is unsurprizing that IL-6 blockade is an attractive target for the amelioration of inflammatory disease given that IL-6 is also established as a pleiotropic cytokine (as we have noted for TNFα). IL-6 is a proinflammatory cytokine secreted by several immune cells including T-cells, monocytes and fibroblasts at the sites of inflammation (Hunter & Jones, 2015). The complex pathway for IL-6 is not yet completely understood, but IL-6 was first identified to stimulate the activity of T-cells and trigger the differentiation of B-cells by binding exclusively to the IL-6 receptor. As such, IL-6 is involved in mediation of both the innate and adaptive arms of the immune response.

The first approved IL-6 blocking bDMARD was Tocilizumab, which is a humanized monoclonal anti-IL6 receptor (IL-6R) antibody, that has been shown to bind to soluble IL6-R (sIL6-R) and membrane bound IL6-R (mIL6-R). Toculizumab was first developed in 1993 (Sato et al., 1993) and subsequent results from clinical trials showed to improve the appearance of bone erosion, anemia and disease activity in RA and JIA patients (Bongartz, 2008) and was later approved for use in Europe in 2009 for intravenous administration and in 2014 for subcutaneous administration. It can be used in both monotherapy or in combination therapy with MTX. The second anti-IL-6 receptor antibody that was approved for use in rheumatic disease in 2017 is Sarilumab. Similarly to Toculizumab, Sarilumab is fully humanized, monoclonal and binds directly to mIL-6R and sIL-6R. Sarilumab has been shown to bind to its epitope with a higher affinity and a longer half-life, with a comparable safety profile (Kim et al., 2015). As such, drug stability is enhanced with Sarilumab, and premature termination of treatment by a patient is usually due to an adverse event, rather than lack of efficacy.

5.5 The IL-17/IL-23 axis

Although NSAIDs are the first line therapy for many patients with inflammatory arthritis, and many options of bDMARD therapy exist for those with poor prognosis or severe disease, there remains an unmet need for treatment of SpA. Due to the systemic and multifactorial nature of disease, both innate and adaptive cells are involved in pathogenesis of SpA. The IL-17/IL-23 axis is an attractive therapeutic target for inhibiting both arms of the immune response as these cytokines serve as a bridge between the adaptive and innate immune pathways. IL-23 and IL-17 are proinflammatory cytokines that have been directly measured in psoriatic lesions of patients (Piskin, Sylva-Steenland, Bos, & Teunissen, 2006). As such, a targeted approach to inhibit the IL23-IL17 axis was suggested and has been a successful option for mediation of PsA and AS. IL-17 triggers the release of other proinflammatory cytokines, and IL-23 has been shown to be critical for joint damage in SpA (Sherlock et al., 2012).

IL-17 includes a family of 6 proteins secreted by Th17 cells, γδT cells, NKT cells, NK cells and 3 types of lymphoid cell. The function of IL-17 is to enhance production of other proinflammatory cytokines (including IL-6 and IL-1) in order to recruit lymphocytes to the site of release. Studies have shown that IL-17 can directly induce the activation of keratinocytes and endothelial cells (Skepner et al., 2014), it has also been shown to work in synergy with TNFα (Wang et al., 2013) and thus enhances the inflammatory cascade.

There are currently two EU bDMARDs approved for the targeted treatment of the IL17 axis. Ustekinumab is a human MAb that binds to the p40 subunit of either IL12 or IL23 to thus will directly inhibit Th17 cell expansion. Ustekinumab was approved in 2013 for the treatment of PsA having demonstrated efficacy in 2 phase III trials for psoriasis (Leonardi et al., 2008; Papp et al., 2008) and in later trials for PsA (McInnes et al., 2013; Ritchlin et al., 2014). In these Phoenix I and II trials, Ustekinumab showed improved dactylitis, improved enthesitis and improved spondylitis in PsA patients. Furthermore the safety profile has been shown to be good, with very few serious adverse events (López-Ferrer, Laiz, & Puig, 2017). Secukinumab is the second fully humanized MAb approved by the FDA for PsA in 2015. Secukinumab binds to IL-17A in order to reduce inflammation and has proven efficacy for both TNFα naïve and TNFα experienced patients, with no significant difference observed between the effectiveness of Secukinumab and Infliximab. Furthermore, Secukinumab has been shown to be the most cost efficient bDMARD in a Finnish study (Purmonen, Puolakka, Bhattacharyya, Jain, & Martikainen, 2018) and is therefore a highly attractive alternative treatment option for the amelioration of PsA.

5.6 Small molecule inhibitors

5.6.1 JAK inhibitors

The final group of DMARDs to be discussed are the synthetic targeted DMARDs (tDMARDS) which include the newer, orally administered drugs that inhibit the JAK-STAT signaling pathway. These are small molecule inhibitors, referred to as JAK inhibitors (JAKis), that have provided exciting results with excellent efficacy and safety profiles (Keystone et al., 2017), rivalling that of MTX and the anti-TNFα drugs. Tofacitinib and

Baricitinib are two small molecule inhibitors which are demonstrating impressive efficacy across a spectrum of rheumatic diseases.

The JAK/STAT pathway is a critical signaling pathway that involves the transduction of signals across the plasma membrane into the nucleus upon binding to a receptor. A cytokine will bind selectively to its transmembrane receptor which will bear an enzymatic JAK domain. This will then trigger a downstream signal to stimulate the transcription of the target genes in order to exhibit a response to the cytokine. In mammals there are a limited number of JAK/STAT proteins; the JAK family comprises JAK1, JAK2, JAK3 and Tyk3 where there are 7 members of the STAT proteins. It has already been shown that the JAK/STAT pathway acts downstream of over 40 cytokines and it is therefore clear that the JAK/STAT pathway has a critical role in cytokine signaling, and targeting any JAK protein will have a profound impact on multiple signaling pathways. Cytokines known to be affected include IL-23 and IL-17, which were discussed above.

JAKi are quite different molecules to the bDMARDs discussed previously. Importantly, JAKi are small molecules of only 400Da in size. This is significantly smaller than the $\geq$100,000Da biological molecules that have so far dominated the market. Their small size allows for oral administration and subsequent rapid entrance into the cells. In addition, the molecules have a shorter half-life compared to the bDMARDs, with fast clearance via the liver and kidney (Dowty et al., 2014). This can prove beneficial to any patient experiencing an adverse event as the effects can be limited and recovery will be accelerated.

Tofacitinib was the first JAKi approved as a treatment for RA patients that had already failed treatment with MTX or other cDMARDs. Tofacitinib specifically targets JAK1 and JAK3 and may be used as a monotherapy or as a combination therapy with MTX and this treatment regimen demonstrates a similar response to anti-TNF therapies (Fleischmann et al., 2017). Baricitinib was approved more recently than Tofacitinib, targeting JAK1 and JAK2, and has proved to be even more effective than MTX in clinical trials, where clinical improvement was observed as early as week 1 of trials (Keystone et al., 2017).

6. Limitations to current treatments and the era of biosimilars

Despite the many available treatment options discussed for inflammatory arthritis, it remains impossible to predict how a patient will respond to any particular DMARD. It is without a doubt that bDMARDs have revolutionized the treatment regimens for almost all arthritic conditions, however, there remain some clear limitations to their use. Predominantly, the high percentage of both primary and secondary non-responders means that they do not work for every patient and so there is therefore good justification for continued research into their mechanisms of action. In particular, identification of predictive biomarkers would be gigantic leap for the efficiency of biological therapy. Specifically, the ability to predict whether a patient will respond positively to a treatment prior to administration would prove to enhance remission rates and also save vast amounts of money on unsuccessful treatments and wasted time. This would ultimately lead to the ability to correctly treat more patients in a shorter amount of time. However, this route has not yet provided and clinically significant data (Ferreiro-Iglesias et al., 2019; Lopez-Rodriguez et al., 2018; Márquez et al., 2014; Roszkiewicz & Smolewska, 2017) but is still an active area of research. The reported adverse systemic events and potential immunogenicity also remain a cause for concern. With some systemically presenting rheumatological diseases, such as SLE and RA, these adverse effects may be difficult to avoid at present, as the effect of the biologicals are required throughout the body; however in some more localized arthritic conditions such as monoarthritis or localized psoriasis it may be possible to target the diseased areas more effectively as has been shown for the use of oral biologicals for diseases of the gut (Crowe et al., 2019; Tjellström, Stenhammar, Magnusson, & Sandqvist, 1997; Worledge, Godiska, Barrett, & Kink, 2000). For example, as opposed to the local injection of steroids, local injections of biologicals have shown significant improvement for monoarthritis (Conti et al., 2005); whereby biologicals may be targeted directly to the joints. This practice has not been adopted in clinics, despite encouraging results. This concept could be further investigated by assessing whether application, topically, to inflamed skin would provide relief for patients. By utilizing a localized delivery method for biologicals, there would be the benefit of administration at a lower total dose, lower systemic effects and of course a lower overall cost to the healthcare systems.

Another limitation to the use of bDMARDs compared to cDMARDs is the high cost of production and later requirement of cold storage. As such, there remains room on the market for new biologicals or other DMARDs that would be more cost effective. This relatively high cost of biological therapy, and the increasing prevalence of rheumatological disease are clear driving factors in the development of other similar molecules with comparable activities (the biosimilars). A biosimilar (bsDMARD) is a biological treatment that has a highly similar biological activity, physiochemical characteristics, efficacy and safety profile to an already approved biological therapeutic. Such proteins will not be exact replicas of the original MAbs, but will have been rigorously tested and demonstrate similar results from at least one randomized clinical trial. They will demonstrate a similar safety profile to their reference biological (Lie, Sciascia, & Cuadrado, 2015). The main purpose of developing such molecules is to reduce the cost burden to healthcare services. At the moment, the anti-TNF molecules are the most expensive treatments in the world with adalimumab remaining the World's bestselling drug earning close to $20 billion USD in 2018. As the patents for these costly biosimilars are expiring, biosimilars are beginning to make their presence known on the market. Several biosimilars, predominantly anti-TNF biosimilars, have been approved and recommended by EULAR for use in rheumatic disease, and many more are in development.

The infliximab biosimilars CT-P13 and SB2 were on the market prior to 2017 (Smolen et al., 2017; Yoo, 2017), and data has shown that these treatments have indeed allowed for a greater access to therapy for a higher number of patients with a 13% drop in price per treatment observed in the EU from 2016 to 2017.

Despite these encouraging figures, it is vital to ensure careful and delicate communication with the patients in order to prevent any negative preconceptions of biosimilars prior to administration (Tweehuysen et al., 2018). If a patient is already receiving a biological therapy, successfully, he or she may be reluctant to switch to what they perceive as a "cheaper" therapy (Glintborg et al., 2017). If a patient feels uncertain or believes that a new treatment will provide no benefit, a nocebo effect may manifest (Planès, Villier, & Mallaret, 2016). Here, an increase in stress and anxiety may trigger neurochemical networks that could increase pain signaling and may have an overall negative impact on treatment adherence and treatment results (Rezk & Pieper, 2017). The overall impact of the nocebo effect is significant, it has been shown to increase the number of reported adverse events, the perceived efficacy of the drug and ultimately the quality of life for that patient and pressure on healthcare systems. As such, the full benefit of bDMARDs is not yet in full effect. In order to truly reap the benefits of the biosimilar agents, there needs to be strong confidence in these agents and a better understanding of the clinical data which can then be directly delivered to the patients in order to give them good assurance that these treatments will succeed.

7. The future

The inflammatory forms of arthritis discussed in this review are no longer considered universally debilitating diseases, rather, manageable and treatable conditions with the realistic goal of patient remission. In this review we have discussed the cDMARDs, the bDMARDs and the tDMARDs as invaluable treatment options for the control of disease. As there are many drug targets to choose from and the option of multiple routes of administration it is somewhat possible to tailor treatment to a patient's needs. However, as we have discussed, some patients do not respond to any of the available agents and it is internationally accepted that there is no clear strategy for bDMARD selection after the failure of an anti-TNF treatment. There have been no complete head-to-head clinical trials to directly compare all of the bDMARDs on offer for arthritis and it is still not possible to accurately prescribe the drug with the best likelihood of success for any one patient.

A personalized approach to drug selection is, of course, the most desirable scenario and may be slowly becoming achievable. Some studies have shown that different environmental factors are able to predict a response to treatments. Factors including BMI, smoking, the use of previous drugs and the disease serotype can all influence the response to a particular DMARD. It is also clear that adherence to the drug and patient attitude can impact the success of treatment. Such factors must therefore be taken into consideration during patient assessment and drug selection. Although these factors are certainly helpful, a truly personalized regimen does not yet exist due to the complex and multifactorial nature of these inflammatory disorders. Many of the inflammatory arthritides have been linked to hundreds of genes in GWAS, and have differentially expressed proteins involving multiple pathways. These factors, as well as the incomplete understanding of pathogenesis of many of the diseases, make it incredibly difficult to predict the success of different treatments.

At present, studies involving SNP data, expression abundance and DNA methylation status are being investigated and should provide more predictive treatment strategies for the rheumatic diseases (Wu, Sheng, Sheng, Lu, & Xu, 2019). Together, with the development of biosimilars, we should look forward to more affordable, predictive and successful treatment strategies for inflammatory forms of arthritis in the future.

References

Abd-Allah, S. H., El-Shal, A. S., Shalaby, S. M., Pasha, H. F., Abou El-Saoud, A. M., Abdel Galil, S. M., et al. (2015). Influence of matrix metalloproteinase 1 and 3 genetic variations on susceptibility and severity of juvenile idiopathic arthritis: MMP-1 and MMP-3 polymorphisms in JIA. *IUBMB Life, 67*(12), 934–942. http://dx.doi.org/10.1002/iub.1446.

Álvarez-Quiroga, C., Abud-Mendoza, C., Doníz-Padilla, L., Juárez-Reyes, A., Monsiváis-Urenda, A., Baranda, L., et al. (2011). CTLA-4-Ig therapy diminishes the frequency but enhances the function of Treg cells in patients with rheumatoid arthritis. *Journal of Clinical Immunology, 31*(4), 588–595. http://dx.doi.org/10.1007/s10875-011-9527-5.

Ansari, A., Hassan, C., Duley, J., Marinaki, A., Shobowale-Bakre, E.-M., Seed, P., et al. (2002). Thiopurine methyltransferase activity and the use of azathioprine in inflammatory bowel disease. *Alimentary Pharmacology & Therapeutics, 16*(10), 1743–1750.

Atkinson, H. J., Slater, M. A., Patterson, T. L., Grant, I., & Garfin, S. R. (1991). Prevalence, onset, and risk of psychiatric disorders in men with chronic low back pain: A controlled study. *The Journal of Pain, 45*(2), 111–121. http://dx.doi.org/10.1016/0304-3959(91)90175-W.

Barton, A., John, S., Ollier, W. E., Silman, A., & Worthington, J. (2001). Association between rheumatoid arthritis and polymorphism of tumor necrosis factor receptor II, but not tumor necrosis factor receptor I, in Caucasians. *Arthritis & Rheumatism, 44*(1), 61–65. http://dx.doi.org/10.1002/1529-0131(200101)44:1<61::AID-ANR9>3.0.CO;2-Q.

Berger, M. (2002). A history of immune globulin therapy, from the harvard crash program to monoclonal antibodies. *Current Allergy and Asthma Reports, 2*(5), 368–378. http://dx.doi.org/10.1007/s11882-002-0069-z.

Beutler, B., Milsark, I., & Cerami, A. (1985). Passive immunization against cachectin/tumor necrosis factor protects mice from lethal effect of endotoxin. *Science, 229*(4716), 869–871. http://dx.doi.org/10.1126/science.3895437.

Bleyer, W. A. (1978). The clinical pharmacology of methotrexate: New applications of an old drug. *Cancer, 41*(1), 36–51. http://dx.doi.org/10.1002/1097-0142(197801)41:1<36::aid-cncr2820410108>3.0.co;2-i.

Bonelli, M., & Scheinecker, C. (2018). How does abatacept really work in rheumatoid arthritis? *Current Opinion in Rheumatology, 30*(3), 295–300. http://dx.doi.org/10.1097/BOR.0000000000000491.

Bongartz, T. (2008). Tocilizumab for rheumatoid and juvenile idiopathic arthritis. *The Lancet, 371*(9617), 961–963. http://dx.doi.org/10.1016/S0140-6736(08)60428-6.

Braun, J., Brandt, J., Listing, J., Zink, A., Alten, R., Burmester, G., et al. (2005). Two year maintenance of efficacy and safety of infliximab in the treatment of ankylosing spondylitis. *Annals of the Rheumatic Diseases, 64*(2), 229–234. http://dx.doi.org/10.1136/ard.2004.025130.

Brenner, D., Blaser, H., & Mak, T. W. (2015). Regulation of tumour necrosis factor signalling: Live or let die. *Nature Reviews Immunology, 15*(6), 362–374. http://dx.doi.org/10.1038/nri3834.

Brewerton, D. A., Hart, F. D., Nicholls, A., Caffrey, M., James, D. C. O., & Sturrock, R. D. (1973). Ankylosing spondylitis and HL-A 27. *The Lancet, 301*(7809), 904–907. http://dx.doi.org/10.1016/S0140-6736(73)91360-3.

Brunner, H. I., Johnson, A. L., Barron, A. C., Passo, M. H., Griffin, T. A., Graham, T. B., et al. (2005). Gastrointestinal symptoms and their association with health-related quality of life of children with juvenile rheumatoid arthritis: Validation of a gastrointestinal symptom questionnaire. *Journal of Clinical Rheumatology: Practical Reports on Rheumatic & Musculoskeletal Diseases, 11*(4), 194–204.

Buss, N. A., Henderson, S. J., McFarlane, M., Shenton, J. M., & de Haan, L. (2012). Monoclonal antibody therapeutics: History and future. *Current Opinion in Pharmacology, 12*(5), 615–622. http://dx.doi.org/10.1016/j.coph.2012.08.001.

Chan, E. S. L., & Cronstein, B. N. (2010). Methotrexate—how does it really work? *Nature Reviews Rheumatology, 6*(3), 175–178. http://dx.doi.org/10.1038/nrrheum.2010.5.

Choi, H. K., Mount, D. B., & Reginato, A. M. (2005). Pathogenesis of gout. *Annals of Internal Medicine, 143*(7), 499. http://dx.doi.org/10.7326/0003-4819-143-7-200510040-00009.

Choy, E. H. S., & Panayi, G. S. (2001). Cytokine pathways and joint inflammation in rheumatoid arthritis. *New England Journal of Medicine, 344*(12), 907–916. http://dx.doi.org/10.1056/NEJM200103223441207.

Clark, E. A. (2005). How does B cell depletion therapy work, and how can it be improved? *Annals of the Rheumatic Diseases, 64*(Suppl. 1_4), iv77–iv80. http://dx.doi.org/10.1136/ard.2005.042507.

Conti, F., Priori, R., Chimenti, M. S., Coari, G., Annovazzi, A., Valesini, G., et al. (2005). Successful treatment with intraarticular infliximab for resistant knee monarthritis in a patient with spondylarthropathy: A role for scintigraphy with^{99m}Tc-infliximab. *Arthritis & Rheumatism, 52*(4), 1224–1226. http://dx.doi.org/10.1002/art.20979.

Cronstein, B. N. (2005). Low-dose methotrexate: A mainstay in the treatment of rheumatoid arthritis. *Pharmacological Reviews, 57*(2), 163–172. http://dx.doi.org/10.1124/pr.57.2.3.

Crowe, J. S., Roberts, K. J., Carlton, T. M., Maggiore, L., Cubitt, M. F., Ray, K. P., et al. (2019). Oral delivery of the anti-tumor necrosis factor α domain antibody, V565, results in high intestinal and fecal concentrations with minimal systemic exposure in cynomolgus monkeys. *Drug Development and Industrial Pharmacy, 45*(3), 387–394. http://dx.doi.org/10.1080/03639045.2018.1542708.

Dall'Era, M., Bruce, I. N., Gordon, C., Manzi, S., McCaffrey, J., & Lipsky, P. E. (2019). Current challenges in the development of new treatments for lupus. *Annals of the Rheumatic Diseases, 78*(6), 729–735. http://dx.doi.org/10.1136/annrheumdis-2018-214530.

Dass, S., Vital, E. M., & Emery, P. (2007). Development of psoriasis after B cell depletion with rituximab. *Arthritis & Rheumatism, 56*(8), 2715–2718. http://dx.doi.org/10.1002/art.22811.

Dean, L. E., Jones, G. T., MacDonald, A. G., Downham, C., Sturrock, R. D., & Macfarlane, G. J. (2014). Global prevalence of ankylosing spondylitis. *Rheumatology, 53*(4), 650–657. http://dx.doi.org/10.1093/rheumatology/ket387.

Dowty, M. E., Lin, J., Ryder, T. F., Wang, W., Walker, G. S., Vaz, A., et al. (2014). The pharmacokinetics, metabolism, and clearance mechanisms of tofacitinib, a Janus kinase inhibitor, in humans. *Drug Metabolism and Disposition, 42*(4), 759–773. http://dx.doi.org/10.1124/dmd.113.054940.

Elkayam, O., & Elalouf, O. (2015). Long-term safety and efficacy of infliximab for the treatment of ankylosing spondylitis. *Therapeutics and Clinical Risk Management, 1719*. http://dx.doi.org/10.2147/TCRM.S55928.

Ellerin, T., Rubin, R. H., & Weinblatt, M. E. (2003). Infections and anti-tumor necrosis factor ? Therapy. *Arthritis & Rheumatism, 48*(11), 3013–3022. http://dx.doi.org/10.1002/art.11301.

Elliott, M. J., Maini, R. N., Feldmann, M., Long-Fox, A., Charles, P., Bijl, J. A., et al. (1994). Repeated therapy with monoclonal antibody to tumour necrosis factor α (cA2) in patients with rheumatoid arthritis. *The Lancet, 344*(8930), 1125–1127. http://dx.doi.org/10.1016/S0140-6736(94)90632-7.

Ferreiro-Iglesias, A., Montes, A., Perez-Pampin, E., Cañete, J. D., Raya, E., Magro-Checa, C., et al. (2019). Evaluation of 12 GWAS-drawn SNPs as biomarkers of rheumatoid arthritis response to TNF inhibitors. A potential SNP association with response to etanercept. *PLoS One, 14*(2), e0213073. http://dx.doi.org/10.1371/journal.pone.0213073.

Firestein, G. S. (2003). Evolving concepts of rheumatoid arthritis. *Nature, 423*(6937), 356–361. http://dx.doi.org/10.1038/nature01661.

Fleischmann, R., Mysler, E., Hall, S., Kivitz, A. J., Moots, R. J., Luo, Z., et al. (2017). Efficacy and safety of tofacitinib monotherapy, tofacitinib with methotrexate, and adalimumab with methotrexate in patients with rheumatoid arthritis (oral strategy): A phase 3b/4, double-blind, head-to-head, randomised controlled trial. *The Lancet, 390*(10093), 457–468. http://dx.doi.org/10.1016/S0140-6736(17)31618-5.

Gattorno, M., Facchetti, P., Ghiotto, F., Vignola, S., Buoncompagni, A., Prigione, I., et al. (1997). Synovial fluid T cell clones from oligoarticular juvenile arthritis patients display a prevalent Th1/Th0-type pattern of cytokine secretion irrespective of immunophenotype. *Clinical and Experimental Immunology, 109*(1), 4–11. http://dx.doi.org/10.1046/j.1365-2249.1997.4331330.x.

Gattorno, M., Vignola, S., Falcini, F., Sabatini, F., Buoncompagni, A., Simonini, G., et al. (2002). Serum and synovial fluid concentrations of matrix metalloproteinases 3 and its tissue inhibitor 1 in juvenile idiopathic arthritides. *Journal of Rheumatology, 29*(4), 826–831.

Ghodke-Puranik, Y., & Niewold, T. B. (2015). Immunogenetics of systemic lupus erythematosus: A comprehensive review. *Journal of Autoimmunity, 64*, 125–136. http://dx.doi.org/10.1016/j.jaut.2015.08.004.

Giles, J., Shaw, J., Piper, C., Wong-Baeza, I., McHugh, K., Ridley, A., et al. (2012). HLA-B27 homodimers and free H chains are stronger ligands for leukocyte Ig-like receptor B2 than classical HLA class I. *The Journal of Immunology, 188*(12), 6184–6193. http://dx.doi.org/10.4049/jimmunol.1102711.

Glintborg, B., Sørensen, I. J., Loft, A. G., Lindegaard, H., Linauskas, A., Hendricks, O., et al. (2017). A nationwide non-medical switch from originator infliximab to biosimilar CT-P13 in 802 patients with inflammatory arthritis: 1-year clinical outcomes from the DANBIO registry. *Annals of the Rheumatic Diseases, 76*(8), 1426–1431. http://dx.doi.org/10.1136/annrheumdis-2016-210742.

Grell, M. (1995). Tumor necrosis factor (TNF) receptors in cellular signaling of soluble and membrane-expressed TNF. *Journal of Inflammation, 47*(1–2), 8–17.

Gridley, G., McLaughlin, J. K., Ekbom, A., Klareskog, L., Adami, H.-O., Hacker, D. G., et al. (1993). Incidence of cancer among patients with rheumatoid arthritis. *Journal of the National Cancer Institute, 85*(4), 307–311. http://dx.doi.org/10.1093/jnci/85.4.307.

Gumpel, J. M. (1976). Cyclophosphamide, gold and penicillamine—disease-modifying drugs in rheumatoid arthritis—tailored dosage and ultimate success. *Rheumatology and Rehabilitation, 15*(3), 217–220.

Haq, I. (2003). Osteoarthritis. *Postgraduate Medical Journal, 79*(933), 377–383. http://dx.doi.org/10.1136/pmj.79.933.377.

Hayter, S. M., & Cook, M. C. (2012). Updated assessment of the prevalence, spectrum and case definition of autoimmune disease. *Autoimmunity Reviews, 11*(10), 754–765. http://dx.doi.org/10.1016/j.autrev.2012.02.001.

Heuft-Dorenbosch, L. (2003). Assessment of enthesitis in ankylosing spondylitis. *Annals of the Rheumatic Diseases, 62*(2), 127–132. http://dx.doi.org/10.1136/ard.62.2.127.

Hoffbrand, A. V., & Tripp, E. (1972). Unbalanced deoxyribonucleotide synthesis caused by methotrexate. *The BMJ, 2*(5806), 140–142. http://dx.doi.org/10.1136/bmj.2.5806.140.

Hootman, J. M., Helmick, C. G., Barbour, K. E., Theis, K. A., & Boring, M. A. (2016). Updated projected prevalence of self-reported doctor-diagnosed arthritis and arthritis-attributable activity limitation among US adults, 2015–2040. Projected prevalence of arthritis in the US, 2015–2040 *Arthritis & Rheumatology, 68*(7), 1582–1587. http://dx.doi.org/10.1002/art.39692.

Hopkins, S. J., & Meager, A. (1988). Cytokines in synovial fluid: II. The presence of tumour necrosis factor and interferon. *Clinical and Experimental Immunology, 73*(1), 88–92.

Houssiau, F. A., Devogelaer, J.-P., Damme, J. V., Deuxchaisnes, C. N. D., & Snick, J. V. (1988). Interleukin-6 in synovial fluid and serum of patients with rheumatoid arthritis and other inflammatory arthritides. *Arthritis & Rheumatism, 31*(6), 784–788. http://dx.doi.org/10.1002/art.1780310614.

Hunter, C. A., & Jones, S. A. (2015). IL-6 as a keystone cytokine in health and disease. *Nature Immunology, 16*(5), 448–457. http://dx.doi.org/10.1038/ni.3153.

International Genetics of Ankylosing Spondylitis Consortium (IGAS). (2013). Identification of multiple risk variants for ankylosing spondylitis through high-density genotyping of immune-related loci. *Nature Genetics, 45*(7), 730–738. http://dx.doi.org/10.1038/ng.2667.

Jafarzadeh, S. R., & Felson, D. T. (2018). Updated estimates suggest a much higher prevalence of arthritis in United States adults than previous ones. *Arthritis & Rheumatology, 70*(2), 185–192. http://dx.doi.org/10.1002/art.40355.

Karami, J., Aslani, S., Jamshidi, A., Garshasbi, M., & Mahmoudi, M. (2019). Genetic implications in the pathogenesis of rheumatoid arthritis; an updated review. *Gene, 702*, 8–16. http://dx.doi.org/10.1016/j.gene.2019.03.033.

Kean, W. F., Forestier, F., Kassam, Y., Buchanan, W. W., & Rooney, P. J. (1985). The history of gold therapy in rheumatoid disease. *Seminars in Arthritis and Rheumatism, 14*(3), 180–186. http://dx.doi.org/10.1016/0049-0172(85)90037-X.

Keystone, E. C., Taylor, P. C., Tanaka, Y., Gaich, C., DeLozier, A. M., Dudek, A., et al. (2017). Patient-reported outcomes from a phase 3 study of baricitinib versus placebo or adalimumab in rheumatoid arthritis: Secondary analyses from the RA-BEAM study. *Annals of the Rheumatic Diseases, 76*(11), 1853–1861. http://dx.doi.org/10.1136/annrheumdis-2017-211259.

Khare, S. D. (1995). Spontaneous inflammatory arthritis in HLA-B27 transgenic mice lacking beta 2-microglobulin: A model of human spondyloarthropathies. *Journal of Experimental Medicine, 182*(4), 1153–1158. http://dx.doi.org/10.1084/jem.182.4.1153.

Khare, S. D., Hansen, J., Luthra, H. S., & David, C. S. (1996). HLA-B27 heavy chains contribute to spontaneous inflammatory disease in B27/human beta2-microglobulin (beta2m) double transgenic mice with disrupted mouse beta2m. *Journal of Clinical Investigation, 98*(12), 2746–2755. http://dx.doi.org/10.1172/JCI119100.

Kim, G. W., Lee, N. R., Pi, R. H., Lim, Y. S., Lee, Y. M., Lee, J. M., et al. (2015). IL-6 inhibitors for treatment of rheumatoid arthritis: Past, present, and future. *Archives of Pharmacal Research, 38*(5), 575–584. http://dx.doi.org/10.1007/s12272-015-0569-8.

Kitasato, S. (1890). Ueber das Zustandekommen der Diphtherie-Immunität und der Tetanus-Immunität bei Thieren. *DMW – Deutsche Medizinische Wochenschrift, 16*(49), 1113–1114. http://dx.doi.org/10.1055/s-0029-1207589.

Köhler, G., & Milstein, C. (1975). Continuous cultures of fused cells secreting antibody of predefined specificity. *Nature, 256*(5517), 495–497. http://dx.doi.org/10.1038/256495a0.

Kremer, J. M., Genant, H. K., Moreland, L. W., Russell, A. S., Emery, P., Abud-Mendoza, C., et al. (2006). Effects of abatacept in patients with methotrexate-resistant active rheumatoid arthritis: A randomized trial. *Annals of Internal Medicine, 144*(12), 865. http://dx.doi.org/10.7326/0003-4819-144-12-200606200-00003.

Kriegler, M., Perez, C., DeFay, K., Albert, I., & Lu, S. D. (1988). A novel form of TNF/cachectin is a cell surface cytotoxic transmembrane protein: Ramifications for the complex physiology of TNF. *Cell, 53*(1), 45–53. http://dx.doi.org/10.1016/0092-8674(88)90486-2.

Leonardi, C. L., Kimball, A. B., Papp, K. A., Yeilding, N., Guzzo, C., Wang, Y., et al. (2008). Efficacy and safety of ustekinumab, a human interleukin-12/23 monoclonal antibody, in patients with psoriasis: 76-week results from a randomised, double-blind, placebo-controlled trial (PHOENIX 1). *The Lancet, 371*(9625), 1665–1674. http://dx.doi.org/10.1016/S0140-6736(08)60725-4.

Lie, G., Sciascia, S., & Cuadrado, M. J. (2015). Biosimilar vs biological agents in rheumatology: When are biosimilar agents similar enough? *International Immunopharmacology, 27*(2), 220–223. http://dx.doi.org/10.1016/j.intimp.2015.04.022.

Liu, J.-T. (2014). Psoriatic arthritis: Epidemiology, diagnosis, and treatment. *World Journal of Orthopedics, 5*(4), 537. http://dx.doi.org/10.5312/wjo.v5.i4.537.

Liu, Z., & Davidson, A. (2012). Taming lupus—a new understanding of pathogenesis is leading to clinical advances. *Nature Medicine, 18*(6), 871–882. http://dx.doi.org/10.1038/nm.2752.

López-Ferrer, A., Laiz, A., & Puig, L. (2017). The safety of ustekinumab for the treatment of psoriatic arthritis. *Expert Opinion on Drug Safety, 16*(6), 733–742. http://dx.doi.org/10.1080/14740338.2017.1323864.

Lopez-Rodriguez, R., Perez-Pampin, E., Marquez, A., Blanco, F. J., Joven, B., Carreira, P., et al. (2018). Validation study of genetic biomarkers of response to TNF inhibitors in rheumatoid arthritis. *PLoS One, 13*(5), e0196793. http://dx.doi.org/10.1371/journal.pone.0196793.

Luchetti, M. M., Benfaremo, D., & Gabrielli, A. (2018). Biologics in inflammatory and immunomediated arthritis. *Current Pharmaceutical Biotechnology, 18*(12), 989–1007. http://dx.doi.org/10.2174/1389201019666171226151852.

Lybecker, K. M. (July 12, 2016). *The biologics revolution in the production of drugs.* Retrieved from https://www.fraserinstitute.org/studies/biologics-revolution-in-the-production-of-drugs.

Maggi, L., Cosmi, L., Simonini, G., Annunziato, F., & Cimaz, R. (2016). T cell subpopulations in juvenile idiopathic arthritis and their modifications after biotherapies. *Autoimmunity Reviews, 15*(12), 1141–1144. http://dx.doi.org/10.1016/j.autrev.2016.09.012.

Maini, R., St Clair, E. W., Breedveld, F., Furst, D., Kalden, J., Weisman, M., et al. (1999). Infliximab (chimeric anti-tumour necrosis factor alpha monoclonal antibody) versus placebo in rheumatoid arthritis patients receiving concomitant methotrexate: A randomised phase III trial. ATTRACT study group. *Lancet (London, England), 354*(9194), 1932–1939. http://dx.doi.org/10.1016/s0140-6736(99)05246-0.

Malaviya, A. N. (2016). Landmark papers on the discovery of methotrexate for the treatment of rheumatoid arthritis and other systemic inflammatory rheumatic diseases: A fascinating story. *International Journal of Rheumatic Diseases, 19*(9), 844–851. http://dx.doi.org/10.1111/1756-185X.12862.

Márquez, A., Ferreiro-Iglesias, A., Dávila-Fajardo, C. L., Montes, A., Pascual-Salcedo, D., Perez-Pampin, E., et al. (2014). Lack of validation of genetic variants associated with anti–tumor necrosis factor therapy response in rheumatoid arthritis: A genome-wide association study replication and meta-analysis. *Arthritis Research and Therapy, 16*(2), R66. http://dx.doi.org/10.1186/ar4504.

Marra, C. A., Bansback, N., Anis, A. H., & Shojania, K. (2011). Introduction to economic modeling for clinical rheumatologists: Application to biologic agents in rheumatoid arthritis. *Clinical Rheumatology, 30*(S1), 9–18. http://dx.doi.org/10.1007/s10067-010-1635-8.

Maxwell, L. J., Zochling, J., Boonen, A., Singh, J. A., Veras, M. M., Tanjong Ghogomu, E., et al. (2015). TNF-alpha inhibitors for ankylosing spondylitis. *Cochrane Database of Systematic Reviews.* http://dx.doi.org/10.1002/14651858.CD005468.pub2.

McBain, H., Shipley, M., Newman, S., & members of the Appearance Research Collaboration (ARC). (2013). The impact of appearance concerns on depression and anxiety in rheumatoid arthritis: Concerns about appearance and their impact on mood in RA. *Musculoskeletal Care, 11*(1), 19–30. http://dx.doi.org/10.1002/msc.1020.

McHugh, K., & Bowness, P. (2012). The link between HLA-B27 and SpA—new ideas on an old problem. *Rheumatology, 51*(9), 1529—1539. http://dx.doi.org/10.1093/rheumatology/kes061.

McInnes, I. B., Buckley, C. D., & Isaacs, J. D. (2016). Cytokines in rheumatoid arthritis—shaping the immunological landscape. *Nature Reviews Rheumatology, 12*(1), 63—68. http://dx.doi.org/10.1038/nrrheum.2015.171.

McInnes, I. B., Kavanaugh, A., Gottlieb, A. B., Puig, L., Rahman, P., Ritchlin, C., et al. (2013). Efficacy and safety of ustekinumab in patients with active psoriatic arthritis: 1 year results of the phase 3, multicentre, double-blind, placebo-controlled PSUMMIT 1 trial. *Lancet (London, England), 382*(9894), 780—789. http://dx.doi.org/10.1016/S0140-6736(13)60594-2.

Mikkelsen, T. S., Thorn, C. F., Yang, J. J., Ulrich, C. M., French, D., Zaza, G., et al. (2011). PharmGKB summary: Methotrexate pathway. *Pharmacogenetics and Genomics, 21*(10), 679—686. http://dx.doi.org/10.1097/FPC.0b013e328343dd93.

Mitoma, H., Horiuchi, T., Tsukamoto, H., & Ueda, N. (2018). Molecular mechanisms of action of anti-TNF-α agents — comparison among therapeutic TNF-α antagonists. *Cytokine, 101*, 56—63. http://dx.doi.org/10.1016/j.cyto.2016.08.014.

Munn, D. H., Sharma, M. D., & Mellor, A. L. (2004). Ligation of B7-1/B7-2 by human $CD4^+$ T cells triggers indoleamine 2,3-dioxygenase activity in dendritic cells. *The Journal of Immunology, 172*(7), 4100—4110. http://dx.doi.org/10.4049/jimmunol.172.7.4100.

Murphy, L. B., Sacks, J. J., Brady, T. J., Hootman, J. M., & Chapman, D. P. (2012). Anxiety is more common than depression among US adults with arthritis. *Arthritis Care & Research*. http://dx.doi.org/10.1002/acr.21685.

Murray, K. J., Luyrink, L., Grom, A. A., Passo, M. H., Emery, H., Witte, D., et al. (1996). Immunohistological characteristics of T cell infiltrates in different forms of childhood onset chronic arthritis. *Journal of Rheumatology, 23*(12), 2116—2124.

Nadler, L. M., Stashenko, P., & Hardy, R. (1980). Serotherapy of a patient with a monoclonal antibody directed against a human lymphoma-associated antigen. *Cancer Research, 40*(9), 3147—3154.

Niewold, T. B. (2015). Advances in lupus genetics. *Current Opinion in Rheumatology, 1*. http://dx.doi.org/10.1097/BOR.0000000000000205.

Nisar, M. K. (2019). Early arthritis clinic is cost-effective, improves outcomes and reduces biologic use. *Clinical Rheumatology, 38*(6), 1555—1560. http://dx.doi.org/10.1007/s10067-019-04515-3.

Nistala, K., Adams, S., Cambrook, H., Ursu, S., Olivito, B., de Jager, W., et al. (2010). Th17 plasticity in human autoimmune arthritis is driven by the inflammatory environment. *Proceedings of the National Academy of Sciences, 107*(33), 14751—14756. http://dx.doi.org/10.1073/pnas.1003852107.

Ober, R. J., Radu, C. G., Ghetie, V., & Ward, E. S. (2001). Differences in promiscuity for antibody—FcRn interactions across species: Implications for therapeutic antibodies. *International Immunology, 13*(12), 1551—1559. http://dx.doi.org/10.1093/intimm/13.12.1551.

Palylyk-Colwell, E., & McGahan, L. (2006). Rituximab for rheumatoid arthritis. *Issues in Emerging Health Technologies*, (89), 1—4.

Papp, K. A., Langley, R. G., Lebwohl, M., Krueger, G. G., Szapary, P., Yeilding, N., et al. (2008). Efficacy and safety of ustekinumab, a human interleukin-12/23 monoclonal antibody, in patients with psoriasis: 52-week results from a randomised, double-blind, placebo-controlled trial (PHOENIX 2). *The Lancet, 371*(9625), 1675—1684. http://dx.doi.org/10.1016/S0140-6736(08)60726-6.

Peake, N. J., Khawaja, K., Myers, A., Jones, D., Cawston, T. E., Rowan, A. D., et al. (2005). Levels of matrix metalloproteinase (MMP)-1 in paired sera and synovial fluids of juvenile idiopathic arthritis patients: Relationship to inflammatory activity, MMP-3 and tissue

inhibitor of metalloproteinases-1 in a longitudinal study. *Rheumatology, 44*(11), 1383–1389. http://dx.doi.org/10.1093/rheumatology/kei025.
Piskin, G., Sylva-Steenland, R. M. R., Bos, J. D., & Teunissen, M. B. M. (2006). In vitro and in situ expression of IL-23 by keratinocytes in healthy skin and psoriasis lesions: Enhanced expression in psoriatic skin. *The Journal of Immunology, 176*(3), 1908–1915. http://dx.doi.org/10.4049/jimmunol.176.3.1908.
Planès, S., Villier, C., & Mallaret, M. (2016). The nocebo effect of drugs. *Pharmacology Research & Perspectives, 4*(2), e00208. http://dx.doi.org/10.1002/prp2.208.
Purmonen, T., Puolakka, K., Bhattacharyya, D., Jain, M., & Martikainen, J. (2018). Cost--effectiveness analysis of secukinumab versus other biologics and apremilast in the treatment of active psoriatic arthritis: A Finnish perspective. *Cost Effectiveness and Resource Allocation, 16*(1), 56. http://dx.doi.org/10.1186/s12962-018-0162-3.
Ravelli, A., & Martini, A. (2007). Juvenile idiopathic arthritis. *The Lancet, 369*(9563), 767–778. http://dx.doi.org/10.1016/S0140-6736(07)60363-8.
Reichert, J. M., & Paquette, C. (2003). Clinical development of therapeutic recombinant proteins. *BioTechniques, 35*(1), 176–185. http://dx.doi.org/10.2144/03351dd01.
Reynolds, T., de Zafra, C., Kim, A., & Gelzleichter, T. R. (2013). Overview of biopharmaceuticals and comparison with small-molecule drug development. In *Nonclinical development of novel biologics, biosimilars, vaccines and specialty biologics* (pp. 3–33). http://dx.doi.org/10.1016/B978-0-12-394810-6.00001-0.
Rezk, M. F., & Pieper, B. (2017). Treatment outcomes with biosimilars: Be aware of the nocebo effect. *Rheumatology and Therapy, 4*(2), 209–218. http://dx.doi.org/10.1007/s40744-017-0085-z.
Ritchlin, C., Rahman, P., Kavanaugh, A., McInnes, I. B., Puig, L., Li, S., et al. (2014). Efficacy and safety of the anti-IL-12/23 p40 monoclonal antibody, ustekinumab, in patients with active psoriatic arthritis despite conventional non-biological and biological anti-tumour necrosis factor therapy: 6-month and 1-year results of the phase 3, multicentre, double-blind, placebo-controlled, randomised PSUMMIT 2 trial. *Annals of the Rheumatic Diseases, 73*(6), 990–999. http://dx.doi.org/10.1136/annrheumdis-2013-204655.
Roddy, E., & Doherty, M. (2010). Gout. Epidemiology of gout. *Arthritis Research and Therapy, 12*(6), 223. http://dx.doi.org/10.1186/ar3199.
Roszkiewicz, J., & Smolewska, E. (2017). In the pursuit of methotrexate treatment response biomarker in juvenile idiopathic arthritis—are we getting closer to personalised medicine? *Current Rheumatology Reports, 19*(4), 19. http://dx.doi.org/10.1007/s11926-017-0646-8.
Ruperto, N., Murray, K. J., Gerloni, V., Wulffraat, N., de Oliveira, S. K. F., Falcini, F., et al. (2004). A randomized trial of parenteral methotrexate comparing an intermediate dose with a higher dose in children with juvenile idiopathic arthritis who failed to respond to standard doses of methotrexate. *Arthritis & Rheumatism, 50*(7), 2191–2201. http://dx.doi.org/10.1002/art.20288.
Saag, K. G., Teng, G. G., Patkar, N. M., Anuntiyo, J., Finney, C., Curtis, J. R., et al. (2008). American college of rheumatology 2008 recommendations for the use of nonbiologic and biologic disease-modifying antirheumatic drugs in rheumatoid arthritis. *Arthritis & Rheumatism, 59*(6), 762–784. http://dx.doi.org/10.1002/art.23721.
Sanz, I., Yasothan, U., & Kirkpatrick, P. (2011). Belimumab. *Nature Reviews Drug Discovery, 10*(5), 335–336. http://dx.doi.org/10.1038/nrd3436.
Sarzi-Puttini, P., Ceribelli, A., Marotto, D., Batticciotto, A., & Atzeni, F. (2019). Systemic rheumatic diseases: From biological agents to small molecules. *Autoimmunity Reviews, 18*(6), 583–592. http://dx.doi.org/10.1016/j.autrev.2018.12.009.

Sato, K., Tsuchiya, M., Saldanha, J., Koishihara, Y., Ohsugi, Y., Kishimoto, T., et al. (1993). Reshaping a human antibody to inhibit the interleukin 6-dependent tumor cell growth. *Cancer Research, 53*(4), 851–856.

Saxne, T., Palladino, M. A., Heinegård, D., Talal, N., & Wollheim, F. A. (1988). Detection of tumor necrosis factor alpha but not tumor necrosis factor beta in rheumatoid arthritis synovial fluid and serum. *Arthritis & Rheumatism, 31*(8), 1041–1045. http://dx.doi.org/10.1002/art.1780310816.

Schiff, M., Keiserman, M., Codding, C., Songcharoen, S., Berman, A., Nayiager, S., et al. (2008). Efficacy and safety of abatacept or infliximab vs placebo in ATTEST: A phase III, multi-centre, randomised, double-blind, placebo-controlled study in patients with rheumatoid arthritis and an inadequate response to methotrexate. *Annals of the Rheumatic Diseases, 67*(8), 1096–1103. http://dx.doi.org/10.1136/ard.2007.080002.

Schlosstein, L., Terasaki, P. I., Bluestone, R., & Pearson, C. M. (1973). High association of an HL-A antigen, W27, with ankylosing spondylitis. *New England Journal of Medicine, 288*(14), 704–706. http://dx.doi.org/10.1056/NEJM197304052881403.

Sherlock, J. P., Joyce-Shaikh, B., Turner, S. P., Chao, C.-C., Sathe, M., Grein, J., et al. (2012). IL-23 induces spondyloarthropathy by acting on ROR-γt^+ $CD3^+CD4^-CD8^-$ entheseal resident T cells. *Nature Medicine, 18*(7), 1069–1076. http://dx.doi.org/10.1038/nm.2817.

Shi, Y., Mucsi, A. D., & Ng, G. (2010). Monosodium urate crystals in inflammation and immunity. *Immunological Reviews, 233*(1), 203–217. http://dx.doi.org/10.1111/j.0105-2896.2009.00851.x.

Skepner, J., Ramesh, R., Trocha, M., Schmidt, D., Baloglu, E., Lobera, M., et al. (2014). Pharmacologic inhibition of RORγt regulates Th17 signature gene expression and suppresses cutaneous inflammation in vivo. *The Journal of Immunology, 192*(6), 2564–2575. http://dx.doi.org/10.4049/jimmunol.1302190.

Smolen, J. S., Aletaha, D., Barton, A., Burmester, G. R., Emery, P., Firestein, G. S., et al. (2018). Rheumatoid arthritis. *Nature Reviews Disease Primers, 4*(1), 18001. http://dx.doi.org/10.1038/nrdp.2018.1.

Smolen, J. S., Choe, J.-Y., Prodanovic, N., Niebrzydowski, J., Staykov, I., Dokoupilova, E., et al. (2017). Comparing biosimilar SB2 with reference infliximab after 54 weeks of a double-blind trial: Clinical, structural and safety results. *Rheumatology, 56*(10), 1771–1779. http://dx.doi.org/10.1093/rheumatology/kex254.

Stern, M., & Herrmann, R. (2005). Overview of monoclonal antibodies in cancer therapy: Present and promise. *Critical Reviews in Oncology/Hematology, 54*(1), 11–29. http://dx.doi.org/10.1016/j.critrevonc.2004.10.011.

Strand, V., Kimberly, R., & Isaacs, J. D. (2007). Biologic therapies in rheumatology: Lessons learned, future directions. *Nature Reviews Drug Discovery, 6*(1), 75–92. http://dx.doi.org/10.1038/nrd2196.

Tjellström, B., Stenhammar, L., Magnusson, K. E., & Sandqvist, T. (1997). Oral immunoglobulin treatment in Crohn's disease. *Acta Paediatrica, 86*(2), 221–223. http://dx.doi.org/10.1111/j.1651-2227.1997.tb08873.x.

Tracey, K. J., Fong, Y., Hesse, D. G., Manogue, K. R., Lee, A. T., Kuo, G. C., et al. (1987). Anti-cachectin/TNF monoclonal antibodies prevent septic shock during lethal bacteraemia. *Nature, 330*(6149), 662–664. http://dx.doi.org/10.1038/330662a0.

Tweehuysen, L., van den Bemt, B. J. F., van Ingen, I. L., de Jong, A. J. L., van der Laan, W. H., van den Hoogen, F. H. J., et al. (2018). Subjective complaints as the main reason for biosimilar discontinuation after open-label transition from reference infliximab to biosimilar infliximab. *Arthritis & Rheumatology, 70*(1), 60–68. http://dx.doi.org/10.1002/art.40324.

Ursini, F., Russo, E., De Giorgio, R., De Sarro, G., & D'Angelo, S. (2018). Current treatment options for psoriatic arthritis: Spotlight on abatacept. *Therapeutics and Clinical Risk Management, 14*, 1053–1059. http://dx.doi.org/10.2147/TCRM.S148586.

Wang, C. Q. F., Akalu, Y. T., Suarez-Farinas, M., Gonzalez, J., Mitsui, H., Lowes, M. A., et al. (2013). IL-17 and TNF synergistically modulate cytokine expression while suppressing melanogenesis: Potential relevance to psoriasis. *Journal of Investigative Dermatology, 133*(12), 2741–2752. http://dx.doi.org/10.1038/jid.2013.237.

Webb, L. M. C., Walmsley, M. J., & Feldmann, M. (1996). Prevention and amelioration of collagen-induced arthritis by blockade of the CD28 co-stimulatory pathway: Requirement for both B7-1 and B7-2. *European Journal of Immunology, 26*(10), 2320–2328. http://dx.doi.org/10.1002/eji.1830261008.

Weiner, G. J. (2010). Rituximab: Mechanism of action. *Seminars in Hematology, 47*(2), 115–123. http://dx.doi.org/10.1053/j.seminhematol.2010.01.011.

Weiss, J. E., & Ilowite, N. T. (2007). Juvenile idiopathic arthritis. *Rheumatic Disease Clinics of North America, 33*(3), 441–470. http://dx.doi.org/10.1016/j.rdc.2007.07.006.

Wong-Baeza, I., Ridley, A., Shaw, J., Hatano, H., Rysnik, O., McHugh, K., et al. (2013). KIR3DL2 binds to HLA-B27 dimers and free H chains more strongly than other HLA class I and promotes the expansion of T cells in ankylosing spondylitis. *The Journal of Immunology, 190*(7), 3216–3224. http://dx.doi.org/10.4049/jimmunol.1202926.

Worledge, K. L., Godiska, R., Barrett, T. A., & Kink, J. A. (2000). Oral administration of avian tumor necrosis factor antibodies effectively treats experimental colitis in rats. *Digestive Diseases and Sciences, 45*(12), 2298–2305. http://dx.doi.org/10.1023/a:1005554900286.

Wu, X., Sheng, X., Sheng, R., Lu, H., & Xu, H. (2019). Genetic and clinical markers for predicting treatment responsiveness in rheumatoid arthritis. *Frontiers of Medicine, 13*(4), 411–419. http://dx.doi.org/10.1007/s11684-018-0659-3.

Yocum, D. E., Esparza, L., Dubry, S., Benjamin, J. B., Volz, R., & Scuderi, P. (1989). Characteristics of tumor necrosis factor production in rheumatoid arthritis. *Cellular Immunology, 122*(1), 131–145. http://dx.doi.org/10.1016/0008-8749(89)90154-8.

Yoo, D. H. (2017). CT-P13 in the treatment of rheumatoid arthritis. *Expert Review of Clinical Immunology, 13*(7), 653–666. http://dx.doi.org/10.1080/1744666X.2017.1337510.

CHAPTER EIGHT

Comprehensive *in silico* screening and molecular dynamics studies of missense mutations in Sjogren-Larsson syndrome associated with the *ALDH3A2* gene

S. Udhaya Kumar[a], D. Thirumal Kumar[a], Pinky D. Mandal[a], Srivarshini Sankar[a], Rishin Haldar[b], Balu Kamaraj[c], Charles Emmanuel Jebaraj Walter[d], R. Siva[a], C. George Priya Doss[a,*] and Hatem Zayed[e,*]

[a]School of BioSciences and Technology, Vellore Institute of Technology, Vellore, Tamil Nadu, India
[b]School of Computer Science and Engineering, Vellore Institute of Technology, Vellore, Tamil Nadu, India
[c]Department of Neuroscience Technology, College of Applied Medical Sciences, Imam Abdulrahman Bin Faisal University, Jubail, Saudi Arabia
[d]Department of Biotechnology, Sri Ramachandra Institute of Higher Education and Research (Deemed to be University), Chennai, India
[e]Department of Biomedical Sciences, College of Health and Sciences, Qatar University, Doha, Qatar
*Corresponding authors: E-mail: hatem.zayed@qu.edu.qa, georgepriyadoss@vit.ac.in

Contents

Advances in Protein Chemistry and Structural Biology, Volume 120
ISSN 1876-1623
https://doi.org/10.1016/bs.apcsb.2019.11.004

Abstract

Sjögren–Larsson syndrome (SLS) is an autoimmune disorder inherited in an autosomal recessive pattern. To date, 80 missense mutations have been identified in association with the *Aldehyde Dehydrogenase* 3 Family Member A2 (*ALDH3A2*) gene causing SLS. Disruption of the function of *ALDH3A2* leads to excessive accumulation of fat in the cells, which interferes with the normal function of protective membranes or materials that are necessary for the body to function normally. We retrieved 54 missense mutations in the ALDH3A2 from the OMIM, UniProt, dbSNP, and HGMD databases that are known to cause SLS. These mutations were examined with various *in silico* stability tools, which predicted that the mutations p.S308N and p.R423H that are located at the protein-protein interaction domains are the most destabilizing. Furthermore, to determine the atomistic-level differences within the protein-protein interactions owing to mutations, we performed macromolecular simulation (MMS) using GROMACS to validate the motion patterns and dynamic behavior of the biological system. We found that both mutations (p.S380N and p.R423H) had significant effects on the protein-protein interaction and disrupted the dimeric interactions. The computational pipeline provided in this study helps to elucidate the potential structural and functional differences between the ALDH3A2 native and mutant homodimeric proteins, and will pave the way for drug discovery against specific targets in the SLS patients.

1. Introduction

An autoimmune disorder is a condition caused by an impaired immune response to a healthy body part (Hayter & Cook, 2012; Lerner, Jeremias, & Matthias, 2015; Manzel et al., 2013). Sjögren–Larsson syndrome (SLS; MIM #270200) is inherited neurocutaneously; this rare autoimmune disorder is characterized by ichthyosis, spastic diplegia or tetraplegia and mental retardation (Gordon, 2007; Rizzo & Carney, 2005; Sjogren & Larsson, 1957). SLS is a progressive autoimmune condition that often leads to a severe inflammatory illness that could also impair certain healthy organs and tissues, such as the nervous system, skin, lungs and joints (Ramos-Casals, Tzioufas, & Font, 2005; Skopouli, Dafni, Ioannidis, & Moutsopoulos, 2000). The cutaneous symptoms of ichthyosis are usually seen in a newborn child as dry, scaly skin. Dry skin is due to a build-up of skin to form a scale-like covering. Spasticity blocks motor development, thus making the patient physically impaired in some cases. The neurological symptoms are observed later in the first or second year of life. Mental retardation may range from mild to profound (Carney, Wei, & Rizzo, 2004). Additional clinical manifestations include short height, macular degeneration in the eyes, upper eyelid ichthyosis, thoracic kyphosis, seizures, photophobia, speech abnormalities, and premature birth (Willemsen et al., 1999). SLS is inherited in an autosomal recessive pattern. The leading cause of this disease is the mutation in the gene *ALDH3A2*, which results in a toxic accumulation of fatty alcohol. The location of the gene is in the short arm of chromosome number 17 (17p11.2), and it consists of 11 exons (Chang & Yoshida, 1997; Rogers, Markova, De Laurenzi, Rizzo, & Compton, 1997), encoding a protein measuring 485 amino acids (Chang & Yoshida, 1997). When the gene undergoes alternative splicing, it produces a second minor transcript, which is translated into a FALDH protein of 508 amino acids possessing a unique carboxy-terminus (Rogers et al., 1997).

This gene is responsible for producing enzyme fatty aldehyde dehydrogenase. In humans, the gene *ALDH3A2* codes for a particular membrane-bound fatty aldehyde dehydrogenase enzyme, which is responsible for the conversion from long-chain fatty aldehydes to their respective fatty alcohols (Nakahara et al., 2012). *ALDH3A2* is also involved in the conversion of the sphingosine-1-phosphate (SIP) degradation product hexadecenal to hexadecenoic acid. The function of ALDH3A2 is not well understood, and most of the known functions of ALDH3A2 have been elucidated by studying SLS patients (Lloyd et al., 2007). ALDH3A2 contributes to detoxification processes in various lipid degradation pathways. This protein acts against oxidative stress-induced lipid peroxidation, and its expression is regulated by proliferator-activated receptor alpha (Ashibe, Hirai, Higashi, Sekimizu, & Motojima, 2007; Demozay, Mas, Rocchi, & Obberghen, 2008; Hait et al., 2009; Kihara, Mitsutake, Mizutani, & Igarashi, 2007). The prevalence of SLS is estimated to occur at a frequency of 6 per million in the Swedish population. However, in northern Sweden, the incidence rate of SLS was estimated to be approximately 164 per million (Sillén, Jagell, & Wadelius, 1997).

ALDH3A2 is a homodimeric protein (homo-oligomeric) that contains a catalytic domain, a NAD-binding domain, and a C-terminal oligomerization domain. These domains are involved in the process of turning a monomer or a mixture of monomers into an oligomer (oligomerization)

(Liu et al., 1997). Based on the alternatively spliced hydrophobic C-terminal transmembrane domain, ALDH3A2 is embedded in the membrane of the endoplasmic reticulum or peroxisome (Rogers et al., 1997). At the interface of the catalytic and cofactor binding domain, a deep funnel-like structure is established that gives access to catalytic cysteine (Liu et al., 1997). The amino acids present in this funnel are entailed in substrate selectivity (Khanna et al., 2011; Liu et al., 1997). The protein subunits in the asymmetric structure assume the canonical aldehyde dehydrogenase fold, which includes an amino-terminal (N- terminal), a cofactor binding domain/catalytic domain (residues 1–79 and 103–208), a catalytic domain (residues 29–419) and an oligomerization domain (residues 82–102 and 420–443) that connects two subunits of the homodimer (Keller et al., 2014). To date, 72 different mutations have been reported in SLS patients (Rizzo & Carney, 2005). The interactions of proteins could be interrupted by chemical changes and alterations in amino acids, and the structural context behind this effect has not been determined (Chan et al., 2015, 2016; Landersdorfer, Caliph, Shackleford, Ascher, & Kaminskas, 2015). Therefore, it is vital to understand the implications of mutations in a structural context at protein-protein interface regions (Jubb et al., 2017). Studying these mutations would enable us to understand the disease's molecular mechanism to be linked to their inheritance modes and illustrate how genetic mutations in the same trait can exhibit different clinical features by interfering with various protein interaction interfaces (Wang et al., 2012; Zhong et al., 2009).

This study seeks to determine the role of missense mutations in ALDH3A2, present in the PPI, causing changes in the protein's native structure and function. In the past few years, computational pipelines have been used to interpret the effects of these alterations and unravel the ramification of mutations in various genes (Doss et al., 2016; Sneha et al., 2018; Udhaya Kumar, Priyanka, Sneha, & George Priya Doss, 2015). Integrating several algorithms to determine the stability change provides more precise results (George et al., 2014; Sneha et al., 2017; Thirumal Kumar & George Priya Doss, 2016). It is important to grasp the relationship between mutation and disease by studying the effects of such variations on biological functions. Moreover, nsSNPs associated with disease have been shown to appear at PPI hotspots, and they have also been shown to alter the stability of proteins and interactions with other proteins, ligands, and DNA (David, Razali, Wass, & Sternberg, 2012; Doss et al., 2012; George Priya Doss, Nagasundaram, Chakraborty, Chen, & Zhu, 2013; Kucukkal, Petukh, Li, & Alexov, 2015; Pires, Chen, Blundell, & Ascher, 2016). Additionally, some prior studies showed the best influence of molecular dynamics simulation and its association with the experimental findings (Agrahari, Sneha, George Priya Doss, Siva, & Zayed, 2018, p. 20; Sneha et al., 2017; Tanwar, Kumar, Doss, & Zayed, 2019). The mutations obtained in this study were subjected to stability and conservation *in silico* prediction tools. Although ALDH3A2 is a homodimeric protein, we performed protein-protein docking using the ZDOCK web server for the wild-type and mutant proteins. Finally, the macromolecular simulation was conducted for the protein complexes using a package from GROMACS to determine the atomistic-level significance of the mutation. In this study, the negative impact of nsSNPs on the stabilization and role of fatty aldehyde dehydrogenase was extended and explored.

2. Materials and methods

2.1 Datasets

For our *in silico* analysis, the human *ALDH3A2* gene was retrieved from OMIM (OMIM #270200), UniProt (Uniprot ID: P51648), dbSNP (Ref. No: NP_000373.1), and HGMD databases. (Amberger & Hamosh, 2017; Bateman et al., 2017; Sherry et al., 2001; Stenson et al., 2017). The RCSB Protein Data Bank was used to obtain a human ALDH3A2 dimeric protein structure with PDB ID: 4QGK (https://www.rcsb.org/) (Berman et al., 2000; Keller et al., 2014). We retrieved 54 missense mutations associated with SLS from public databases. The protein-protein interface interacting amino acids were obtained from PDBsum for the ALDH3A2 homodimeric protein (Laskowski, 2001; Laskowski, Jabłońska, Pravda, Vařeková, & Thornton, 2018). For further investigation, we selected 10 missense mutations located in the PPI of ALDH3A2.

2.2 ConSurf prediction for conserved residues

Of using empirical Bayesian induction, ConSurf accessible at http://consurf.tau.ac.il/computes the evolutionary conservation of amino acids in a given protein sequence. Together with the color palette, ConSurf provides a conservation score ranging from 9 (most conserved amino acids) to 0 (least conserved amino acids) (Ashkenazy, Erez, Martz, Pupko, & Ben-Tal, 2010; Ashkenazy et al., 2016).

2.3 Determining protein stability changes

The preceding algorithms, such as I-Mutant2.0 (Capriotti, Fariselli, & Casadio, 2005), MUpro (Cheng, Randall, & Baldi, 2006), iStable (Chen, Lin, & Chu, 2013), DynaMut (Rodrigues, Pires, & Ascher, 2018), mCSM (Pires, Ascher, & Blundell, 2014b), SDM (Pandurangan, Ochoa-Montaño, Ascher, & Blundell, 2017), and DUET (Pires, Ascher, & Blundell, 2014a), were used to determine the stability and functional consequences of p.S380N and p.R423H mutations. I-Mutant2.0 is an online tool based on neural networks that automatically detects changes in protein stability in point mutations. I-Mutant2.0 can compute a mutation's severity effect on the stability of the folded protein (Capriotti et al., 2005). MUpro is a collection of programs with machine learning that precisely computes the stability of proteins that changes based on prime sequence data, particularly where the tertiary structure is not exposed. This approach overcomes significant restrictions on previous approaches based on the tertiary structure (Cheng et al., 2006). The integrated predictor iStable was constructed using information from sequence and data prediction from different predictors of elements. Two different types of input are required for iStable: sequence and structure (Chen et al., 2013). The DynaMut combines graph-based signatures with normal mode parameters to produce a collective assessment of the effect of a mutation on the stability of proteins (Rodrigues et al., 2018). The mCSM depends on a graph-based signature principle and computes the consequences of missense mutations on protein stability, protein-protein binding, and protein-DNA interaction (Pires et al., 2014b). The SDM, a statistical method, and many other systems are the most effective approaches to use. SDM examines the substitution of amino acids in different structural conditions tolerated in the families of homologous proteins of specified 3D structures and transforms them into probability tables for substitution (Pandurangan et al., 2017). The DUET is an online tool that predicts the impact of point mutations on the stability of protein using an embedded computational approach. The computation of DUET depends on additional mutation information, including a pharmacophore vector and secondary structure (Pires et al., 2014a). The retrieved missense mutation (p.S380N and p.R423H) and the ALDH3A2 FASTA sequence were provided as input for all prediction tools mentioned above.

2.4 Physicochemical properties of amino acids

The amino acid features of wild-type and mutant proteins were distinguished using the Project HOPE server (HOPE, http://www.cmbi.ru.nl/hope/). Project HOPE is an automatic program that utilizes individual servers (FASTA, UniProt, DSSP, DAS, BLAST, YASARA, PDB-3D structure, ClustalW, and HSSP) to analyze the consequences of point mutation on the protein structure and function (Venselaar, te Beek, Kuipers, Hekkelman, & Vriend, 2010). The FASTA format protein sequence was given as an input to assess the physicochemical features of wild-type and mutant amino acids.

2.5 Molecular docking and preparation of mutant protein models

To initiate protein-protein docking, we employed a Fast Fourier Transform (FFT)-based protein docking online server called ZDOCK (http://zdock.umassmed.edu/) to understand the improvement of the binding interface of the 4QGK homodimer in the case of mutation (Pierce et al., 2014). 4QGK homodimer-interacting residues were retrieved from the PDBsum database (Laskowski et al., 2018). The wild-type protein was retrieved from Protein Data Bank (PDB) with ID: 4QGK (Keller et al., 2014) through the X-Ray diffraction model. At a defined amino acid position of a native ALDH3A2 protein, the induction of mutations was performed using Swiss PDB Viewer (Guex & Peitsch, 1997). The energy-minimized model of the mutant protein was obtained by performing the GROMOS96 force field, which is embedded in the Swiss PDB Viewer (van Gunsteren, 1996). The interactions, such as hydrogen bonds, hydrophobicity, and other molecules, between the wild-type and mutant protein were studied and visualized using the UCSF Chimera Molecular Modeling System (Meng, Pettersen, Couch, Huang, & Ferrin, 2006; Pettersen et al., 2004).

2.6 Macromolecular simulation

For analysis, GROMACS software (Abraham et al., 2015) was used to conduct molecular dynamics simulations for ALDH3A2 wild-type and mutant complex proteins. The inbuilt force field GROMOS 54A7 parameter was utilized from the Gromacs package to perform system energy minimization, and it explicitly incorporates each atom (Huang, Lin, & van Gunsteren, 2011; Schmid et al., 2011). For ALDH3A2 native and mutant protein complexes, the system was energy minimized to eliminate the local strains and used as the entry point for MD simulations. The solvation of protein models was performed using Simple Point Charge water models in a cubic box, which places the protein complex in 1.2 nm of the box's edge (Jorgensen, Chandrasekhar, Madura, Impey, & Klein, 1983). The MD system was neutralized by using the GENION method by adding ions that were required by the system. Then, all protein complexes were energy-minimized using the steepest descent method. Successful energy minimization obtains a negative potential energy (Epot) value (10^5-10^6). After energy minimization of the system, we conducted three essential steps for the ALDH3A2 wild-type and mutant protein complexes, such as NPT at 300 K for 100 ps, NVT at 300 K for 100 ps, and MD production for 30 ns at 300 K. For all MD simulations, we preferred a time step of 0.002 ps, and for every 2.0 ps of the entire simulation, the energies, coordinates, and velocities were saved.

2.7 Trajectory analysis of macromolecular simulation

We intended to find the differences in protein flexibility, protein compactness change, and stability loss between the ALDH3A2 wild-type and mutant protein complexes using trajectory files from MD. MD inbuilt parameters, such as gmx rms, gmx h_bond, gmx covar, gmx anaeig, gmx sham, and gmx mdmat, were used in this analysis to identify the differences between protein complexes. The resulting values were plotted with the use of Xmgrace to visualize the differences between these protein complexes (Turner, 2005).

3. Results

3.1 Dataset

Mutations (p.R228H, p.R228C, p.S365L, p.S380N, p.N386S, p.G401E, p.G406R, p.R423H, p.R423C, and p.K447E) that were present in the protein-protein interface were retrieved from the UniProt, dbSNP and HGMD databases. The PPI interacting amino acids from PDBsum are indexed in Table 2. We selected only protein-protein interface residue mutations for further investigation.

3.2 PPI interaction

We used PDBsum to retrieve the residues involved in the protein-protein interaction of the 4QGK homodimer. This analysis showed that 86 amino

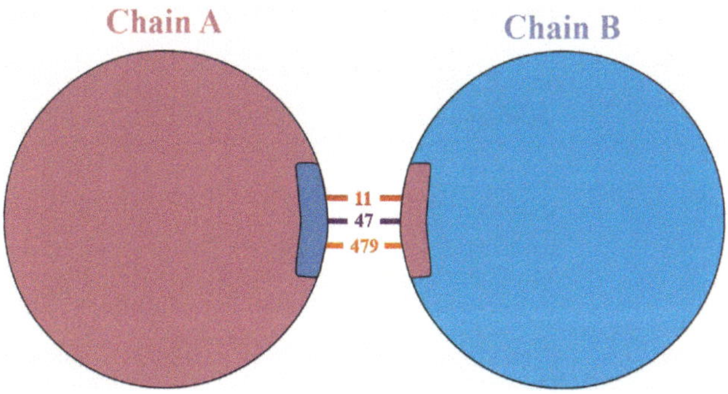

Fig. 1 Illustration of the PDBsum shows the number of interactive amino acids from the homodimeric ALDH3A2 chain A and B. Colored lines form interacting chains; red, violet, and mustard represent a type of interaction salt bridges, H—bonds, and non-bonded contacts, respectively.

acids from chain A and chain B of 4QGK are involved in the protein-protein interaction network (Fig. 1 and Tables 1 and 2). Among these residues, it was reported that 10 mutations (p.R228H, p.R228C, p.S365L, p.S380N, p.N386S, p.G401E, p.G406R, p.R423H, p.R423C, and p.K447E) were present in the protein-protein interface region and involved in the dimeric interaction. The variant groups alter the dimeric interactions, such as inter-chain salt bridges, H—bonds, and non-bonded contacts. To understand the molecular basis of the binding interface of the 4QGK homodimer, we initiated molecular docking analysis. The mutation positions 380 and 423 were found in the ALDH3A2-interacting chains A and B. The mutations p.S380N and p.R423H were focused on further investigations. The prediction of evolutionary conservation analysis, molecular docking, alterations in physicochemical features, and macromolecular simulations were implemented for mutations p.S380N and p.R423H. The essential domains and the position of native and mutant residues of the ALDH3A2 homodimeric 3D complex protein structures are depicted in Fig. 2.

Table 1 Interactions between the residues across the interfaces of 4QGK.

Chain	No. of interface residues	Interface area (Å^2)	No. of salt bridges	No. of H—bonds	No. of non-bonded contacts
A	86	4486	11	47	479
B	86	4473			

Table 2 The amino acid residues present in the protein-protein interface region from PDBsum.

S.No	Amino acid residues in chain A	Amino acid residues in chain B
1.	Ser444	Ser58
2.	Val448	Phe54
3.	Lys452	Tyr57
4.	Asn443	Asp278
5.	Gln445	Leu65
6.	Pro441	Asn55
7.	Pro442	Phe268
8.	Leu456	Trp231
9.	Tyr440	Arg227
10.	Asn436	Arg228
11.	Lys431	Asn386
12.	Ser429	Ile224
13.	Glu433	Asp387
14.	Arg439	His391
15.	Leu430	His366
16.	Lys428	Met390
17.	Leu426	Gly385
18.	Gln100	Ser379
19.	Arg423	Thr384
20.	His421	Ile375
21.	Pro424	Ser380
22.	Thr90	Thr378
23.	Asn87	Gly381
24.	Cys425	Leu361
25.	Ile97	Val383
26.	Val84	Gly401
27.	Arg177	Val402
28.	Lys85	Thr393
29.	Leu427	Gly382
30.	Glu94	Leu394
31.	Leu102	Pro398
32.	Tyr96	Ser404
33.	Met194	Phe397
34.	Lys191	Ser396
35.	Met407	Glu356
36.	Thr187	Ile371
37.	Glu195	Lys372
38.	Ala198	Ala197
39.	Leu208	Pro203
40.	Thr202	Lys357

Table 2 The amino acid residues present in the protein-protein interface region from PDBsum.—cont'd

S.No	Amino acid residues in chain A	Amino acid residues in chain B
41.	Leu206	Ala198
42.	Lys357	Leu201
43.	Lys199	Thr202
44.	Leu201	Lys199
45.	Pro203	Met194
46.	Glu356	Glu195
47.	Val402	Leu208
48.	Ser396	Lys191
49.	Lys372	Leu102
50.	Ser404	Met407
51.	Asp376	Leu206
52.	Ile371	Gln98
53.	Gln98	Arg177
54.	Leu394	Gly190
55.	Thr378	Val184
56.	Ser380	Lys85
57.	Thr393	Asp376
58.	Ile375	Thr90
59.	Ser379	Glu94
60.	Gly381	Leu427
61.	Phe397	Gln100
62.	His391	Tyr96
63.	Val383	Asn87
64.	Gly385	Arg423
65.	Met390	His421
66.	Gly382	Pro424
67.	Thr384	Cys425
68.	Asn386	Leu426
69.	Asp387	Lys428
70.	Asp221	Leu430
71.	Pro398	Arg439
72.	His366	Asn436
73.	Trp231	Ile97
74.	Arg228	Lys431
75.	Arg227	Ser429
76.	Cys220	Glu433
77.	Ile224	Tyr440
78.	Phe268	Pro442
79.	Leu65	Pro441
80.	Asp278	Lys457

(*Continued*)

Table 2 The amino acid residues present in the protein-protein interface region from PDBsum.—cont'd

S.No	Amino acid residues in chain A	Amino acid residues in chain B
81.	Asp69	Gln445
82.	Tyr57	Lys452
83.	Phe54	Leu456
84.	Asn55	Asn443
85.	Gly66	Ser444
86.	Ser58	Val448

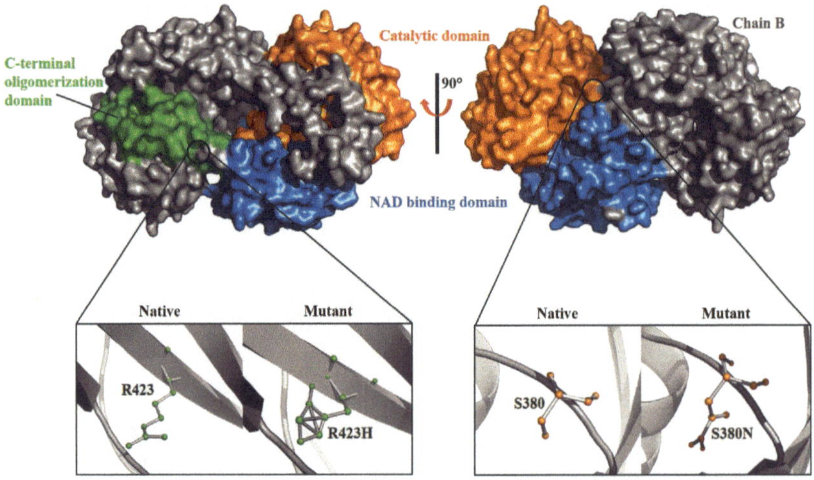

Fig. 2 Orientation of ALDH3A2 homodimeric 3D complex protein structure and position of the mutation site. The homodimeric protein depicts the essential domains involved in the protein-protein interaction, and only chain A contains colored domains. The left panel shows the C-terminal oligomerization domain (green; AA82-102, AA420-443), the catalytic domain (orange; AA209-419), and NAD-binding domain or co-factor binding domain (blue; AA1-79, AA103-208) of chain A in ALDH3A2 homodimeric protein and also shows the location of wild (p.R423) and mutant (p.R423H) residues. The right panel is the rear view of the left angle rotated, depicting chain B (gray) and the location of wild-type (p.S380) and mutant (p.S380N) residues.

3.3 *In silico* methods for screening missense mutations

The retrieved mutations that are involved in the protein-protein interaction were subjected to a variety of stability prediction tools to identify their effects on the interacting sites of the protein chains. Tools such as mCSM, iStable, SDM, DUET, MUpro, DynaMut, and I-Mutant2.0 showed that these mutations reduce the stability of the protein by indicating a decrease or destabilizing as an output. All these tools indicated that p.R228H, p.R228C, p.S380N and p.R423H was either decreasing or destabilizing the protein's stability upon amino acid substitution (Table 3). The ΔΔG values of p.S380N are −0.41 kcal/mol, −0.0382 kcal/mol, 0.842 kcal/mol, −1.404 kcal/mol, −1.02 kcal/mol, −1.154 kcal/mol, and −0.506 kcal/mol for I-Mutant2.0, MUpro, iStable, mCSM, SDM, DUET, and DynaMut, respectively. The ΔΔG values of p.R423H are −0.96, −1, 0.874, −1.588 kcal/mol, −0.56 kcal/mol, −1.637 kcal/mol, and −0.278 kcal/mol for I-Mutant2.0, MUpro, iStable, mCSM, SDM, DUET, and DynaMut, respectively. The negative ΔΔG value represents the destabilizing effect of the mutation with the tools I-Mutant2.0, MUpro, mCSM, SDM, DUET, and DynaMut, whereas the positive ΔΔG values ($\Delta\Delta G < 0$) represent a decrease in stability with the tool iStable. These computed results led us to employ further analysis for the mutations to identify how they affect the protein-protein interactions of the homodimeric ALDH3A2 protein.

3.4 Physicochemical properties of amino acids

We used the HOPE web server to classify the physicochemical features of the amino acids for mutations exhibiting destabilizing effects on the protein-protein interaction of native ALDH3A2. The amino acids contain unique characteristics, such as charge, hydrophobicity, polarity, and size that play prominent roles in the function of the protein. The results from the HOPE web server are indexed in Table 4. The p.S380N was predicted to be less hydrophobic (0.448) than the native residue (0.601), and the charge of this mutation remained unchanged, but we observed an increase in the size of the amino acid compared to the native residue. However, the p.R423H was computed to have more hydrophobic (0.548) than the native residue (0.000), and their charge changed from positive to neutral. We also observed that the amino acid size decreased at this position compared to the native residue. Ultimately, these variations occurred due to mutations that cause significant alterations in the protein-protein interacting sites.

3.5 Evolutionary conservation analysis

The ConSurf tool was utilized to analyze the conservation level of each amino acid shown at its defined position. Based on the conservation level, a Multiple Sequence Alignment (MSA) was drawn across species of the

Table 3 Stability prediction of mutations using various prediction tools.

Accession ID	Variants	I-Mutant2.0	MUpro	iStable	mCSM	SDM	DUET	DynaMut
CM122178	**R228H**	**Decrease**	**Decrease**	**Decrease**	**Destabilizing**	**Destabilizing**	**Destabilizing**	**Destabilizing**
CM993310	**R228C**	**Decrease**	**Decrease**	**Decrease**	**Destabilizing**	**Destabilizing**	**Destabilizing**	**Destabilizing**
CM980052	S365L	Increase	Decrease	Increase	Stabilizing	Stabilizing	Stabilizing	Stabilizing
CM042002	**S380N**	**Decrease**	**Decrease**	**Decrease**	**Destabilizing**	**Destabilizing**	**Destabilizing**	**Destabilizing**
CM001057	N386S	Increase	Decrease	Increase	Destabilizing	Destabilizing	Destabilizing	Destabilizing
CM083802	G401E	Increase	Increase	Increase	Destabilizing	Destabilizing	Destabilizing	Stabilizing
CM993315	G406R	Increase	Decrease	Increase	Destabilizing	Destabilizing	Destabilizing	Stabilizing
VAR_017527	**R423H**	**Decrease**	**Decrease**	**Decrease**	**Destabilizing**	**Destabilizing**	**Destabilizing**	**Destabilizing**
500063	R423C	Increase	Decrease	Decrease	Destabilizing	Destabilizing	Destabilizing	Destabilizing
CM993320	K447E	Decrease	Increase	Increase	Destabilizing	Stabilizing	Stabilizing	Destabilizing

The bold represents a decrease or destabilizing mutation by all the prediction tools.

Table 4 Structural visualization of missense mutations based on the Physiochemical properties of amino acids.

Variants	Change in charge	Change in hydrophilicity/ hydrophobicity	Change in size	Mutation effect
S380N	Unchanged	Less (0.448) hydrophobic than native (0.601)	Small to Large	The native residue is located on the surface of the protein, mutation on this residue can disturb interactions with other molecules and might also cause loss of hydrophobic interactions on the surface of the protein
R423H	Positive to Neutral	More (0.548) hydrophobic than native (0.000)	Large to Small	The charge of the native residue is lost in the case of mutation. This may cause a possible loss of external interactions

same protein. The ConSurf computed that S380 and R423 residues attained the highest conservation score, '9'. Aside from establishing the conservation level, ConSurf also computed the structural/functional effect of the amino acid on the protein by showing buried and exposed regions. In this study, both mutations were predicted to be functionally significant and exposed (Fig. 3). Therefore, the conserved residues at protein binding regions are essential for the normal functioning of homodimeric proteins, and the alterations at these residue positions affect the protein-protein interactions.

3.6 PPI and molecular docking analysis

For molecular docking, we performed a freely available ZDOCK online server that predicts protein-protein complexes based on rigid-body docking programs. Docking was performed to characterize the ALDH3A2 homodimer's alteration in binding interactions. The mutant model (p.R423H) acquired the highest ZDOCK score of 4508.859 compared with the native model 4180.225, and we obtained the lowest ZDOCK score of 4170.336

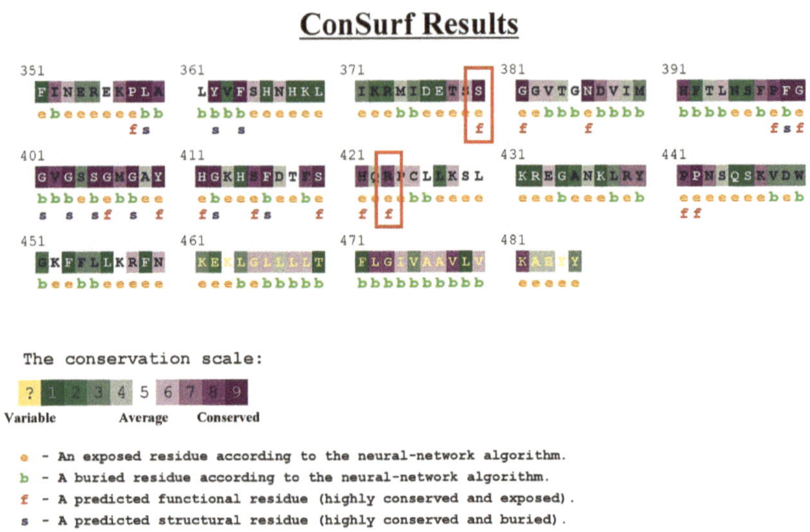

Fig. 3 ConSurf computed the conservation level of the ALDH3A2 protein FASTA sequence (1–485 amino acid residues). In ConSurf, the conservation level is situated between the scale of 0–9, and residues are coded with conservation color grades. The amino acid at a particular position with a score of '9' determines the highest conservation that is indicated in violet color, and a score of '0' predicts the least conserved amino acid indicated in green color. Both mutations (p.S380N and p.R423H) are highly conserved, functionally significant, and located in the exposed region of homodimeric protein (red box).

for the mutant model p.S380N. The UCSF Chimera tool was used to visualize the native protein-docked complex and is shown in Fig. 4.

3.7 Macromolecular simulations

The native and mutant (S135L, Q188R, K285N, and N314D) docked complexes were subjected to molecular dynamics simulation to understand alterations at the atomistic level of ALDH3A2 homodimeric protein for 30 ns. The analysis of RMSD from the MD trajectory was performed to identify the alterations of divergence patterns between the complexes of native and mutant models. The last 10 ns of RMSD were plotted in Fig. 5A for the native and mutant models. As a result, both the native and mutant models were converged at the completion of 30 ns simulation with a ~0.28 nm RMSD value. We validated the intramolecular H—bonds that formed the protein complexes in both ALDH3A2 native and mutant proteins. The maximum amount of intramolecular H—bonds between the ALDH3A2 native protein complex was obtained at approximately 780 followed by

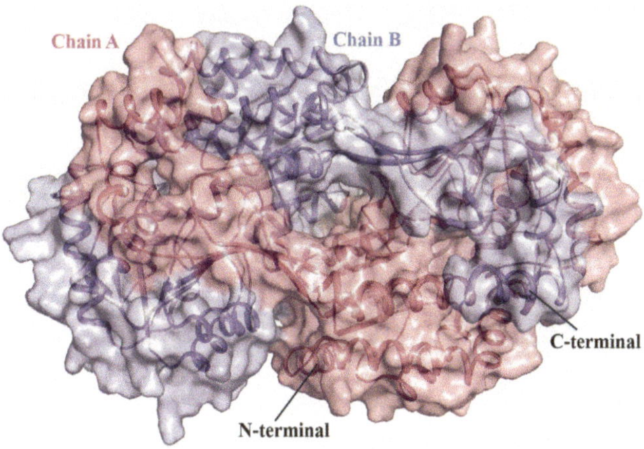

Fig. 4 Depiction of the native protein-docked complex marked with the N-terminal and C-terminal. Color scheme as follows: pink represents chain A and blue represents chain B.

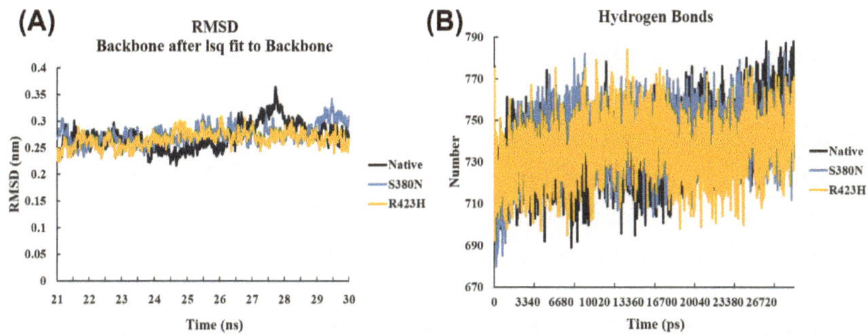

Fig. 5 (A) Backbone RMSD plot representing 30 ns MD simulation of ALDH3A2 native and mutant protein complexes. The x-axis indicates the time (ns), and the y-axis indicates the RMSD (nm). (B) Analysis of intramolecular H—bonds of ALDH3A2 native and mutant protein complexes at 30 ns. The x-axis indicates the time (ps), and the y-axis indicates the number of H—bonds. The native, p.S380N, and p.R423H is indicated in black, blue, and yellow, respectively.

the mutant protein complexes p.S380N and p.R423H, obtained at approximately 765 and 770, respectively (Fig. 5B). The lowest number of H—bonds was formed in the p.R423H protein complex, which shows the detrimental effect of this mutant in a structural context. The contact maps were analyzed to explore the specific atomistic interactions between residues throughout the entire MD simulation. The results from the contact maps revealed that both mutant complexes (p.S380N and p.R423H) lose an

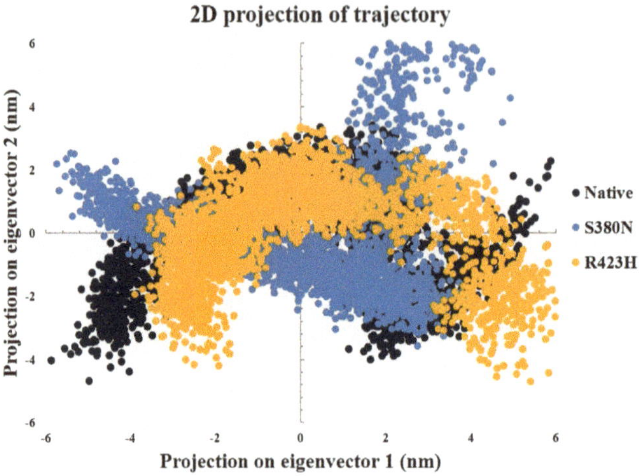

Fig. 6 Comparative analysis of contact maps between the ALDH3A2 native complex, p.S380N complex, and p.R423H complex.

interaction in the region of anti-parallel beta-sheet and alpha helices, leading to increased destabilization owing to a mutation in comparison with the ALDH3A2 native complex (Fig. 6). With this phenomenon, we implemented gmx anaeig and gmx covar parameters from Gromacs to determine the combined dynamics of atoms on the largest scale and the interrelation between the ALDH3A2 native and mutant atomic motions (Fig. 7). The two mutant complexes showed differences in atomic motion in comparison with the ALDH3A2 native complex. Both mutant complexes (p.S380N and p.R423H) was found to have a greater deviation pattern. This deviation could be attributed to the structure unfolding as observed in the mutant complexes. Furthermore, we analyzed this unfolding/folding structural pattern by employing a Gibbs free energy landscape analysis (FEL). For an appropriate function of a protein, the pattern of folding and proper conformation is essential. We examined the FEL against the two PC1 and PC2 (principal component) for the ALDH3A2 native and mutant complexes to govern the different (folded and unfolded) states of a protein. The unfolded-state proteins obtained an immense Gibbs free energy and appeared as blue spots, while the folded-state proteins obtained a lower Gibbs free energy and appeared as red spots. As a result, p.S380N and p.R423H gained a larger Gibbs free energy with an increase in scattered blue spots in comparison with the ALDH3A2 native complex (Fig. 8). Therefore, the protein's folding pattern adversely affects the stability, and

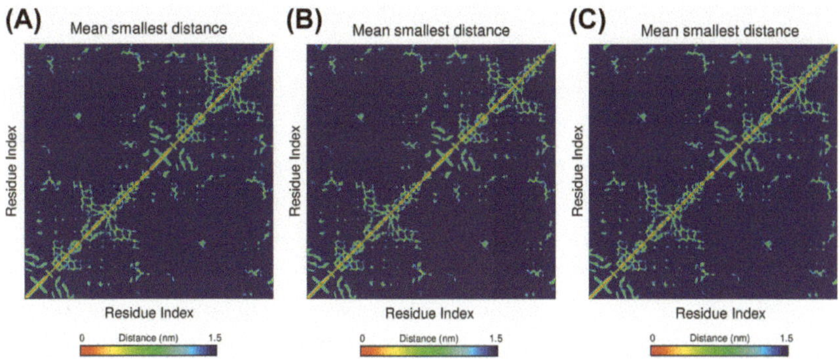

Fig. 7 (A) Projection of protein motion in eigenvalues at 300 K for the first two principal eigenvectors of ALDH3A2 native and mutant (p.S380N and p.R423H) protein complexes. The native, p.S380N, and p.R423H are indicated in black, blue, and yellow, respectively.

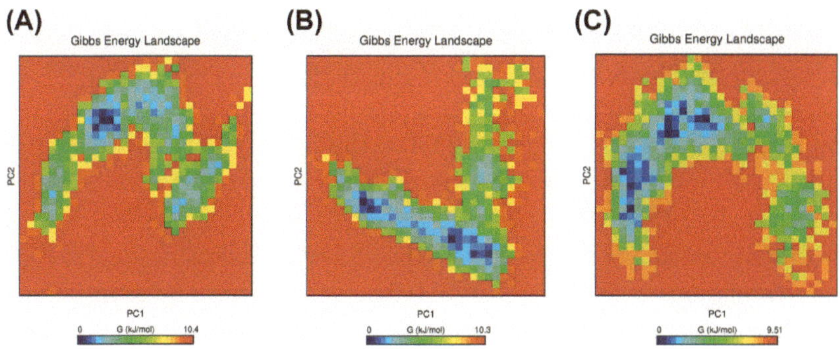

Fig. 8 Plot representing the Gibbs free energy landscape of principal components 1 and 2 for (A) the ALDH3A2 native complex, (B) p.S380N complex, and (C) p.R423H complex.

the decrease in H—bonds corresponds well with the stability decrease in the mutant complexes.

4. Discussion

Previous studies have emphasized the importance of simple and complex disease-causing non-synonymous single nucleotide polymorphisms and *in silico* prediction tools to analyze the how these mutations affect protein structure and function (Hicks, Wheeler, Plon, & Kimmel, 2011; Rajasekaran, Sudandiradoss, Doss, & Sethumadhavan, 2007; Sneha & George Priya Doss, 2016; Udhaya Kumar et al., 2015; Yates & Sternberg,

2013). The activity and stable function of a protein complex depend on the protein-protein interaction, which contributes to many molecular features, such as cell signaling and cell regulation. The stable heterodimers and homodimers formed from the interactions within the protein, presence of single amino acid substitutions at the interfaces of PPI results in the disruption of the shape, size, secondary structure, hydrophobicity, and conformational change on the protein complex formation (Jones & Thornton, 1996; Zhanhua, Gan, lei, Sakharkar, & Kangueane, 2005). The stable interaction between two or more proteins is important for the specific function of homo- or heterodimeric proteins (Nooren & Thornton, 2003). In addition, the mutation of a small amino acid residue into a larger one leads to intermolecular clashes, whereas the reverse case might lead to gaps, which might hinder protein folding and function (Yates & Sternberg, 2013). The aldehyde dehydrogenase enzymes play a crucial role in directing toxic accumulates to restore networks for the biosynthesis of lipids (Nakahara et al., 2012). The deficient ALDH3A2 enzyme leads to a severe form of a genetic disorder called Sjögren—Larsson syndrome. The amino acid residues (p.S380 and p.R423) form two identical interactions at the site of the dimeric interface, and mutations (p.S380N and p.R423H) at this site result in dimeric interface destabilization by disturbing the crucial H—bonds involved in dimerization (Keller et al., 2014).

In a prevailing study, the PPI of the ALDH3A2 dimeric interface was due to mutations (p.S380N, p.R423H) that were classified by using computational pipelines, such as stability predictors, physiochemical features, and molecular dynamics simulations, to understand their motion patterns and functions. The presence of mutations at the highly conserved site of amino acids has a more significant effect than the less conserved site of amino acids (Kumar & Doss, 2017; Mathew et al., 2015; Shastry, 2007). Our study focused on examining the consequences of mutations in the interacting region of the ALDH3A2 homodimeric protein (Fig. 1 and Table 1). In this study, we utilized the ConSurf web server to compute the conservation level of amino acid residues at 380 and 423 in which the mutations were reported in HGMD and UniProt. With this analysis, we determined that both the amino acid residues 380 and 423, which have serine and arginine in the ALHD3A2 native type, are strongly conserved with the conservation scale of '9' (Fig. 3). In addition, it was found that both residues are functionally significant and present in the homodimer's exposed region. The change in stability due to mutation was evaluated using the webservers DynaMut and iStable. The output of these servers computed that both mutations (p.S380N and p.R423H) either decrease or destabilize the protein's stability with significant ΔΔG scores (Table 3). Thus, we selected the mutations p.S380N and p.R423H for further theoretical molecular analyses. The dimeric core region accommodates amino acids that help to sustain the protein's stability, while the interface region contains amino acids that help to maintain the interaction between dimeric proteins (Kumar et al., 2017; Tina, Bhadra, & Srinivasan, 2007). In general, the protein-protein interaction site is disrupted due to amino acid mutations, resulting in a diseased state in patients (Petukh, Kucukkal, & Alexov, 2015). To analyze this phenomenon, we initiated protein-protein docking using the ZDOCK online server for the ALDH3A2 native and mutant proteins. This analysis revealed that p.R423H and p.S380N mutant models acquired the ZDOCK highest and lowest scores of 4508.859 and 4170.336, respectively, compared with the ALDH3A2 native model (4180.225). With a high docking score, the mutant (p.R423H) is supposed to have an increase in stability, but it showed a destabilizing effect from all the stability predictor tools. The p.S380N mutation showed the lowest docking score compared to the ALDH3A2 native model, thereby indicating a disruption in the interaction between chain A and chain B owing to mutation. This finding ensures that the mutation disrupts the chain interactions of the ALDH3A2 homodimer.

MD simulations, a versatile computational tool, enabled us to illustrate the complex nature of protein-protein interactions between and to understand atomistic-level differences (Sneha et al., 2017; Padhi, Jayaram, & Gomes, 2013). The native and mutant docked complexes were subjected to MD simulations. Simulations for 30 ns were conducted to examine changes in the pattern of convergence, intramolecular H—bonds, contact maps, and atomic motions using gmx rms, gmx hbond, gmx anaeig, gmx covar, and gmx mdmat, respectively. Fig. 5A shows the last 10 ns of RMSD for native and mutant complexes (p.S380N and p.R423H) that converged after simulation with an RMSD value of ~0.28 nm. The computed results for these mutant complexes from the stability tools correlate well with the results from RMSD (Table 3). These results indicated that protein stability was affected in the case of mutant complexes (p.S380N and p.R423H). Protein stability and function are the two crucial interrelated features that must be taken into account when investigating the structure of a protein, the genetic variations that suppress the protein's properties directly influence all other properties, and protein stability thus plays a vital role in maintaining the function of the protein (Sneha et al., 2017; Shoichet, Baase, Kuroki, & Matthews, 1995; Sudhakar et al., 2016). As per the neural-network algorithm, which was obtained from the ConSurf server, both mutant residues were located in the exposed region of the protein complex (Fig. 3). A previous study suggested that the exposed arginine residues have evolved to obtain favored protein folding and to stabilize the protein structures (Strub et al., 2004). From FEL analysis, we observed decreased stability and increased Gibbs free energy due to the substitution of serine to asparagine and arginine to histidine (Fig. 8). The folding/unfolding patterns of protein can be demonstrated by the Gibbs free energy, and they contribute to the protein's structure and function (Bloom, Wilke, Arnold, & Adami, 2004; Studer, Dessailly, & Orengo, 2013). Depending on the outcome from a series of MD trajectory analyses, the mutant complex functionality was affected by variations in their stability. In comparison with the ALDH3A2 native complex, we defined that the mutants have a more significant impact on protein-protein interaction, protein stability, and folding patterns.

The amino acid size change was observed in the mutations (Table 4); this finding emphasized that the side chain length is required for the interaction of various geometric conformations by introducing new H—bonds (Luscombe & Thornton, 2002). Protein stability is maintained by maintaining the H—bonds between the amino acid residues. Higher H—bonds in the protein lead to increased stability. In addition, a reduction in the number of H—bonds can result in a loss of protein stabilization (Sneha et al., 2017; Pace et al., 2014). Consequently, we plotted the contact maps and intramolecular H—bonds in comparison with the native complex (Fig. 6) to analyze these phenomena; thus, we observed the loss of protein-protein intramolecular contacts in both mutant complexes. The H—bonds are usually formed at 90° angles in the beta sheets (anti-parallel), and this region might have increased dimeric destabilization due to lack of contacts (Strub et al., 2004). A significant number of intramolecular H—bonds were observed in the ALDH3A2 native complex, whereas the mutant complexes formed the lowest number of H—bonds (Fig. 6). Conclusively, we observed significant differences in the RMSD, intramolecular H—bonds, contact maps, principal eigenvectors, and FEL analyses, suggesting that these amino acid substitutions have a detrimental impact on the dimeric interface region. The core amino acid residues present in the PPI location play an essential part in sustaining the protein's stability; the residues located at the PPI interface are responsible for functional and structural alterations (Guharoy & Chakrabarti, 2005; Kumar et al., 2017). This property contributes to making our study essential to characterize the mutation located at the protein-

protein interface. It is labor-intensive to examine a PPI using an experimental strategy (Jordán, Nguyen, & Liu, 2012). This vital tool has become a useful alternative to traditional methods in biomedical research. Computational methods, such as molecular docking and molecular dynamics, enable us to comprehend protein-protein interactions at the molecular level. The results of our study may not only help to characterize the mutations in ALDH3A2 but may also substantially demonstrate the functions of the PPI interface mutations that are conserved.

5. Conclusion

The function and activity of a protein complex depend on protein-protein interactions, which play an essential role in the molecular process. In some instances, the interaction of stable homodimers and heterodimers is disrupted due to mutations located at the most conserved residues that affect the function and protein complex formation. Our study explored mutations (p.S380N and p.R423H) situated at the PPI interface that are found to be present at the highest conserved sites of amino acids and therefore disrupt the protein-protein interactions in comparison with the ALDH3A2 native protein, leading to protein destabilization. Determining the precise crystal structure of ALDH3A2 mutations, which causes Sjögren–Larsson syndrome, is necessary to characterize the dynamic behavior of the protein complex within the system. Additionally, our study suggests that bioinformatic methods can help to characterize mutations in the homo/heterodimeric protein complex at the molecular level.

Conflict of interest

No conflicts of interest exist.

Acknowledgments

The authors thank the management of VIT for providing the facilities and the encouragement to carry out this work.

References

Abraham, M. J., Murtola, T., Schulz, R., Páll, S., Smith, J. C., Hess, B., et al. (2015). GROMACS: High performance molecular simulations through multi-level parallelism from laptops to supercomputers. *SoftwareX, 1–2*, 19–25. http://dx.doi.org/10.1016/j.softx.2015.06.001.

Agrahari, A. K., Sneha, P., George Priya Doss, C., Siva, R., & Zayed, H. (2018). A profound computational study to prioritize the disease-causing mutations in PRPS1 gene. *Metabolic Brain Disease, 33*(2), 589–600. http://dx.doi.org/10.1007/s11011-017-0121-2.

Amberger, J. S., & Hamosh, A. (2017). Searching online Mendelian inheritance in man (OMIM): A knowledgebase of human genes and genetic phenotypes. *Current Protocols in Bioinformatics, 58*(1), 1.2.1–1.2.12. http://dx.doi.org/10.1002/cpbi.27.

Ashibe, B., Hirai, T., Higashi, K., Sekimizu, K., & Motojima, K. (2007). Dual subcellular localization in the endoplasmic reticulum and peroxisomes and a vital role in protecting against oxidative stress of fatty aldehyde dehydrogenase are achieved by alternative splicing. *Journal of Biological Chemistry, 282*(28), 20763–20773. http://dx.doi.org/10.1074/jbc.M611853200.

Ashkenazy, H., Abadi, S., Martz, E., Chay, O., Mayrose, I., Pupko, T., et al. (2016). ConSurf 2016: An improved methodology to estimate and visualize evolutionary conservation in macromolecules. *Nucleic Acids Research, 44*(W1), W344–W350. http://dx.doi.org/10.1093/nar/gkw408.

Ashkenazy, H., Erez, E., Martz, E., Pupko, T., & Ben-Tal, N. (2010). ConSurf 2010: Calculating evolutionary conservation in sequence and structure of proteins and nucleic acids. *Nucleic Acids Research, 38*(Suppl. l_2), W529–W533. http://dx.doi.org/10.1093/nar/gkq399.

Bateman, A., Martin, M. J., O'Donovan, C., Magrane, M., Alpi, E., Antunes, R., et al. (2017). UniProt: The universal protein knowledgebase. *Nucleic Acids Research, 45*(D1), D158–D169. http://dx.doi.org/10.1093/nar/gkw1099.

Berman, H. M., Westbrook, J., Feng, Z., Gilliland, G., Bhat, T. N., Weissig, H., et al. (2000). The protein data bank. *Nucleic Acids Research, 28*(1), 235–242. http://dx.doi.org/10.1093/nar/28.1.235.

Bloom, J. D., Wilke, C. O., Arnold, F. H., & Adami, C. (2004). Stability and the evolvability of function in a model protein. *Biophysical Journal, 86*(5), 2758–2764.

Capriotti, E., Fariselli, P., & Casadio, R. (2005). I-Mutant2.0: Predicting stability changes upon mutation from the protein sequence or structure. *Nucleic Acids Research, 33*(Suppl. l_2), W306–W310. http://dx.doi.org/10.1093/nar/gki375.

Carney, G., Wei, S., & Rizzo, W. B. (2004). Sjögren-Larsson syndrome: Seven novel mutations in the fatty aldehyde dehydrogenase gene ALDH3A2. *Human Mutation, 24*(2), 186. http://dx.doi.org/10.1002/humu.9262.

Chan, L. J., Ascher, D. B., Yadav, R., Bulitta, J. B., Williams, C. C., Porter, C. J. H., et al. (2016). Conjugation of 10 kDa linear PEG onto trastuzumab fab' is sufficient to significantly enhance lymphatic exposure while preserving in vitro biological activity. *Molecular Pharmaceutics, 13*(4), 1229–1241. http://dx.doi.org/10.1021/acs.molpharmaceut.5b00749.

Chan, L. J., Bulitta, J. B., Ascher, D. B., Haynes, J. M., McLeod, V. M., Porter, C. J. H., et al. (2015). PEGylation does not significantly change the initial intravenous or subcutaneous pharmacokinetics or lymphatic exposure of trastuzumab in rats but increases plasma clearance after subcutaneous administration. *Molecular Pharmaceutics, 12*(3), 794–809. http://dx.doi.org/10.1021/mp5006189.

Chang, C., & Yoshida, A. (1997). Human fatty aldehyde dehydrogenase gene (ALDH10): Organization and tissue-dependent expression. *Genomics, 40*(1), 80–85. http://dx.doi.org/10.1006/geno.1996.4547.

Cheng, J., Randall, A., & Baldi, P. (2006). Prediction of protein stability changes for single-site mutations using support vector machines. *Proteins: Structure, Function, and Bioinformatics, 62*(4), 1125–1132. http://dx.doi.org/10.1002/prot.20810.

Chen, C.-W., Lin, J., & Chu, Y.-W. (2013). iStable: Off-the-shelf predictor integration for predicting protein stability changes. *BMC Bioinformatics, 14*(2), S5. http://dx.doi.org/10.1186/1471-2105-14-S2-S5.

David, A., Razali, R., Wass, M. N., & Sternberg, M. J. E. (2012). Protein–protein interaction sites are hot spots for disease-associated nonsynonymous SNPs. *Human Mutation, 33*(2), 359–363. http://dx.doi.org/10.1002/humu.21656.

Demozay, D., Mas, J.-C., Rocchi, S., & Obberghen, E. V. (2008). FALDH reverses the deleterious action of oxidative stress induced by lipid peroxidation product 4-hydroxynonenal on insulin signaling in 3T3-L1 adipocytes. *Diabetes, 57*(5), 1216–1226. http://dx.doi.org/10.2337/db07-0389.

Doss, C. G. P., Alasmar, D. R., Bux, R. I., Sneha, P., Bakhsh, F. D., Al-Azwani, I., et al. (2016). Genetic epidemiology of glucose-6-phosphate dehydrogenase deficiency in the Arab world. *Scientific Reports, 6*, 37284. http://dx.doi.org/10.1038/srep37284.

Doss, C. G. P., Rajith, B., Garwasis, N., Mathew, P. R., Raju, A. S., Apoorva, K., et al. (2012). Screening of mutations affecting protein stability and dynamics of FGFR1—a simulation analysis. *Applied & Translational Genomics, 1*, 37–43. http://dx.doi.org/10.1016/j.atg.2012.06.002.

George Priya Doss, C., Nagasundaram, N., Chakraborty, C., Chen, L., & Zhu, H. (2013). Extrapolating the effect of deleterious nsSNPs in the binding adaptability of flavopiridol with CDK7 protein: A molecular dynamics approach. *Human Genomics,* 7(1), 10. http://dx.doi.org/10.1186/1479-7364-7-10.

George, D. C. P., Chakraborty, C., Haneef, S. S., NagaSundaram, N., Chen, L., & Zhu, H. (2014). Evolution- and structure-based computational strategy reveals the impact of deleterious missense mutations on MODY 2 (Maturity-onset diabetes of the young, type 2). *Theranostics, 4*(4), 366–385. http://dx.doi.org/10.7150/thno.7473.

Gordon, N. (2007). Sjögren-Larsson syndrome. *Developmental Medicine and Child Neurology, 49*(2), 152–154. http://dx.doi.org/10.1111/j.1469-8749.2007.00152.x.

Guex, N., & Peitsch, M. C. (1997). SWISS-MODEL and the Swiss-Pdb viewer: An environment for comparative protein modeling. *Electrophoresis, 18*(15), 2714–2723. http://dx.doi.org/10.1002/elps.1150181505.

Guharoy, M., & Chakrabarti, P. (2005). Conservation and relative importance of residues across protein-protein interfaces. *Proceedings of the National Academy of Sciences of the United States of America, 102*(43), 15447–15452. http://dx.doi.org/10.1073/pnas.0505425102.

Hait, N. C., Allegood, J., Maceyka, M., Strub, G. M., Harikumar, K. B., Singh, S. K., et al. (2009). Regulation of histone acetylation in the nucleus by sphingosine-1-phosphate. *Science, 325*(5945), 1254–1257. http://dx.doi.org/10.1126/science.1176709.

Hayter, S. M., & Cook, M. C. (2012). Updated assessment of the prevalence, spectrum and case definition of autoimmune disease. *Autoimmunity Reviews, 11*(10), 754–765. http://dx.doi.org/10.1016/j.autrev.2012.02.001.

Hicks, S., Wheeler, D. A., Plon, S. E., & Kimmel, M. (2011). Prediction of missense mutation functionality depends on both the algorithm and sequence alignment employed. *Human Mutation, 32*(6), 661–668. http://dx.doi.org/10.1002/humu.21490.

Huang, W., Lin, Z., & van Gunsteren, W. F. (2011). Validation of the GROMOS 54A7 force field with respect to β-peptide folding. *Journal of Chemical Theory and Computation,* 7(5), 1237–1243. http://dx.doi.org/10.1021/ct100747y.

Jones, S., & Thornton, J. M. (1996). Principles of protein-protein interactions. *Proceedings of the National Academy of Sciences, 93*(1), 13–20. http://dx.doi.org/10.1073/pnas.93.1.13.

Jordán, F., Nguyen, T.-P., & Liu, W. (2012). Studying protein–protein interaction networks: A systems view on diseases. *Briefings in Functional Genomics, 11*(6), 497–504. http://dx.doi.org/10.1093/bfgp/els035.

Jorgensen, W. L., Chandrasekhar, J., Madura, J. D., Impey, R. W., & Klein, M. L. (1983). Comparison of simple potential functions for simulating liquid water. *The Journal of Chemical Physics, 79*(2), 926–935. http://dx.doi.org/10.1063/1.445869.

Jubb, H. C., Pandurangan, A. P., Turner, M. A., Ochoa-Montaño, B., Blundell, T. L., & Ascher, D. B. (2017). Mutations at protein-protein interfaces: Small changes over big surfaces have large impacts on human health. *Progress in Biophysics and Molecular Biology, 128*, 3–13. http://dx.doi.org/10.1016/j.pbiomolbio.2016.10.002.

Keller, M. A., Zander, U., Fuchs, J. E., Kreutz, C., Watschinger, K., Mueller, T., et al. (2014). A gatekeeper helix determines the substrate specificity of Sjögren–Larsson syndrome enzyme fatty aldehyde dehydrogenase. *Nature Communications, 5*(1), 1–12. http://dx.doi.org/10.1038/ncomms5439.

Khanna, M., Chen, C.-H., Kimble-Hill, A., Parajuli, B., Perez-Miller, S., Baskaran, S., et al. (2011). Discovery of a novel class of covalent inhibitor for aldehyde dehydrogenases. *Journal of Biological Chemistry, 286*(50), 43486–43494. http://dx.doi.org/10.1074/jbc.M111.293597.

Kihara, A., Mitsutake, S., Mizutani, Y., & Igarashi, Y. (2007). Metabolism and biological functions of two phosphorylated sphingolipids, sphingosine 1-phosphate and ceramide 1-phosphate. *Progress in Lipid Research, 46*(2), 126–144. http://dx.doi.org/10.1016/j.plipres.2007.03.001.

Kucukkal, T. G., Petukh, M., Li, L., & Alexov, E. (2015). Structural and physico-chemical effects of disease and non-disease nsSNPs on proteins. *Current Opinion in Structural Biology, 32*, 18–24. http://dx.doi.org/10.1016/j.sbi.2015.01.003.

Kumar, D. T., & Doss, C. G. P. (2017). Role of E542 and E545 missense mutations of PIK3CA in breast cancer: A comparative computational approach. *Journal of Biomolecular Structure and Dynamics, 35*(12), 2745–2757. http://dx.doi.org/10.1080/07391102.2016.1231082.

Kumar, D. T., Doss, C. G. P., Sneha, P., Tayubi, I. A., Siva, R., Chakraborty, C., et al. (2017). Influence of V54M mutation in giant muscle protein titin: A computational screening and molecular dynamics approach. *Journal of Biomolecular Structure and Dynamics, 35*(5), 917–928. http://dx.doi.org/10.1080/07391102.2016.1166456.

Landersdorfer, C. B., Caliph, S. M., Shackleford, D. M., Ascher, D. B., & Kaminskas, L. M. (2015). PEGylated interferon displays differences in plasma clearance and bioavailability between male and female mice and between female immunocompetent C57Bl/6J and athymic nude mice. *Journal of Pharmaceutical Sciences, 104*(5), 1848–1855. http://dx.doi.org/10.1002/jps.24412.

Laskowski, R. A. (2001). PDBsum: Summaries and analyses of PDB structures. *Nucleic Acids Research, 29*(1), 221–222. http://dx.doi.org/10.1093/nar/29.1.221.

Laskowski, R. A., Jabłońska, J., Pravda, L., Vařeková, R. S., & Thornton, J. M. (2018). PDBsum: Structural summaries of PDB entries. *Protein Science, 27*(1), 129–134. http://dx.doi.org/10.1002/pro.3289.

Lerner, A., Jeremias, P., & Matthias, T. (2015). The world incidence and prevalence of autoimmune diseases is increasing. *International Journal of Celiac Disease, 3*(4), 151–155. http://dx.doi.org/10.12691/ijcd-3-4-8.

Liu, Z.-J., Sun, Y.-J., Rose, J., Chung, Y.-J., Hsiao, C.-D., Chang, W.-R., et al. (1997). The first structure of an aldehyde dehydrogenase reveals novel interactions between NAD and the Rossmann fold. *Nature Structural Biology, 4*(4), 317–326. http://dx.doi.org/10.1038/nsb0497-317.

Lloyd, M. D., Boardman, K. D. E., Smith, A., van den Brink, D. M., Wanders, R. J. A., & Threadgill, M. D. (2007). Characterisation of recombinant human fatty aldehyde dehydrogenase: Implications for Sjögren-Larsson syndrome. *Journal of Enzyme Inhibition and Medicinal Chemistry, 22*(5), 584–590. http://dx.doi.org/10.1080/14756360701425360.

Luscombe, N. M., & Thornton, J. M. (2002). Protein–DNA interactions: Amino acid conservation and the effects of mutations on binding specificity. *Journal of Molecular Biology, 320*(5), 991–1009. http://dx.doi.org/10.1016/S0022-2836(02)00571-5.

Manzel, A., Muller, D. N., Hafler, D. A., Erdman, S. E., Linker, R. A., & Kleinewietfeld, M. (2013). Role of "western diet" in inflammatory autoimmune diseases. *Current Allergy and Asthma Reports, 14*(1), 404. http://dx.doi.org/10.1007/s11882-013-0404-6.

Mathew, J., Jagadeesh, S. M., Bhat, M., Udhaya Kumar, S., Thiyagarajan, S., & Srinivasan, S. (2015). Mutations in ARSB in MPS VI patients in India. *Molecular Genetics and Metabolism Reports, 4*, 53–61. http://dx.doi.org/10.1016/j.ymgmr.2015.06.002.

Meng, E. C., Pettersen, E. F., Couch, G. S., Huang, C. C., & Ferrin, T. E. (2006). Tools for integrated sequence-structure analysis with UCSF Chimera. *BMC Bioinformatics, 7*(1), 339. http://dx.doi.org/10.1186/1471-2105-7-339.

Nakahara, K., Ohkuni, A., Kitamura, T., Abe, K., Naganuma, T., Ohno, Y., et al. (2012). The Sjögren-Larsson syndrome gene encodes a hexadecenal dehydrogenase of the sphingosine 1-phosphate degradation pathway. *Molecular Cell, 46*(4), 461–471. http://dx.doi.org/10.1016/j.molcel.2012.04.033.

Nooren, I. M. A., & Thornton, J. M. (2003). Diversity of protein–protein interactions. *The EMBO Journal, 22*(14), 3486–3492. http://dx.doi.org/10.1093/emboj/cdg359.

Pace, C. N., Fu, H., Fryar, K. L., Landua, J., Trevino, S. R., Schell, D., et al. (2014). Contribution of hydrogen bonds to protein stability. *Protein Science: A Publication of the Protein Society, 23*(5), 652–661. http://dx.doi.org/10.1002/pro.2449.

Padhi, A. K., Jayaram, B., & Gomes, J. (2013). Prediction of functional loss of human angiogenin mutants associated with ALS by molecular dynamics simulations. *Scientific Reports, 3*, 1225. http://dx.doi.org/10.1038/srep01225.

Pandurangan, A. P., Ochoa-Montaño, B., Ascher, D. B., & Blundell, T. L. (2017). SDM: A server for predicting effects of mutations on protein stability. *Nucleic Acids Research, 45*(W1), W229–W235. http://dx.doi.org/10.1093/nar/gkx439.

Pettersen, E. F., Goddard, T. D., Huang, C. C., Couch, G. S., Greenblatt, D. M., Meng, E. C., et al. (2004). UCSF Chimera—a visualization system for exploratory research and analysis. *Journal of Computational Chemistry, 25*(13), 1605–1612. http://dx.doi.org/10.1002/jcc.20084.

Petukh, M., Kucukkal, T. G., & Alexov, E. (2015). On human disease-causing amino acid variants: Statistical study of sequence and structural patterns. *Human Mutation, 36*(5), 524–534. http://dx.doi.org/10.1002/humu.22770.

Pierce, B. G., Wiehe, K., Hwang, H., Kim, B.-H., Vreven, T., & Weng, Z. (2014). ZDOCK server: Interactive docking prediction of protein–protein complexes and symmetric multimers. *Bioinformatics, 30*(12), 1771–1773. http://dx.doi.org/10.1093/bioinformatics/btu097.

Pires, D. E. V., Ascher, D. B., & Blundell, T. L. (2014a). DUET: A server for predicting effects of mutations on protein stability using an integrated computational approach. *Nucleic Acids Research, 42*(W1), W314–W319. http://dx.doi.org/10.1093/nar/gku411.

Pires, D. E. V., Ascher, D. B., & Blundell, T. L. (2014b). mCSM: Predicting the effects of mutations in proteins using graph-based signatures. *Bioinformatics, 30*(3), 335–342. http://dx.doi.org/10.1093/bioinformatics/btt691.

Pires, D. E. V., Chen, J., Blundell, T. L., & Ascher, D. B. (2016). In silico functional dissection of saturation mutagenesis: Interpreting the relationship between phenotypes and changes in protein stability, interactions and activity. *Scientific Reports, 6*, 19848. http://dx.doi.org/10.1038/srep19848.

Rajasekaran, R., Sudandiradoss, C., Doss, C. G. P., & Sethumadhavan, R. (2007). Identification and in silico analysis of functional SNPs of the *BRCA1* gene. *Genomics, 90*(4), 447–452. http://dx.doi.org/10.1016/j.ygeno.2007.07.004.

Ramos-Casals, M., Tzioufas, A. G., & Font, J. (2005). Primary Sjögren's syndrome: New clinical and therapeutic concepts. *Annals of the Rheumatic Diseases, 64*(3), 347–354. http://dx.doi.org/10.1136/ard.2004.025676.

Rizzo, W. B., & Carney, G. (2005). Sjögren-Larsson syndrome: Diversity of mutations and polymorphisms in the fatty aldehyde dehydrogenase gene (ALDH3A2). *Human Mutation, 26*(1), 1–10. http://dx.doi.org/10.1002/humu.20181.

Rodrigues, C. H., Pires, D. E., & Ascher, D. B. (2018). DynaMut: Predicting the impact of mutations on protein conformation, flexibility and stability. *Nucleic Acids Research, 46*(W1), W350–W355. http://dx.doi.org/10.1093/nar/gky300.

Rogers, G. R., Markova, N. G., De Laurenzi, V., Rizzo, W. B., & Compton, J. G. (1997). Genomic organization and expression of the human fatty aldehyde dehydrogenase gene (FALDH). *Genomics, 39*(2), 127–135. http://dx.doi.org/10.1006/geno.1996.4501.

Schmid, N., Eichenberger, A. P., Choutko, A., Riniker, S., Winger, M., Mark, A. E., et al. (2011). Definition and testing of the GROMOS force-field versions 54A7 and 54B7. *European Biophysics Journal, 40*(7), 843. http://dx.doi.org/10.1007/s00249-011-0700-9.

Shastry, B. S. (2007). SNPs in disease gene mapping, medicinal drug development and evolution. *Journal of Human Genetics, 52*(11), 871–880. http://dx.doi.org/10.1007/s10038-007-0200-z.

Sherry, S. T., Ward, M.-H., Kholodov, M., Baker, J., Phan, L., Smigielski, E. M., et al. (2001). dbSNP: The NCBI database of genetic variation. *Nucleic Acids Research, 29*(1), 308–311. http://dx.doi.org/10.1093/nar/29.1.308.

Shoichet, B. K., Baase, W. A., Kuroki, R., & Matthews, B. W. (1995). A relationship between protein stability and protein function. *Proceedings of the National Academy of Sciences, 92*(2), 452–456. http://dx.doi.org/10.1073/pnas.92.2.452.

Sillén, A., Jagell, S., & Wadelius, C. (1997). A missense mutation in the FALDH gene identified in Sjögren-Larsson syndrome patients originating from the northern part of Sweden. *Human Genetics, 100*(2), 201–203. http://dx.doi.org/10.1007/s004390050490.

Sjogren, T., & Larsson, T. (1957). Oligophrenia in combination with congenital ichthyosis and spastic disorders; a clinical and genetic study. *Acta Psychiatrica et Neurologica Scandinavica Supplementum, 113*, 1–112.

Skopouli, F. N., Dafni, U., Ioannidis, J. P. A., & Moutsopoulos, H. M. (2000). Clinical evolution, and morbidity and mortality of primary Sjögren's syndrome. *Seminars in Arthritis and Rheumatism, 29*(5), 296–304. http://dx.doi.org/10.1016/S0049-0172(00)80016-5.

Sneha, P., & George Priya Doss, C. (2016). Chapter seven – molecular dynamics: New frontier in personalized medicine. In R. Donev (Ed.), *Advances in protein chemistry and structural Biology* (pp. 181–224). http://dx.doi.org/10.1016/bs.apcsb.2015.09.004.

Sneha, P., Thirumal Kumar, D., George Priya Doss, C., Siva, R., & Zayed, H. (2017). Determining the role of missense mutations in the POU domain of HNF1A that reduce the DNA-binding affinity: A computational approach. *PLoS One, 12*(4), e0174953. http://dx.doi.org/10.1371/journal.pone.0174953.

Sneha, P., Kumar Thirumal, D., Himani, T., Siva, R., George Priya Doss, C., & Zayed, H. (2017). Structural analysis of G1691S variant in the human filamin B gene responsible for Larsen syndrome: A comparative computational approach. *Journal of Cellular Biochemistry, 118*(7), 1900–1910. http://dx.doi.org/10.1002/jcb.25920.

Sneha, P., Thirumal Kumar, D., Saini, S., Kajal, K., Magesh, R., Siva, R., et al. (2017). Chapter four – analyzing the effect of V66M mutation in BDNF in causing mood disorders: A computational approach. In R. Donev (Ed.), *Advances in protein chemistry and structural biology* (pp. 85–103). http://dx.doi.org/10.1016/bs.apcsb.2017.01.006.

Sneha, P., Ebrahimi, E. A., Ghazala, S. A., Thirumal Kumar, D., Siva, R., George Priya Doss, C., et al. (2018). Structural analysis of missense mutations in galactokinase 1 (GALK1) leading to galactosemia type-2. *Journal of Cellular Biochemistry, 119*(9), 7585–7598. http://dx.doi.org/10.1002/jcb.27097.

Stenson, P. D., Mort, M., Ball, E. V., Evans, K., Hayden, M., Heywood, S., et al. (2017). The human gene mutation database: Towards a comprehensive repository of inherited mutation data for medical research, genetic diagnosis and next-generation sequencing studies. *Human Genetics, 136*(6), 665–677. http://dx.doi.org/10.1007/s00439-017-1779-6.

Strub, C., Alies, C., Lougarre, A., Ladurantie, C., Czaplicki, J., & Fournier, D. (2004). Mutation of exposed hydrophobic amino acids to arginine to increase protein stability. *BMC Biochemistry, 5*(1), 9. http://dx.doi.org/10.1186/1471-2091-5-9.

Studer, R. A., Dessailly, B. H., & Orengo, C. A. (2013). Residue mutations and their impact on protein structure and function: Detecting beneficial and pathogenic changes. *Biochemical Journal, 449*(3), 581–594. http://dx.doi.org/10.1042/BJ20121221.

Sudhakar, N., Doss, C. G. P., Kumar, D. T., Chakraborty, C., Anand, K., & Suresh, M. (2016). Deciphering the impact of somatic mutations in exon 20 and exon 9 of PIK3CA gene in breast tumors among Indian women through molecular dynamics approach. *Journal of Biomolecular Structure and Dynamics, 34*(1), 29–41. http://dx.doi.org/10.1080/07391102.2015.1007483.

Tanwar, H., Kumar, D. T., Doss, C. G. P., & Zayed, H. (2019). Bioinformatics classification of mutations in patients with Mucopolysaccharidosis IIIA. *Metabolic Brain Disease*. http://dx.doi.org/10.1007/s11011-019-00465-6.

Thirumal Kumar, D., & George Priya Doss, C. (2016). Chapter nine – investigating the inhibitory effect of wortmannin in the hotspot mutation at codon 1047 of PIK3CA kinase domain: A molecular docking and molecular dynamics approach. In R. Donev (Ed.), *Advances in protein chemistry and structural biology* (pp. 267–297). http://dx.doi.org/10.1016/bs.apcsb.2015.09.008.

Tina, K. G., Bhadra, R., & Srinivasan, N. (2007). PIC: Protein interactions calculator. *Nucleic Acids Research, 35*(Suppl. 1_2), W473–W476. http://dx.doi.org/10.1093/nar/gkm423.

Turner, P. J. (2005). *Grace home*. Retrieved August 8, 2019, from XMGRACE, Version 5.1.19. website: http://plasma-gate.weizmann.ac.il/Grace/.

Udhaya Kumar, S., Priyanka, N., Sneha, P., & George Priya Doss, C. (2015). Functional and structural characterization of missense mutations in PAX6 gene. *Frontiers in Biology, 10*(4), 377–385. http://dx.doi.org/10.1007/s11515-015-1346-2.

van Gunsteren, W. F. (1996). *Biomolecular simulation: The GROMOS96 manual and user guide*. Zürich: Biomos .

Venselaar, H., te Beek, T. A., Kuipers, R. K., Hekkelman, M. L., & Vriend, G. (2010). Protein structure analysis of mutations causing inheritable diseases. An e-Science approach with life scientist friendly interfaces. *BMC Bioinformatics, 11*(1), 548. http://dx.doi.org/10.1186/1471-2105-11-548.

Wang, X., Wei, X., Thijssen, B., Das, J., Lipkin, S. M., & Yu, H. (2012). Three-dimensional reconstruction of protein networks provides insight into human genetic disease. *Nature Biotechnology, 30*(2), 159–164. http://dx.doi.org/10.1038/nbt.2106.

Willemsen, M. A. A. P., Rotteveel, J. J., van Domburg, P. H. M. F., Gabreëls, F. J. M., Mayatepek, E., & Sengers, R. C. A. (1999). Preterm birth in Sjögren-Larsson syndrome. *Neuropediatrics, 30*(6), 325–327. http://dx.doi.org/10.1055/s-2007-973513.

Yates, C. M., & Sternberg, M. J. E. (2013). The effects of non-synonymous single nucleotide polymorphisms (nsSNPs) on protein–protein interactions. *Journal of Molecular Biology, 425*(21), 3949–3963. http://dx.doi.org/10.1016/j.jmb.2013.07.012.

Zhanhua, C., Gan, J. G.-K., lei, L., Sakharkar, M. K., & Kangueane, P. (2005). Protein subunit interfaces: Heterodimers versus homodimers. *Bioinformation, 1*(2), 28–39.

Zhong, Q., Simonis, N., Li, Q.-R., Charloteaux, B., Heuze, F., Klitgord, N., et al. (2009). Edgetic perturbation models of human inherited disorders. *Molecular Systems Biology, 5*(1), 321. http://dx.doi.org/10.1038/msb.2009.80.

CHAPTER NINE

Computational model to analyze and characterize the functional mutations of NOD2 protein causing inflammatory disorder – Blau syndrome

D. Thirumal Kumar[a], S. Udhaya Kumar[a], Ahmed Shaikh Nishaat Laeeque[a], Shivalkar Apurva Abhay[a], R. Bithia[a], R. Magesh[b], Maignana Kumar[c], Hatem Zayed[d,]* and C. George Priya Doss[a,]*

[a]School of BioSciences and Technology, Vellore Institute of Technology, Vellore, Tamil Nadu, India
[b]Department of Biotechnology, College of Biomedical Sciences Technology & Research, Sri Ramachandra Institute of Higher Education and Research (DU), Chennai, Tamil Nadu, India
[c]Department of Pharmacology, Chettinad Academy of Research and Education, Kelambakkam, Tamil Nadu, India
[d]Department of Biomedical Sciences, College of Health and Sciences, Qatar University, Doha, Qatar
*Corresponding authors: E-mail: hatem.zayed@qu.edu.qa, georgepriyadoss@vit.ac.in

Contents

Advances in Protein Chemistry and Structural Biology, Volume 120
ISSN 1876-1623
https://doi.org/10.1016/bs.apcsb.2019.11.005

Abstract

Blau syndrome (BS), which affects the eyes, skin, and joints, is an autosomal dominant genetic inflammatory disorder. BS is caused by mutations in the *NOD2* gene. However, there are no direct treatments, and treatment with conventional anti-inflammatory drugs such as adrenal glucocorticoids, anti-metabolites, and biological agents such as anti-TNF and infliximab have all been attempted with varying degrees of success. In this study, we tried to identify all the reported mutations in the NOD2 protein that cause BS. Collectively, 114 missense mutations were extracted from the UniProt, ClinVar, and HGMD databases. The mutations were further subjected to pathogenic, stability, and conservation analyses. According to these computational analyses, six missense mutations (R334Q, R334W, E383G, E383K, R426H, and T605P) were found to be highly deleterious, destabilizing, and positioned in the conserved position. ADP to ATP conversion plays a crucial role in switching the closed-form of NOD2 protein to the open-form, thus activating the protein. Accordingly, the mutations in the ADP binding sites have received more attention in comparison to the mutations in the non-ADP binding positions. Interestingly, the W490L mutation is positioned in the ADP binding site and exhibits highly deleterious and destabilizing properties. Additionally, W490L was also found to be conserved, with a ConSurf score of 7. Therefore, we further performed homology modeling to determine the 3D structure of native NOD2 and the W490L mutant. Molecular docking analysis was carried out to understand the change in the interaction of ADP with the NOD2 protein. We observed that ADP had a stronger interaction with the native NOD2 protein compared to the W490L mutant. Finally, ADP complexed with native NOD2 and W490L mutant were subjected to molecular dynamics simulations, and the trajectories were analyzed. In the simulations, we observed decreased deviation and fluctuations in native NOD2, whereas decreased compactness and inter- and intramolecular hydrogen bonds were observed in the W490L mutant. This study is expected to serve as a platform for developing targeted drug therapy for BS.

1. Introduction

BS (OMIM #186580) is a rare autosomal dominant inflammatory disorder that is also known as arthrocutaneous granulomatosis. It was first discovered in 1985 by the pediatricians Jabs and Blau, who treated two separate families exhibiting a similar clinical phenotype, including severe uveitis (Blau, 1985; Jabs, Houk, Bias, & Arnett, 1985). The family reported by Blau also displayed a dark rash on the skin, which was absent in Jabs' report, whereas the second family exhibited cranial neuropathy (Blau, 1985; Rosé et al., 2009). It is indicated by three conditions: rash and granulomatous inflammation of the skin, uveitis (eye inflammation), and arthritis. The general symptoms of the disorder are skin granulomas, which may invade joints, eyes, and other organs (Punzi et al., 2009). This syndrome is often referred to as juvenile or pediatric, as it most often affects children. The signs and symptoms of the disorder are widespread in different parts of the body. It causes uveitis (redness and inflammation), cataracts, optic disc edema, glaucoma, iritis, and other conditions in the eyes. The skin is affected by different symptoms, such as granulomatous dermatitis, skin ulceration, and cysts over the wrists and ankle joints (Pastores et al., 1990). Other symptoms include synovial cysts, tendonitis, joint swelling, joint contractures, camptodactyly (bent fingers), flexion and contractures in the fingers and toes, etc. (Raphael, Blau, Zhang, & Hsu, 1993). The primary cause of BS is mutations in the *NOD/CARD15* gene. Mutations in the nucleotide-binding oligomerization domain 2 (NOD2), which is also known as the caspase recruitment domain-containing protein 15 (CARD15), leads to Blau syndrome and Crohn's disease by playing a vital role in innate immune reactions. NOD2 is present on the long 'q-arm' of chromosome 16 at position 12.1. The NOD2 protein contains three domains, which include the nucleotide-binding oligomerization domain (NOD), leucine-rich repeat (LRR), and the caspase activation and recruitment domain (CARD) (Maekawa, Ohto, Shibata, Miyake, & Shimizu, 2016) (Fig. 1A).

The *NOD2* gene is located on the q arm of chromosome 16 and encodes a 110 kDa protein of 1040 amino acids (Sterka, Rati, & Marriott, 2006; Tromp et al., 1996). Multiple exons and introns are incorporated throughout the 39.5 kb of the gene. Two binding sites for NF-κB are present at 26 and 301 bp in the promoter region upstream of the NOD2 transcription start site. The NOD2 gene also contains the binding site for vitamin D receptors within it (Salem et al., 2015). NOD2 protein has been found to be located chiefly in the cytoplasm but also in mitochondrial NLRs. NOD-like receptors (NLRs) are receptors comprising a subfamily called NLRC that encompasses NOD2 (Grube et al., 2015; Moreira & Zamboni, 2012). NOD2 is composed of 3 domains (Philpott, Sorbara, Robertson, Croitoru, & Girardin, 2014). There are two CARDs (caspase activation and recruitment domains) that act as effector domains near the N terminus, which is followed by a central NOD (nucleotide-binding and oligomerization) domain comprising 5 subdomains, namely, NBD

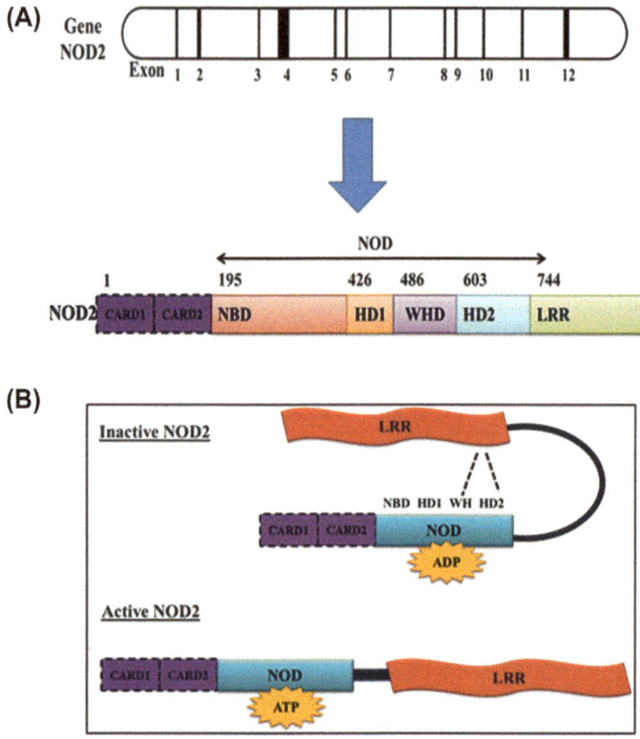

Fig. 1 (A) Graphical illustration of the gene location in the chromosome and the domains of the NOD2 protein. (B) Graphical illustration of the inactive and active forms of NOD2 protein.

(nucleotide-binding domain), HD1 (helical domain 1), WHD (winged helical domain), HD2 (helical domain 2), which plays a significant role in ATP hydrolysis, and the C terminal domain, which consists of 10 LRRs (leucine-rich repeats) that have the function of sensing the molecular patterns associated with pathogens (Maekawa et al., 2016; Moreira & Zamboni, 2012) (Fig. 1B).

The structure of NOD2 in the inactive state is folded due to the binding of the central NOD domain to the C-terminal LRR domain, restricting self-dimerization. Once it is stimulated by bacteria, there is a conformational change in the NOD2 protein structure, converting it to its active form and allowing it to self-dimerize, and downstream signaling is activated (Lechtenberg, Mace, & Riedl, 2014). The transformation of the NOD2 protein from an inactive state to an active state is due to the conformational change caused by the exchange of ADP for ATP. ADP plays a vital role in retaining the closed form of NOD2 protein, thus preserving the inactive state (Philpott et al., 2014). The inactive form is achieved by the binding of the ADP molecule to the center of the NOD domain, forming a bridge between the NBD, HD1, and WHD and thus leading to the formation of a closed compact tertiary structure that prevents self-oligomerization. Stabilization of the inactive form of NOD2 is also carried out by the subdomain interactions. There is the formation of interdomain interactions between the NBD and HD1 (Lechtenberg et al., 2014). The interdomain interactions between NBD and HD2 are exclusive to NOD2. Interdomain interactions between NOD2 and WHD are also responsible for the inactivation of NOD2 (Maekawa et al., 2016; Zhang et al., 2015). Activation of NOD2 occurs by the binding of the ligand. The ligand, when bound to the LRR domain, undergoes a conformational change mediated by NBD, WH, and HD1 to allow the exchange of bound ADP for ATP, which leads to the formation of active self-dimers carry out further downstream signaling. Thus, the release of ADP and binding of ATP to the NOD region causes the closed conformation to relax and allows the domains to become available for dimerization, resulting in NOD activation (Maekawa et al., 2016; Moreira & Zamboni, 2012). Once the LRR domain senses the MDP from bacteria, the initiation of the signaling cascade occurs, which leads to the activation of MAPK and NF-κB (Hasegawa et al., 2008). The binding of a ligand to NOD2 causes the protein to unfold, and there is an exchange of ADP to ATP, which stabilizes the unfolded structure. This conformation of NOD2 has the capability to self-oligomerize, and the resulting structure has its CARD domains exposed. The exposed CARD domains have the capability to recruit and bind to RIP2 due to the interactions between CARD-CARD27 (Hasegawa et al., 2008). The activation of NF-κB is due to this binding. 6-E3 ubiquitin-protein ligase, which is a TNF receptor-associated factor, is recruited by RIP2; this molecule has the ability to trigger self-ubiquitination and polyubiquitinate downstream proteins. The RIP2 molecule is equipped with a kinase domain that is associated with cellular inhibitors of apoptosis (cIAP1/2), which comprise an E3 ubiquitin ligase. Transforming growth factor beta-activated kinase 1 complex (TAK1) is recruited by the polyubiquitination of RIP2, which results in the activation of IkK kinases (Jakopin, Gobec, Mlinarič-Raščan, & Sollner Dolenc, 2012). IkK kinases phosphorylate IkB, which is degraded, and this triggers the activation of NF-κB1 (p50) and RelA (p65), which enter the nucleus and bind to DNA to enhance the production of inflammatory cytokines (Wertz & Dixit, 2010). Another pathway that is activated by

NOD2 is the mitogen-activated protein kinase (MAPK) pathway. NOD2, once it is activated by peptidoglycan, causes the activation of p38 MAPK and extracellular signal-regulated kinase (ERK). At present, the mechanism is not well known, but it has been observed that NOD2 binds to procaspase-1 through the CARD–CARD interaction and is converted into caspase, leading to the secretion of interleukin-1β (IL-1β) (Sabbah et al., 2009). In the case of BS, mutations in the gene result in the instability of the inactive form of NOD2, and there is constitutive activation of the protein and thus an increase in the secretion of inflammatory cytokines, ultimately leading to the granulomas that are characteristic of the disease (Fig. 2).

In this study, we used a series of computational tools, such as PredictSNP, iStable, ConSurf, etc., to determine the impact of NOD2 protein mutations causing BS (Bendl et al., 2014; Chen, Lin, & Chu, 2013; Glaser et al., 2003). Protein changes occur due to mutations, which often cause the loss of protein function. To predict protein functional changes, the synchronism classifier PredictSNP performs the prediction of disease-linked mutations. PredictSNP presents a consensus score that is generated based on the output of six different amino acid-based predictors. PredictSNP is thus utilized to evaluate substitutions in an amino acid sequence positioned in any region of the genome. The prediction of the protein stability changes helps with certain predictions regarding protein design. One relevant tool is

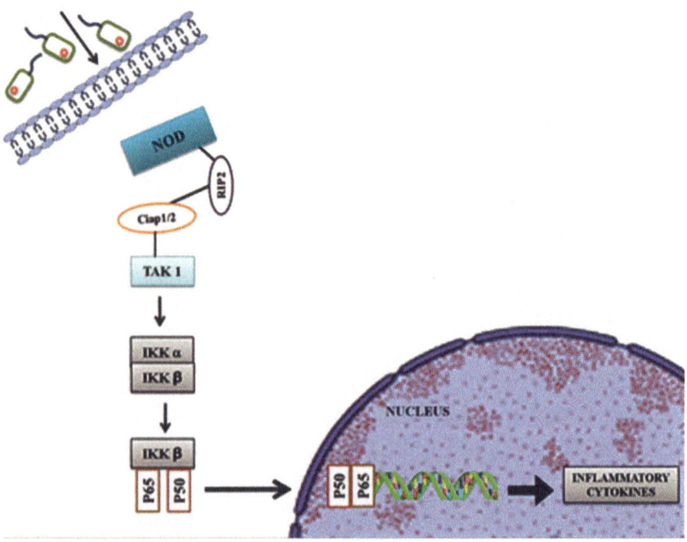

Fig. 2 Graphical illustration of the NOD2 signaling transduction pathway.

iStable (Integrated Predictor for Protein Stability Change upon Single Mutation). iStable is an integrated predictor that is constructed using the sequence information and the results of predictions obtained from different sources, such as element predictors. iStable is available in two types based on the input option: sequential and structural. The ConSurf web server scrutinizes the evolutionary pattern of the amino acids of the macromolecule to determine sites that are important for structure and/or function. Starting from a structure or query sequence, ConSurf collects the homologs of the sequence and applies a multiple sequence alignment and, with the aid of algorithms, reconstructs a phylogenetic tree to understand the evolutionary relationship between the sequences. These data are then used in accordance with a framework for the estimation of the evolutionary rates of each sequence position. Using such collective algorithms to filter the most significant mutations helps to attain more confidence in the significance of mutations (Bendl et al., 2014; Capriotti, Altman, & Bromberg, 2013; Sneha, Thirumal Kumar, George Priya Doss, Siva, & Zayed, 2017; Thirumal Kumar, Jerushah Emerald, et al., 2018). Finally, we performed molecular docking and used dynamic approaches to understand the impact of mutations on the ADP interaction with the NOD2 protein. A similar computational pipeline has been shown to be more accurate and realistic in several previous studies (George Priya Doss & Zayed, 2017; Thirumal Kumar, Sneha, Uppin, Usha, & George Priya Doss, 2018; Thirumal Kumar, Judith et al., 2019, Thirumal Kumar, Mendonca, Priyadharshini Christy, George Priya Doss, & Zayed, 2019; Zaki et al., 2017).

2. Materials and methods

2.1 Data set

In regards to the missense mutations observed in the NOD2 protein, data associated with BS were studied through a literature survey of recent articles. The mutation data were collected from UniProt, Clinvar, and HGMD (Apweiler et al., 2004; Landrum et al., 2014; Stenson et al., 2017). UniProtKB and Swiss-Prot contain high-quality manual illustrations and concise records of protein sequences. Manual illustration consists of the comparison, merging, and analysis of all available sequences of a given protein and a review of experimental and predicted data (UniProtKB-Q9HC29; NOD2_HUMAN). ClinVar is a freely available collection of reports of the interactions among human variants and phenotypes based on clinical or experimental evidence. ClinVar uses a similar query interface to that used in PubMed or other databases at NCBI. HGMD creates an extensive collection of published germ-line mutations that underlie or are closely associated with inherited human disease. The amino acids in FASTA format were obtained from the UniProt database, and ADP binding was assessed according to earlier reports by Maekawa et al. in 2016 (Maekawa et al., 2016). Collectively, 114 mutations were retrieved from these databases.

2.2 Pathogenicity

Several tools are available online to predict amino acid mutations, which may lead to specific functional changes. The PredictSNP tool was used to analyze the pathogenicity of the collected mutations, which indicated that the mutations were neutral or deleterious. PredictSNP uses a combination of the various pathogenicity predictor tools (Bendl et al., 2014). nsSNPAnalyzer is an online software that predicts the structural and evolutionary information for a queried nsSNP through a method called Random Forest that calculates the phenotypic expression of nsSNPs. The nsSNPAnalyzer server incorporates multiple sequence alignment and protein structure analysis to recognize nsSNPs that are related to a disease. nsSNPAnalyzer analyzes a protein sequence and the supplementary nsSNPs to determine whether the nsSNP is likely to be disease-associated or functionally neutral (Bao, Zhou, & Cui, 2005). PolyPhen is an acronym for the Polymorphism Phenotyping Server. PolyPhen-1 and PolyPhen-2 are the two tools included in PolyPhen. PolyPhen-2 calculates the naïve Bayes posterior probability, which indicates the damaging capacity of mutations (Adzhubei, Jordan, & Sunyaev, 2013). SNAP is a screening tool for non-acceptable polymorphisms that uses a neural network-centered method to estimate the functional expression of nonsynonymous SNPs (Bromberg & Rost, 2007). The remaining tools that assess mutations according to the PredictSNP dataset sort the data by the continuous value, indicating the confidence score.

2.3 Stability of the mutants

The iStable tool was used to check the stability of the protein polymorphisms. iStable is available with structural and sequential input protocols. iStable generates a combined result by using the I-Mutant2.0 and MUpro tools. I-Mutant2.0 implements an SVM model to predict the free energy ddG value of the protein and predicts the direction of the stability change. iStable analysis uses both the sequence (I-Mutant_SEQ) and the structure

(I-Mutant_PDB). MUPRO devises an SVM model to predict stability changes that occur due to single-site mutations predominantly by using successive information; it also uses the provided structural description. The result indicates only whether the alterations will cause destabilization but does not provide an actual ddG value (Chen et al., 2013).

2.4 Conservation analysis

The ConSurf tool determined the conservation of the amino acid at each position. ConSurf involves an examination of similar homologous sequences using PSI-BLAST, CSI-BLAST, or BLAST (Ashkenazy et al., 2016). ConSurf uses a color code, with green-through-purple indicating variable (grade 1)-through-conserved (grade 9) positions. The level of conservation is indicated by a scale from 0 to 9 in ConSurf, with '9' indicating the highest conservation of the amino acid and '0' indicating the least conservation.

2.5 Protein structural analysis

The protein structure was modeled using the Swiss Model Server (Guex & Peitsch, 1997). The FASTA sequence of NOD2 protein was chosen from UniProtKB for further analysis. The numbering was performed manually to match the amino acids of the protein structure with the amino acids of the protein sequence. The most deleterious mutation found in the ADP-protein interaction was modeled using SwissPDB Viewer (Guex & Peitsch, 1997). The active site amino acid numbering of the protein sequence from the UniProt database and the modeled structure in PDB format were mapped manually and are shown in Table 1.

2.6 Molecular docking

AutoDock v4.2 software was used to carry out the molecular docking (Morris et al., 1998). All the nonpolar hydrogen atoms were merged after the addition of all hydrogen molecules while docking the native and mutant protein structures in the AutoDock tool. The Gasteiger method was used to designate the charges. The torsions were set for the ligand (ADP-C10H15N5O10P2-, PubChem CID: 6022). The grid box was placed around the active sites for the interaction of the ligands. The residues L237, T239, I251, Y252, T253, E254, N255, E300, A301, G302, S303, G304, K305, S306, T307, L308, F447, P486, V487, W490, and H603 were set within a grid box of 60x60x60 with 3.75 Å. Auto grid was used to complete the rigid box, which was followed by molecular docking with AutoDock using the Lamarckian genetic algorithm, which helped to

Table 1 The active site amino acids numbering of the protein sequence from the UniProt database and the modeled structure in PDB.

Rabbit (template structure)	Human (modeled structure)
LEU217	LEU237
THR219	THR239
ILE231	ILE251
TYR232	TYR252
THR233	THR253
GLU234	GLU254
ASN235	ASN255
GLU280	GLU300
ALA281	ALA301
GLY282	GLY302
SER283	SER303
GLY284	GLY304
LYS285	LYS305
SER286	SER306
THR287	THR307
LEU288	LEU308
PHE427	PHE447
PRO466	PRO486
VAL467	VAL487
TRP470	TRP490
HIS583	HIS603

generate the elite docking conformation (Morris et al., 1998). The conformation with the best binding energy was used for the interaction analysis and molecular dynamics simulations.

2.7 Molecular dynamics simulations

Molecular dynamics simulations were performed using the GROMACS package with the latest GROMOS force field to derive the protein parameters (Abraham et al., 2015). The proteins were solvated in a cubical box with a simple point charge water molecule. The simulation systems were energy-minimized by adding the counterions (NA^+ or CL^-). The equilibration of the systems was performed using the NVT and NPT ensembles for 50,000 steps each, and molecular dynamics simulations were finally performed for 50 ns. The built-in packages in GROMACS were used to generate the outputs from the 50 ns trajectories, and the plots were generated.

3. Results

3.1 Data set

Biological databases such as UniProt, ClinVar, and HGMD containing the most common mutation reports were used to retrieve the mutation reports for the NOD2 protein. Collectively, 114 missense mutations were extracted from these databases, and these were subjected to further computational analysis.

3.2 Pathogenicity

The PredictSNP server was used to determine the pathogenicity of all 114 NOD2 missense mutations. Based on the overall screening, 17 missense mutations were found to be 'deleterious' by all the pathogenicity predicting algorithms embedded within the PredictSNP server (Table 2). The 17 mutations were further studied to determine the influence of the mutations on the stability of the protein.

3.3 Stability of the mutants and conservation analysis

Seventeen missense mutations in NOD2 protein that showed deleteriousness according to the pathogenicity predictors were subjected to stability analysis using the iStable server. From the predictions, 11 mutations were identified that destabilized the protein structure (Table 3). The 17 highly pathogenic mutations were also subjected to conservation analysis to understand the conservation range using the ConSurf server. From the ConSurf analysis, nine mutations (R334Q, R334W, E383G, E383K, R426H, C495Y, T605N, T605P, and S900F) were found to be highly conserved residues with a ConSurf score of 9 (Table 4). In the overall stability and conservation analysis, six mutations (R334Q, R334W, E383G, E383K, R426H, and T605P) were found to be both destabilizing and highly conserved.

3.4 Protein structural analysis

Homology modeling was performed using the Swiss Model server. The FASTA sequence of NOD2 protein was retrieved from the UniProt database and was used as the input for modeling. The protein structure with

Table 2 Screening of NOD2 mutations for the pathogenicity using PredictSNP server.

S. No	Feature Key	Variant	PredictSNP prediction	PredictSNP expected accuracy	MAPP prediction	MAPP expected accuracy	PhD-SNP prediction	PhD-SNP expected accuracy	PolyPhen-1 prediction	Poly Phen-1 expected accuracy	PolyPhen-2 prediction	Poly Phen-2 expected accuracy	SIFT prediction	SIFT expected accuracy	SNAP prediction	SNAP expected accuracy
1	NM_022162.2	M1T	NEUTRAL	0.62831343	NA	NA	NEUTRAL	0.7828765	UNKNOWN	0	UNKNOWN	0	DELETERIOUS	0.79280784	NEUTRAL	0.66524217
2	NM_022162.2	E3Q	NEUTRAL	0.62831343	NA	NA	NEUTRAL	0.89245476	NEUTRAL	0.66884082	DELETERIOUS	0.39845758	DELETERIOUS	0.79280784	NEUTRAL	0.55439739
3	NM_022162.2	V18I	NEUTRAL	0.82622462	NA	NA	NEUTRAL	0.7828765	NEUTRAL	0.66884082	NEUTRAL	0.8730899	NEUTRAL	0.67141585	NEUTRAL	0.61167883
4	NM_022162.2	E43K	NEUTRAL	0.82622462	NA	NA	NEUTRAL	0.7828765	NEUTRAL	0.66884082	NEUTRAL	0.63363363	NEUTRAL	0.82295374	NEUTRAL	0.76823399
5	NM_022162.2	S47L	NEUTRAL	0.82622462	NA	NA	NEUTRAL	0.58230958	NEUTRAL	0.66884082	NEUTRAL	0.62762763	NEUTRAL	0.68210151	NEUTRAL	0.55439739
6	NM_022162.2	R87C	DELETERIOUS	0.71871275	NA	NA	NEUTRAL	0.55202703	DELETERIOUS	0.74491225	DELETERIOUS	0.81142888	DELETERIOUS	0.79280784	DELETE-RIOUS	0.55551884
7	NM_022162.2	V92I	NEUTRAL	0.75203963	NA	NA	NEUTRAL	0.71871412	NEUTRAL	0.66884082	DELETERIOUS	0.45308483	NEUTRAL	0.66429207	NEUTRAL	0.50014676
8	NM_022162.2	A105V	NEUTRAL	0.6025641	NEUTRAL	0.6778169	NEUTRAL	0.83210379	DELETERIOUS	0.59445019	DELETERIOUS	0.64974293	DELETERIOUS	0.42969871	NEUTRAL	0.76823399
9	NM_022162.2	P125L	NEUTRAL	0.75291375	NA	NA	NEUTRAL	0.68183996	DELETERIOUS	0.59445019	NEUTRAL	0.60960961	NEUTRAL	0.60819234	NEUTRAL	0.70878378
10	NM_022162.2	R138Q	NEUTRAL	0.63151762	NEUTRAL	0.65022026	NEUTRAL	0.71871412	NEUTRAL	0.66884082	DELETERIOUS	0.55077121	DELETERIOUS	0.52778524	NEUTRAL	0.50014676
11	NM_022162.2	A140T	NEUTRAL	0.82622462	NEUTRAL	0.72246696	NEUTRAL	0.71871412	NEUTRAL	0.66884082	NEUTRAL	0.74974975	NEUTRAL	0.79073909	NEUTRAL	0.61167883
12	NM_022162.2	D154 N	NEUTRAL	0.82622462	NEUTRAL	0.78854626	NEUTRAL	0.83210379	NEUTRAL	0.66884082	NEUTRAL	0.74074074	NEUTRAL	0.76067616	NEUTRAL	0.76823399
13	NM_022162.2	R159Q	NEUTRAL	0.82622462	NEUTRAL	0.85110132	NEUTRAL	0.89245476	NEUTRAL	0.66884082	NEUTRAL	0.67667668	NEUTRAL	0.76936776	NEUTRAL	0.55439739
14	NM_022162.2	V162I	NEUTRAL	0.82622462	NEUTRAL	0.85110132	NEUTRAL	0.83210379	NEUTRAL	0.66884082	NEUTRAL	0.74074074	NEUTRAL	0.75800712	NEUTRAL	0.61167883
15	NM_022162.2	T189 M	NEUTRAL	0.6025641	NEUTRAL	0.73480176	NEUTRAL	0.71871412	DELETERIOUS	0.74491225	DELETERIOUS	0.60089974	DELETERIOUS	0.42969871	NEUTRAL	0.61167883
16	NM_022162.2	P209S	NEUTRAL	0.82622462	NEUTRAL	0.65903084	NEUTRAL	0.68183996	NEUTRAL	0.66884082	NEUTRAL	0.78928929	NEUTRAL	0.76224399	NEUTRAL	0.58359402
17	NM_022162.2	Q233P	DELETERIOUS	0.86908365	DELETE-RIOUS	0.58770615	DELETERIOUS	0.67620995	DELETERIOUS	0.59445019	DELETERIOUS	0.64717224	DELETERIOUS	0.45232975	DELETE-RIOUS	0.55551884
18	NM_022162.2	R235C	NEUTRAL	0.65307311	DELETE-RIOUS	0.46176912	DELETERIOUS	0.57795276	NEUTRAL	0.66884082	NEUTRAL	0.69369369	NEUTRAL	0.71326803	NEUTRAL	0.55439739
19	NM_022162.2	S238C	NEUTRAL	0.82622462	NEUTRAL	0.73480176	NEUTRAL	0.71871412	NEUTRAL	0.66884082	NEUTRAL	0.68268268	NEUTRAL	0.66429207	NEUTRAL	0.76823399
20	NM_022162.2	T245 M	NEUTRAL	0.63151762	NEUTRAL	0.71013216	NEUTRAL	0.71871412	DELETERIOUS	0.74491225	DELETERIOUS	0.54177378	NEUTRAL	0.60819234	NEUTRAL	0.61167883
21	NM_022162.2	L248R	DELETERIOUS	0.78903456	DELETE-RIOUS	0.87706147	DELETERIOUS	0.87523992	DELETERIOUS	0.74491225	DELETERIOUS	0.81142888	DELETERIOUS	0.79280784	NEUTRAL	0.50014676
22	NM_022162.2	E249G	DELETERIOUS	0.71871275	NEUTRAL	0.64229075	DELETERIOUS	0.57795276	DELETERIOUS	0.59445019	DELETERIOUS	0.54177378	DELETERIOUS	0.79280784	DELETE-RIOUS	0.55551884
23	NM_022162.2	P268S	NEUTRAL	0.82622462	NEUTRAL	0.73480176	NEUTRAL	0.7828765	NEUTRAL	0.66884082	NEUTRAL	0.72472472	NEUTRAL	0.75800712	NEUTRAL	0.61167883

24	NM_022162.2	L281F	DELETERIOUS	0.60548272	DELETE-RIOUS	0.57314329	NEUTRAL	0.58230958	DELETERIOUS	0.59445019	DELETERIOUS	0.55077121	DELETERIOUS	0.79280784	NEUTRAL	0.70878378
25	NM_022162.2	H287Y	NEUTRAL	0.82622462	NEUTRAL	0.63348018	NEUTRAL	0.83210379	NEUTRAL	0.66884082	NEUTRAL	0.74074074	NEUTRAL	0.76067616	NEUTRAL	0.50014676
26	NM_022162.2	N289S	NEUTRAL	0.82622462	NEUTRAL	0.73480176	NEUTRAL	0.83210379	NEUTRAL	0.66884082	NEUTRAL	0.70240481	NEUTRAL	0.64501779	NEUTRAL	0.66524217
27	NM_022162.2	A292V	NEUTRAL	0.63151762	NEUTRAL	0.7596831	NEUTRAL	0.660879	NEUTRAL	0.66884082	DELETERIOUS	0.50257069	DELETERIOUS	0.79280784	NEUTRAL	0.66524217
28	VAR_012676	R334Q	DELETERIOUS	0.86908365	DELETE-RIOUS	0.40929535	DELETERIOUS	0.67620995	DELETERIOUS	0.74491225	DELETERIOUS	0.81142888	DELETERIOUS	0.79280784	DELETE-RIOUS	0.80510276
29	VAR_012677	R334W	DELETERIOUS	0.86908365	DELETE-RIOUS	0.63268366	DELETERIOUS	0.85822785	DELETERIOUS	0.74491225	DELETERIOUS	0.81142888	DELETERIOUS	0.79280784	DELETE-RIOUS	0.84845361
30	NM_022162.2	L343P	DELETERIOUS	0.86908365	DELETE-RIOUS	0.75112444	DELETERIOUS	0.7733853	DELETERIOUS	0.74491225	DELETERIOUS	0.60089974	DELETERIOUS	0.79280784	DELETE-RIOUS	0.72038776
31	NM_022162.2	H352R	DELETERIOUS	0.60548272	DELETE-RIOUS	0.46176912	NEUTRAL	0.50824588	DELETERIOUS	0.59445019	DELETERIOUS	0.47429306	DELETERIOUS	0.79280784	NEUTRAL	0.61167883
32	NM_022162.2	I363F	NEUTRAL	0.6025641	NA	NA	DELETERIOUS	0.7733853	NEUTRAL	0.66884082	NEUTRAL	0.63363363	DELETERIOUS	0.79280784	NEUTRAL	0.50014676
33	VAR_023822	D382E	DELETERIOUS	0.7556615	DELETE-RIOUS	0.81934033	DELETERIOUS	0.81731169	DELETERIOUS	0.59445019	DELETERIOUS	0.81142888	DELETERIOUS	0.79280784	NEUTRAL	0.66524217
34	VAR_073231	E383G	DELETERIOUS	0.86908365	DELETE-RIOUS	0.85682159	DELETERIOUS	0.81731169	DELETERIOUS	0.59445019	DELETERIOUS	0.81142888	DELETERIOUS	0.79280784	DELETE-RIOUS	0.55551884
35	VAR_023823	E383K	DELETERIOUS	0.86908365	DELETE-RIOUS	0.77494374	DELETERIOUS	0.87523992	DELETERIOUS	0.59445019	DELETERIOUS	0.81142888	DELETERIOUS	0.79280784	DELETE-RIOUS	0.80510276
36	CM1414276	E383Q	DELETERIOUS	0.7556615	DELETE-RIOUS	0.75112444	DELETERIOUS	0.7733853	DELETERIOUS	0.74491225	DELETERIOUS	0.81142888	DELETERIOUS	0.79280784	NEUTRAL	0.61167883
37	NM_022162.2	F384Y	NEUTRAL	0.63151762	DELETE-RIOUS	0.65667166	NEUTRAL	0.71871412	NEUTRAL	0.66884082	NEUTRAL	0.60960961	DELETERIOUS	0.52778524	NEUTRAL	0.58359402
38	CM149331	D390V	NEUTRAL	0.6025641	NEUTRAL	0.73480176	DELETERIOUS	0.58885542	NEUTRAL	0.66884082	DELETERIOUS	0.55077121	DELETERIOUS	0.52778524	NEUTRAL	0.61167883
39	NM_022162.2	R391H	NEUTRAL	0.82622462	NEUTRAL	0.7596831	NEUTRAL	0.71871412	NEUTRAL	0.66884082	NEUTRAL	0.72472472	NEUTRAL	0.72662511	NEUTRAL	0.70878378
40	NM_022162.2	R393H	NEUTRAL	0.6025641	NEUTRAL	0.7596831	NEUTRAL	0.71871412	DELETERIOUS	0.74491225	DELETERIOUS	0.67524116	DELETERIOUS	0.45949821	NEUTRAL	0.58359402
41	NM_022162.2	P397L	NEUTRAL	0.62831343	NEUTRAL	0.6778169	NEUTRAL	0.58230958	NEUTRAL	0.66884082	DELETERIOUS	0.59318766	DELETERIOUS	0.79280784	NEUTRAL	0.61167883
42	NM_022162.2	N414S	NEUTRAL	0.65307311	NEUTRAL	0.71013216	DELETERIOUS	0.60798122	NEUTRAL	0.66884082	DELETERIOUS	0.39845758	NEUTRAL	0.7747106	NEUTRAL	0.61167883
43	NM_022162.2	R426H	DELETERIOUS	0.86908365	DELETE-RIOUS	0.76536732	DELETERIOUS	0.81731169	DELETERIOUS	0.74491225	DELETERIOUS	0.81142888	DELETERIOUS	0.79280784	DELETE-RIOUS	0.86956522
44	NM_022162.2	P427S	NEUTRAL	0.6025641	DELETE-RIOUS	0.42653673	DELETERIOUS	0.7733853	NEUTRAL	0.66884082	DELETERIOUS	0.59318766	NEUTRAL	0.6841637	NEUTRAL	0.61167883
45	NM_022162.2	P427L	DELETERIOUS	0.7556615	DELETE-RIOUS	0.7711928	DELETERIOUS	0.87523992	DELETERIOUS	0.74491225	DELETERIOUS	0.81142888	DELETERIOUS	0.79280784	NEUTRAL	0.61167883

(Continued)

Table 2 Screening of NOD2 mutations for the pathogenicity using PredictSNP server.—cont'd

S. No	Feature Key	Variant	PredictSNP prediction	PredictSNP expected accuracy	MAPP prediction	MAPP expected accuracy	PhD-SNP prediction	PhD-SNP expected accuracy	PolyPhen-1 prediction	Poly Phen-1 expected accuracy	PolyPhen-2 prediction	Poly Phen-2 expected accuracy	SIFT prediction	SIFT expected accuracy	SNAP prediction	SNAP expected accuracy
46	NM_022162.2	R439H	NEUTRAL	0.68365861	NEUTRAL	0.79030837	NEUTRAL	0.50824588	NEUTRAL	0.66884082	DELETERIOUS	0.39845758	DELETERIOUS	0.45232975	NEUTRAL	0.66524217
47	NM_022162.2	E441K	NEUTRAL	0.65307311	NEUTRAL	0.74185022	DELETERIOUS	0.60798122	NEUTRAL	0.66884082	DELETERIOUS	0.39845758	NEUTRAL	0.77560107	NEUTRAL	0.58359402
48	NM_022162.2	R468S	DELETERIOUS	0.50595948	NEUTRAL	0.63348018	DELETERIOUS	0.7733853	NEUTRAL	0.66884082	DELETERIOUS	0.55077121	DELETERIOUS	0.79280784	NEUTRAL	0.58359402
49	VAR_012685	L469F	DELETERIOUS	0.86908365	DELETE-RIOUS	0.5089955	DELETERIOUS	0.73260309	DELETERIOUS	0.74491225	DELETERIOUS	0.60089974	DELETERIOUS	0.52778524	DELETE-RIOUS	0.55551884
50	NM_022162.2	R471C	NEUTRAL	0.63151762	NEUTRAL	0.65286344	DELETERIOUS	0.58885542	NEUTRAL	0.66884082	NEUTRAL	0.74074074	NEUTRAL	0.72662511	DELETE-RIOUS	0.62208185
51	NM_022162.2	T476 N	NEUTRAL	0.82622462	NEUTRAL	0.85110132	NEUTRAL	0.71871412	NEUTRAL	0.66884082	NEUTRAL	0.74974975	NEUTRAL	0.89670526	NEUTRAL	0.55439739
52	VAR_073235	G481D	DELETERIOUS	0.86908365	DELETE-RIOUS	0.9197901	DELETERIOUS	0.60798122	DELETERIOUS	0.74491225	DELETERIOUS	0.64717224	DELETERIOUS	0.79280784	DELETE-RIOUS	0.80510276
53	VAR_073236	W490L	DELETERIOUS	0.86908365	DELETE-RIOUS	0.58770615	DELETERIOUS	0.85822785	DELETERIOUS	0.59445019	DELETERIOUS	0.81142888	DELETERIOUS	0.79280784	DELETE-RIOUS	0.72038776
54	CM1414277	M491L	NEUTRAL	0.73834499	NEUTRAL	0.64229075	NEUTRAL	0.55202703	NEUTRAL	0.66884082	NEUTRAL	0.78928929	DELETERIOUS	0.42969871	NEUTRAL	0.66524217
55	VAR_073237	C495Y	DELETERIOUS	0.86908365	DELETE-RIOUS	0.58889722	DELETERIOUS	0.73260309	DELETERIOUS	0.74491225	DELETERIOUS	0.67524116	DELETERIOUS	0.79280784	DELETE-RIOUS	0.72038776
56	VAR_023824	H496L	NEUTRAL	0.82622462	NEUTRAL	0.78325991	NEUTRAL	0.71871412	NEUTRAL	0.66884082	NEUTRAL	0.7027027	NEUTRAL	0.52894034	NEUTRAL	0.58359402
57	CM1414294	E498G	NEUTRAL	0.6025641	DELETE-RIOUS	0.5089955	NEUTRAL	0.55202703	NEUTRAL	0.66884082	DELETERIOUS	0.60089974	DELETERIOUS	0.52778524	NEUTRAL	0.55439739
58	NM_022162.2	E498D	NEUTRAL	0.82622462	NEUTRAL	0.63348018	NEUTRAL	0.89245476	NEUTRAL	0.66884082	NEUTRAL	0.60960961	NEUTRAL	0.60819234	NEUTRAL	0.55439739
59	VAR_073180	P507S	NEUTRAL	0.63151762	NEUTRAL	0.63348018	DELETERIOUS	0.73260309	NEUTRAL	0.66884082	DELETERIOUS	0.47429306	NEUTRAL	0.67141585	NEUTRAL	0.70878378
60	VAR_073238	M513T	DELETERIOUS	0.54946365	DELETE-RIOUS	0.57121439	DELETERIOUS	0.81731169	NEUTRAL	0.66884082	DELETERIOUS	0.54177378	DELETERIOUS	0.52778524	NEUTRAL	0.70878378
61	CM114213	M513R	DELETERIOUS	0.86908365	DELETE-RIOUS	0.87706147	DELETERIOUS	0.88474971	DELETERIOUS	0.74491225	DELETERIOUS	0.64717224	DELETERIOUS	0.79280784	DELETE-RIOUS	0.55551884
62	CM149332	H520Y	DELETERIOUS	0.50595948	NEUTRAL	0.6778169	NEUTRAL	0.55202703	DELETERIOUS	0.74491225	DELETERIOUS	0.64717224	NEUTRAL	0.71326803	DELETE-RIOUS	0.55551884
63	NM_022162.2	A525T	NEUTRAL	0.82622462	NA	NA	NEUTRAL	0.71871412	NEUTRAL	0.66884082	NEUTRAL	0.76252505	NEUTRAL	0.7747106	NEUTRAL	0.76823399
64	NM_022162.2	R543H	NEUTRAL	0.82622462	NEUTRAL	0.78325991	NEUTRAL	0.660879	NEUTRAL	0.66884082	NEUTRAL	0.68268268	NEUTRAL	0.76067616	NEUTRAL	0.55439739

65	NM_022162.2	P545L	NEUTRAL	0.75203963	DELETE-RIOUS	0.62143928	NEUTRAL	0.58230958	NEUTRAL	0.66884082	NEUTRAL	0.76252505	NEUTRAL	0.72662511	NEUTRAL	0.58359402
66	CM1412976	Y563S	NEUTRAL	0.6025641	NEUTRAL	0.64229075	DELETERIOUS	0.60798122	NEUTRAL	0.66884082	DELETERIOUS	0.43123393	DELETERIOUS	0.79280784	NEUTRAL	0.66524217
67	VAR_073240	R587C	NEUTRAL	0.75291375	NEUTRAL	0.73127753	DELETERIOUS	0.81731169	NEUTRAL	0.66884082	NEUTRAL	0.7027027	NEUTRAL	0.7366548	NEUTRAL	0.61167883
68	NM_022162.2	R587H	NEUTRAL	0.73688811	NEUTRAL	0.78325991	DELETERIOUS	0.60798122	NEUTRAL	0.66884082	NEUTRAL	0.68268268	NEUTRAL	0.77560107	NEUTRAL	0.70878378
69	CM127716	E600K	DELETERIOUS	0.60548272	NEUTRAL	0.65022026	DELETERIOUS	0.67620995	DELETERIOUS	0.59445019	DELETERIOUS	0.64974293	DELETERIOUS	0.79280784	NEUTRAL	0.55439739
70	VAR_065228	T605 N	DELETERIOUS	0.86908365	DELETE-RIOUS	0.76086957	DELETERIOUS	0.87523992	DELETERIOUS	0.74491225	DELETERIOUS	0.81142888	DELETERIOUS	0.79280784	DELETE-RIOUS	0.55551884
71	VAR_073241	T605P	DELETERIOUS	0.86908365	DELETE-RIOUS	0.9137931	DELETERIOUS	0.87523992	DELETERIOUS	0.74491225	DELETERIOUS	0.81142888	DELETERIOUS	0.79280784	DELETE-RIOUS	0.80510276
72	VAR_012686	A612T	DELETERIOUS	0.7556615	DELETE-RIOUS	0.77494374	DELETERIOUS	0.87523992	DELETERIOUS	0.74491225	DELETERIOUS	0.81142888	DELETERIOUS	0.79280784	NEUTRAL	0.70878378
73	NM_022162.2	A612S	DELETERIOUS	0.7556615	DELETE-RIOUS	0.76611694	DELETERIOUS	0.85822785	DELETERIOUS	0.74491225	DELETERIOUS	0.81142888	DELETERIOUS	0.79280784	NEUTRAL	0.70878378
74	NM_022162.2	A619V	NEUTRAL	0.82622462	NEUTRAL	0.65022026	NEUTRAL	0.660879	NEUTRAL	0.66884082	NEUTRAL	0.76252505	NEUTRAL	0.75690116	NEUTRAL	0.82941176
75	NM_022162.2	S657G	NEUTRAL	0.82622462	NEUTRAL	0.76563877	NEUTRAL	0.83210379	NEUTRAL	0.66884082	NEUTRAL	0.8730899	NEUTRAL	0.89145907	NEUTRAL	0.61167883
76	NM_022162.2	A661P	NEUTRAL	0.65307311	DELETE-RIOUS	0.46176912	DELETERIOUS	0.60798122	NEUTRAL	0.66884082	NEUTRAL	0.72745491	NEUTRAL	0.76224399	NEUTRAL	0.61167883
77	NM_022162.2	P668L	NEUTRAL	0.65307311	NEUTRAL	0.74823944	DELETERIOUS	0.58885542	NEUTRAL	0.66884082	DELETERIOUS	0.39845758	NEUTRAL	0.60819234	NEUTRAL	0.58359402
78	CM1414278	H669R	NEUTRAL	0.6025641	NEUTRAL	0.65286344	NEUTRAL	0.55202703	DELETERIOUS	0.59445019	DELETERIOUS	0.50257069	NEUTRAL	0.60819234	DELETE-RIOUS	0.62208185
79	VAR_073242	N670K	DELETERIOUS	0.54946365	DELETE-RIOUS	0.48350825	NEUTRAL	0.58230958	DELETERIOUS	0.59445019	DELETERIOUS	0.59318766	DELETERIOUS	0.52778524	NEUTRAL	0.66524217
80	NM_022162.2	G680R	DELETERIOUS	0.7556615	DELETE-RIOUS	0.58770615	DELETERIOUS	0.87523992	DELETERIOUS	0.74491225	DELETERIOUS	0.81142888	DELETERIOUS	0.79280784	NEUTRAL	0.66524217
81	NM_022162.2	L682F	DELETERIOUS	0.71871275	NEUTRAL	0.70484581	DELETERIOUS	0.60798122	DELETERIOUS	0.59445019	DELETERIOUS	0.81142888	DELETERIOUS	0.79280784	DELETE-RIOUS	0.62208185
82	NM_022162.2	R684W	DELETERIOUS	0.54946365	DELETE-RIOUS	0.46176912	DELETERIOUS	0.85822785	NEUTRAL	0.66884082	NEUTRAL	0.7027027	DELETERIOUS	0.42969871	DELETE-RIOUS	0.55551884
83	NM_022162.2	R702W	DELETERIOUS	0.7556615	DELETE-RIOUS	0.63268366	NEUTRAL	0.50824588	DELETERIOUS	0.74491225	DELETERIOUS	0.67524116	DELETERIOUS	0.79280784	DELETE-RIOUS	0.62208185
84	NM_022162.2	R703C	DELETERIOUS	0.86908365	DELETE-RIOUS	0.58889722	DELETERIOUS	0.73260309	DELETERIOUS	0.74491225	DELETERIOUS	0.63431877	DELETERIOUS	0.79280784	DELETE-RIOUS	0.62208185

(Continued)

Table 2 Screening of NOD2 mutations for the pathogenicity using PredictSNP server.—cont'd

S. No	Feature Key	Variant	PredictSNP prediction	PredictSNP expected accuracy	MAPP prediction	MAPP expected accuracy	PhD-SNP prediction	PhD-SNP expected accuracy	PolyPhen-1 prediction	Poly Phen-1 expected accuracy	PolyPhen-2 prediction	Poly Phen-2 expected accuracy	SIFT prediction	SIFT expected accuracy	SNAP prediction	SNAP expected accuracy
85	NM_022162.2	R708H	NEUTRAL	0.63151762	NEUTRAL	0.73480176	NEUTRAL	0.83210379	DELETERIOUS	0.74491225	DELETERIOUS	0.54177378	NEUTRAL	0.69991095	NEUTRAL	0.55439739
86	NM_022162.2	R716H	NEUTRAL	0.82622462	NEUTRAL	0.65286344	NEUTRAL	0.44670846	NEUTRAL	0.66884082	NEUTRAL	0.70240481	NEUTRAL	0.67141585	NEUTRAL	0.58359402
87	NM_022162.2	A725G	NEUTRAL	0.82622462	NEUTRAL	0.74185022	NEUTRAL	0.68183996	NEUTRAL	0.66884082	NEUTRAL	0.78928929	NEUTRAL	0.72662511	NEUTRAL	0.61167883
88	NM_022162.2	A755V	NEUTRAL	0.63151762	NEUTRAL	0.71742958	NEUTRAL	0.660879	NEUTRAL	0.66884082	DELETERIOUS	0.54177378	DELETERIOUS	0.79280784	NEUTRAL	0.66524217
89	NM_022162.2	R756W	DELETERIOUS	0.86908365	DELETE-RIOUS	0.65667166	DELETERIOUS	0.87523992	DELETERIOUS	0.74491225	DELETERIOUS	0.59318766	DELETERIOUS	0.79280784	DELETE-RIOUS	0.62208185
90	NM_022162.2	R756Q	NEUTRAL	0.73688811	NEUTRAL	0.76563877	DELETERIOUS	0.58885542	NEUTRAL	0.66884082	NEUTRAL	0.72745491	NEUTRAL	0.86910062	NEUTRAL	0.66524217
91	NM_022162.2	R760H	DELETERIOUS	0.52145411	NEUTRAL	0.65903084	NEUTRAL	0.660879	DELETERIOUS	0.59445019	DELETERIOUS	0.39845758	DELETERIOUS	0.52778524	DELETE-RIOUS	0.55551884
92	NM_022162.2	T777N	NEUTRAL	0.75291375	NEUTRAL	0.79559471	NEUTRAL	0.68183996	NEUTRAL	0.66884082	NEUTRAL	0.72272272	DELETERIOUS	0.79280784	NEUTRAL	0.58359402
93	NM_022162.2	E778K	DELETERIOUS	0.52145411	NEUTRAL	0.70484581	DELETERIOUS	0.58885542	NEUTRAL	0.66884082	DELETERIOUS	0.47429306	DELETERIOUS	0.52778524	DELETE-RIOUS	0.62208185
94	NM_022162.2	L789F	NEUTRAL	0.68365861	NEUTRAL	0.76123348	DELETERIOUS	0.60798122	NEUTRAL	0.66884082	DELETERIOUS	0.39845758	NEUTRAL	0.77560107	NEUTRAL	0.76823399
95	NM_022162.2	R790Q	NEUTRAL	0.82622462	NEUTRAL	0.76563877	NEUTRAL	0.71871412	NEUTRAL	0.66884082	NEUTRAL	0.64364364	NEUTRAL	0.81300089	NEUTRAL	0.66524217
96	NM_022162.2	R791Q	NEUTRAL	0.82622462	NEUTRAL	0.77004405	NEUTRAL	0.71871412	NEUTRAL	0.66884082	NEUTRAL	0.67667668	NEUTRAL	0.69991095	NEUTRAL	0.66524217
97	NM_022162.2	V807 M	NEUTRAL	0.6025641	DELETE-RIOUS	0.58889722	NEUTRAL	0.68183996	NEUTRAL	0.66884082	DELETERIOUS	0.47429306	DELETERIOUS	0.79280784	NEUTRAL	0.66524217
98	CM149329	Q809K	NEUTRAL	0.75291375	NEUTRAL	0.71013216	NEUTRAL	0.7828765	NEUTRAL	0.66884082	NEUTRAL	0.62762763	DELETERIOUS	0.79280784	NEUTRAL	0.76823399
99	NM_022162.2	R830Q	NEUTRAL	0.82622462	NEUTRAL	0.79559471	NEUTRAL	0.71871412	NEUTRAL	0.66884082	NEUTRAL	0.69369369	NEUTRAL	0.76067616	NEUTRAL	0.61167883
100	NM_022162.2	I832V	NEUTRAL	0.82622462	NEUTRAL	0.85110132	NEUTRAL	0.89245476	NEUTRAL	0.66884082	NEUTRAL	0.78928929	NEUTRAL	0.83526269	NEUTRAL	0.58359402
101	NM_022162.2	A849V	DELETERIOUS	0.52145411	DELETE-RIOUS	0.57121439	NEUTRAL	0.660879	NEUTRAL	0.66884082	DELETERIOUS	0.40745501	DELETERIOUS	0.42969871	DELETE-RIOUS	0.72038776
102	NM_022162.2	N852S	NEUTRAL	0.73834499	NEUTRAL	0.74185022	NEUTRAL	0.71871412	NEUTRAL	0.66884082	NEUTRAL	0.62762763	DELETERIOUS	0.45232975	NEUTRAL	0.61167883
103	NM_022162.2	M863V	NEUTRAL	0.73834499	DELETE-RIOUS	0.5089955	NEUTRAL	0.58230958	NEUTRAL	0.66884082	NEUTRAL	0.8730899	NEUTRAL	0.89670526	NEUTRAL	0.76823399
104	NM_022162.2	S900F	DELETERIOUS	0.86908365	DELETE-RIOUS	0.65667166	DELETERIOUS	0.73260309	DELETERIOUS	0.59445019	DELETERIOUS	0.50257069	DELETERIOUS	0.79280784	DELETE-RIOUS	0.62208185
105	NM_022162.2	Q902K	NEUTRAL	0.82622462	NEUTRAL	0.74185022	NEUTRAL	0.71871412	NEUTRAL	0.66884082	NEUTRAL	0.68268268	NEUTRAL	0.76224399	NEUTRAL	0.70878378

106	NM_022162.2	G908R	DELETERIOUS	0.71871275	DELETE-RIOUS	0.5089955	DELETERIOUS	0.81731169	NEUTRAL	0.66884082	DELETERIOUS	0.43123393	DELETERIOUS	0.79280784	DELETE-RIOUS	0.55551884
107	NM_022162.2	E914K	NEUTRAL	0.73688811	NEUTRAL	0.74185022	DELETERIOUS	0.57795276	NEUTRAL	0.66884082	NEUTRAL	0.70240481	NEUTRAL	0.7747106	NEUTRAL	0.66524217
108	NM_022162.2	A918D	DELETERIOUS	0.71871275	DELETE-RIOUS	0.63268366	DELETERIOUS	0.67620995	NEUTRAL	0.66884082	DELETERIOUS	0.59318766	DELETERIOUS	0.79280784	DELETE-RIOUS	0.72038776
109	NM_022162.2	D925G	NEUTRAL	0.75291375	DELETE-RIOUS	0.71814093	NEUTRAL	0.68183996	NEUTRAL	0.66884082	NEUTRAL	0.78928929	NEUTRAL	0.69991095	NEUTRAL	0.58359402
110	NM_022162.2	V955I	NEUTRAL	0.73834499	NEUTRAL	0.72246696	NEUTRAL	0.71871412	NEUTRAL	0.66884082	NEUTRAL	0.64364364	DELETERIOUS	0.45949821	NEUTRAL	0.83058824
111	NM_022162.2	E963G	NEUTRAL	0.65307311	NEUTRAL	0.71742958	NEUTRAL	0.7828765	NEUTRAL	0.66884082	NEUTRAL	0.72472472	DELETERIOUS	0.45949821	DELETE-RIOUS	0.55551884
112	NM_022162.2	S974F	NEUTRAL	0.75203963	NEUTRAL	0.73127753	DELETERIOUS	0.60798122	NEUTRAL	0.66884082	NEUTRAL	0.72745491	NEUTRAL	0.69991095	NEUTRAL	0.58359402
113	NM_022162.2	A976T	DELETERIOUS	0.54946365	DELETE-RIOUS	0.48350825	DELETERIOUS	0.67620995	NEUTRAL	0.66884082	DELETERIOUS	0.47429306	DELETERIOUS	0.45232975	NEUTRAL	0.61167883
114	NM_022162.2	G1032S	NEUTRAL	0.82622462	NEUTRAL	0.79559471	NEUTRAL	0.660879	NEUTRAL	0.66884082	NEUTRAL	0.74974975	NEUTRAL	0.89670526	NEUTRAL	0.83058824

Table 3 Screening of pathogenic NOD2 mutations for the stability changes using iStable server.

S.No	Feature Key	Variant	i-Mutant2.0 SEQ	DDG	MUpro	Conf. Score	iStable	Conf. Score
1	NM_022162.2	Q233P	Decrease	−0.58	Decrease	−0.1508532	Decrease	0.789505
2	VAR_012676	R334Q	Decrease	−0.17	Decrease	−0.32867187	Decrease	0.711317
3	VAR_012677	R334W	Decrease	0.07	Decrease	−0.033876936	Decrease	0.702579
4	NM_022162.2	L343P	Decrease	−1.55	Decrease	−1	Decrease	0.900319
5	VAR_073231	E383G	Decrease	−1.38	Decrease	−1	Decrease	0.817872
6	VAR_023823	E383K	Decrease	−1.03	Decrease	−1	Decrease	0.770885
7	NM_022162.2	R426H	Decrease	−1.49	Decrease	−0.4268387	Decrease	0.757283
8	VAR_012685	L469F	null	−0.91	Decrease	−1	Decrease	0.801367
9	VAR_073235	G481D	Decrease	−0.91	Increase	0.47785331	Decrease	0.551764
10	VAR_073236	W490L	Decrease	−0.89	Decrease	−0.25221905	Decrease	0.813972
11	VAR_073237	C495Y	Increase	−0.04	Decrease	−0.14810199	Decrease	0.656363
12	CM114213	M513R	Increase	−0.34	Decrease	−1	Decrease	0.520911
13	VAR_065228	T605N	Increase	−0.09	Decrease	−0.45686193	Decrease	0.506883
14	VAR_073241	T605P	Decrease	−0.45	Decrease	−0.6040249	Decrease	0.732037
15	NM_022162.2	R703C	Decrease	−0.25	Decrease	−0.7420117	Decrease	0.716146
16	NM_022162.2	R756W	Decrease	−0.42	Decrease	−1	Decrease	0.793283
17	NM_022162.2	S900F	null	0.47	Increase	0.25722331	Decrease	0.521354

Table 4 Screening of pathogenic NOD2 mutations for the conservation nature using ConSurf server.

S.No	Feature Key	Variant	Consurf score
1	NM_022162.2	Q233P	8
2	VAR_012676	R334Q	9
3	VAR_012677	R334W	9
4	NM_022162.2	L343P	7
5	VAR_073231	E383G	9
6	VAR_023823	E383K	9
7	NM_022162.2	R426H	9
8	VAR_012685	L469F	7
9	VAR_073235	G481D	7
10	VAR_073236	W490L	7
11	VAR_073237	C495Y	9
12	CM114213	M513R	6
13	VAR_065228	T605 N	9
14	VAR_073241	T605P	9
15	NM_022162.2	R703C	7
16	NM_022162.2	R756W	2
17	NM_022162.2	S900F	9

PDB ID 5IRM (crystal structure of rabbit NOD2 in an ADP-bound state (crystal form2)) was used as the template for building the 3D model. This template had an identity of 86.25% with the queried NOD2 sequence. Furthermore, the modeled protein structure was used to generate a Ramachandran plot using the Rampage server. The model showed that approximately 99% of the residues were in the favored and allowed regions (Fig. 3).

3.5 Molecular docking

Molecular docking was performed using AutoDock software to understand the influence of the W490L mutation, which showed high deleteriousness and a destabilizing effect and is positioned in the ADP binding site of the NOD2 protein. ADP complexed with the native NOD2 protein showed a binding energy of −9.34 kcal/mol, whereas ADP complexed with the mutant NOD2 protein showed a binding energy of −5.44 kcal/mol. The interacting amino acids and the hydrogen bonds are shown in Fig. 4A and B.

3.6 Molecular dynamics simulations

Molecular dynamics simulations were performed using the GROMACS simulation package for 50 ns. The in-built analysis packages were used to

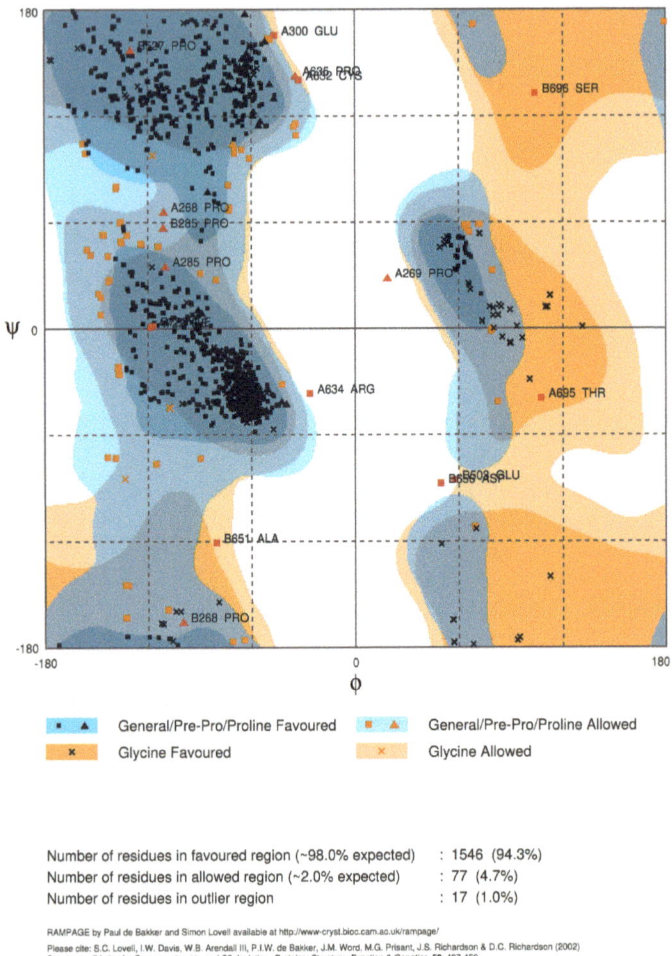

Fig. 3 Modeled 3D structure of NOD2 protein and the Ramachandran plot evaluation obtained by using the Rampage server.

analyze the 50 ns trajectories obtained at the end of the molecular dynamics simulations. The RMSD, RMSF, GYRATE, and HBOND packages of GROMACS were used to analyze the root mean square deviation (RMSD), root mean square fluctuation (RMSF), radius of gyration (Rg) and the number of inter/intramolecular hydrogen bonds of the native and mutant NOD2 proteins, respectively. In the RMSD plot, we observed a lower deviation in the native NOD2 protein molecule (deviation between ~0.2 nm and ~0.42 nm) compared to that in the NOD2 protein molecule with the W490L mutation (deviation between ~0.2 nm and ~0.63 nm).

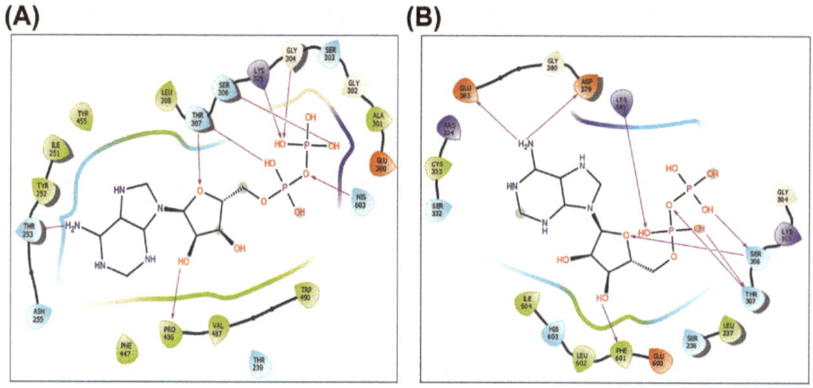

Fig. 4 (A) Amino acids in the NOD2 native protein that interact with the ADP molecule. (B) Amino acids in the NOD2 mutant protein that interact with the ADP molecule.

Both the native NOD2 protein and protein with the W490L mutation showed convergence after ~47 ns (Fig. 5A). Further analyses were performed according to the confidence of the convergence. To understand the actions of each amino acid, the RMSF plot was analyzed. From the analysis, we observed overall higher fluctuations in the residues in the NOD2 protein with the W490L mutation when compared to those in the native NOD2 protein (Fig. 5B). A radius of gyration analysis was performed to analyze the change in the compactness of the NOD2 protein due to the W490L mutation. The overall radius of gyration of native NOD2 deviated between ~3.05 nm and ~3.21 nm, whereas the overall radius of gyration of mutant NOD2 with the W490L mutation deviated between ~3.13 nm and ~3.3 nm (Fig. 6). The change in a number of intermolecular and

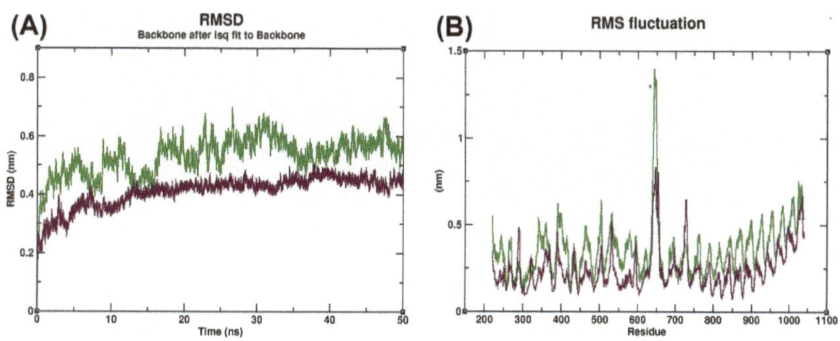

Fig. 5 (A) Root mean square deviation (RMSD) of the native and mutant NOD2 proteins complexed with ADP. Color scheme: maroon – native; green – mutant. (B) Root mean square fluctuation (RMSF) of the native and mutant NOD2 proteins complexed with ADP. Color scheme: maroon – native; green – mutant.

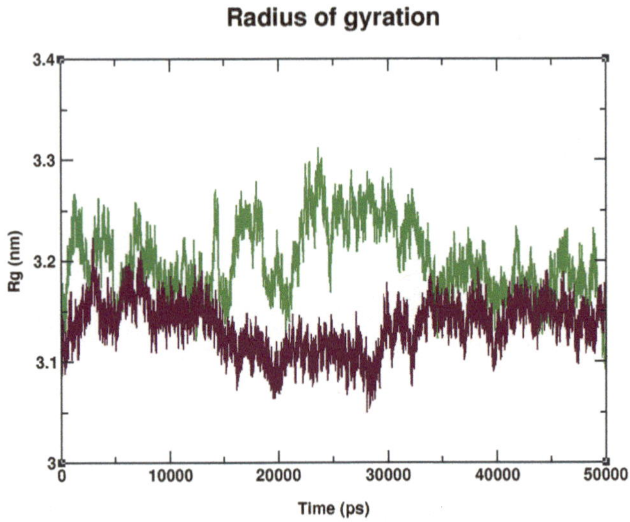

Fig. 6 Radius of gyration (Rg) of the native and mutant NOD2 proteins complexed with ADP. Color scheme: maroon – native; green – mutant.

intramolecular hydrogen bonds was calculated to understand the role of hydrogen bonds. According to the intramolecular hydrogen bond analysis, there was an overall higher number of hydrogen bonds in the native NOD2 when compared to that in the NOD2 protein with the W490L mutation (Fig. 7). We also calculated the number of intermolecular hydrogen

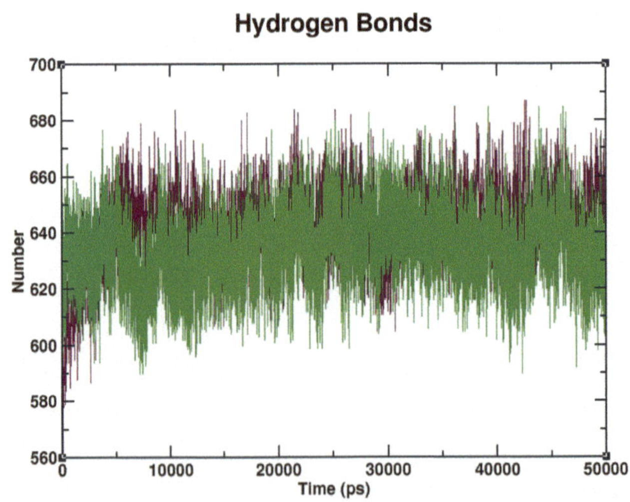

Fig. 7 Number of intramolecular hydrogen bonds in the native and mutant NOD2 proteins complexed with ADP. Color scheme: maroon – native; green – mutant.

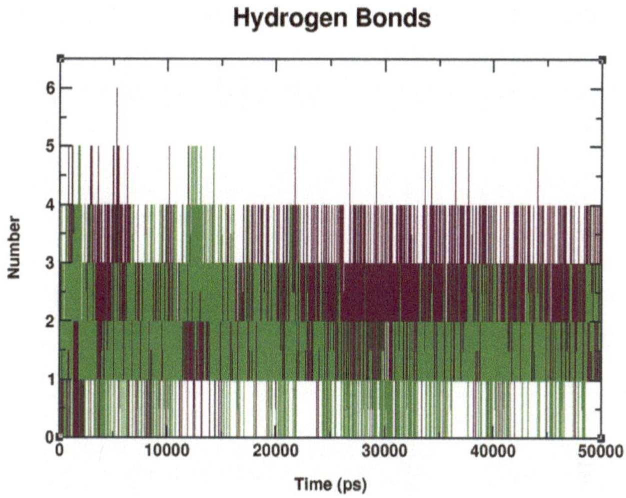

Fig. 8 Number of intermolecular hydrogen bonds between ADP and the native and mutant NOD2 proteins. Color scheme: maroon – native; green – mutant.

bonds between the ADP molecule and native NOD2 and NOD2 with the W490L mutation. From the analysis, we observed an overall higher number of intermolecular hydrogen bonds between ADP and native NOD2 compared with that between ADP and NOD2 with the W490L mutation. The number of intermolecular hydrogen bonds between ADP and native NOD2 was 6; on the other hand, a maximum of only five intermolecular hydrogen bonds was observed between ADP and NOD2 with the W490L mutation (Fig. 8).

4. Discussion

Several inflammation conditions result in a wide range of disorders and the development of inflammatory diseases. The most common inflammatory diseases include various allergies, asthma, and several autoimmune diseases (Inflammatory diseases - Latest research and news | Nature, n.d.). BS is an unusual autosomal dominant inflammatory disorder that was first discovered by the pediatricians Jabs and Blau in 1985 (Becker & Rose, 2005; Blau, 1985). This syndrome is caused by missense mutations in the NOD2 protein (Borzutzky et al., 2010). In this study, we formulated a computational pipeline to identify the most significant mutations in the NOD2 protein that cause BS. We have used these computational approaches in several previous studies and obtained promising results (George Priya Doss, Chakraborty, Monford Paul Abishek, Kumar, & Narayanan, 2014; George Priya Doss & Zayed, 2017; Sneha et al., 2017; Thirumal Kumar, Eldous, Mahgoub, George Priya Doss, & Zayed, 2018; Thirumal Kumar, Judith, et al., 2019). Besides facilitating the computational screening of mutations, these approaches have also been shown to obtain results that correlate with experimental pieces of evidence (Mosaeilhy et al., 2017; Moseilhy et al., 2017; Zaki et al., 2017). The list of missense mutations in NOD2 protein responsible for BS was retrieved from public databases such as UniProt, ClinVar, and HGMD. We retrieved 114 missense mutations from these databases. During the initial screening, mutations were screened for pathogenicity using the PredictSNP server. This server utilizes several other common pathogenicity predictors in addition to its prediction algorithm. Based on the predictions, we identified 17 mutations that were 'deleterious' according to the algorithm embedded within the PredictSNP server (Table 2). Furthermore, these 17 mutations were used for stability and conservation analysis using the iStable and ConSurf servers, respectively. Nine out of 17 mutations were found to be destabilizing by all the predictors embedded in the iStable server (R334Q, R334W, E383G, E383K, R426H, C495Y, T605N, T605P, and S900F). We also observed 6 out of 17 mutations to be both destabilizing and positioned in the highly conserved site (R334Q, R334W, E383G, E383K, R426H, and T605P) (Tables 3 and 4). Since there were no 3D structures deposited in the public databases such as the Protein Data Bank (PDB), we modeled the 3D structure of NOD2 protein using the Swiss Model server with the template PDB ID 5IRM. The template structure contains ADP, which plays a crucial role in maintaining the closed-form of NOD2 protein, thus maintaining the inactive state (Philpott et al., 2014). Interestingly, we found the W490L mutation to be positioned in the ADP interaction site in the NOD2 protein and found it to be highly pathogenic and destabilizing according to the PredictSNP and iStable predictions (Tables 2 and 3). W490L was also found to be positioned in a comparatively conserved site, with a ConSurf score of 7 (Table 4). Based on these interesting features, the W490L mutation was used for our structural analysis using molecular docking and dynamics simulations with native NOD2 as the control. In addition to BS, the W490L mutation was also found to be clinically reported in other inflammatory disease conditions, such as familial cold autoinflammatory syndrome (FCAS) and arthritis (Galozzi, Punzi, & Sfriso, 2017; Rosé et al., 2009). Additionally, a report by Parkhouse et al. in 2014 reported that the level of signaling response was slightly increased due to the W490L mutation (Parkhouse, Boyle, & Monie, 2014). From the molecular docking analysis, we observed that the binding affinity between the NOD2 protein and the ADP molecule was comparatively reduced when compared to that of the native NOD2 protein and the ADP molecule. Additionally, NOD2 interacted with ADP through 20 amino acids, whereas the NOD2 protein with the W490L mutation showed two fewer interactions, which resulted in 18 amino acid being involved in the interaction (Fig. 4A and B). The loss of interactions might be highly involved in destabilizing the protein structure (Warren et al., 2006). Finally, molecular dynamics simulations were performed with native NOD2 and mutant NOD2 complexed with ADP molecules. Molecular dynamics simulations are one of the modern higher-end approaches used to understand the behavior of proteins in real-time simulations (Agrahari, Muskan, George Priya Doss, Siva, & Zayed, 2018; Thirumal Kumar, Iyer, et al., 2019, 2016). From the plots of the RMSD and RMSF, we observed decreased deviation and fluctuation in the native NOD2 protein when compared to that of the NOD2 protein with the W490L mutation. The final convergence was observed at the end of the simulation period (Fig. 5A and B). The radius of gyration was inversely proportional to the compactness of the protein (Lobanov et al., 2008; Thirumal Kumar, Susmita et al., 2019). The increased radius of gyration of the protein with the W490L mutation, when compared to that of the native NOD2 protein, reveals that the compactness of the protein is lost due to the mutation (Fig. 6). The decrease in hydrogen bonds plays a crucial role in affecting the structural stability and compactness of the protein (Kumar & Doss, 2016; Perczel, Gáspári, & Csizmadia, 2005). From the intramolecular and intermolecular hydrogen bond analyses, we observed a decrease in the number of hydrogen bonds in both cases due to the W490L mutation, indicating that hydrogen bonds play a crucial role in maintaining the compactness of the protein (Figs. 7 and 8).

5. Conclusion

The study aims to understand the role of missense mutations that cause BS in NOD2 protein. We used the computational pipeline to understand the structural and functional impacts of the missense mutations. Overall, 144 missense mutations were retrieved from databases such as UniProt, ClinVar, and HGMD. Out of 114 mutations, 17 mutations were found to be the most pathogenic mutations. Furthermore, out of these 17 mutations, nine mutations were found to destabilize the protein structure, and six mutations were found to be destabilizing and positioned in a highly conserved position. ADP plays a crucial role in NOD2 protein by affecting the open-closed conformation of the protein that causes BS, so we were interested in the missense mutations occurring in the ADP binding sites of the NOD2 protein. The W490L mutation was found to be the most deleterious and destabilizing mutation and was positioned in a conserved region in the ADP binding site. From the molecular docking analysis, we observed that the binding of ADP was affected by the W490L mutation. Finally, from the molecular dynamics simulations, we observed that the NOD2 protein with the W490L mutation possessed increased deviation and fluctuation. It also showed lower compactness and the loss of inter- and intramolecular hydrogen bonds. To the best of our knowledge, no structural studies have been carried out in this area to understand the functional impact of mutations and the role of the ADP interaction with the NOD2 protein. We believe that this study will aid in the understanding of the role of mutations and switching between ADP and ATP in the binding of ADP to the NOD2 protein, which causes BS.

Acknowledgments

The authors acknowledge the aid of the Vellore Institute of Technology, Vellore, India, and BRAF @ CDAC for providing the facilities required to perform this work.

Disclosure statement

No potential conflict of interest exists.

References

Abraham, M. J., Murtola, T., Schulz, R., Páll, S., Smith, J. C., Hess, B., et al. (2015). GROMACS: High performance molecular simulations through multi-level parallelism from laptops to supercomputers. *SoftwareX, 1–2*, 19–25. http://dx.doi.org/10.1016/j.softx.2015.06.001.

Adzhubei, I., Jordan, D. M., & Sunyaev, S. R. (2013). Predicting functional effect of human missense mutations using PolyPhen-2. *Current Protocols in Human Genetics*. http://dx.doi.org/10.1002/0471142905.hg0720s76.

Agrahari, A. K., Muskan, M., George Priya Doss, C., Siva, R., & Zayed, H. (2018). Computational insights of K1444N substitution in GAP-related domain of NF1 gene associated with neurofibromatosis type 1 disease: A molecular modeling and dynamics approach. *Metabolic Brain Disease, 33*, 1443–1457. http://dx.doi.org/10.1007/s11011-018-0251-1.

Apweiler, R., Bairoch, A., Wu, C. H., Barker, W. C., Boeckmann, B., Ferro, S., et al. (2004). UniProt: The universal protein knowledgebase. *Nucleic Acids Research, 32*, 115D–119. http://dx.doi.org/10.1093/nar/gkh131.

Ashkenazy, H., Abadi, S., Martz, E., Chay, O., Mayrose, I., Pupko, T., et al. (2016). ConSurf 2016: An improved methodology to estimate and visualize evolutionary conservation in macromolecules. *Nucleic Acids Research, 44*, W344–W350. http://dx.doi.org/10.1093/nar/gkw408.

Bao, L., Zhou, M., & Cui, Y. (2005). nsSNPAnalyzer: identifying disease-associated nonsynonymous single nucleotide polymorphisms. *Nucleic Acids Research, 33*, W480–W482. http://dx.doi.org/10.1093/nar/gki372.

Becker, M. L., & Rose, C. D. (2005). Blau syndrome and related genetic disorders causing childhood arthritis. *Current Rheumatology Reports, 7*, 427–433. http://dx.doi.org/10.1007/s11926-005-0046-3.

Bendl, J., Stourac, J., Salanda, O., Pavelka, A., Wieben, E. D., Zendulka, J., et al. (2014). PredictSNP: Robust and accurate consensus classifier for prediction of disease-related mutations. *PLoS Computational Biology, 10*, e1003440. http://dx.doi.org/10.1371/journal.pcbi.1003440.

Blau, E. B. (1985). Familial granulomatous arthritis, iritis, and rash. *The Journal of Pediatrics, 107*, 689–693. http://dx.doi.org/10.1016/S0022-3476(85)80394-2.

Borzutzky, A., Fried, A., Chou, J., Bonilla, F. A., Kim, S., & Dedeoglu, F. (2010). NOD2-associated diseases: Bridging innate immunity and autoinflammation. *Journal of Clinical Immunology, 134*, 251–261. http://dx.doi.org/10.1016/j.clim.2009.05.005.

Bromberg, Y., & Rost, B. (2007). SNAP: Predict effect of non-synonymous polymorphisms on function. *Nucleic Acids Research, 35*, 3823–3835. http://dx.doi.org/10.1093/nar/gkm238.

Capriotti, E., Altman, R. B., & Bromberg, Y. (2013). Collective judgment predicts disease-associated single nucleotide variants. *BMC Genomics, 14*, S2. http://dx.doi.org/10.1186/1471-2164-14-S3-S2.

Chen, C. W., Lin, J., & Chu, Y. W. (2013). iStable: Off-the-shelf predictor integration for predicting protein stability changes. *BMC Bioinformatics, 14*, S5. http://dx.doi.org/10.1186/1471-2105-14-S2-S5.

Galozzi, P., Punzi, L., & Sfriso, P. (2017). Clinical overlapping in autoinflammatory diseases: The role of gene duplication. *Frontiers in Immunology, 8*, 392. http://dx.doi.org/10.3389/fimmu.2017.00392.

George Priya Doss, C., Chakraborty, C., Monford Paul Abishek, N., Kumar, D., & Narayanan, V. (2014). Application of evolutionary based in silico methods to predict the impact of single amino acid substitutions in vitelliform macular dystrophy. *Advances in Protein Chemistry and Structural Biology, 94*. http://dx.doi.org/10.1016/B978-0-12-800168-4.00006-8.

George Priya Doss, C., & Zayed, H. (2017). Comparative computational assessment of the pathogenicity of mutations in the Aspartoacylase enzyme. *Metabolic Brain Disease, 32*, 2105–2118. http://dx.doi.org/10.1007/s11011-017-0090-5.

Glaser, F., Pupko, T., Paz, I., Bell, R. E., Bechor-Shental, D., Martz, E., et al. (2003). ConSurf: Identification of functional regions in proteins by surface-mapping of phylogenetic information. *Bioinformatics, 19*, 163–164.

Grube, M., Brenmoehl, J., Rogler, G., Hahn, J., Herr, W., & Holler, E. (2015). Donor nucleotide-binding oligomerization-containing protein 2 (NOD2) single nucleotide polymorphism 13 is associated with Septic Shock after Allogeneic Stem Cell Transplantation. *Biology of Blood and Marrow Transplantation, 21*, 1399–1404. http://dx.doi.org/10.1016/j.bbmt.2015.05.011.

Guex, N., & Peitsch, M. C. (1997). SWISS-MODEL and the Swiss-Pdb Viewer: An environment for comparative protein modeling. *Electrophoresis, 18*, 2714–2723. http://dx.doi.org/10.1002/elps.1150181505.

Hasegawa, M., Fujimoto, Y., Lucas, P. C., Nakano, H., Fukase, K., Núñez, G., et al. (2008). A critical role of RICK/RIP2 polyubiquitination in Nod-induced NF-kappaB activation. *The EMBO Journal, 27*, 373–383. http://dx.doi.org/10.1038/sj.emboj.7601962.

Inflammatory diseases - Latest research and news | Nature. n.d. https://www.nature.com/subjects/inflammatory-diseases (accessed October 28, 2019).

Jabs, D. A., Houk, J. L., Bias, W. B., & Arnett, F. C. (1985). Familial granulomatous synovitis, uveitis, and cranial neuropathies. *The American Journal of Medicine, 78*, 801–804. http://dx.doi.org/10.1016/0002-9343(85)90286-4.

Jakopin, Ž., Gobec, M., Mlinarič-Raščan, I., & Sollner Dolenc, M. (2012). Immunomodulatory properties of novel nucleotide oligomerization domain 2 (Nod2) agonistic desmuramyldipeptides. *Journal of Medicinal Chemistry, 55*, 6478–6488. http://dx.doi.org/10.1021/jm300503b.

Kumar, D. T., & Doss, C. G. P. (2016). Investigating the inhibitory effect of Wortmannin in the Hotspot mutation at Codon 1047 of PIK3CA kinase domain: A molecular docking and molecular dynamics approach. *Advances in Protein Chemistry and Structural Biology, 102*, 267–297. http://dx.doi.org/10.1016/bs.apcsb.2015.09.008.

Landrum, M. J., Lee, J. M., Riley, G. R., Jang, W., Rubinstein, W. S., Church, D. M., et al. (2014). ClinVar: Public archive of relationships among sequence variation and human phenotype. *Nucleic Acids Research, 42*. http://dx.doi.org/10.1093/nar/gkt1113.

Lechtenberg, B. C., Mace, P. D., & Riedl, S. J. (2014). Structural mechanisms in NLR inflammasome signaling. *Current Opinion in Structural Biology, 29*, 17–25. http://dx.doi.org/10.1016/j.sbi.2014.08.011.

Lobanov, M. I., Bogatyreva, N. S., & Galzitskaia, O. V. (2008). Radius of gyration is indicator of compactness of protein structure. *Molecular Biology (Mosk), 42*, 701–706.

Maekawa, S., Ohto, U., Shibata, T., Miyake, K., & Shimizu, T. (2016). Crystal structure of NOD2 and its implications in human disease. *Nature Communications, 7*, 1–11. http://dx.doi.org/10.1038/ncomms11813.

Moreira, L. O., & Zamboni, D. S. (2012). NOD1 and NOD2 signaling in infection and inflammation. *Frontiers in Immunology, 3*, 328. http://dx.doi.org/10.3389/fimmu.2012.00328.

Morris, G. M., Goodsell, D. S., Halliday, R. S., Huey, R., Hart, W. E., Belew, R. K., et al. (1998). Automated docking using a Lamarckian genetic algorithm and an empirical binding free energy function. *Journal of Computational Chemistry, 19*, 1639–1662. http://dx.doi.org/10.1002/(SICI)1096-987X(19981115)19:14<1639::AID-JCC10>3.0.CO;2-B.

Mosaeilhy, A., Mohamed, M. M., C, G. P. D., El Abd, H. S. A., Gamal, R., Zaki, O. K., et al. (2017). Genotype-phenotype correlation in 18 Egyptian patients with glutaric acidemia type I. *Metabolic Brain Disease, 32*, 1417–1426. http://dx.doi.org/10.1007/s11011-017-0006-4.

Moseilhy, A., Hassan, M. M., El Abd, H. S. A., Mohammad, S. A., El Bekay, R., Abdel-Motal, U. M., et al. (2017). Severe neurological manifestations in an Egyptian patient with a novel frameshift mutation in the Glutaryl-CoA dehydrogenase gene. *Metabolic Brain Disease, 32*, 35–40. http://dx.doi.org/10.1007/s11011-016-9879-x.

Parkhouse, R., Boyle, J. P., & Monie, T. P. (2014). Blau syndrome polymorphisms in NOD2 identify nucleotide hydrolysis and helical domain 1 as signalling regulators. *FEBS Letters, 588*, 3382–3389. http://dx.doi.org/10.1016/j.febslet.2014.07.029.

Pastores, G. M., Michels, V. V., Stickler, G. B., Daniel Su, W. P., Nelson, A. M., & Bovenmyer, D. A. (1990). Autosomal dominant granulomatous arthritis, uveitis, skin rash, and synovial cysts. *The Journal of Pediatrics, 117*, 403–408. http://dx.doi.org/10.1016/S0022-3476(05)81080-7.

Perczel, A., Gáspári, Z., & Csizmadia, I. G. (2005). Structure and stability of beta-pleated sheets. *Journal of Computational Chemistry, 26*, 1155–1168. http://dx.doi.org/10.1002/jcc.20255.

Philpott, D. J., Sorbara, M. T., Robertson, S. J., Croitoru, K., & Girardin, S. E. (2014). NOD proteins: Regulators of inflammation in health and disease. *Nature Reviews Immunology, 14*, 9–23. http://dx.doi.org/10.1038/nri3565.

Punzi, L., Furlan, A., Podswiadek, M., Gava, A., Valente, M., De Marchi, M., et al. (2009). Clinical and genetic aspects of Blau syndrome: A 25-year follow-up of one family and a literature review. *Autoimmunity Reviews, 8*, 228–232. http://dx.doi.org/10.1016/j.autrev.2008.07.034.

Raphael, S. A., Blau, E. B., Zhang, W. H., & Hsu, S. H. (1993). Analysis of a large kindred with Blau syndrome for HLA, autoimmunity, and sarcoidosis. *American Journal of Diseases of Children, 147*, 842–848. http://dx.doi.org/10.1001/archpedi.1993.02160320044017.

Rosé, C. D., Aróstegui, J. I., Martin, T. M., Espada, G., Scalzi, L., Yagüe, J., et al. (2009). NOD2-associated pediatric granulomatous arthritis, an expanding phenotype: Study of an international registry and a national cohort in Spain. *Arthritis & Rheumatism, 60*, 1797–1803. http://dx.doi.org/10.1002/art.24533.

Sabbah, A., Te Hung Chang, R. H., Frohlich, V., Tominaga, K., Dube, P. H. P. H., Xiang, Y., et al. (2009). Activation of innate immune antiviral responses by Nod2. *Nature Immunology, 10*, 1073–1080. http://dx.doi.org/10.1038/ni.1782.Activation.

Salem, M., Seidelin, J. B., Eickhardt, S., Alhede, M., Rogler, G., & Nielsen, O. H. (2015). Species-specific engagement of human nucleotide oligomerization domain 2 (NOD)2 and Toll-like receptor (TLR) signalling upon intracellular bacterial infection: Role of Crohn's associated NOD2 gene variants. *Clinical and Experimental Immunology, 179*(3), 426–434. http://dx.doi.org/10.1111/cei.12471.

Sneha, P., Thirumal Kumar, D., George Priya Doss, C., Siva, R., & Zayed, H. (2017). Determining the role of missense mutations in the POU domain of HNF1A that reduce the DNA-binding affinity: A computational approach. *PLoS One, 12*, e0174953. http://dx.doi.org/10.1371/journal.pone.0174953.

Sneha, P., Kumar Thirumal, D., Himani, T., Siva, R., George Priya Doss, C., & Zayed, H. (2017). Structural analysis of G1691S variant in the human *Filamin B* gene responsible for Larsen syndrome: A comparative computational approach. *Journal of Cellular Biochemistry, 118*, 1900–1910. http://dx.doi.org/10.1002/jcb.25920.

Stenson, P. D., Mort, M., Ball, E. V., Evans, K., Hayden, M., Heywood, S., et al. (2017). The human gene mutation database: Towards a comprehensive repository of inherited mutation data for medical research, genetic diagnosis and next-generation sequencing studies. *Human Genetics, 136*, 665–677. http://dx.doi.org/10.1007/s00439-017-1779-6.

Sterka, D., Rati, D. M., & Marriott, I. (2006). Functional expression of NOD2, a novel pattern recognition receptor for bacterial motifs, in primary murine astrocytes. *Glia, 53*, 322–330. http://dx.doi.org/10.1002/glia.20286.

Thirumal Kumar, D., Eldous, H. G., Mahgoub, Z. A., George Priya Doss, C., & Zayed, H. (2018). Computational modelling approaches as a potential platform to understand the molecular genetics association between Parkinson's and Gaucher diseases. *Metabolic Brain Disease, 33*, 1835–1847. http://dx.doi.org/10.1007/s11011-018-0286-3.

Thirumal Kumar, D., George Priya Doss, C., Sneha, P., Tayubi, I. A., Siva, R., Chakraborty, C., et al. (2016). Influence of V54M mutation in giant muscle protein titin: A computational screening and molecular dynamics approach. *Journal of Biomolecular Structure and Dynamics, 1102*, 1–12. http://dx.doi.org/10.1080/07391102.2016.1166456.

Thirumal Kumar, D., Iyer, S., Christy, J. P., Siva, R., Tayubi, I. A., George Priya Doss, C., et al. (2019). A comparative computational approach toward pharmacological chaperones (NN-DNJ and ambroxol) on N370S and L444P mutations causing Gaucher's disease. *Advances in Protein Chemistry and Structural Biology, 114*, 315–339. http://dx.doi.org/10.1016/bs.apcsb.2018.10.002.

Thirumal Kumar, D., Jerushah Emerald, L., George Priya Doss, C., Sneha, P., Siva, R., Charles Emmanuel Jebaraj, W., et al. (2018). Computational approach to unravel the impact of missense mutations of proteins (D2HGDH and IDH2) causing D-2-hydroxyglutaric aciduria 2. *Metabolic Brain Disease*. http://dx.doi.org/10.1007/s11011-018-0278-3.

Thirumal Kumar, D., Judith, E., Priyadharshini Christy, J., Siva, R., Tayubi, I. A., Chakraborty, C., et al. (2019). Computational and modeling approaches to understand the impact of the Fabry's disease causing mutation (D92Y) on the interaction with pharmacological chaperone 1-deoxygalactonojirimycin (DGJ). *Advances in Protein Chemistry and Structural Biology, 114*, 341–407. http://dx.doi.org/10.1016/bs.apcsb.2018.10.009.

Thirumal Kumar, D., Mendonca, E., Priyadharshini Christy, J., George Priya Doss, C., & Zayed, H. (2019). A computational model to predict the structural and functional consequences of missense mutations in O6-methylguanine DNA methyltransferase. *Advances in Protein Chemistry and Structural Biology, 115*, 351–369. http://dx.doi.org/10.1016/bs.apcsb.2018.11.006.

Thirumal Kumar, D., Sneha, P., Uppin, J., Usha, S., & George Priya Doss, C. (2018). Investigating the influence of hotspot mutations in protein–protein interaction of IDH1 homodimer protein: A computational approach. *Advances in Protein Chemistry and Structural Biology, 111*, 243–261. http://dx.doi.org/10.1016/BS.APCSB.2017.08.002.

Thirumal Kumar, D., Susmita, B., Judith, E., Priyadharshini Christy, J., George Priya Doss, C., & Zayed, H. (2019). Elucidating the role of interacting residues of the MSH2-MSH6 complex in DNA repair mechanism: A computational approach. *Advances in Protein Chemistry and Structural Biology, 115*, 325–350. http://dx.doi.org/10.1016/bs.apcsb.2018.11.005.

Tromp, G., Kuivaniemi, H., Raphael, S., Ala-Kokko, L., Christiano, A., Considine, E., et al. (1996). Genetic linkage of familial granulomatous inflammatory arthritis, skin rash, and uveitis to chromosome 16. *The American Journal of Human Genetics, 59*, 1097–1107.

Warren, G. L., Andrews, C. W., Capelli, A.-M., Clarke, B., LaLonde, J., Lambert, M. H., et al. (2006). A critical assessment of docking programs and scoring functions. *Journal of Medicinal Chemistry, 49*, 5912–5931. http://dx.doi.org/10.1021/jm050362n.

Wertz, I. E., & Dixit, V. M. (2010). Signaling to NF-kappaB: Regulation by ubiquitination. *Cold Spring Harbor Perspectives in Biology, 2*, a003350. http://dx.doi.org/10.1101/cshperspect.a003350.

Zaki, O. K., Krishnamoorthy, N., El Abd, H. S., Harche, S. A., Mattar, R. A., Al disi, R. S., et al. (2017). Two patients with Canavan disease and structural modeling of a novel mutation. *Metabolic Brain Disease, 32*, 171–177. http://dx.doi.org/10.1007/s11011-016-9896-9.

Zhang, P., Wang, F., Hu, J., Sorrentino, R., Percha, B., Altman, R. B., et al. (2015). Label propagation prediction of drug-drug interactions based on clinical side effects. *Scientific Reports, 5*, 12339. http://dx.doi.org/10.1038/srep12339.

CHAPTER TEN

Structural sequence evolution and computational modeling approaches of the complement system in leishmaniasis

Prajakta Ingale, Ritika Kabra and Shailza Singh*
National Centre for Cell Science, NCCS Complex, SP Pune University Campus, Pune, Maharashtra, India
*Corresponding author: E-mail: singhs@nccs.res.in

Contents

Abstract

The complement system is one of the first barriers and consists of well-balanced cascades of reactions which generates anaphylatoxins such as C5a and C3a. A G-protein coupled receptor C5a anaphylatoxin chemotactic receptor 1 (C5AR1, also known as CD88) is the receptor for C5a which is present on cells of myeloid origin. Owing to difficulty in obtaining crystal structures of GPCRs in either inactive or active state, accurate structural modeling is still highly desirable for the majority of GPCRs. In an attempt to

Advances in Protein Chemistry and Structural Biology, Volume 120
ISSN 1876-1623
https://doi.org/10.1016/bs.apcsb.2019.12.004

dissect the conformational changes associated with GPCR activation, computational modeling approaches is being pursued in this paper along with the evolutionary divergence to deal with the structural variability.

1. Introduction

Leishmaniasis consists of wide spectrum of diseases ranging from simple asymptomatic cutaneous forms to a dreadful, disfiguring mucocutaneous form and a fatal, if untreated, visceral form of leishmaniasis. (Peter J. Myler, book cha. 1). Leishmaniasis has been persistent for centuries as life threating and disfiguring parasitic diseases affecting millions of people across the subtropics. Currently, 98 countries are listed as having endemic diseases amounting to an estimated 12 million cases with annual incidences of 2 million more cases (Hartley, Drexler, Ronet, Beverley, & Fasel, 2014). The three clinical syndromes classification are namely cutaneous leishmaniasis (Oriental sore), mucocutaneous leishmaniasis (espundia) and visceral leishmaniasis (kala-azar). Because of the interplay of numerous factors in the parasites, vectors, hosts and environments involved, the epidemiology and clinical manifestation of the disease are highly variable specifically when resources are limited and species identification is required. Human disease is caused by species of *Leishmania* protozoan parasites and is cycled among hosts through the bite of phlebotomine sand flies. Symptoms range from single self-healing cutaneous lesions to fatal visceralization or chronic metastatic dissemination throughout the skin (Hartley et al., 2014).

1.1 Epidemiology

Almost 20 *Leishmania* species are responsible to cause disease in humans and 30 species are proven vectors during blood meals. There are mainly two epidemiological entities: (1) zoonotic, which includes animal reservoir hosts in the transmission cycle (2) anthroponotic, in which man is the sole source of infection for the vector (Desjeux, 2004).

Leishmania is generally transmitted through the bite of an infected sand fly. However, from this common origin, the same sp. can cause widely different outcomes. In most instances, disease is 'asymptomatic', without any obvious pathology, although still able to support life-long infection. The *Leishmania* parasite multiplies in the gut of the sand fly and becomes infective 8–20 days later. Biting an animal or a human affected by the disease first infects the sand fly. The organisms are passed on when the sandfly bites its next victim. Sandflies are most active from dusk to dawn. The disease endemicity spreads to over 88 countries, 72 being in the developing world while 13 belong to the category of least developed countries, still its public health impact remained neglected. (Peter J. Myler, book cha. 1). This disease is also classified according to its occurrence in the world regions. Old World (OW) leishmaniasis exists in the Eastern Hemisphere and is endemic in Asia, Africa and southern Europe. New World (NW) leishmaniasis is endemic to the Western Hemisphere, extending from south-central Texas to Central and South America (except Chile and Uruguay) (Kevric, Cappel, & Keeling, 2015; Bailey & Lockwood, 2007).

1.2 Life cycle

The life cycle of *leishmania* is dimorphic, shifting between the female *Phlebotomus* sandfly vector and mammalian host. There are two morphologies of *leishmania*, first is infective, extracellular, flagellated and metacyclic promastigotes shifting between the alimentary tract of their sand fly vector and other one is intracellular amastigotes residing in phagolysosomal vacuoles of their mammalian host mononuclear phagocytes (Inbar et al., 2017). Inside gut of the fly various developmental forms of *leishmania* can be observed. *Leishmania* parasites are taken up by the sandfly during its bloodmeal. The first transformation occurs is conversion from amastigotes to promastigotes. These early promastigotes are ellipsoid in shape with flagella and body length of 6–8 μm (Pan et al., 1993). This stage of promastigotes is referred as procyclic stage (replicating and non-infectious). This event occurs in the peritrophic membrane (matrix), (Mosser and Brittingham, 1997). Approximately 3 days later these promastigotes move towards the anterior portion of the midgut. Promastigotes at this stage are attached to midgut epithelium with their flagella inserted between microvilli. After nearly 5 days, increasing number of ellipsoid, non-replicating, infectious and rapidly moving metacyclic staged promastigotes are seen in the lumen of the anterior midgut and foregut (Mosser and Brittingham, 1997). During vector's blood meal it creates a small blood pool through which infectious promastigotes are delivered to the vertebrate host. The damage created by sandfly feeding invites tissue phagocytes, monocytes, neutrophils to the site to quickly swallow up newly administered promastigotes. Within macrophages, the promastigotes transform into non-motile, round shaped amastigotes. These amastigotes can survive and replicate in the acidic environment created by phagolysosome, eventually lysing the infected cell and freely move in the outer space of cells infecting other phagocytic cells. So, when sandfly acquires blood meal it sucks either free amastigotes or amastigote-infected mononuclear cells and the cycle continues.

1.3 The complement cascade after *Leishmania* infection

The major hurdle with Leishmania infection is being able to evade and subvert host immune responses. These qualities allow them to establish chronic infection (Gurung & Kanneganti, 2015). During feeding process of sandfly promastigotes are regurgitated into blood pool and immediately encounters complement proteins of the host. The complement system of the host is a group of more than 30 soluble proteins work in cascade reaction for clearance of invading pathogens.

Three major pathways are present in complement system: classical, lectin and alternative pathway. All of these pathways are aimed for activation of C3 convertase that cleaves C3 to generate C3b. C3b deposition on to the pathogen surface induces deposition of C5b-C6-C7-C8-C9 to form membrane attack complex and sort of makes holes on the target pathogen surface and promotes lysis of it. Also, C3b acts as marker for promoting phagocytosis of the pathogen by neutrophils and macrophages (Gurung & Kanneganti, 2015). It has been seen that *L.major* and *L. enrietti* were capable of activating complement system via the alternative pathway, which proceeds in the absence of antibody (Mosser & Edelson, 1984). *L. donovani* and metacyclic promastigotes of *Leishmania major* activate complement system by both classical pathway and alternative pathway (Mosser and Brittingham, 1997). The two major virulence factors of *Leishmania* are glycocalyx component lipophosphoglycan (LPG) and the metalloprotease leishmanolysin GP63 (Gurung & Kanneganti, 2015). Lacking these glycocalyx receptors makes *L.major* highly susceptible for killing by complement proteins. Lipophosphoglycan is present on all over surface of Leishmania including the flagella and the major surface protein expressed on the surface of Leishmania is 63 KDa glycoprotein, also known as GP63 (Mosser and Brittingham, 1997; Mosser et al., 1987).

The first step in the activation of alternate pathway of complement activation is the deposition of the C3b onto the surface of target pathogen and it requires the formation of amide or ester linkage to be formed between the thioester of C3b and a free amino or hydroxyl group on the activating surface (Mosser and Brittingham, 1997). Knowing this there is no wonder that the most abundant surface structures on promastigotes, gp63 and LPG are acceptors of C3b (Russell and Wright, 1988). Therefore, LPG in metacyclic promastigotes is essential to save themselves from killing by complement pathway. Even though promastigote bearing LPG bound C3b initiated the MAC (Membrane Attack Complex), the complex does not reach the leishmania membrane and is immediately released as soluble C5b-9 complex (Gurung & Kanneganti, 2015). Similarly, GP63 accepts C3 deposition on the surface of Leishmania but inactivates C3b (iC3b), completely inhibiting the formation of the subsequent C5b-9 MAC complex (Brittingham et al., 1995). This inactivated form of C3 is unable to help formation of C5 convertase and thus formation of MAC. LPG and GP63 facilitate C3b opsonization, which promotes CD11b dependent phagocytosis which is required for parasite persistence within the host. Thus, GP63 and LPG mediated entry of Leishmania promastigotes into host cells protects them from complement cascade killing (Fig. 1). Because of the selective pressure to survive the highly sophisticated complement system mediated killing has resulted in the evolution of the several immune escape strategies that are adapted by *Leishmania* parasite.

Along with the complement system second central arm of the immunity are TLRs that have the ability to recognize PAMPs (Pathogen associated

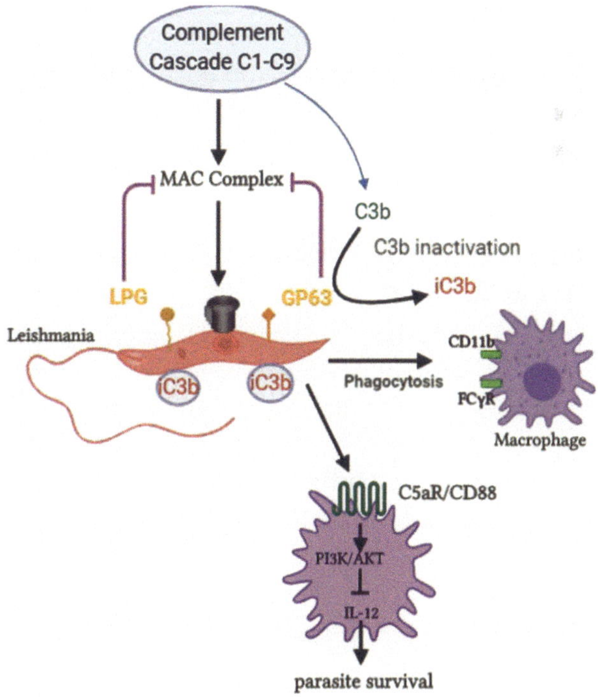

Fig. 1 Survival of *Leishmania* parasite overcoming complement system of host.

molecular patterns) and to destroy microbial invaders. The C5a receptor is a 74 amino acid peptide, known as complement component 5a receptor 1 (C5aR1). It's a G protein coupled receptor for C5a, classically described as an anaphylatoxin because of its vasodilatation, histamine release, and smooth muscle contraction ability (Wetsel et al., 2000). C5a is also considered widely as a proinflammatory molecule because of its anaphylactic and chemotactic properties, as well as its ability to enhance the expression of inflammatory cytokines like TNF-α, IL-6 and IL-1b (Daniel Calame, Mueller-Ortiz, Morales, & Wetsel, 2014). C5a is a negative regulator of IL-12 family cytokines through activation of PI3K and ERK1/2 (Fig. 1), (Hawlisch et al., 2005). A relatively low homology level has been observed across the mammalian C5AR1 sequences (Gerard & Gerard, 1991; Hani Boshra et al., 2004). All C5aR molecules possess an extra-cellular N-terminal region, seven helical hydrophobic TM regions and an intracellular C-terminal domain. C5a interact with receptor protein C5a or C5aR on the surface of target cells such as macrophages, neutrophils and endothelial cells. It is one of the major chemoattractant receptors. To date C5AR1 has only been identified and cloned in mammalian species, and its evolutionary history remains ill-defined.

To gain insights into the evolution, the divergence over the period of time between species and conservation of amino acids we have carried out the molecular clock analysis and statistical coupling analysis (SCA).

2. Materials and methods

2.1 Data acquisition

Total 149 sequences (both predicted and annotated) were retrieved from NCBI database across all the species and saved in fasta format. All the sequences were merged to single file to form multifasta file.

2.2 Sequence alignment and phylogenetic reconstruction

Multiple sequence alignment was performed using ClustalW online server and saved in nexus format which is an input for Bayesian inference phylogeny and molecular clock. Phylogeny was created using MrBayes version 3.2 x64 and the resultant tree was visualized in FigTree software (v 1.4.3).

2.3 Bayesian inference phylogeny and relaxed clock model

For estimation of Bayesian inference phylogeny, GTR+I+G model was used. Markov chain Monte Carlo (MCMC) simulation was run for 75,00,000 generations until split frequency is less than 0.01. The tree was visualized in FigTree (v1.4.3) and estimated outliers in the tree. Obtained outliers were kept as root to construct relaxed molecular clock phylogeny and run for 75,00,000 generations using MCMC simulation. The model used for constructing the molecular clock was the independent gamma rates model (IGR model) (Fig. 2).

2.4 Molecular modeling of C5aR1

Ab initio modeling was carried out to identify structural homologs through proteinBLAST search against PDB database. A primary amino acid sequence was retrieved from NCBI (NP_001727.2). Structure prediction was done by online Robetta software (Kim et al, 2004) Further loop refinement was made through loop.py file in modeller software (v9.18) while Model structure was validated using online servers such as RAMPAGE, PDBsum and ProsA web.

2.5 Statistical coupling analysis

Ranganathan et al. in 1999 generated random protein sequences through Monte Carlo simulations as a part of statistical coupling analysis (SCA) (Russ et al., 2005). Teşileanu, Colwell, Leibler (2015) assessed the relative importance of factors contributing to SCA viz., single site amino acid frequencies and pairwise correlations. SCA was done for conservation of amino acids over multiple aligned protein sequences in the form of sectors. All the sequences available for C5aR1 (both annotated and predicted) were converted into required input format for SCA analysis to be done and program files were run using MATAB (v2019a). SCA gives coupling energy which gives degree of evolutionary dependence between amino acid residues which further indicates higher coupling energy corresponding to increased dependence.

2.6 Selection pressure assessment of the protein families

To study the effect of selection pressure on the conservedness of the protein, MSA was submitted to the CONSURF server (Ashkenazy et al., 2016) which classifies the residues based on their conservation in the MSA.

(A)

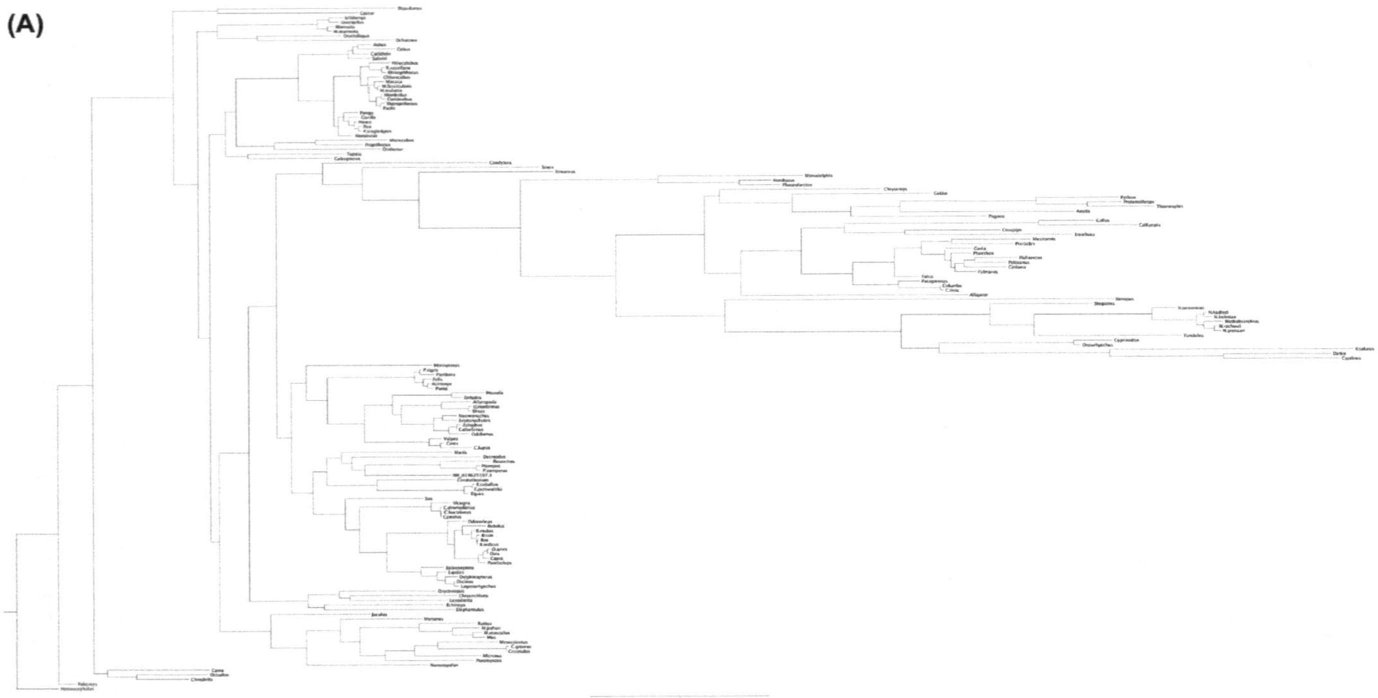

Fig. 2 ***Bayesian inference phylogeny and Relaxed clock model.*** (A). Phylogram based on Bayesian analysis of 149 taxa of C5AR1 genes of different organisms. The horizontal branches represent evolutionary lineages changing over time and the longer the horizontal branch, the larger the amount of change. The scale bar at the bottom of the phylogram showing value of 0.2 is the amount of genetic change over the period. The species *Fucomys* and *Heterocephalus* were considered as out-group. (B). Relaxed clock model based on the Bayesian tree. The scale bar at the bottom of the figure represents the amount of change that is 0.04.

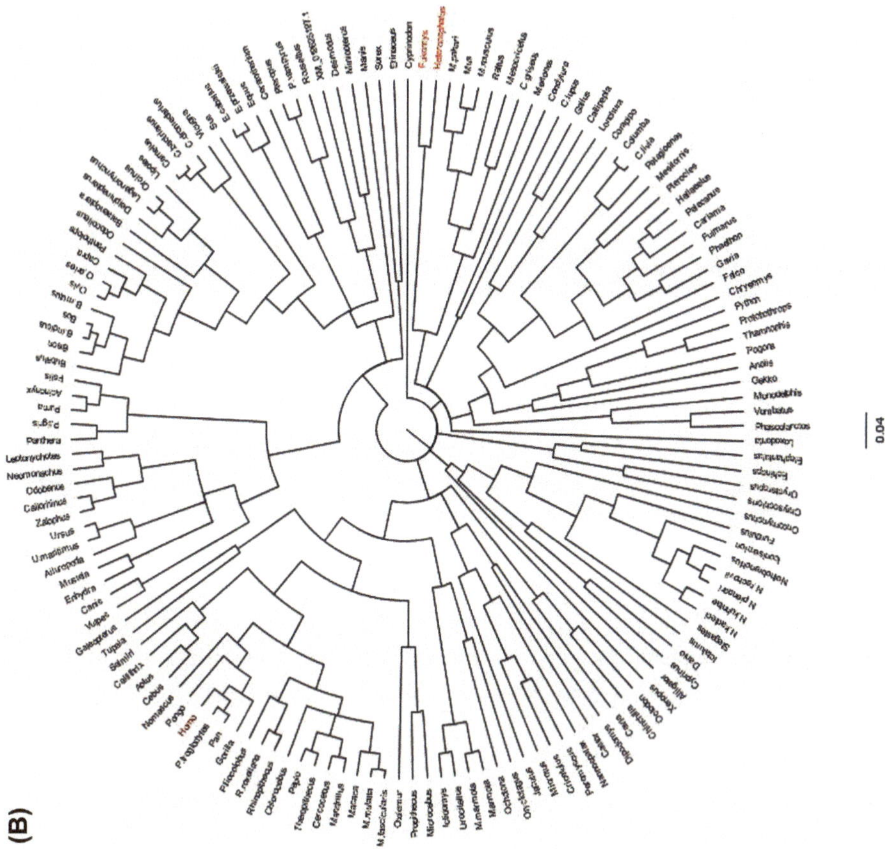

Fig. 2 (continued).

3. Result and discussion

Table 1 represents the posterior distribution of IGR branch rates for outgroups and their respective sister lineages 95% HPD (Highest Posterior Density) interval.

In particular, the evolutionary rate is decelerated in the case of *Fucomys* (bipartition ID 113) and that of its sister lineage with bipartition ID 113. Also, in case of *Heterocephalus* (bipartition ID 307) rate is decelerated and so is the case with its sister lineage, bipartition ID 329. Thus, it indicates as rate is decelerating, the tree is getting converged over the time. On the contrary, both the mean and median indicate that this rate is more likely to be below than 1.0. The potential scale reduction factor (PSRF) value should be reasonably closure to 1.0 as runs converge. It indicates heterogeneity between the two runs.

3.1 Molecular Modeling of C5aR1

In order to carry out homology modeling of C5aR1 using MODELLER (v9.18), best template (maximum sequence coverage, x-ray crystal structure available) selection was performed through protein-protein blast against the PDB database. 5o9h (it's a crystal structure ID of thermostabilised human anaphylatoxin chemotactic receptor 1) was found to be best template for C5aR1 with >90% sequence similarity. After generating models for C5aR1 using target protein it was observed that N-terminal and C-terminal ends are not able to span over the 5o9h and no proper secondary structure for initial 36 residues and at the end nearly 20 residues are formed. Thus, instead of using homology models, online server named Robetta (http://robetta.bakerlab.org) was used, it's a full chain protein structure prediction server.

Table 1 95% HPD interval values: If the clock is accelerated, the branch rate should be above 1.0; if the rate is decelerated, the branch rate should be less than 1.0.

			95% HPD interval			
Parameter	**Mean**	**Variance**	**Lower**	**Upper**	**Median**	**PSRF+**
Length[1]	0.081235	0.001379	0.020550	0.149170	0.077495	1.016
Length[113]	0.081235	0.001379	0.020550	0.149170	0.077495	1.016
Length [307]	0.007960	0.000033	0.000117	0.019358	0.006654	1.001
Length[329]	0.005135	0.000013	0.000235	0.012653	0.004299	1.001

Table 2 Model validation.

Robetta models	**Model 1**	**Model 2**	**Model 3**	**Model 4**	**Model 5**
PDBsum (No. of residues in disallowed region)	2	0	0	0	1
Prosaweb (Z-score)	−5.84	−5.74	−5.64	−5.51	−5.26
RAMPAGE (No. of residues in outlier region)	4	2	2	0	0

Robetta generates best 5 models, out of which one possible best model has to be chosen. Model validation is performed based on criterions like PDBsum (shows disallowed region), ProsaWeb (Z-score), Ramachandran plot by RAMPAGE (shows no. of residues in outlier region).

Based on the values shown in Table 2 model 1 has more outlier residues, thus to reduce it loop refinement was performed by using loop.py program of modeller and again three models were generated. These models had to again go through validation process as mentioned earlier. Table 3 enlists the values of outlier residues for respective models.

After model validation it's clear that model 2 is the best out of three. The structure of model 2 is given in Fig. 3.

3.2 Statistical Coupling Analysis

Statistical coupling analysis (SCA) analyzes multiple sequence alignments and identifies groups of coevolving residues termed 'sectors'. The sectors are important and controls different biochemical functions of the protein. This method applies spectral analysis to a matrix pursued by combining correlation information with sequence conservation (Teşileanu et al., 2015). Fig. 4 shows the three diagonal sectors identified by SCA distinguishingly in three colors viz; red, blue and green.

It implies more the red color in sector, more conserved is the protein. So, C5aR1 shows the conservation of amino acids, precisely 49 amino acids in red color whereas 74 and 36 amino acids in green and blue region

Table 3 Model validation after loop refinement.

Robetta models	**Model 1**	**Model 2**	**Model 3**
ProsAweb	−5.03	−5.28	−5.38
Rampage	6	1	3
PDBsum	3	0	1

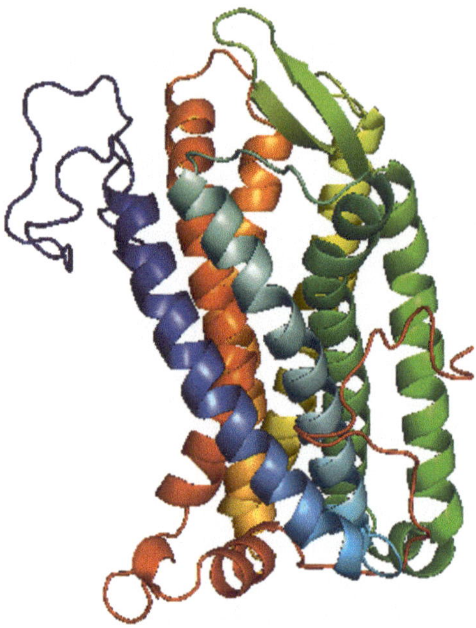

Fig. 3 Final model (Model 2) of C5aR1 visualized through Pymol. Initial blue part is N-terminal and end of red part is C-terminal., whereas middle region consists of 7 helical domains.

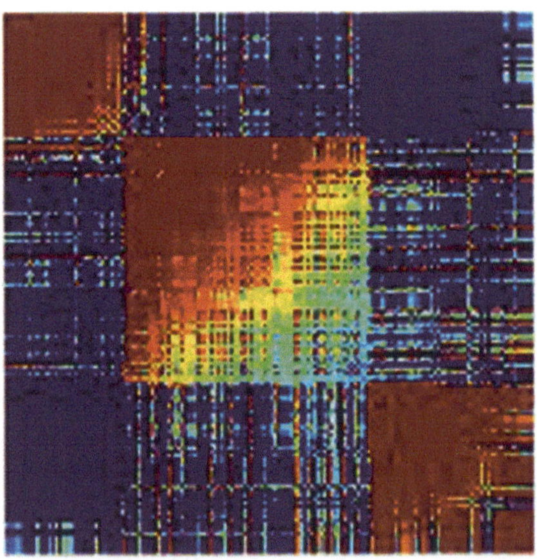

Fig. 4 Three sectors obtained from SCA analysis.

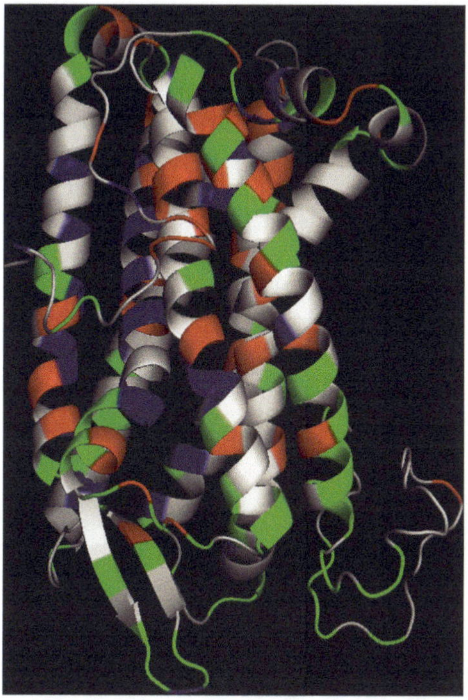

Fig. 5 Distribution of Conservedness of amino acids.

respectively. When we map this conserved amino acids over the 3D structure of C5aR1 it is observed that distribution of conservedness as shown in Fig. 5.

3.3 Conservation over the sequence

Phylogenetic analysis doesn't provide much information for genes that have accumulated over few mutations. Analysis of functional divergence provides sequence variations across a set of homologous gene sequences (Mandlik et al. 2014). Functionally significant amino acids are evolutionary conserved, also evolutionary pressure over protein sequence vary from one amino to another. Amino acids from middle domains of C5aR1 protein are significantly conserved than end terminals can be visualized in Fig. 6.

In nutshell, the developed model of C5aR1 shows overall good structural quality and was confirmed using several different validation tools. It can act as a good chemo attractant target as C5a interacts with receptor protein C5a1 or C5aR on the surface of target cells such as macrophages, neutrophils and endothelial cell. In Leishmania infection, the amastigote

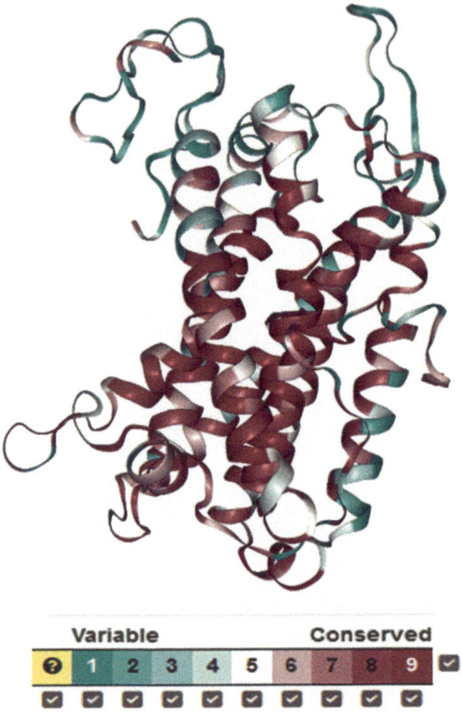

Fig. 6 ***Conservation profile of the C5aR1 protein.*** Residues marked in green are variable and that of in pink color are conserved. No yellow colored residue observed meaning least variable residue is absent.

form of the pathogen is found present in human macrophages. So by targeting this C5AR receptor which is one of the most potent inflammatory chemoattractant and has been implicated in pathogenesis of numerous inflammatory diseases, we can further stop the cell division of amastigote form of Leishmania. Our study provides the molecular model and its *in-silico* structural stability but still further study needs to be done in order to understand the biochemical nature of the molecule. C5aR1 has raised the intriguing possibility of the use of this receptor having possible role in inflammation.

Acknowledgments

The authors thank Director, National Center for Cell Science (NCCS), Pune for supporting the Bioinformatics and High Performance Computing Facility (BHPCF) at NCCS, Pune, India. The authors would also like to thank the Intramural fund from Department of Biotechnology (DBT). Ministry of Science and Technology, Government of India.

References

Ashkenazy, H., Abadi, S., Martz, E., Chay, O., Mayrose, I., Pupko, T., et al. (May 10, 2016). ConSurf 2016: An improved methodology to estimate and visualize evolutionary conservation in macromolecules. *Nucleic Acids Research, 44*(W1), W344–W350.

Bailey, M. S., & Lockwood, D. N. (March 1, 2007). Cutaneous leishmaniasis. *Clinics in Dermatology, 25*(2), 203–211.

Boshra, H., Li, J., Peters, R., Hansen, J., Matlapudi, A., & Sunyer, J. O. (April 1, 2004). Cloning, expression, cellular distribution, and role in chemotaxis of a C5a receptor in rainbow trout: The first identification of a C5a receptor in a nonmammalian species. *The Journal of Immunology, 172*(7), 4381–4390.

Brittingham, A., Morrison, C. J., McMaster, W. R., McGwire, B. S., Chang, K. P., & Mosser, D. M. (September 15, 1995). Role of the Leishmania surface protease gp63 in complement fixation, cell adhesion, and resistance to complement-mediated lysis. *The Journal of Immunology, 155*(6), 3102–3111.

Calame, D. G., Mueller-Ortiz, S. L., Morales, J. E., & Wetsel, R. A. (November 15, 2014). The C5a anaphylatoxin receptor (C5aR1) protects against *Listeria* monocytogenes infection by inhibiting type 1 IFN expression. *The Journal of Immunology, 193*(10), 5099–5107.

Desjeux, P. (September 1, 2004). Leishmaniasis: Current situation and new perspectives. *Comparative Immunology, Microbiology and Infectious Diseases, 27*(5), 305–318.

Gerard, N. P., & Gerard, C. (February 1991). The chemotactic receptor for human C5a anaphylatoxin. *Nature, 349*(6310), 614.

Gurung, P., & Kanneganti, T. D. (September 2015). Innate immunity against Leishmania infections. *Cellular Microbiology, 17*(9), 1286–1294.

Hartley, M. A., Drexler, S., Ronet, C., Beverley, S. M., & Fasel, N. (August 1, 2014). The immunological, environmental, and phylogenetic perpetrators of metastatic leishmaniasis. *Trends in Parasitology, 30*(8), 412–422.

Hawlisch, H., Belkaid, Y., Baelder, R., Hildeman, D., Gerard, C., & Köhl, J. (April 1, 2005). C5a negatively regulates toll-like receptor 4-induced immune responses. *Immunity, 22*(4), 415–426.

Inbar, E., Hughitt, V. K., Dillon, L. A., Ghosh, K., El-Sayed, N. M., & Sacks, D. L. (May 3, 2017). The transcriptome of *Leishmania major* developmental stages in their natural sand fly vector. *mBio, 8*(2). e00029-17.

Kevric, I., Cappel, M. A., & Keeling, J. H. (July 1, 2015). New world and old world *Leishmania* infections: A practical review. *Dermatologic Clinics, 33*(3), 579–593.

Kim, D. E., Chivian, D., & Baker, D. (July 1, 2004). Protein structure prediction and analysis using the Robetta server. *Nucleic Acids Research, 32*(Suppl. 1_2), W526–W531.

Mandlik, V., Shinde, S., & Singh, S. (December 2014). Molecular evolution of the enzymes involved in the sphingolipid metabolism of Leishmania: Selection pressure in relation to functional divergence and conservation. *BMC Evolutionary Biology, 14*(1), 142.

Mosser, D. M., & Brittingham, A. (December 1997). Leishmania, macrophages and complement: A tale of subversion and exploitation. *Parasitology, 115*(7), 9–23.

Mosser, D. M., & Edelson, P. J. (March 1, 1984). Activation of the alternative complement pathway by Leishmania promastigotes: Parasite lysis and attachment to macrophages. *The Journal of Immunology, 132*(3), 1501–1505.

Mosser, D. M., Vlassara, H. E., Edelson, P. J., & Cerami, A. (January 1, 1987). Leishmania promastigotes are recognized by the macrophage receptor for advanced glycosylation endproducts. *Journal of Experimental Medicine, 165*(1), 140–145.

Pan, A., Duboise, S., Eperon, S., Rivas, L., Hodgkinson, V., Traub-Cseko, Y., et al. (March 1993). Developmental life cycle of Leishmania—cultivation and characterization of cultured extracellular amastigotes 1. *The Journal of Eukaryotic Microbiology, 40*(2), 213–223.

Russell, D. G., & Wright, S. D. (July 1, 1988). Complement receptor type 3 (CR3) binds to an Arg-Gly-Asp-containing region of the major surface glycoprotein, gp63, of Leishmania promastigotes. *Journal of Experimental Medicine, 168*(1), 279–292.

Russ, W. P., Lowery, D. M., Mishra, P., Yaffe, M. B., & Ranganathan, R. (September 2005). Natural-like function in artificial WW domains. *Nature, 437*(7058), 579.

Teşileanu, T., Colwell, L. J., & Leibler, S. (February 27, 2015). Protein sectors: Statistical coupling analysis versus conservation. *PLoS Computational Biology, 11*(2), e1004091.

Wetsel, R. A., Kildsgaard, J., & Haviland, D. L. (2000). Complement anaphylatoxins (C3a, C4a, C5a) and their receptors (C3aR, C5aR/CD88) as therapeutic targets in inflammation. In J. D. Lambris, & V. M. Holers (Eds.), *Contemporary immunology: Therapeutic interventions in the complement system* (pp. 113–154). Totowa, NJ: Humana Press.

FSC
www.fsc.org
MIX
Papier | Fördert
gute Waldnutzung
FSC® C083411

Kolibri 360 GmbH
Ferdinand-Jühlke-Straße 7
99095 Erfurt, Deutschland
produktsicherheit@kolibri360.de